Interdisciplinary Biotechnological Advances

This series is an authoritative resource that explains recent advances and emerging directions in biotechnology, reflecting the forefront of research clearly and reliably, without excessive hype. Each volume is written by authors with excellent reputations and acknowledged expertise in the topic under discussion. The volumes span the entire field from an interdisciplinary perspective, covering everything from biotechnology principles and methods to applications in areas including genetic engineering, transgenic plants and animals, environmental problems, genomics, proteomics, diagnosis of disease, gene therapy, and biomedicine. The significance of these applications for the achievement of UN Sustainable Development Goals is highlighted. The series will be highly relevant for Master's and PhD students in Biotechnology, Nanochemistry, Biochemical Engineering, and Microbiology, medical students, academic and industrial researchers, agricultural scientists, farmers, clinicians, industry personnel, and entrepreneurs.

Santosh Kumar Singh • Emanuel Vamanu
Sachchida Nand Rai • Anurag Kumar Singh
Editors

Microbiota Profiling for Precision Medicine

Biomarker-Targeted Drug Delivery Strategies

Springer

Editors
Santosh Kumar Singh
Centre of Experimental Medicine and Surgery, Institute of Medical Sciences
Banaras Hindu University
Varanasi, Uttar Pradesh, India

Sachchida Nand Rai
Centre of Experimental Medicine and Surgery, Institute of Medical Sciences
Banaras Hindu University
Varanasi, Uttar Pradesh, India

Emanuel Vamanu
Department of Industrial Biotechnology
Faculty of Biotechnology
University of Agricultural Sciences and Veterinary Medicine
Bucharest, Romania

Anurag Kumar Singh
Department of Pharmaceutical Engineering and Technology,
Indian Institute of Technology (Banaras Hindu University)
Varanasi, Uttar Pradesh, India

ISSN 2730-7069 ISSN 2730-7077 (electronic)
Interdisciplinary Biotechnological Advances
ISBN 978-981-95-2103-6 ISBN 978-981-95-2104-3 (eBook)
https://doi.org/10.1007/978-981-95-2104-3

This Springer imprint is published by the registered company Springer Nature Singapore Pte Ltd.
The registered company address is: 152 Beach Road, #21-01/04 Gateway East, Singapore 189721, Singapore

This book, Microbiota Profiling for Precision Medicine: Biomarker-Targeted Drug Delivery Strategies, is respectfully dedicated to ***Pt. Madan Mohan Malaviya Ji****, the visionary founder of Banaras Hindu University. His steadfast dedication to education, innovation, and the quest for knowledge has inspired countless scholars and researchers. It is his pioneering vision that established a world-class institution, nurturing excellence in science, technology, and the humanities. May this work, born from the very university he created, serve as a tribute to his enduring legacy and his commitment to societal progress through education and research.*

With profound respect and gratitude.

Santosh Kumar Singh
Emanuel Vamanu
Sachchida Nand Rai
Anurag Kumar Singh

Preface

The dynamic convergence of genomics, bioinformatics, and microbiome research is revolutionizing the landscape of modern healthcare, laying the foundation for precision medicine an innovative approach that customizes medical treatment based on individual patient characteristics. At the core of this transformation is the expanding knowledge of the human microbiota, a diverse and interactive community of microorganisms that significantly influences health, disease progression, and therapeutic outcomes. This book, *Microbiota Profiling for Precision Medicine: Biomarker-Targeted Drug Delivery Strategies*, offers a comprehensive exploration of how microbiota-based profiling is paving new pathways for personalized medical interventions.

Starting with an introduction to the principles of precision medicine and its growing role in personalized health management, the book delves into recent technological advancements in microbiota profiling. These techniques have enabled the identification of novel biomarkers critical for disease diagnosis, therapeutic monitoring, and precise drug delivery. The chapters progress to examine the microbiota's influence on drug metabolism, toxicity, immune modulation, and its involvement in neurodegenerative diseases such as Alzheimer's and Parkinson's, underlining the therapeutic promise of microbiome-targeted strategies.

In addition, this volume highlights the integration of microbiota-derived data with cutting-edge tools like nanotechnology and machine learning to enhance drug delivery systems. Case studies provide practical insights into microbiota-informed therapies for cancer, autoimmune conditions, gastrointestinal disorders, and neurological diseases. Later chapters address essential topics such as biomarker validation, data integration, regulatory landscapes, and the ethical considerations required for the clinical application of microbiome-based therapeutics.

This book is intended for researchers, clinicians, academicians, and students engaged in microbiome science, pharmacology, biotechnology, and precision medicine. By bridging multiple disciplines, it aims to inspire innovation and support the development of individualized, microbiota-informed therapeutic strategies.

We would like to extend our deepest gratitude to our families, mentors, colleagues, and the publishing team for their unwavering support throughout the creation of this book. We are also sincerely thankful to the contributors whose valuable insights and expertise have shaped the content presented here. It is our hope that this work serves as a valuable resource and a catalyst for future advancements in microbiota-based precision medicine.

Varanasi, Uttar Pradesh, India	Santosh Kumar Singh
Bucharest, Romania	Emanuel Vamanu
Varanasi, Uttar Pradesh, India	Sachchida Nand Rai
Varanasi, Uttar Pradesh, India	Anurag Kumar Singh

Acknowledgments

Santosh Kumar Singh gratefully acknowledges funding from the Institution of Eminence Incentive Grant Scheme (1209; R/Dev/D/IoE/2021-22/32115), the Trans-Disciplinary Research Scheme (R/Dev/D/IoE/TDR/2023-24/69559), the Department of Science and Technology (SEED grant, New Delhi), and the Institute of Medical Sciences Intramural Grant, Banaras Hindu University, Varanasi, India. Sachchida Nand Rai sincerely acknowledges the Indian Council of Medical Research (ICMR), Government of India, for support through the Research Associate (ICMR-RA) Award (No.: 3/1/3/196Neuro/2021-NCD-I). Anurag Kumar Singh extends his gratitude to the Indian Council of Medical Research (ICMR), Government of India, for support through the Research Associate (ICMR-RA) Award (No.: 3/1/2/110Neuro/2019-NCD-I). We extend our heartfelt gratitude to all contributors whose quality material has enriched this book. We are deeply thankful to our beloved family members for their joyful support and understanding during the countless hours of our absence while completing this project. Our sincere thanks also go to our friends and colleagues for their encouragement and assistance throughout the writing process. Finally, we acknowledge the publishing team at Springer for their professionalism, enthusiasm, and steadfast belief in this project. Their expertise in design, production, and marketing has been invaluable in bringing this book to fruition.

Santosh Kumar Singh
Emanuel Vamanu
Sachchida Nand Rai
Anurag Kumar Singh

Contents

About the Editors

Santosh Kumar Singh is a Senior Scientist at the Centre of Experimental Medicine & Surgery, Institute of Medical Sciences, Banaras Hindu University (BHU), Varanasi. He completed his Ph.D. at Banaras Hindu University in 1998 and pursued postdoctoral research at the same institution. He also served as a Visiting Scientist at the Department of Photosynthesis, Institute of Physical Biology, University of South Bohemia, Czech Republic. He has presented his scientific work internationally, including in the United States and other countries. With more than 25 years of professional experience, he has supervised over 12 PhD scholars, 2 postdoctoral fellows, and more than 10 MD/MS/DM/MCh theses. He has published over 120 research articles in internationally reputed journals and has edited more than 5 books. He also serves on the editorial boards of various scientific journals. He has led numerous research projects funded by esteemed agencies such as the Department of Science and Technology (DST), the Department of Biotechnology (DBT), and the Science and Engineering Research Board (SERB). His contributions to science earned him membership in the National Academy of Sciences, India (NASI), Prayagraj. He is a recipient of the National Award for Science and Technology Communication, presented by the Ministry of Science & Technology, Government of India, in recognition of his exceptional contributions to the field.

Emanuel Vamanu (March 18, 1976, Bucharest) is a Romanian biotechnologist and Professor in the Faculty of Biotechnologies at University of Agronomic Sciences and Veterinary Medicine of Bucharest (USAMV) Bucharest, where he has taught and led research in pharmaceutical biotechnology, cellular pathology, microbial bioremediation, and industrial biotechnology practices, since March 2022. After earning his BSc (1999) and MSc in Genetics (2001) at the University of Bucharest, he completed his doctoral studies in Horticulture at USAMV Bucharest in 2006 with a thesis on probiotic microbial biomass production. Between 1999 and 2003, he served as lab preparator and assistant, advancing through research and teaching roles, including head of works and full professor, before joining the Applied Biochemistry and Biotechnology Center BIOTEHNOL (2002–2012). He holds a Publons Academy certification in peer review (2019). He has completed numerous advanced training courses across Europe in gene manipulation, lipid metabolism, animal science, and EU structural funds.

Sachchida Nand Rai is a research associate (ICMR-RA) at the Centre of Experimental Medicine and Surgery, Institute of Medical Sciences, Banaras Hindu University (BHU), Varanasi, since 27 January 2023. He also completed his PhD from the Biochemistry Department (Institute of Science) of this same university in 2018. The BHU consistently ranks among the top universities in the India in terms of research output and quality in the life sciences. Before joining the BHU as research associate, he was post-doctoral fellow at the Centre of Biotechnology, University of Allahabad, Prayagraj, India, from 27 January 2020 to 26 January 2023 (3 years). He has impressive publications in several reputed journals. He serves as Academic Editor for the journals *Mediators of Inflammation*, *International Journal of Inflammation*, and *BMC Research Notes* and Associate Editor for *3-Biotech* (Springer Nature), *Molecular Neurobiology*, *Oxidative Medicine and Cellular Longevity* along with the *Journal of Inflammation Research* and *Degenerative Neurological and Neuromuscular Disease* (Dove Press). He has edited more than 100 manuscripts as editor and reviewed more than 900 manuscripts as reviewer for high-impact journals. He has published two books with the renowned publishers (Bentham). His current Google scholar citation is 2701 with 29 h-indexes. He investigating the neuroprotective activity of common Indian medicinal plants in toxin-induced Parkinsonian mouse model. He is also member of the Indian Academy of Neurosciences (India).

Anurag Kumar Singh is currently working as a research associate (ICMR-RA) in the Department of Pharmaceutical Engineering and Technology at the Indian Institute of Technology (Banaras Hindu University), Varanasi, India. Previously, he served as a postdoctoral research assistant in Cancer Biology at the Department of Biological Sciences, Alabama State University, USA, and at the Institute of Nanotechnology & Advanced Materials, Department of Chemistry, Bar-Ilan University, Israel. He brings over 2.5 years of post-doctoral research experience to his current role. He earned his doctoral degree in pharmaceutical chemistry from the Centre of Experimental Medicine and Surgery, Institute of Medical Sciences, Banaras Hindu University, Varanasi, and his MPharm degree from the School of Chemical Sciences and Pharmacy, Central University of Rajasthan, Ajmer, India. His research focuses on the development of innovative nano-particulate systems for applications in chronic pain management, neurodegenerative disorders, and brain-targeted drug delivery. This includes expertise in dendrimers, nanoporous silica/silicon materials, and polymeric nanoparticles for controlled drug delivery, diagnostics, and nanotheranostics. His work aims to integrate engineering, pharmaceutical, and medical sciences to advance nanomedicine for healthcare and biomedical applications. He has an impressive academic record with over 50 research articles, reviews, and journal editorials published in esteemed national and international journals. He has edited or authored more than 8 books and contributed over 15 book chapters published by internationally renowned presses.

Introduction to Precision Medicine and Personalized Therapies in Health and Diseases

Manoleena Sarkar and Rajnandan Borah

Abstract Two healthcare approaches that seek to customize medical interventions and therapies for specific patients are precision medicine and personalized medicine. These methods are predicated on the notion that each patient's genetic composition, way of life and surroundings significantly influence their health and reaction to therapy. By examining a patient's DNA, RNA and protein expression, precision medicine uses genetic and molecular data to pinpoint disease subtypes and provide customized treatments. In contrast, personalized medicine develops a comprehensive treatment plan that is specific to each patient by considering not only genetic information but also other characteristics like the patient's lifestyle, environment and medical history. Both of these have significant success in the treatment of certain types of cancer (breast and lung cancer), and neurodegenerative diseases like Alzheimer's disease and Parkinson's disease. Both of these play a significant role in disease treatment but they also have significant effects on early detection and disease prevention. All things considered, personalized and precision medicine are becoming increasingly important in the treatment of diseases. These methods are being utilized to create new medicines for different illnesses and have already resulted in notable advancements in the treatment of several forms of cancer. The purpose of this chapter is to examine the potential, constraints and opportunities of precision or personalized medicine by reviewing current research that places the topic at the centre of the discussion.

M. Sarkar (✉)
Faculty of Pharmacognosy, School of Pharmacy, Himalayan University, Itanagar, Arunachal Pradesh, India

R. Borah
School of Pharmacy, Himalayan University, Itanagar, Arunachal Pradesh, India

Institute of Pharmacy, Assam Don Bosco University, Tepesia, Kamarkuchi, Assam, India

S. K. Singh et al. (eds.), *Microbiota Profiling for Precision Medicine*, Interdisciplinary Biotechnological Advances,
https://doi.org/10.1007/978-981-95-2104-3_1

Keywords Precision medicine · Personalized medicine · Genetic composition · Cancer · Neurodegenerative diseases · Disease prevention

1 Introduction: Precision Medicine

After spending more than 10 years and almost US$3 billion, the first human genome was sequenced in 2001, and since then, the technology has advanced significantly in speed and affordability. Nowadays, genomes can be sequenced in a single day for about $1000. Consequently, genome sequencing is becoming a part of medicine, especially in cases where traditional methods have not been successful in diagnosing rare disorders (Hodson 2016).

The phrase 'precision medicine' has gained significant attraction in recent years. In 2008, Clayton Christensen, a business strategist, coined the term 'precision medicine' to describe how molecular diagnostics enables doctors to definitively diagnose a patient's cause of disease without relying solely on intuition (Katsnelson 2013).

2 Evolution of Precision Medicine

'To wrest from nature the secrets which have perplexed philosophers in all ages, to track to their sources the causes of disease, to correlate the vast stores of knowledge, that they may be quickly available for the prevention and cure of disease—these are our ambitions', stated Sir William Osler, summarizing the objectives of medicine.

Precision medicine (PM) aims to use preventative measures or therapies that are specific to the illness course or symptoms of a certain person. The management of communicable illnesses has historically aligned with PM's objectives in many aspects, since it aims to identify the organisms causing the sickness and establishes data archives to guide targeted therapy for infections (Brownstein et al. 2010). Particularly, DNA sequencing techniques have made a significant contribution to the availability of data that might help prevent and treat illness. It took $95 million to sequence the first human genome in 2001. Since then, costs have decreased in tandem with sequencing technologies' automation, mirroring Moore's laws in the electronics industry. These days, sequencing a whole exosome costs less than $1000. The goal of PM's technological integration with medicine is to establish a data ecosystem that will enable more accurate diagnosis and treatment of each patient's unique condition (Ramaswami et al. 2018).

Theoretically, PM approaches will contribute to the improvement and knowledge of disease taxonomy, increasing specificity regarding the aetiology of complicated illnesses like obesity, cancer and heart disease. They may also cause a paradigm shift in the thinking behind possible therapeutic interventions, maximizing disease treatment while minimizing unfavourable outcomes (Collins and Varmus 2015).

3 Precision Medicine and Drug Development

Precision medicine is fundamentally the pursuit of comprehending the root cause of disease in unique people. By definition, this drug-related opportunity development is not the same as any other mentioned stages. Finding the underlying cause of a patient's illness can provide fresh perspectives on the basic biology and pathophysiology of the condition. In the long run, this will help create medications that specifically address the disease's root cause. The medication will next be developed for the patient groups that stand to gain the most (Raviña Rubira 2011). Examples are taken specifically from the therapeutic areas that are anticipated to move toward a precision medicine framework the fastest: cancer and hereditary illnesses with a significant genetic component, especially neurological disorders like epilepsy.

3.1 Cancer

With 9.6 million fatalities globally, cancer is one of the main causes of mortality (Freddie Bray et al. 2018; Hanahan and Weinberg 2000). This illness is typified by unchecked cell proliferation, resistance to antigrowth agents, apoptosis avoidance, persistent angiogenesis invasion and organ metastasis. Chemotherapy medications have a tendency to target rapidly proliferating cells, which includes healthy cells like blood, digestive tract cells and hair follicles. Toxic effects on healthy tissues result from this. Through improved tumour targeting and careful avoidance of healthy cells, targeted cancer therapies seek to reduce or eliminate these adverse effects (Hanahan and Weinberg 2011).

Precision medicine was undoubtedly pioneered by the discipline of oncology. These days, therapies can be created to more precisely target cancer cells using two main techniques:

(a) Creating responses to cancer cells by artificially altering patients' immune systems (immunotherapy) (Kosorok and Laber 2019).
(b) Selectively interfering with processes necessary for the growth or survival of cancer cells (Naithani et al. 2021).

For both approaches, access to tumour cells enables an investigation into tumour biomarkers using various tools, including companion diagnostics, gene expression profiling and proteomics (Stegmeier et al. 2010).

Limited success has been observed in the results of one randomized clinical trial and a few feasibility and/or tumour response studies centred around the idea of precision medicine.

These biomarker investigations are used to identify master regulators and important drivers of networks and pathways that support cancer survival and proliferation in order to develop pathway-based targeted therapies. While it was widely believed that a pathway-based strategy would be the mainstay of cancer treatment, the need of acquired drug resistance in patients has cast doubt on this notion (Friedman et al. 2015). The USFDA-approved medications that specifically target these pathways can then be found, or chances for novel target discovery or repurposing of existing medications can be investigated.

Recent advancements in cancer treatment have brought immunotherapy to the forefront thanks to decades of improvement in our understanding about tumour immunology. Strategies based on immunotherapy are distinct from those based on chemotherapy and radiation in that the latter directly target the growth of the cancer cell, while the former indirectly target the tumour by boosting the immune response against the malignancy. Through a variety of cutting-edge techniques, immunotherapies seek to maximize and strengthen the power of the cancer immunity cycle by stimulating immune cells to strengthen and augment the host response or by combating immunosuppression (Yang 2015; Farkona et al. 2016).

3.2 Epilepsy

Over 50 million individuals worldwide suffer from epilepsy, a common neurological illness. The aim of treating epilepsy is to enhance the patients' overall quality of life and induce seizure remission without creating any negative side effects. The aetiology and clinical manifestation of epilepsy vary greatly, making seizure prevention and treatment more challenging. Approximately one-third of patients still have uncontrollable seizures despite the availability of over 24 anti-epileptic medications on the market (Brodie et al. 2012; Sankaraneni and Lachhwani 2015). Precision medicine pertains to a customized approach to medical care that takes into account the unique attributes of every patient (Bruxel et al. 2021) (Table 1).

4 Precision Medicine and Novel Therapies

The objective that precision medicine potentially discovers novel cures for illnesses that were previously incurable, is a significant portion of its promise. Originally, the phrase was used to describe the development of targeted treatment against organ-specific tumour features. Therapies like erlotinib or gefitinib for lung cancer have been developed as a result of the change from organ type to molecular identification. These treatments have provided an alternative to chemotherapy regimens (Lynch et al. 2004) (Fig. 1).

The potential to tailor treatment plans grounded on the information gathered regarding the individual's unique traits and the biologic composition of the tumour, independent of the organ where the cancer started, has demonstrated the promise of PM in recent times. Certain medications have been licensed to use in some circumstances in people who inherit the *BRCA* mutation (Tutt et al. 2010).

Table 1 Broad-spectrum function of precision medicine in various disease management

Title	Major findings	Reference
Precision medicine and management of rheumatoid arthritis	A highly personalized and customized approach to patient care is implied by the term precision medicine (PM). Prognostic precision for rheumatoid arthritis (RA) is well established and includes indicators of genetics, autoantibodies and disease activity or structure. Subtypes of RA may go unnoticed and PM markers for RA diagnosis are usually indicators of RA categorization rather than diagnosis.	Aletaha (2020)
Precision medicine in chronic disease management: The multiple sclerosis BioScreen	The innovative tool from BioScreen tackles the difficulties of dynamically managing a complicated chronic illness that is multiple sclerosis; the way patients and professionals interact with it shows how much translational digital medicine has advanced. It is an application for precision medicine designed to treat multiple sclerosis.	Gourraud et al. (2014)
Precision medicine in the management of type 2 diabetes	The development and use of this novel knowledge to clinical practice were critically addressed in this review, along with potential avenues for supporting type 2 diabetes diagnosis and treatment through the application of precision medicine.	Gloyn and Drucker (2018)
Precision medicine and personalized medicine in cardiovascular disease	Precision medicine has potential in cardiovascular medicine that extends to all phases of disease progression and includes focused therapy methods, risk prediction and prevention interventions. New advances in genomes and other omics will be helpful to precision medicine, but it also largely relies on imaging, functional testing and well-established biomarkers.	Currie and Delles (2018)
The development of precision medicine in clinical practice	Personalized therapy is the goal of precision medicine, which bases choices and procedures on each patient's individual biology and clinical 'signs and indications'. Precision therapy has advanced considerably in recent years. Clinical trials have demonstrated an improvement in the therapeutic efficacy of newly discovered and produced medicinal medicines.	He et al. (2015)
A precision medicine approach to SARS-CoV-2 pandemic management	Numerous methods to stop the spread of the virus have been developed as a result of the coronavirus disease 2019 (COVID-19) pandemic, which has affected people's daily habits and ways of living. The diagnosis, prevention and treatment of diseases are being revolutionized by precision medicine (PM). Recent advancements in PM's approach to COVID-19 disease prevention, assessment and management are highlighted, along with current information.	Crisci et al. (2020)

(continued)

Table 1 (continued)

Title	Major findings	Reference
Emerging role of precision medicine in cardiovascular disease	An individual's genetics, lifestyle and exposures are taken into consideration in precision medicine as factors influencing their heart health and disease characteristics. This review examined the extreme phenotypes' reaction to population-based vs. high-risk treatment approaches as a pretext, as well as the existing spectrum of both cardiovascular wellness and illness.	Leopold and Loscalzo (2018)
Precision medicine and sickle cell disease	The clinical consequences of sickle cell disease (SCD) vary with time as a significant potential role for precision medicine, which selects specific treatments for certain patient groups, arises from this variability. However, the lack of effective treatments and the challenges in distinguishing between various prognostic categories restrict the use of PM in SCD.	El Hoss et al. (2022)
The role of precision medicine in interstitial lung disease	A change in perspective could be beneficial for the treatment of interstitial lung disease (ILD). Precision medicine in ILD is still in initial stages of development when compared to therapeutic fields like oncology.	Maher et al. (2022)
Precision medicine for multiple sclerosis promotes preventative medicine	Young individuals who have multiple sclerosis (MS), a chronic or a lifelong illness for which there is now no cure, later tend to suffer from severe brain damage. Using clinical status, disease subtype and other variables to inform recommendations, PM for MS seeks to offer a more rigorous and sophisticated approach to care.	Hansen and Okuda (2018)
Precision medicine in cardiology	By means of the advancement and implementation of precision medicine, cardiovascular disease study and therapeutic communities are in a prime position to address further our comprehension of human health and malady by addressing the non-communicable reasons for mortality epidemic. Precision medicine will transform healthcare into a more dynamic, interconnected system where patients actively participate in shared decision-making and provide data, rather than being viewed as a passive object on which measurements are made.	Antman and Loscalzo (2016)

(continued)

Table 1 (continued)

Title	Major findings	Reference
Precision medicine in the care of rheumatoid arthritis: Focus on prediction and prevention of future clinically-apparent disease	There is growing knowledge that a number of indicators can be used to predict the probability of an individual contracting rheumatoid arthritis, which is high in the future. It was determined from this analysis that in order to identify people for intervention, the most precise illness prognosis models will require the integration of clinical and demographic information, biomarkers and potentially medical imaging data. The main points of precision medicine in RA were outlined in this review, with an emphasis on illness prevention and prediction.	Mahler et al. (2020)
Biomarkers for precision medicine in airways disease	The clinical condition known as chronic obstructive pulmonary disease (COPD) is complicated. The care could be revolutionized by the possible stratification of this heterogeneous patient population using predictive biomarkers. In order to monitor and enhance clinical outcomes in this illness, as well as to optimize diagnosis and therapy options, precision medicine was the focus of this review.	Rooney and Sethi (2015)
Precision medicine in chronic obstructive pulmonary disease	In managerial decision-making, the fundamental principle of precision medicine is to account for individual variability. Research is currently underway to demonstrate the effectiveness of biomarker-based enrichment of clinical trials in COPD.	Hurst (2016)
Toward precision medicine in Parkinson's disease	With new physiological and pathological understandings of a particular disease, precision medicine aims to develop prevention and therapy plans tailored to each patient. In healthcare, a multifaceted data-driven strategy is poised to elevate 'precision medicine' to a higher level of more individualization, moving away from treating specific diseases like Parkinson's disease (PD) and toward treating individual patients. This review focused on the prospects of precision medicine in PD, using multifaceted data regarding deep brain stimulation (DBS), molecular imaging and omics. It is anticipated that precision medicine will include the finest available evidence-based information to help people with PD receive the best possible care in the future.	Bu et al. (2016)

(continued)

Table 1 (continued)

Title	Major findings	Reference
Precision medicine in kidney disease: The patient's view	The knowledge of the processes that cause kidney disease is constantly being enhanced by nephrology research. Informing and facilitating the application of precision medicine—in which patient care is customized for each patient based on the mechanisms underlying their illness to improve the likelihood of treatment achievement—is one of the objectives of these research initiatives.	Brown et al. (2020)
A precision medicine initiative for Alzheimer's disease: The road ahead to biomarker-guided integrative disease modelling	Alzheimer's disease (AD) continues to be a deadly worldwide epidemic. According to this review, a prototype transfer toward the application using precision medicine (PM) and its enhanced risk screening, detection and treatment of Alzheimer's disease are long overdue. The convergence of cutting-edge technology advancements, including big data science and systems biology, will offer the framework for how PM for AD can be accomplished.	Hampel et al. (2017)
Precision medicine in diabetes prevention, classification and management	Abnormal β-cell biology is a key factor in diabetes mellitus. The foundation for precision medicine, which aims to tailor diabetes prevention and treatment, has been established by recent developments.	Xie et al. (2018)
Application of precision medicine in neurodegenerative diseases	The primary elements that could be included in targeted and individualized treatment plans for neurodegenerative diseases are discussed in this overview. According to this viewpoint, the development of web-based networks can play a key role in supporting clinical and therapeutic decisions, advancing a more preventive and participatory approach to treating neurodegenerative diseases and implementing precision medicine strategies across various specialized centres.	Strafella et al. (2018)
Precision medicine for managing diabetes	How precision medicine promises to integrate biomarker data on diabetes risks and responses, physiologic and behavioural data from novel sensors, and individual data about genetic predispositions to diabetes has been discussed in this review.	Klonoff (2015)

(continued)

Table 1 (continued)

Title	Major findings	Reference
Precision medicine: Diagnosis and management of obesity	PM is medical care that targets the needs of individual patients. The effectiveness of the various treatments for obesity and associated comorbidities varies greatly from person to person. In addition to the lack of specialized medication candidates to enable innovative and tailored treatments for obese subpopulations, there are few appropriate biomarkers which predict a patient's response to treatment.	Frühbeck et al. (2018)
Personalized management of bone and mineral disorders and precision medicine in end-stage kidney disease	End-stage renal disease frequently has prolonged kidney problems causing disturbance of the mineral bones, which is connected to a higher risk of death. The incorporation of precision medicine to CKD-MBD therapy and its potential to enhance clinical-therapeutic outcomes were covered in this review.	Jovanovich and Kendrick (2018)
Precision medicine in non-communicable diseases	Cardiovascular diseases account for the majority of non-communicable disease (NCD)-related mortality, which are followed by diabetes, COPD and various forms of cancer. In order to optimize advantages, the purpose of precision medicine as a predictive and preventive technique is to improve patient outcomes.	Hasanzad et al. (2019)
Feasibility of precision medicine in hypertension management—Scope and technological aspects	Giving the patient the greatest care possible with the fewest possible side effects is the cornerstone of clinical disease management. For more accurate and individualized hypertension therapy, a new photoplethysmography-based method was recently discovered to be more accurate. In the review, the scope and the technological aspects of precision medicine in the management of hypertension were successfully described.	Nitzan and Nitzan (2022)
Perspective for precision medicine for tuberculosis	Worldwide, tuberculosis is the most fatal reason of death. Patients with tuberculosis have been treated using a systematic approach over the past 50 years. A set of precision medicine-guided treatments was suggested in this review, which could influence clinical management choices by taking into account the genetic makeup of *Mycobacterium tuberculosis* strains as well as host immunity.	Lange et al. (2020)

(continued)

Table 1 (continued)

Title	Major findings	Reference
Precision medicine in Fabry disease	A shortage in α-galactosidase A causes Fabry disease (FD). This article examines the clinical presentation, diagnosis and interdisciplinary clinical management of FD. With a particular emphasis on precision medicine, this article takes into account individual variability in genetic mutations enabling doctors to more correctly predict which prevention and treatment is suitable for each individual.	Lenders and Brand (2021)
Achieving precision medicine in allergic disease: Progress and challenges	The idea behind precision medicine is to use various levels of patient-specific data to customize available therapies and diagnoses for each patient. Although, despite the recent achievements of biological products and the development of technology, personalized medicine in allergy remains in its infancy, it aims at integrating big datasets, which are encouraging and show how useful it is to better understand allergic endotypes in order to build better treatments.	Proper et al. (2021)
Precision medicine in oral health and diseases: A systematic review	Personalized medicine that may create customized medical treatments for each patient is known as precision medicine (PM). One interesting application that requires more research is 'precision dentistry' (DP). Through the identification of risk factors and deformities including orofacial clefts, PM seeks to provide insight into cancer preventive techniques. The potential for building a global database network will help with disease outbreak diagnosis, forecasting and prevention, saving the world's healthcare systems a substantial amount of money.	Malcangi et al. (2023)
Precision medicine in pancreatic cancer: Treating every patient as an exception	Recent advances in overall survival in other cancers due to precision medicine have not helped patients with pancreatic cancer. The present work focuses on identification of exceptional patients of pancreatic cancer and management or control of the disease by precision medicine treatment.	Herbst and Zheng (2019)
Precision medicine approach to Alzheimer's disease: Successful pilot project	There are currently no viable treatments for neurodegenerative disorders like Alzheimer's disease. Some studies have shown that a precision medicine approach in oncology, which targets the presumed genetic drivers of the disease process therapeutically, improves results. This study sought to ascertain whether a proof-of-concept experiment for disease impairment was effective enough for a precise medicine approach to Alzheimer's.	Toups et al. (2022)

(continued)

Table 1 (continued)

Title	Major findings	Reference
Precision medicine for lysosomal disorders	About 70 monogenic ailments resulting from a malfunction in lysosomal function are included in the group of hereditary metabolic disorders known as lysosomal diseases (LDs). A developing strategy for treating and preventing disease is precision medicine (PM). PM can significantly enhance LD management, but it will take some time for the necessary improvements to become commonplace.	Pinto e Vairo et al. (2020)
Precision medicine in rhinosinusitis	The presentation of precision medicine research and clinical advancements that are anticipated to change the clinical approach to chronic rhinosinusitis (CRS) is the scope. This review mainly stated that the precision medicine or therapy is able to treat not only severe chronic upper airways diseases but also milder and more prevalent cases that are being seen in the community.	Vlastos et al. (2019)
Precision medicine in fatty liver disease/non-alcoholic fatty liver disease	Excess fat deposition in the liver is a hallmark of non-alcoholic fatty liver disease (NAFLD). It is anticipated that a precision medicine approach to fatty liver disease (FLD) will enhance patient care and create more effective treatments. Based on the recently suggested subcategorization, a precision medicine approach for FLD was given in this publication.	Valenzuela-Vallejo et al. (2023)
Precision medicine in atopic diseases	It is becoming more well acknowledged that atopic disorders are diverse and can range widely in severity. The goal of precision medicine is to direct targeted, customized therapy using known endotype phenotypes. Given the availability of increasingly precise treatments that can target specific pathways, the effort targeted at detailed characterization of diseases to assist disease management is essential.	Cianferoni and Annesi-Maesano (2019)
Treatments for non-small-cell lung cancer: The multiple options for precision medicine	Advanced training for non-small-cell lung cancer (NSCLC) therapy has changed recently due to developments in molecular diagnostics. With the advancement of treatments, it is now more crucial to enhance the early detection of lung lesions. Giving a summary of the developments in precision medicine, lung cancer treatment at various phases of the disease is the aim.	Pons-Tostivint and Bennouna (2022)

(continued)

Table 1 (continued)

Title	Major findings	Reference
Precision medicine approaches to vascular disease: JACC focus seminar 2/5	The most widespread vascular condition in the world, atherosclerosis is the primary cause of coronary artery disease, peripheral vascular disease and other conditions. Precision medicine is being applied to atherosclerosis and Mendelian arteriopathies through integrated approaches, despite the potential differences in their genetic makeup.	Miller et al. (2021)
Towards precision medicine in systemic lupus erythematosus	The chronic, multi-organ, autoimmune disease known as systemic lupus erythematosus (SLE) has no recognized aetiology. In SLE, precision medicine aims to meet the needs of ideal patient assessment, disease course prediction and therapy response upon diagnosis. Although there is significant potential for increasing precision in lupus care, this aim is unlikely to be universally achieved in the near future.	Lever et al. (2020)
Precision medicine in breast cancer	Future developments in detection and treatment of breast cancer rely on the practical application of precision medicine and genomics. The article also discusses the difficulties in implementing precision medicine in cancer treatment as well as the identification and treatment of cancer of the breast utilizing precision medicine employing imaging.	Odle (2017)
Precision medicine approaches for diabetic kidney disease: Opportunities and challenges	Chronic kidney disease (CKD) affects about 700 million individuals globally. This review described recent developments of precision medication for kidney disease caused by diabetes as well as potential research opportunities.	Tye et al. (2021)
Precision medicine in lupus nephritis	A common form of systemic lupus erythematosus (SLE), lupus nephritis (LN) is marked by a variety of clinical and histological characteristics and places a significant burden on patients. The main focus of this research was how multi-omics methods can help direct precision pharmacotherapeutics and precision medicine in LN.	Sharafaldin et al. (2024)
Integrating precision medicine into the contemporary management of gynaecologic cancers	Patients with advanced gynaecologic cancers are still difficult to treat. The current developments in precision medicine for the treatment of gynaecologic cancers are reviewed in this article. Future research must find, identify and validate reliable biomarkers with significant prognostic/predictive capacities because there are obstacles to the widespread use of precision medicine.	Wolford et al. (2022)

(continued)

Table 1 (continued)

Title	Major findings	Reference
Personalized treatment for multiple sclerosis: The role of precision medicine	Multiple sclerosis (MS), a chronic, autoimmune disease that affects the central nervous system, is a complex and multifaceted sickness. A promising strategy to treat multiple sclerosis is precision medicine, which uses cutting-edge technologies like genomes, metabolomics and imaging to find particular biomarkers. This review paper aims to provide an overview of the current and future directions of precise medicine in MS.	Pathak (2023)

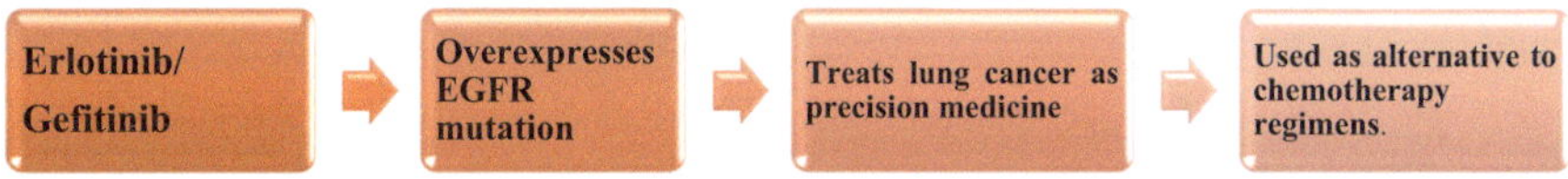

Fig. 1 Demonstration of precision medicine as an alternative to chemotherapy regimen

Targeting the defective gene is the particular goal of gene editing methods and gene therapy, which have developed into a treatment option for individuals with some uncommon disorders like haemophilia B. Novel treatments targeting the *CFTR* gene, such as ivacaftor and lumacaftor, have been authorized for use in other uncommon disorders, like cystic fibrosis. Patients with certain *CFTR* mutations may benefit from a combination of these therapies; it has been shown that these treatments improve respiratory function (Fig. 2).

There are several obstacles to the effectiveness of PM techniques in this way:

- First, it is far from certain that these methods have broad applicability. Patients have not benefited from cancer trials intended to precisely explore the use of genetic alterations to direct therapy as defined by PM (Prasad et al. 2016).
- Second, although PM approaches simultaneously build prognostic and predictive biomarkers, these can be used to develop novel medicines or assist doctors in directing specific therapy for a patient's ailment or avoiding hazardous therapies (Parkinson et al. 2012).
- Third, a significant amount of money has been invested in PM by the public and commercial sectors with an eye toward helping sick people.

5 Introduction: Personalized Therapies

The phrase 'personalized therapy' was first used in published works in 1999; however, some of the fundamental ideas in the subject date all the way back to the early 1960s (Jain 2002). Instead of employing 'established norms of care' that are established by averaging responses from sizable cohorts of clinical trial participants,

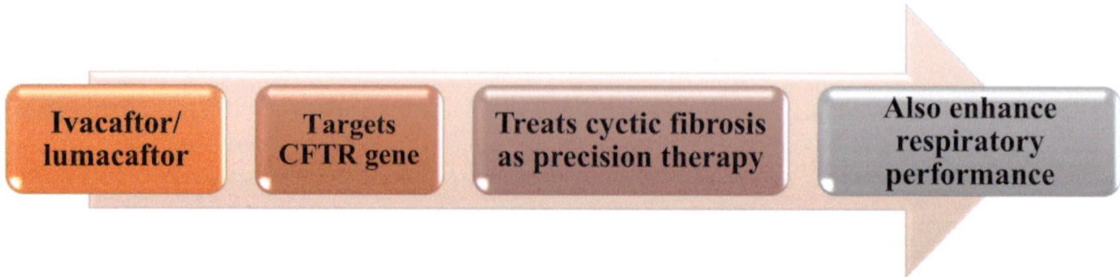

Fig. 2 Demonstration of how precision medicine treats uncommon disorders

personalized medicine was founded on the notion that treatment decisions should be made based on each patient's particular characteristics, such as their gender, age, height/weight and surroundings (Jain 2015) (Table 2).

6 Conclusion

Two healthcare philosophies that seek to customize medical interventions and therapies for specific patients are precision medicine and personalized medicine. These methods are predicated on the notion that each patient's genetic composition, way of life and surroundings significantly influence their health and reaction to therapy (Farrokhi et al. 2023). Although they are somewhat different, these two are frequently used. To understand the underlying processes of disease and customize treatments for each patient, precision medicine analyses a patient's DNA, RNA and protein expression. For creating a thorough treatment plan that is specific to each patient, personalized medicine considers not just genetic information but also other elements including the patient's lifestyle, surroundings and medical history. A number of elements will come together for the broad use of precision and customized medicine technologies, including changing medical–engineering education and regulations that encourage innovative clinical trial designs (Ho et al. 2020). The demand for personalized therapies along with precision medicines is growing day by day and in near future these two therapies will occupy a large portion of medical world in terms of disease identification, management and prevention (Figs. 3 and 4).

7 Future Research

After observing lots of research on precision medicine and personalized therapies, it is confirmed that though they play a very vital role in many kinds of disease management, still they often face challenges in some specific type of disease prevention or management. Therefore, to use these techniques wholly in near future without facing any kind of unusual or unnecessary hurdles, more experiments should be done to check their functionality and authenticity.

Table 2 Broad-spectrum functions of personalized therapies in various disease management

Title	Major findings	Reference
Goals and targets for personalized therapy for HCC	The incidence of hepatocellular carcinoma (HCC) is steadily increasing. In order to provide individualized care targeted at enhancing overall survival, future therapy should ideally focus on numerous pathways in the HCC cascade with combination medicines.	Couri and Pillai (2019)
Personalized therapy for breast cancer	Breast cancer is a multifaceted illness with numerous morphological, clinical and molecular characteristics. Our knowledge of the disease's processes has increased thanks to contemporary methods for assessing breast tumours employing transcriptomics and genomes, which has shown to be more useful for more individualized patient care than more conventional laboratory techniques for evaluating tumours.	De Abreu et al. (2014)
Future of personalized medicine in oncology: A systems biology approach	A multidisciplinary systems biology endeavour will be necessary to determine the factors influencing tumour behaviour that are the most promising targets for treatment. Finding molecular drivers with systems biology techniques will enable smaller, quicker, less expensive and more customized clinical trials, which will boost success rates and speed the introduction of successful treatments into the clinical toolbox.	Gonzalez-Angulo et al. (2010)
Personalized therapy for lung cancer	When a sensitizing mutation in the receptor for epidermal growth factor gene was discovered, it spurred a thorough investigation. A small bit of tissue can be used to evaluate several molecular changes in the same sample according to a technique called next-generation sequencing (NGS). With the use of this technology, the majority of molecular changes can be found in a single test, sparing tissue for further research.	Moreira and Eng (2014)
Heterogeneity of liver cancer and personalized therapy	Chronic liver inflammation and fibrosis are the most common conditions under which hepatocellular carcinoma (HCC) develops. The complicated mechanism and phenotype of liver cancer heterogeneity will be explained in this review paper, which will also assist us in implementing precision medicine in a highly customized way.	Li and Wang (2016)

(continued)

Table 2 (continued)

Title	Major findings	Reference
Personalized therapy for lung cancer: Striking a moving target	The deadliest illness in the world, lung cancer claims around 1.6 million lives annually. For patients with NSCLC driven by oncogenes, molecular targeted therapy marked the beginning of a new era in treatment. Lung cancer patients' resistance to treatment, resistance mechanisms and current developments in targeted therapy are the main topics of this review.	Pakkala and Ramalingam (2018)
Personalized therapy in inflammatory bowel disease	Several well-known disease phenotypes comprise the spectrum of inflammatory bowel disorders (IBD), including Crohn's disease (CD) and ulcerative colitis (UC). The need to choose which drug will be most beneficial for each patient and how quickly it should be begun is highlighted by the variety of therapeutic choices available.	Calderón et al. (2024)
Personalized medicine in Crohn's disease	Although the specific course of Crohn's disease varies greatly from one to another individual, it usually follows a relapsing-remitting pattern. Precision Medicine Initiative has recently fuelled a rising push toward personalized or precision medicine across all areas. The scientific community's support and related funding requests will be crucial in propelling additional advancements in personalized medicine.	Noor et al. (2020)
Personalized therapy in pain management: Where do we stand?	Currently, research is being done to determine how genomic variants affect and how well pain medications work. In addition to genetic variations, other factors may affect pharmacotherapy and should be taken into account when pain therapists prescribe medication, including age, disease, concurrent medication and patient compliance.	Stamer et al. (2010)
Personalized type 2 diabetes management: An update on recent advances and recommendations	Newer recommendations advocate for a customized approach to diabetes care to improve overall health outcomes, medication adherence, patient satisfaction and quality of life. Together with patients, clinicians should create individualized treatment plans that address issues including weight control, targeted glycaemic control and avoiding consequences like hypoglycaemia.	Williams et al. (2022)

(continued)

Table 2 (continued)

Title	Major findings	Reference
Perspectives for personalized therapy for patients with multidrug-resistant tuberculosis	Tuberculosis is the most common reason of death globally from a single microbiological infection. The individualized care, significantly affects how well multidrug-resistant tuberculosis (MDR-TB) patients respond to treatment. Getting these advancements to the patients who need them the most is now a difficulty.	Lange et al. (2018)
Predictive genomic tools in disease stratification and targeted prevention: A recent update in personalized therapy advancements	Genomic testing has led the medical community to create customized, preventive and predictive treatment in the present medical revolution. The use of artificial intelligence (AI) and machine learning must be in tune with the advancement of preventative and personalized medicine in order to meet the demands of modern clinical practice.	Jain et al. (2022)
Current therapy of the patients with MDS: Walking towards personalized therapy	Dysplasia, peripheral cytopenias and inefficient haematopoiesis are the hallmarks of myelodysplastic syndromes (MDS). Numerous hypomethylating agent-based treatment combinations (eprenetapopt, venetoclax etc.) have emerged in recent years and have shown promise in treating MDS, even individuals with high-risk characteristics.	Palacios-Berraquero and Alfonso-Piérola (2021)
Towards personalized therapy for multiple sclerosis: Prediction of individual treatment response	Effective therapy must be started to prevent multiple sclerosis-related impairment. This review paper concludes that individual response to disease-modifying medications from the onset can be predicted with the use of demographic, clinical and paraclinical data.	Kalincik et al. (2017)
Personalized therapy of neurological disorders	The use of personalized medicine concepts to increase the effectiveness and lessen the side effects of different treatments is known as personalized neurology. This chapter provides examples of individualized methods to the treatment of certain neurologic illnesses, some of which are integrated into management algorithms.	Jain and Jain (2021a)
Colorectal cancer: Personalized therapy	The second most common kind of cancer is colorectal cancer (CRC). Particularly for patients with metastatic cancer, the way this disease is treated has changed significantly. In an effort to further individualize treatment and find new treatment targets, several studies have lately attempted to further profile patients with colorectal cancer.	Leichsenring et al. (2015)

(continued)

Table 2 (continued)

Title	Major findings	Reference
Personalized medicine: Marking a new epoch in cancer patient management	Personalized medicine has a broad range of clinical uses in cancer, like screening, diagnosis and therapy efficacy prediction. Additionally, PM is reorienting the focus of patient care from initial care to early intervention and prevention for high-risk patients. High-throughput techniques can be utilized for PM in addition to traditional single molecular markers.	Diamandis et al. (2010)
Advances in personalized therapy for lung cancer	The management of lung cancer is quickly evolving due to advances in genomic medicine that can discover important genetic changes and then produce tailored treatments that match those changes. An overview of clinically authorized methods for tailoring lung cancer treatment is given in this article, along with a look at developing methods at different phases of clinical development.	Kim and Pandya (2013)
Immunomicelles for advancing personalized therapy	Personalized medicine aims to provide individualized treatment plans for each patient. In order to provide a customized approach to disease management, this review emphasizes polymeric immunomicelles (nanocarriers) that may be able to combine targeted medication therapy and therapeutic response monitoring with diagnostics.	Sawant et al. (2012)
Personalized management approach for OSA	OSA is a diverse disorder. Using respiratory phenotyping tools, recent breakthroughs in our understanding of the various causes of OSA have revealed novel targets or 'treatable traits' that can guide treatment. The most recent advancements and cutting-edge treatments for individualized OSA management strategies are highlighted in this succinct review.	Carberry et al. (2018)
Personalized therapy algorithms for type 2 diabetes: A phenotype-based approach	The biology of type 2 diabetes mellitus (DM) is complicated and it progresses over time. For the treatment of type 2 DM, a novel tailored algorithm was created. Six algorithms are suggested based on the primary characteristics displayed by the patient. Physicians may find this easily navigable algorithm useful in selecting and expeditiously optimizing a customized treatment strategy for each patient.	Ceriello et al. (2014)

(continued)

Table 2 (continued)

Title	Major findings	Reference
Personalized therapy in chronic viral hepatitis	At least 600 million people worldwide are chronic carriers of the primary hepatitis viruses, HBV and HCV. The combination of molecular biology and bio-mathematical modelling may provide a better solution for individualized antiviral therapy.	Brunetto et al. (2008)
Promises and challenges of personalized medicine to guide ARDS therapy	Finding novel, efficient therapies for acute respiratory distress is still difficult. Clinical trials can be adaptively enhanced by identifying biological and clinical traits that may respond differently to current and experimental treatments. It is still difficult to target pharmacologic and supportive therapies based on the stage of the respiratory failure syndrome.	Wick et al. (2021)
Personalized therapy approach for hospitalized patients with Coronavirus Disease 2019	There are several consequences for hospitalized patients with respiratory symptoms caused by the COVID-19. According to the study's findings, COVID-19 individuals experience a variety of issues that can be identified by analytics patterns. Every circumstance calls for a unique therapeutic strategy. An independent risk factor for survival is personalized therapy.	Garcia-Vidal et al. (2022)
Personalized therapy in hepatocellular carcinoma: Molecular markers of prognosis and therapeutic response	Hepatocellular carcinoma (HCC) is becoming a major health issue globally. Researches have shown a gene profile that appears to enhance responsiveness and is involved in medication absorption, processing and excretion. Overall, there is little usage of biomarkers for HCC in clinical settings, despite encouraging clinical results.	Burkhart et al. (2017)
Personalized therapy for common variable immunodeficiency	In the 1970s, the clinical descriptive diagnosis of common variable immunodeficiency (CVID) was established. Incremental targeted therapy for these individuals is anticipated in future due to continued growth of biologics and small-molecule immune modulators as well as incremental advancements in our understanding of the pathophysiology of CVID.	Bonilla (2020)

(continued)

Table 2 (continued)

Title	Major findings	Reference
Novel biomarkers for asthma stratification and personalized therapy	All asthma patients are currently advised to follow a step-by-step pharmacological treatment plan that is tailored primarily to the severity of their condition and focus on the shortest disease-controlling step. This review's objective is to provide the reader with an update on the various assays that have been suggested in the most current literature on asthma for assessing possible biomarkers for patient classification and treatment customization.	Bartminski et al. (2015)
Personalized approaches to clopidogrel therapy: Are we there yet?	One of the most often given drugs in the world is clopidogrel. The body's reaction to clopidogrel is also influenced by concurrent drug use. The USFDA has recently issued recommendations that highlight the potential for individualized decision-making for clopidogrel candidates.	Anderson et al. (2010)
Personalized therapy of non-small cell lung cancer (NSCLC)	NSCLC makes up more than 85% of all lung malignancies. In the treatment of advanced stage NSCLC, the shift from chemotherapy to a more individualized strategy has taken place. All NSCLC patients should benefit from such a tailored strategy, which is anticipated to apply to all stages of the disease.	Gadgeel (2016)
Personalized medicine—one size fits one: Tailoring pain therapy to individuals' needs	Oncology is where this medical approach is more well established, although customized pain management is showing promise. To find phenotypic subgroups of neuropathic pain patients, researchers use quantitative sensory testing. According to imaging studies, a distinct 'signature' of cerebral pain may be created by a combination of genetic, environmental and injury-specific factors. There is a continuous search for biomarkers of chronic pain in the central nervous system (CNS).	Ahmedzai (2013)
Personalized therapy of sarcomas: Integration of biomarkers for improved diagnosis, prognosis and therapy selection	There are over 50 distinct forms of 'sarcoma' and each is as distinct from the others. The use of biomarkers in oncology is discussed in this article, which primarily focuses on diagnosis, staging and grading. The kinds of biomarkers that are developing to help choose a treatment for sarcoma patients are also covered.	Ludwig (2008)

(continued)

Table 2 (continued)

Title	Major findings	Reference
Personalized therapy for metastatic melanoma: Could timing be everything?	Strong evidence was found that human immune-related functions are regulated during time. In patients with metastatic melanoma, it is proposed that coordinating the administration of lymphodepleting chemotherapy with biorhythms improves clinical outcomes when compared to using temozolomide in a traditional 'random delivery' fashion.	Dronca et al. (2012)
Personalized therapy of infectious diseases	With a focus on HIV, several examples of individualized management are shown to manage the disease progression. Nanobiotechnology has improved molecular diagnostics and it has also improved the formulation and administration of antimicrobial medicines for more individualized treatment of infectious disorders.	Jain and Jain (2021a, b)
Towards personalized therapy of aortic stenosis	The most prevalent cause of acquired valvular heart disease is calcific aortic stenosis (CAS). In patients with mild-to-moderate CAS, lipid-lowering treatments have proved to give positive effects in case of CAS. In the future, individualized pharmaceutical care planned to consider concurrent conditions may slow the progression of CAS.	Mazur et al. (2021)
Evolving personalized therapy for castration-resistant prostate cancer	Detailed molecular underpinnings for the development of castration-resistant prostate cancer (CRPC) have emerged as a result of advancements in genomic and molecular biologic technologies. A universal approach to curative therapy is still debatable due to patient tumour heterogeneity. Although CRPC patients are still in the early stages of individualized therapy, the best therapeutic drugs and their sequencing will soon be available.	Liu et al. (2014)
Personalized medicine and cancer	It was analysed that the gene expression profiles of hundreds of genes simultaneously differentiate between a cancer-associated gene expression profile and a typical profiling thanks to technologies like gene-expression microarrays. Modern customized medicine considers a patient's gene theory and medical history. This contrasts with traditional individualized medicine, which bases treatment on a patient's day-to-day life, surroundings, socioeconomic situations and family history.	Verma (2012)

(continued)

Table 2 (continued)

Title	Major findings	Reference
Developing adaptive personalized therapy for cystic fibrosis using reinforcement learning	The most prevalent fatal genetic condition affecting Caucasians is cystic fibrosis (CF). This study looked at the issue of determining the best adaptive tailored treatment for patients with cystic fibrosis. According to the results of a simulation study, reinforcement learning can be a useful tool for creating customized treatment plans to enhance long-term outcomes for chronic illnesses.	Tang and Kosorok (2012)
Personalized therapy for advanced breast cancer using molecular signatures	Breast cancer is the second most fatal disease among women in Korea. The most significant predicting indicators for all breast cancers are hormone receptors and HER2 status. This study primarily examined how customized medicine is being utilized to treat breast cancer in Korea by combining several hormonal treatments.	Jung (2009)
Personalized therapy in patients with anaplastic thyroid cancer: Targeting genetic and epigenetic alterations	Anaplastic thyroid cancer is one of the most dangerous human carcinomas and the most deadly type of thyroid cancer. Research into individualized, tailored medicines is showing a lot of promise as a means of finding successful treatment alternatives.	Smith and Nucera (2015)
Personalized matched targeted therapy in advanced pancreatic cancer: A pilot cohort analysis	Results are still poor even with improvements in the treatment of pancreatic cancer. Biomarker-driven clinical studies for this cancer have been attempted more recently. The pilot cohort study about the prevention mechanism of pancreatic cancer using personalized therapy was the primary focus of this particular paper.	Shaya et al. (2023)
Personalized therapy design for liquid tumours via optimal control theory	The creation of trustworthy techniques for defining individualized treatment plans based on the growing amount of experimental data on individual patients is one of the main issues facing cancer research today. Based on population dynamics modelling and optimal control theory, here the Controlled Treatment Theory for Therapy Development, conceptual structure to create customized, ideal cancer treatment regimens was introduced.	Angaroni et al. (2019)
Personalized therapy in endometrial cancer: Challenges and opportunities	The most prevalent gynaecologic cancer in the USA is still endometrial cancer. For women with advanced endometrial cancer, the use of targeted therapy seems to hold the promise of improving response and survival.	Westin and Broaddus (2012)

(continued)

Table 2 (continued)

Title	Major findings	Reference
Personalized medicine for non-small-cell lung cancer	The disease known as non-small-cell lung cancer (NSCLC) is diverse and has a high death rate. By identifying a particular genotypic abnormality and using target-specific therapy, personalized therapy may enhance treatment results. In this research, the available and experimental customized therapeutics for patients were summarized.	Mok et al. (2010)
Personalization of prostate cancer therapy through phosphoproteomics	For castration-resistant prostate cancer (CRPC), a number of novel therapies have been authorized by identifying biomarkers of pathway activity, personalized phosphoproteomics is a promising strategy to develop customized therapy.	Yang et al. (2018)
Recent advances in targeted therapy for inflammatory vascular diseases	Vascular illnesses are a major cause of death rates worldwide. One significant advancement in targeted treatment for inflammatory vascular disorders is personalized therapy. Treatment plans are progressively being tailored to each patient's unique genetic, biomarker and clinical characteristics by utilizing sophisticated modelling algorithms and clinical trial data.	Zhao et al. (2024)
Genomically personalized therapy in head and neck cancer	Upper aerodigestive tract tumours are part of the varied illness known as head and neck cancer, which is intended to enhance clinical outcomes, including survival. Knowing about the biology of cancer of the head and neck can help design tailored therapies for people with various molecular subtypes.	Aung and Siu (2016)
Current approaches for personalized therapy of soft tissue sarcomas	Soft tissue sarcomas (STS) include a range of geometries and clinical characteristics. Current advancements in individualized treatment of advanced and metastatic STS that rely on a combination of genetic testing and chemotherapeutic sensitivity and resistance assays (CSRA) were the focus of the current review.	Kirsanov et al. (2020)
A methodological approach to personalized therapies in metastatic cancer	Cancer that spreads from its primary site to a distant area of body is known as metastatic cancer. A methodical therapeutic strategy ought to be a component of personalized medicine. Therefore, in order to create therapy-linked models and find new target patterns, metastatic tumours must be made for creative approaches.	Reichle et al. (2010)

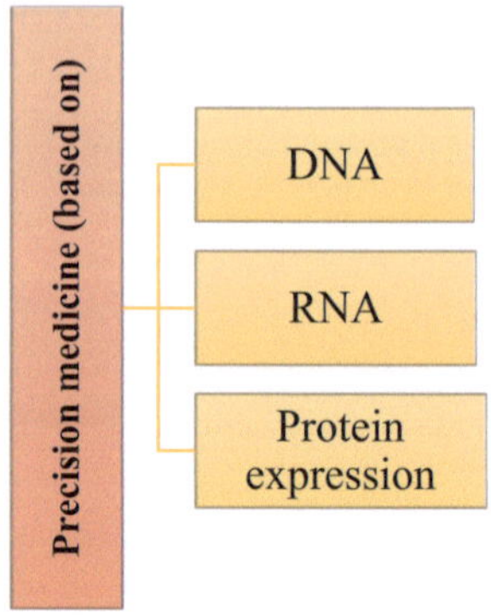

Fig. 3 The main principle behind precision medicine

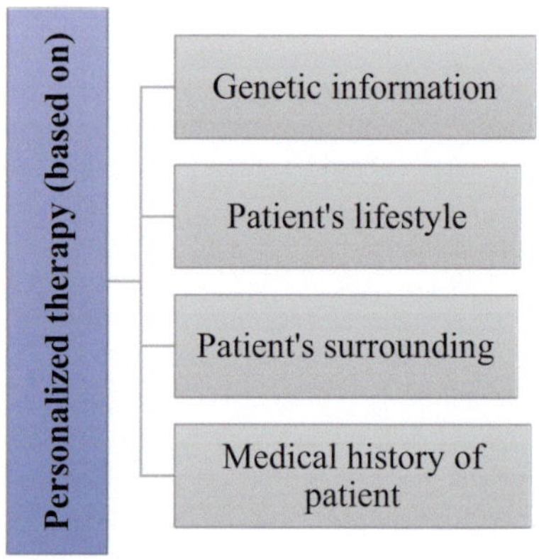

Fig. 4 The main principle behind personalized therapy

References

Ahmedzai SH (2013) Personalized medicine—one size fits one: tailoring pain therapy to individuals' needs. J Pain Palliat Care Pharmacother 27(1):83–85

Aletaha D (2020) Precision medicine and management of rheumatoid arthritis. J Autoimmun 110:102405

Anderson CD, Biffi A, Greenberg SM, Rosand J (2010) Personalized approaches to clopidogrel therapy: are we there yet? Stroke 41(12):2997–3002

Angaroni F, Graudenzi A, Rossignolo M, Maspero D, Calarco T, Piazza R, Montangero S, Antoniotti M (2019) Personalized therapy design for liquid tumors via optimal control theory. bioRxiv:662858

Antman EM, Loscalzo J (2016) Precision medicine in cardiology. Nat Rev Cardiol 13(10):591–602

Aung KL, Siu LL (2016) Genomically personalized therapy in head and neck cancer. Cancers Head Neck 1:1–0

Bartminski G, Crossley M, Turcanu V (2015) Novel biomarkers for asthma stratification and personalized therapy. Expert Rev Mol Diagn 15(3):415–430

Bonilla FA (2020) Personalized therapy for common variable immunodeficiency. Allergy Asthma Proc 41(1):19–25

Bray F, Ferlay J, Soerjomataram I, Siegel RL, Torre LA, Jemal A (2018) Global cancer statistics 2018: GLOBOCAN estimates of incidence and mortality worldwide for 36 cancers in 185 countries. CA Cancer J Clin 68(6):394–424
Brodie MJ, Barry SJ, Bamagous GA, Norrie JD, Kwan YP (2012) Patterns of treatment response in newly diagnosed epilepsy. Neurology 78(20):1548–1554
Brown KD, Campbell C, Roberts GV (2020) Precision medicine in kidney disease: the patient's view. Nat Rev Nephrol 16(11):625–627
Brownstein JS, Freifeld CC, Chan EH, Keller M, Sonricker AL et al (2010) Information technology and global surveillance of cases of 2009 H1N1 influenza. N Engl J Med 362:1731–1735
Brunetto MR, Colombatto P, Bonino F (2008) Personalized therapy in chronic viral hepatitis. Mol Asp Med 29(1–2):103–111
Bruxel EM, Bruno DC, do Canto AM, Geraldis JC, Godoi AB, Martin M, Lopes-Cendes I (2021) Multi-omics in mesial temporal lobe epilepsy with hippocampal sclerosis: clues into the underlying mechanisms leading to disease. Seizure 90:34–50
Bu LL, Yang K, Xiong WX, Liu FT, Anderson B, Wang Y, Wang J (2016) Toward precision medicine in Parkinson's disease. Ann Transl Med 4(2):26
Burkhart RA, Ronnekleiv-Kelly SM, Pawlik TM (2017) Personalized therapy in hepatocellular carcinoma: molecular markers of prognosis and therapeutic response. Surg Oncol 26(2):138–145
Calderón P, Núñez P, Nos P, Quera R (2024) Personalized therapy in inflammatory bowel disease. Gastroenterol Hepatol (Engl Ed) 47:763–770
Carberry JC, Amatoury J, Eckert DJ (2018) Personalized management approach for OSA. Chest 153(3):744–755
Ceriello A, Gallo M, Candido R, De Micheli A, Esposito K, Gentile S, Medea G (2014) Personalized therapy algorithms for type 2 diabetes: a phenotype-based approach. Pharmacogenomics Pers Med 7:129–136
Cianferoni A, Annesi-Maesano I (2019) Precision medicine in atopic diseases. Curr Opin Allergy Clin Immunol 19(6):654–664
Collins FS, Varmus H (2015) A new initiative on precision medicine. N Engl J Med 372:793–795
Couri T, Pillai A (2019) Goals and targets for personalized therapy for HCC. Hepatol Int 13:125–137
Crisci CD, Ardusso LR, Mossuz A, Müller L (2020) A precision medicine approach to SARS-CoV-2 pandemic management. Curr Treat Options Allergy 7:422–440
Currie G, Delles C (2018) Precision medicine and personalized medicine in cardiovascular disease. In: Sex-specific analysis of cardiovascular function. Springer, pp 589–605
De Abreu FB, Schwartz GN, Wells WA, Tsongalis GJ (2014) Personalized therapy for breast cancer. Clin Genet 86(1):62–67
Diamandis M, White NM, Yousef GM (2010) Personalized medicine: marking a new epoch in cancer patient management. Mol Cancer Res 8(9):1175–1187
Dronca RS, Leontovich AA, Nevala WK, Markovic SN (2012) Personalized therapy for metastatic melanoma: could timing be everything? Future Oncol 8(11):1401–1406
El Hoss S, El Nemer W, Rees DC (2022) Precision medicine and sickle cell disease. Hemasphere 6(9):e762
Farkona S, Diamandis EP, Blasutig IM (2016) Cancer immunotherapy: the beginning of the end of cancer? BMC Med 14:73
Farrokhi M, Taheri F, Khouzani PJ, Rahmani E, Tavakoli R, Fard AM, Rajabloo Y, Sadeghniiat-Haghighi A, Shahbazi K, Semnani MA, Manavi MS (2023) Role of precision medicine and personalized medicine in the treatment of diseases. Kindle 3(1):1–64
Friedman AA, Letai A, Fisher DE, Flaherty KT (2015) Precision medicine for cancer with next-generation functional diagnostics. Nat Rev Cancer 15(12):747–756
Frühbeck G, Kiortsis DN, Catalán V (2018) Precision medicine: diagnosis and management of obesity. Lancet Diabetes Endocrinol 6(3):164–166
Gadgeel SM (2016) Personalized therapy of non-small cell lung cancer (NSCLC). In: Lung cancer and personalized medicine: novel therapies and clinical management. Springer, pp 203–222

Garcia-Vidal C, Moreno-García E, Hernández-Meneses M, Puerta-Alcalde P, Chumbita M, Garcia-Pouton N, Linares L, Rico V, Cardozo C, Martínez JA, García F (2022) Personalized therapy approach for hospitalized patients with coronavirus disease 2019. Clin Infect Dis 74(1):127–132

Gloyn AL, Drucker DJ (2018) Precision medicine in the management of type 2 diabetes. Lancet Diabetes Endocrinol 6(11):891–900

Gonzalez-Angulo AM, Hennessy BT, Mills GB (2010) Future of personalized medicine in oncology: a systems biology approach. J Clin Oncol 28(16):2777–2783

Gourraud PA, Henry RG, Cree BA, Crane JC, Lizee A, Olson MP, Santaniello AV, Datta E, Zhu AH, Bevan CJ, Gelfand JM (2014) Precision medicine in chronic disease management: the multiple sclerosis BioScreen. Ann Neurol 76(5):633–642

Hampel H, O'Bryant SE, Durrleman S, Younesi E, Rojkova K, Escott-Price V, Corvol JC, Broich K, Dubois B, Lista S, Alzheimer Precision Medicine Initiative (2017) A precision medicine initiative for Alzheimer's disease: the road ahead to biomarker-guided integrative disease modeling. Climacteric 20(2):107–118

Hanahan D, Weinberg RA (2000) The hallmarks of cancer. Cell 100(1):57–70

Hanahan D, Weinberg RA (2011) Hallmarks of cancer: the next generation. Cell 144(5):646–674

Hansen MR, Okuda DT (2018) Precision medicine for multiple sclerosis promotes preventative medicine. Ann N Y Acad Sci 1420(1):62–71

Hasanzad M, Sarhangi N, Aghaei Meybodi HR, Nikfar S, Khatami F, Larijani B (2019) Precision medicine in non communicable diseases. Int J Mol Cell Med (IJMCM) 8(2):1–8

He M, Xia J, Shehab M, Wang X (2015) The development of precision medicine in clinical practice. Clin Transl Med 4:1–4

Herbst B, Zheng L (2019) Precision medicine in pancreatic cancer: treating every patient as an exception. Lancet Gastroenterol Hepatol 4(10):805–810

Ho D, Quake SR, McCabe ER, Chng WJ, Chow EK, Ding X, Gelb BD, Ginsburg GS, Hassenstab J, Ho CM, Mobley WC (2020) Enabling technologies for personalized and precision medicine. Trends Biotechnol 38(5):497–518

Hodson R (2016) Precision medicine. Nature 537(7619):S49

Hurst JR (2016) Precision medicine in chronic obstructive pulmonary disease. Am J Respir Crit Care Med 193(6):593–594

Jain KK (2002) Personalized medicine. Curr Opin Mol Ther 4:548–558

Jain KK (2015) Textbook of personalized medicine. Springer International Publishing

Jain KK, Jain KK (2021a) Personalized therapy of neurological disorders. In: Textbook of personalized medicine. Springer International Publishing, pp 213–262

Jain KK, Jain KK (2021b) Personalized therapy of infectious diseases. In: Textbook of personalized medicine. Springer International Publishing, pp 325–341

Jain N, Nagaich U, Pandey M, Chellappan DK, Dua K (2022) Predictive genomic tools in disease stratification and targeted prevention: a recent update in personalized therapy advancements. EPMA J 13(4):561–580

Jovanovich A, Kendrick J (2018) Personalized management of bone and mineral disorders and precision medicine in end-stage kidney disease. In: Seminars in nephrology, vol 38, No. 4. WB Saunders, pp 397–409

Jung KH (2009) Personalized therapy for advanced breast cancer using molecular signatures. Korean J Med 77(1):26–34

Kalincik T, Manouchehrinia A, Sobisek L, Jokubaitis V, Spelman T, Horakova D, Havrdova E, Trojano M, Izquierdo G, Lugaresi A, Girard M (2017) Towards personalized therapy for multiple sclerosis: prediction of individual treatment response. Brain 140(9):2426–2443

Katsnelson A (2013) Momentum grows to make 'personalized' medicine more 'precise'. Nat Med 19:249

Kim ES, Pandya KJ (2013) Advances in personalized therapy for lung cancer. Expert Opin Med Diagn 7(5):475–485

Kirsanov KI, Lesovaya EA, Fetisov TI, Bokhyan BY, Belitsky GA, Yakubovskaya MG (2020) Current approaches for personalized therapy of soft tissue sarcomas. Sarcoma 2020(1):6716742
Klonoff DC (2015) Precision medicine for managing diabetes. J Diabetes Sci Technol 9(1):3–7
Kosorok MR, Laber EB (2019) Precision medicine. Annu Rev Stat Appl 6(1):263–286
Lange C, Alghamdi WA, Al-Shaer MH, Brighenti S, Diacon AH, DiNardo AR, Grobbel HP, Gröschel MI, von Groote-Bidlingmaier F, Hauptmann M, Heyckendorf J (2018) Perspectives for personalized therapy for patients with multidrug-resistant tuberculosis. J Intern Med 284(2):163–188
Lange C, Aarnoutse R, Chesov D, Van Crevel R, Gillespie SH, Grobbel HP, Kalsdorf B, Kontsevaya I, van Laarhoven A, Nishiguchi T, Mandalakas A (2020) Perspective for precision medicine for tuberculosis. Front Immunol 11:566608
Leichsenring J, Koppelle A, Reinacher-Schick A (2015) Colorectal cancer: personalized therapy. Gastrointest Tumors 1(4):209–220
Lenders M, Brand E (2021) Precision medicine in Fabry disease. Nephrol Dial Transplan 36(Supplement_2):14–23
Leopold JA, Loscalzo J (2018) Emerging role of precision medicine in cardiovascular disease. Circ Res 122(9):1302–1315
Lever E, Alves MR, Isenberg DA (2020) Towards precision medicine in systemic lupus erythematosus. Pharmgenomics Pers Med 13:39–49
Li L, Wang H (2016) Heterogeneity of liver cancer and personalized therapy. Cancer Lett 379(2):191–197
Liu HH, Tsai YS, Lai CL, Tang CH, Lai CH, Wu HC, Hsieh JT, Yang CR (2014) Evolving personalized therapy for castration-resistant prostate cancer. Biomedicine 4(1):2
Ludwig JA (2008) Personalized therapy of sarcomas: integration of biomarkers for improved diagnosis, prognosis, and therapy selection. Curr Oncol Rep 10:329–337
Lynch TJ, Bell DW, Sordella R, Gurubhagavatula S, Okimoto RA et al (2004) Activating mutations in the epidermal growth factor receptor underlying responsiveness of non–small-cell lung cancer to gefitinib. N Engl J Med 350:2129–2139
Maher TM, Nambiar AM, Wells AU (2022) The role of precision medicine in interstitial lung disease. Eur Respir J 60(3):2102146
Mahler M, Martinez-Prat L, Sparks JA, Deane KD (2020) Precision medicine in the care of rheumatoid arthritis: focus on prediction and prevention of future clinically-apparent disease. Autoimmun Rev 19(5):102506
Malcangi G, Patano A, Guglielmo M, Sardano R, Palmieri G, Di Pede C, de Ruvo E, Inchingolo AD, Mancini A, Inchingolo F, Bordea IR (2023) Precision medicine in oral health and diseases: a systematic review. J Pers Med 13(5):725
Mazur P, Kopytek M, Ząbczyk M, Undas A, Natorska J (2021) Towards personalized therapy of aortic stenosis. J Pers Med 11(12):1292
Miller CL, Kontorovich AR, Hao K, Ma L, Iyegbe C, Björkegren JL, Kovacic JC (2021) Precision medicine approaches to vascular disease: JACC Focus Seminar 2/5. J Am Coll Cardiol 77(20):2531–2550
Mok TS, Zhou Q, Leung L, Loong HH (2010) Personalized medicine for non-small-cell lung cancer. Expert Rev Anticancer Ther 10(10):1601–1611
Moreira AL, Eng J (2014) Personalized therapy for lung cancer. Chest 146(6):1649–1657
Naithani N, Sinha S, Misra P, Vasudevan B, Sahu R (2021) Precision medicine: concept and tools. Med J Armed Forces India 77(3):249–257
Nitzan M, Nitzan I (2022) Feasibility of precision medicine in hypertension management—scope and technological aspects. J Pers Med 12(11):1861
Noor NM, Verstockt B, Parkes M, Lee JC (2020) Personalised medicine in Crohn's disease. Lancet Gastroenterol Hepatol 5(1):80–92
Odle TG (2017) Precision medicine in breast cancer. Radiol Technol 88(4):401M–421M
Pakkala S, Ramalingam SS (2018) Personalized therapy for lung cancer: striking a moving target. JCI Insight 3(15):e120858

Palacios-Berraquero ML, Alfonso-Piérola A (2021) Current therapy of the patients with MDS: walking towards personalized therapy. J Clin Med 10(10):2107

Parkinson DR, Johnson BE, Sledge GW (2012) Making personalized cancer medicine a reality: challenges and opportunities in the development of biomarkers and companion diagnostics. Clin Cancer Res 18:619–624

Pathak L (2023) Personalized treatment for multiple sclerosis: the role of precision medicine. Neurol Lett 2(1):30–34

Pinto e Vairo F, Rojas Málaga D, Kubaski F, Fischinger Moura de Souza C, de Oliveira Poswar F, Baldo G, Giugliani R (2020) Precision medicine for lysosomal disorders. Biomolecules 10(8):1110

Pons-Tostivint E, Bennouna J (2022) Treatments for non-small-cell lung cancer: the multiple options for precision medicine. Curr Oncol 29(10):7106–7108

Prasad V, Fojo T, Brada M (2016) Precision oncology: origins, optimism, and potential. Lancet Oncol 17:e81–e86

Proper SP, Azouz NP, Mersha TB (2021) Achieving precision medicine in allergic disease: progress and challenges. Front Immunol 12:720746

Ramaswami R, Bayer R, Galea S (2018) Precision medicine from a public health perspective. Annu Rev Public Health 39:153–168

Raviña Rubira E (2011) The evolution of drug discovery: from traditional medicines to modern drugs. Wiley-VCH

Reichle A, Vogt T, Hildebrandt GC (2010) A methodological approach to personalized therapies in metastatic cancer. In: From molecular to modular tumor therapy: tumors are reconstructible communicatively evolving systems. Springer Science & Business Media, pp 507–533

Rooney C, Sethi T (2015) Biomarkers for precision medicine in airways disease. Ann N Y Acad Sci 1346(1):18–32

Sankaraneni R, Lachhwani D (2015) Antiepileptic drugs – a review. Pediatr Ann 44:e36–e42

Sawant RR, Jhaveri AM, Torchilin VP (2012) Immunomicelles for advancing personalized therapy. Adv Drug Deliv Rev 64(13):1436–1446

Sharafaldin EN, Sim MS, Lim SK, Alhussieni K, Huri HZ (2024) Precision medicine in lupus nephritis. Clin Chim Acta:119894

Shaya J, Kato S, Adashek JJ, Patel H, Fanta PT, Botta GP, Sicklick JK, Kurzrock R (2023) Personalized matched targeted therapy in advanced pancreatic cancer: a pilot cohort analysis. NPJ Genom Med 8(1):1

Smith N, Nucera C (2015) Personalized therapy in patients with anaplastic thyroid cancer: targeting genetic and epigenetic alterations. J Clin Endocrinol Metabol 100(1):35–42

Stamer UM, Zhang L, Stüber F (2010) Personalized therapy in pain management: where do we stand? Pharmacogenomics 11(6):843–864

Stegmeier F, Warmuth M, Sellers WR, Dorsch M (2010) Targeted cancer therapies in the twenty-first century: lessons from imatinib. Clin Pharmacol Ther 87:543–552

Strafella C, Caputo V, Galota MR, Zampatti S, Marella G, Mauriello S, Cascella R, Giardina E (2018) Application of precision medicine in neurodegenerative diseases. Front Neurol 9:701

Tang Y, Kosorok MR (2012) Developing adaptive personalized therapy for cystic fibrosis using reinforcement learning

Toups K, Hathaway A, Gordon D, Chung H, Raji C, Boyd A, Hill BD, Hausman-Cohen S, Attarha M, Chwa WJ, Jarrett M (2022) Precision medicine approach to Alzheimer's disease: successful pilot project. J Alzheimer's Dis 88(4):1411–1421

Tutt A, Robson M, Garber JE, Domchek SM, Audeh MW et al (2010) Oral poly(ADP-ribose) polymerase inhibitor olaparib in patients with BRCA1 or BRCA2 mutations and advanced breast cancer: a proof-of-concept trial. Lancet 376:235–244

Tye SC, Denig P, Heerspink HJ (2021) Precision medicine approaches for diabetic kidney disease: opportunities and challenges. Nephrol Dial Transplant 36(Supplement_2):ii3–ii9

Valenzuela-Vallejo L, Sanoudou D, Mantzoros CS (2023) Precision medicine in fatty liver disease/non-alcoholic fatty liver disease. J Pers Med 13(5):830

Verma M (2012) Personalized medicine and cancer. J Pers Med 2(1):1–4
Vlastos I, Gkouskou K, Doulaptsi M, Karatzanis A, Prokopakis EP (2019) Precision medicine in rhinosinusitis. Curr Allergy Asthma Rep 19:1–6
Westin SN, Broaddus RR (2012) Personalized therapy in endometrial cancer: challenges and opportunities. Cancer Biol Ther 13(1):1–3
Wick KD, McAuley DF, Levitt JE, Beitler JR, Annane D, Riviello ED, Calfee CS, Matthay MA (2021) Promises and challenges of personalized medicine to guide ARDS therapy. Crit Care 25:1–5
Williams DM, Jones H, Stephens JW (2022) Personalized type 2 diabetes management: an update on recent advances and recommendations. Diabetes Metab Syndr Obes Targets Ther 15:281–295
Wolford JE, Ferrigni E, Margul D, Herzog TJ (2022) Integrating precision medicine into the contemporary management of gynecologic cancers. Curr Oncol Rep 24(7):889–904
Xie F, Chan JC, Ma RC (2018) Precision medicine in diabetes prevention, classification and management. J Diabetes Investig 9(5):998–1015
Yang Y (2015) Cancer immunotherapy: harnessing the immune system to battle cancer. J Clin Invest 125:3335–3337
Yang W, Freeman MR, Kyprianou N (2018) Personalization of prostate cancer therapy through phosphoproteomics. Nat Rev Urol 15(8):483–497
Zhao K, Zeng Z, He Y, Zhao R, Niu J, Sun H, Li S, Dong J, Jing Z, Zhou J (2024) Recent advances in targeted therapy for inflammatory vascular diseases. J Control Release 372:730–750

Microbiota Profiling Techniques: Recent Advancement

Amit Anand, Santhepete Nanjundiah. Manjula, and Kenganora Mruthunjaya

Abstract Microbiota profiling has emerged as a cornerstone of modern microbiological and biomedical research, offering unprecedented insights into the complex communities of the microorganism inhabiting diverse environments, including the human body, soil and aquatic ecosystems. Recent advancements in sequencing technologies, bioinformatics tools and sample preparation techniques have revolutionized our inability to identify, classify and study microbiota composition and function. This chapter explores the latest techniques for microbiota profiling, including next-generation sequencing (NGS), long-read sequencing, shot gun metagenomics and single-cell approaches. Additionally, we discuss advancements in computational pipelines and artificial intelligence driven analytics that facilitate the interpretation of large-scale microbiota datasets. Challenges such as data reproducibility, standardization and clinical translation are also addressed. This chapter concludes by highlighting the future directions of microbiota profiling and its potential applications in personalized medicine, agriculture and environmental monitoring.

Keywords Microbiota profiling · Personalized medicine · Bioinformatics · Artificial intelligence

A. Anand · K. Mruthunjaya (✉)
Department of Pharmacognosy, JSS College of Pharmacy, Mysuru, JSS Academy of Higher Education and Research, Mysuru, India
e-mail: kmruthunjaya@jssuni.edu.in

S. N. Manjula
Department of Pharmacology, JSS College of Pharmacy, Mysuru, JSS Academy of Higher Education and Research, Mysuru, India

S. K. Singh et al. (eds.), *Microbiota Profiling for Precision Medicine*,
Interdisciplinary Biotechnological Advances,
https://doi.org/10.1007/978-981-95-2104-3_2

Abbreviations

AI	Artificial intelligence
DDT	Dichlorodiphenyltrichloroethane
DNA	Deoxyribonucleic acid
FMT	Faecal microbiota transplantation
ML	Machine learning
NGS	Next-generation sequencing
ONT	Oxford Nanopore technology
OTU	Operational taxonomic units
PCR	Polymerase chain reaction
PGPRs	Plant growth promoting rhizobacteria
RNA	Ribonucleic acid
SCNT	Somatic cell nuclear transfer technique
SVM	Support vector machines

1 Introduction

1.1 Definition and Importance of Microbiota Profiling

Extensive research on the human microbiota has uncovered compelling connections between microbial composition and health or disease, generating interest in microbiome-derived biomarkers. However, the complexity of microbial communities and technical limitations significantly influence results, restricting the generalizability of microbiota profiling and raising concerns about its clinical relevance. Additionally, no dedicated quality management framework has been established for microbiota profiling due to inconsistencies in methodologies and findings (Akhatou et al. 2016). This chapter examines the applicability of established quality management strategies from diagnostic laboratories to microbiota profiling and proposes a structured framework (Amaretti et al. 2015). Comprehensive quality controls are essential at every stage from the collection to data processing. Standard operating procedures originally developed for wet lab analysis must be adapted to integrate bioinformatics tools. Moreover, test validation and proficiency testing criteria should account for inherent variations resulting from methodological differences. Implementing the proposed quality management system could enable clinical laboratories to perform routine microbiota profiling more reliably, ultimately enhancing patient care (Belkaid and Hand 2014).

1.2 Comparison of Traditional and Modern Microbiota Profiling Techniques

Traditional and modern microbiota profiling techniques offer a variety of approaches for studying microbial communities. Traditional methods, such as culture-based treatments, require isolating and culturing microorganisms on specific media. While these technologies have provided essential insights into microbial diversity and function, they are confined to culturable species, which account for a small fraction of the total microbiota (Mukhopadhya et al. 2022). Furthermore, while normal microscopy and biochemical techniques are beneficial, they may not provide the resolution required to fully identify microbial species or interactions. In contrast, modern microbiota profiling techniques like sequencing and metagenomics have transformed the discipline by allowing for the analysis of entire microbial communities, including unculturable organisms (Gerasimidis et al. 2016). These methods such as 16S rRNA sequencing and shotgun metagenomics provide precise data on microbial community, taxonomy and functional potential. Furthermore, sophisticated techniques like metabolomics and transcriptomics contribute to this approach by providing system-level understanding of microbiota–host interactions. While contemporary approaches require specialized equipment and bioinformatics expertise, they are more accurate, sensitive and comprehensive than conventional methods. These approaches provide complementary perspectives on microbiological research (Walker 2016). Probiotic strains should be identified using internationally recognized methods, including DNA–DNA hybridization and 16S rRNA sequencing, along with strain-specific techniques such as pulsed field gel electrophoresis or randomly amplified polymorphic DNA analysis. The advent of molecular techniques, particularly polymerase chain reaction (PCR) in the 1980s, significantly enhanced the detection and identification of probiotic bacteria. Furthermore, the limitation of culture-based approaches, which fail to detect non-culturable cells, has been addressed using PCR-based methods that analyse nucleic acid in samples (Donelli et al. 2013).

2 Core Techniques in Microbiota Profiling

2.1 16S rRNA Gene Sequencing

The use of 16S rRNA metagenomic sequencing has greatly improved microbiome studies by allowing for comprehensive bacterial population analysis. Although it is widely used, a standardized method for 16S rRNA-based taxonomic profiling remains to be developed. Originally limited to the laboratory, 16S rRNA sequencing has found increasing application in the clinic, revolutionizing bacterial identification and infection control. This sequencing technique has transformed microbiology through the provision of a precise and effective method for describing bacterial populations (Cuscó et al. 2019). Its most significant clinical application is in identifying antibiotic-resistant infections, improving diagnostic accuracy. Clinicians involved in clinical microbiota research are especially well suited to utilize these technologies in subsequent clinical trials. Conventional identification of bacteria has been based on culture methods, which were unable to identify microorganisms that do not grow on routine media. This limitation is particularly evident when examining heterogeneous microbial populations with different growth characteristics. Some bacteria, like *Burkholderia* spp. and *Mycobacterium tuberculosis*, are resistant to cultivation, and hence alternative identification techniques are necessary. 16S rRNA sequencing is a strong alternative by targeting for a gene coding for a well-conserved bacterial ribosomal element. The gene has hypervariable sections interspersed with conserved regions (Chen et al. 2019). Hypervariable sections are specific to species to enable bacterial grouping, but conserved sections provide the opportunity for the development of universal primers that will attach to common sequences in bacterial species. Short-read sequencing, generally in the range of 200–400 base pairs, can address one or several hypervariable regions by primers based on the conserved flanking sequences. Choice of the proper region and related primers is central to study planning (Walker et al. 2015). As an example, investigators examining the oral microbiome usually favour the V1–V3 region. However, defining the best hypervariable regions to analyse continues to be contentious, with some researchers proposing that multiple regions are sequenced in order to better capture, and without bias, complex microbiomes. Prior to the emergence of next-generation sequencing (NGS), Sanger sequencing was the mainstay for studying bacterial communities, yielding longer reads of over 500 base pairs (Fraher et al. 2012). Nevertheless, NGS technologies have mostly replaced Sanger sequencing because they are faster, cheaper, and have more streamlined workflows, yet with similar read lengths. Irrespective of the platform used for sequencing, bacterial communities are generally described in terms of operational taxonomic units (OTUs), which group organisms to the genus or species level (del Campo-Moreno et al. 2018).

2.2 *Shotgun Metagenomics*

Metagenomics enables the analysis of microbial community composition. While genomics focuses on the genetic makeup of a single organism, metagenomics examines the collective genetic material of entire microbial populations. Following DNA isolation from samples, next-generation sequencing (NGS) is commonly used. NGS generates vast amounts of short-read data, which can be assembled to reconstruct the microbial community profile or other relevant information, much like assembling a puzzle from scattered pieces. Next-generation sequencing technologies enable genetic characterization of all bacteria present within a sample, as opposed to cultivable bacteria (Quince et al. 2017). One such approach is shotgun metagenomics, defined as the untargeted ('shotgun') sequencing of all ('meta-') microbial genomes 'genomics' within a sample. Shotgun sequencing may be applied to elucidate the taxonomic structure and functional potential of microbial communities and to recover whole genome sequences. High-throughput 16S rRNA gene sequencing, which characterizes chosen organisms or individual marker genes, is often called metagenomics, but this is a misnomer since they do not address the complete genomic content of a sample (Sharpton 2014). Subsequent to the first study design, a standard shotgun metagenomics experiment comprises five steps: (i) collection, processing and sequencing of the samples; (ii) pre-processing of the sequencing reads; (iii) analysis of sequences to characterize taxonomic, functional and genomic aspects of the microbiome; (iv) post-processing statistical and biological analysis; and (v) validation. All stages may be performed by utilizing an assortment of experimental and computational approaches, thereby making the investigators face hard choices (Jervis-Bardy et al. 2015). Although it may seem simple, shotgun metagenomics has disadvantages, such as possible experimental bias and the technical difficulties of computer analysis and interpretation. Metagenomic taxonomic profiling determines the presence and relative abundance of microbial species. This can be achieved without assembly from external sequence data resources, for example, reference genomes publicly available. This approach can minimize assembly problems, accelerate computing and allow low-abundance organism profiling that is not possible by building from scratch (Ibrahim and Belheouane 2018). Shotgun metagenomics can reconstruct genomes of diverse skin microbiota, including bacteria, viruses and micro-eukaryotes. Construction of the genome of general skin inhabitants is essential in order to underline strain- and sample-specific contrasts, while with more evasive microbes, the procedure can supply the initial data regarding their traits, taxonomy and functional prospectus. Due to the limited amount of genetic material obtained from skin samples, community DNA is often amplified during metagenomic library construction (Jovel et al. 2016). This amplification can occur either before or as part of the library preparation process. Various amplification techniques are available, including multiple displacement amplification and limited-cycle PCR, both of which are utilized in the widely used Illumina NexteraXT Kit (Illumina®, San Diego, CA). Illumina has become the standard platform for metagenomic sequencing due to its high-throughput capabilities. Typically, metagenomic libraries are multiplexed on the Illumina platform. Since the number and size of organisms in a microbiome are initially unknown, sequencing throughput is generally estimated based on prior research experience and budget constraints (Sato et al. 2019). Deeper sequencing depths provide additional data, aiding in the reconstruction of low-abundance species and the identification of base-calling errors. In skin samples, there is also the problem of human DNA presence, whose proportion is hard to estimate before sequencing, unless qPCR is employed to first analyse human contamination (Santiago-Rodriguez et al. 2023). According to the published literature, the suggested sequencing depth for a skin metagenome is ~5 Gb, comparable with that of well-validated gut metagenomes. This is because species diversity is low in the skin and human DNA is prominently present. To effectively test and validate each sequencing library, multiple samples can be multiplexed and sequenced on a low-throughput platform, such as Illumina MiSeq, before committing the metagenomes to a high-throughput run, like Illumina HiSeq (Schirmer et al. 2016).

2.3 Meta Transcriptomics and Metabolomics Integration

Meta transcriptomics gives an understanding of an active functional profile of a microbial community by considering which genes the whole population is expressing. The meta transcriptome measures total mRNA to provide a snapshot of gene expression in one sample at a point in time and under particular conditions. Early pioneering work to find expressed genes in environmental samples commenced in 2005, with the onset of meta transcriptomics. These were, however, confined to a very small subset of genes (Cavill et al. 2016). For metagenomics, one is now able to perform complete meta transcriptomics shotgun sequencing. This (meta)genome-wide expression defines the expression and functional attributes of a microbiome. In reads processing, a standard meta transcriptomics analysis pipeline will either (1) align them to a reference genome or (2) de novo assemble them into transcript contigs and super contigs. The former, like alignment-based approaches in Widely-Targeted Metabolic Screening (WMS), aligns readings to reference databases, collecting data to ascertain the relative expression of specific genes. The second method makes the same inference, but from built-up sequences. The first method is limited by the amount of information contained within the reference genome database. The second method is limited by the ability of software packages to accurately build contigs and super contigs from short-read data (Fan 2012). Meta transcriptomics is much less widely utilized in microbiome analysis compared to other omics. The majority of the analysis pipelines that have been reported in the literature were made ad hoc. The majority of them employ the previously discussed first method that is based on read mapping. Meta transcriptomic readings in this case are mapped to special databases (usually obtained from the NCBI) by employing tools such as Bowtie2, BWA and BLAST. The resultant reads are thereafter annotated with tools such as GO, Kyoto Encyclopedia of Genes and Genomes (KEGG), COG and Swiss-Prot. Finally, several types of downstream analysis are conducted based on the study's aim (such as PCA-based phylogenetic analysis or enrichment analysis) (Santiago-Rodriguez and Hollister 2021). Most current meta transcriptomics methods incorporate stable isotope probing (SIP), a method that was used to extract specific targeted transcriptomes of aerobes present in lake sediments. Besides facilitating the focusing on specific species, it does significantly contribute towards metabolomics analysis. The second method is to put together meta transcriptomic reads into larger pieces referred to as contigs. There are many software packages that can be used to do this. Celaj et al. compared de novo sequence assemblers and reference-based mapping tools. Trinity, Meta Velvet, Oases, Aby SS, Trans-Abyss and SOAPdenovo were some of the tools that were compared (98–100), in addition to Scripture and Cufflinks (Aguiar-Pulido et al. 2016). Trinity performed better than all the other tools and it appeared to be most appropriate for sensitivity over the widest range of expression levels. This was particularly evident when reconstructing transcripts in the top expression quintiles, whereas other de novo methods did not work well. Li and Dewey developed RNA-Seq using Expectation Maximization (RSEM) (Sheng et al. 2022) (Table 1).

Table 1 Summarizing the core techniques in microbiota profiling, including their principles, advantages and limitations

Profiling techniques	Principle	Advantages	Limitations	References
16s rRNA sequencing	Amplifies and sequences the 16s ribosomal RNA gene to identify bacterial taxa	Well established; cost-effective; allows taxonomic classification at genus level	Limited to bacterial identification; lacks species level resolution	Patel (2001)
Metagenomics (whole genome shotgun sequencing)	Sequences all genetic materials in a sample, identifying both known and unknown microbes	High resolution; detects functional genes; identifies bacteria, fungi, viruses and archaea	Requires complex bioinformatics analysis; high computational demands	Chen and Pachter (2005)
Meta transcriptomics	Analyses RNA transcripts to study active microbial functions and gene expression	Provides insights into microbial activity and functional potential	High cost; RNA is less stable than DNA; complex bioinformatics required	Zhang et al. (2021)
Metabolomics	Identifies and quantifies small molecules (metabolites) produced by bacteria	Can identify viable non-culturable microbes; allows functional studies	Metabolites identification can be challenging; requires sophisticated analytical tools like Liquid chromatography mass spectrometry (LCMS) and Non magnetic resonance (NMR)	Vitale et al. (2024)
Culturomics	Uses multiple culture conditions to isolate and identify microbial species	Can identify viable but non-culturable microbes; allows functional studies	Labour-intensive biases towards cultivable species; difficult for complex microbes	Nowrotek et al. (2019)
Flow cytometry and fluorescence in-situ hybridization	Uses fluorescent probes or staining to identify and quantify bacteria in a sample	Rapid analyses; direct visualization of microbes	Limited resolution; requires specific probes; does not provide full identification of microbes	Prudent and Raoult (2019)
Microarrays	Hybridization-based techniques using DNA/RNA probes to detect microbial sequences	Fast and cost-effective targeted detection	Lower sensitivity; limited to known sequences than sequencing methods	Ventimiglia and Petralia (2013)

3 Recent Technological Advancements

3.1 Long-Read Sequencing (e.g., PacBio, Oxford Nanopore)

Single-molecule, long-read DNA sequencing technologies have emerged as powerful genomic tools over the past decade. These platforms can generate reads ranging from ten to thousands of kilobases with accuracy comparable to short-read sequencing technologies. They have proven instrumental in deciphering complex regions of the human genome, identifying structural variant previously undetectable, and achieving some of the first telomere-to-telomere assemblies of complete chromosomes (Karst et al. 2021). In the near future, long-read sequencing will enable the systematic construction of diploid genomes, advancing genomics to fully capture human genetic variation resolve missing heritability and uncover novel disease mechanisms (Logsdon et al. 2020). Compared to short-read technologies, long-read sequencing can generate continuous reads from 10 kilobases to several mega-bases directly from native DNA. Recent improvements in accuracy and throughput have further enhanced their utility and expanded their applications. PacBio and ONT (Oxford Nanopore Technologies) sequencing platforms both generate long reads that can effortlessly traverse highly repetitive regions of the human genome. However, differences in their sequencing chemistry and detection methods influence read length, base accuracy and throughput. PacBio's SMRT (Single Molecule Real-Time) sequencing technology utilizes a circular DNA template known as an SMRT bell. This structure consists of a double-stranded DNA insert flanked by single-stranded hairpin adapters, and the insert length can range from one kilobase to over a hundred kilobases, enabling the production of exceptionally long sequencing reads. Once assembled, the SMRT bell is anchored by a DNA polymerase (Ardui et al. 2018). During sequencing, the polymerase continuously moves around zero mode circular template, incorporating fluorescently labelled deoxynucleoside triphosphates into the growing strand. After each nucleotide addition, a laser excites the fluorophore and a camera captures the emitted signal. In contrast, ONT (Oxford Nanopore Technologies) long-read sequencing employs linear DNA molecules rather than circular templates. These DNA fragments can range from one kilobase to hundreds of kilobases or even megabases. ONT sequencing begins with a double-stranded DNA molecule binding to a preloaded protein attached to a sequencing adaptor. This setup facilitates controlled movement of the DNA strand through nanopore where changes in ionic current are detected and translated into a nucleotide sequence. A DNA mixture is injected into a flow cell containing hundreds or thousands of nanopores embedded in an artificial membrane. A motor protein unwinds the double-stranded DNA and, with the aid of an electric current, guides the negatively charged DNA through the nanopore at a controlled rate. As the DNA moves through the pore, it disrupts the current in a distinctive manner, allowing real-time decoding of the DNA sequence. ONT sequencing has produced reads exceeding 1 Mb, with the longest reported read surpassing 2.3 Mb when computationally assembled from shorter fragments (Petersen et al. 2019). These advancements have propelled genomics into the era of megabase length sequencing. While ONT read lengths can exceed those of PacBio by at least an order of magnitude producing continuous sequences ranging from several hundred to thousands of kilobases, such long reads constitute only a small portion of the overall read length distribution. These exceptionally long-read lengths are made possible by the unique pore chemistry essential for ONT sequencing, which allows molecules to pass through the nanopore regardless of their length. Several studies have identified the extraction and processing of high molecular weight DNA as the primary constraint on ONT read length (Hu et al. 2021). The preparation methods define two major categories of ONT data: standard long reads (10–100 kb) and specialized ultra-long reads (exceeding 100 kb). Among these, standard ONT long reads are the most commonly produced. Our meta-analysis reveals that these reads typically range from 10 to 100 kb in length with an overall accuracy of 87–98%, though in rare instances, accuracy may drop as low as 69%. Raw ONT long reads correctly identify approximately 91% of homopolymers of at least five bases surpassing PacBio CLRs by 3% points but trailing PacBio HiFi reads by 8% points. The findings align with previous research, reinforcing that ONT raw read accuracy is primarily influenced by the base calling algorithm used. Notably, advancements in these algorithms over the past 5 years have significantly enhanced raw read accuracy (Bowden et al. 2019).

3.2 Single-Cell Microbiology

Microbiology has historically dealt with and is interested in population-level studies. Population-level data inference has been applied mainly to gain insight into how cells react to the environment, communicate with each other, and undergo such complex processes as cellular differentiation and gene expression. Novel appreciation for the ubiquity and significance of cellular heterogeneity, coupled with the latest technological advances, has energized the development of new tools and methods to investigate single microbial cells (Brehm-Stecher and Johnson 2004). It has therefore become possible for researchers to describe microbes and their behaviour with unprecedented specificity. Single-cell techniques have been utilized to precisely characterize the environmental distribution and activities of microorganisms reveal previously undetected processes such as inter-species gene transfer and chemical signalling and clarify specific physicochemical interactions between microbes and the surfaces they inhabit (Ishii et al. 2010). Single-cell methodologies have similarly underpinned knowledge about relationships between cell biochemistry and behaviour and cell-level causes of events at the level of the population. In doing so, the new understanding of the nature of chemical signalling pathways and the underlying mechanisms of coordinating multicellular behaviours has been forthcoming. Single-cell techniques have advanced the direct measurement of individual cell mechanical properties at the micro and nano scales, including turgor pressure, elasticity and bursting force. Additionally, these approaches have been utilized for micro physiological studies, examining metabolite, protein and elemental localization, intracellular water dynamics, host–pathogen interactions and surface-associated redox activity (Balomenos et al. 2017). Plate counting and light microscopy were the initial tools of single-cell analysis. Consequently, they have been very handy for over a century, and they are still sufficient and acceptable for many purposes even in the present time. Nevertheless, the recent decades have brought along with them several technological and methodological advancements, including computerization and imaging techniques, along with the creation of culture-independent techniques like in situ hybridization and PCR. Advances along these lines have hugely enhanced our capability to determine the properties and functions of individual microbial cells (Müller and Nebe-von-Caron 2010).

3.3 Techniques for Isolating and Profiling Individual Cells

Single-cell sorting can be accomplished by separating complete cells, cell-type-specific nuclei, or cell-type-specific organelles and numerous methods have been proposed for sorting the necessary cells. Nonetheless, their decision is dependent on cell type, purchase cost and yield of individual cells (Zeb et al. 2019). Manual isolation is a simple and effective single-cell isolation procedure. The aspirated fluid carrying the cell is distributed into the collection vessel using dispensation. To ensure an unbiased single-cell separation, the separated single cell can be visualized and photographed through the microscope. Micromanipulators are employed for isolating embryonic or live culture cells. Somatic cell nuclear transfer method (SCNT) is employed to clone animals like the sheep Dolly and monkeys Zhong-zhong and Hua-hua, one of the significant phases being the retrieval of donated egg cell through micromanipulation. Micromanipulation is also feasible in the electrophysiology laboratory when applying the patch-clamp technique. Micromanipulation is strong and fast but has low throughput. Moreover, highly skilled professional training is required to apply this method. Cells with complicated structures, such as neuronal cells, cannot be excised completely by micromanipulation and isolated cells may be damaged at times (Lohani et al. 2023).

3.4 Integration of Genomics, Transcriptomics, Proteomics and Metabolomics Data

Omics technologies provide a comprehensive understanding of the chemicals that comprise a cell, tissue, or organism. Omics technologies are primarily concerned with the non-targeted and unbiased identification of genes (genomics), mRNA (transcriptomics), proteins (proteomics) and metabolites (metabolomics) in a biological sample. Over the last decade, fast advances in omics technologies (such as genomics, proteomics and metabolomics) have enabled high-throughput monitoring of a wide range of molecular and organismal processes (Wanichthanarak et al. 2015). Such approaches are commonly used to identify biological variances (e.g. biomarkers), characterize complex biochemical systems, and examine pathophysiological processes. While numerous omics platforms aim to conduct in-depth research of genes (genomics), mRNA (transcriptomics), proteins (proteomics) and metabolites (metabolomics), data integration within and across omics disciplines remains a difficulty. Metabolomics, or the study of small molecules and metabolic intermediates (metabolites), is often used to explore relationships between gene and protein downstream products and environmental stimuli (Cavill et al. 2016). Metabolomics has been extensively used in the past decade to study a range of pathophysiological processes, such as type 1 diabetes and cancer, with shared objectives being the identification of biomarkers predictive of disease onset, prognosis and response to treatment. The metabolome is highly sensitive to both environmental and biological regulatory controls (e.g. epigenetics, transcription and post-translational modification), and its analysis offers a powerful means of characterizing organismal phenotype. But use of metabolomics alone may not be adequate to fully describe complex biological systems or disease (e.g. cancer). Several researchers, for instance, base their analysis on the analysis of circulating metabolites (e.g. serum or plasma), but this pool is the output and input of several biological systems, so it is challenging to obtain insight into tissue- and cellular-level mechanisms. Another issue is seamless integration of metabolomic-based analyses in situations where there is limited biochemical domain knowledge, which could lead to sparse and disconnected biological interpretations (Sanches et al. 2024). There are various software tools designed for integrating various omics datasets according to biochemical route, ontology, network, or empirical association. Genomics describes the scientific investigation of an organism's genome. The genome is a cell or organism's entire DNA. The human genome has 3.2 billion and roughly 30,000–40,000 protein-coding genes. Previously, genes were investigated one at a time, but microarray technology has advanced significantly in recent years. DNA microarrays analyse variances in DNA sequence between individuals, and the expression of hundreds of genes can be studied simultaneously. They are able to identify such abnormalities as chromosomal deletions and insertions, as well as aberrant chromosome numbers, utilizing a procedure known as comparative genomic hybridization. The most prevalent changes in DNA sequences between individuals are single nucleotide polymorphisms (SNPs) (Domenzain et al. 2022), a situation where a single nucleotide is replaced with another; the change might have functional significance if it creates a codon that codes for an alternative amino acid. They become especially intriguing where they are associated with diseases having a genetic foundation. Single nucleotide polymorphism profiling is also used in the field of pharmacogenomics to examine individual patient responses to medications. Transcriptome is an organism's or cell's complete mRNA and is a template for protein synthesis via a process called translation. It indicates what genes are being expressed at a specific point in time. Gene expression microarrays measure packaged mRNA (mRNA with the introns removed) to approximate gene activity (Wanichthanarak et al. 2015) (Fig. 1).

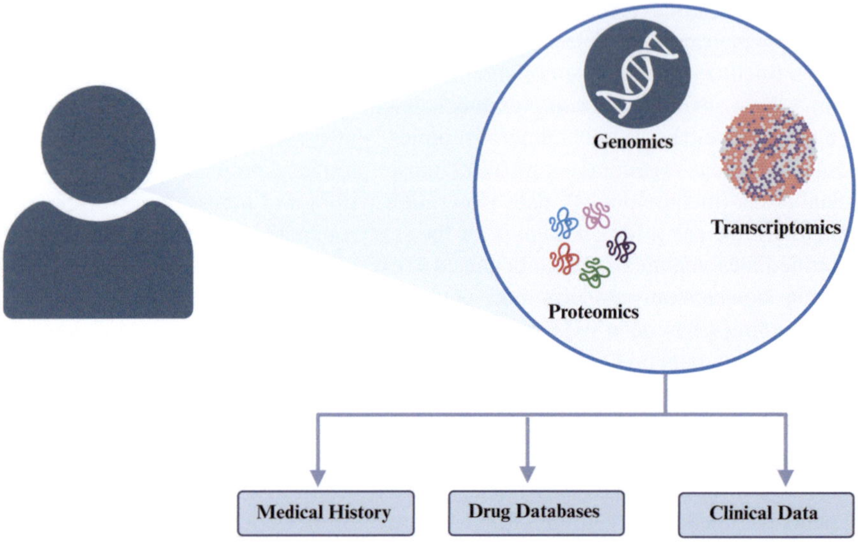

Fig. 1 Integration of genomics, transcriptomics and proteomics with clinical data for better therapeutics

3.5 Artificial Intelligence and Machine Learning

The use of artificial intelligence (AI) in microbiota research has opened up new avenues for investigation and it has proven effective in a variety of scientific disciplines. AI refers to machine learning (ML) algorithms that can learn from data patterns, spot trends and make predictions without direct programming. Because of these characteristics, AI is highly suited to the study of massive and complex microbiome datasets, paving the way for groundbreaking discoveries. The application of AI to microbiota has created new research opportunities and emerged as a valuable tool in a variety of scientific domains (Yilmaz et al. 2020). AI is defined as machine learning algorithms that can learn from data patterns, discover trends and anticipate events without being programmed. Such algorithms can learn, recognize trends, and predict. Because of these characteristics, AI is particularly suited for microbiome investigation on massive and sophisticated datasets, offering unparallel findings (Bahado-Singh et al. 2019). AI tools can effectively analyse large amounts of raw sequencing data obtained from microbiome research. Scientists now have a complete picture of microbial communities and their potential practical applications. They can utilize AI-based predictive models to better understand how changes in the microbiome affect human health. Such models can assess how a certain food intervention changes the microbial population or forecast the likelihood of acquiring a gut infection caused by microbiota (Barberis et al. 2022). AI can identify biomarkers associated with specific disease or disorders, which can then be used to determine the presence of disease, its severity, or therapy response; this will lead to an earlier diagnosis and more targeted treatment. AI-supported medication development platforms can be used to investigate the prospect of curing illnesses caused by bacteria. AI can be used to help develop novel medication capable of healing various ailments by manipulating the microbiome. This is accomplished by studying the relationships between the host and microbiome systems (Srivastava and Uddin 2025). Artificial intelligence algorithms are transforming our understanding of the microbiome's complicated interactions and relationship with the host. This information is crucial for understanding the role of microbial communities in health and disease. However, the quality and AI-readability of this data are critical for reaching accurate conclusions and avoiding misleading outcomes. The effectiveness of AI algorithms is determined by the quantity and diversity of the dataset, and research must involve a broad population to avoid bias. Furthermore, the application of artificial intelligence in microbiome research poses serious ethical questions about data privacy, informed permission, and the potential exploitation of sensitive information (Chi et al. 2024). Machine learning (ML) has become an indispensable technique for microbiota profiling. Scientists can acquire insights into the structure and function of microbial communities by studying massive and complicated microbiome datasets with artificial intelligence (AI) technologies. This chapter looks at numerous machine learning algorithms used in microbiota profiling, including as clustering, classification, and network analysis. K-means clustering is an effective approach for detecting patterns in complex datasets. This algorithm can effectively discriminate between different types of microbial communities, revealing information about their structure and diversity (Woelders et al. 2023). Hierarchical clustering is a useful unsupervised machine learning approach. This method creates a dendrogram, which is a branching tree-like figure that depicts the relationships between different microbiological species or samples. Similar samples or taxa are grouped into hierarchical clusters, allowing for study at various degrees of detail. Hierarchical clustering allows scientists to better visualize the hierarchical order of microbial communities and identify subgroups with comparable taxonomic traits (Aboud et al. 2023). Random Forest is a versatile machine learning technique that is commonly used for both classification and prediction problems. Random Forest performs well in microbiota profiling by classifying samples based on their microbial composition. This capability enables the identification of essential microbial taxa that influence categorization accuracy. Furthermore, Random Forest may be used to identify diseases based on microbiome data, opening up new possibilities in diagnostics and tailored medication (Lin et al. 2014). Support vector machines (SVMs) are a strong supervised machine learning technique that can do binary and multi-class categorization. They are especially well suited for evaluating complex microbiological datasets, such as those used in microbiota profiling, because of their ability to handle high-dimensional data. In this case, SVMs can be used to identify samples as sick or healthy based on the information contained in their microbiomes. The introduction of predictive modelling, a major component of artificial intelligence (AI), transformed the area of microbiota research. Scientists can now use AI techniques like SVMs to anticipate microbial relationships, functional pathways, and disease connections based on empirical evidence. Artificial intelligence is transforming our understanding of the microbiome, paving the way for advances in personalized treatment. Microbiome research has enormous potential for improving human health through the creation of targeted therapeutics. This study investigates the revolutionary significance of AI-powered predictive modelling in this discipline, focusing on its potential to change personalized medicine (Chi et al. 2024). AI systems can evaluate massive volumes of microbiome data to detect patterns of coexistence and competition among microbial species. Understanding these complex ecological interactions is critical for understanding the stability and function of microbial communities. Predictive algorithms can identify keystone species, microorganisms that are essential to the structure and operation of microbial communities. These significant species shape community dynamics and may serve as therapeutic targets for microbiome rebalancing (Cuperlovic-Culf 2018). AI systems that use metagenomics data can anticipate the functional capacities of microbial communities. Metagenomic prediction exposes the functional content of a microbial community's genes based on its taxonomic composition, providing important information about the microbiome's metabolic capability and potential consequences on the host. Predictive modelling can be used to annotate genes in microorganisms and uncover functional pathways linked to various host phenotypes or environmental factors. Understanding the functional features of the microbiome is critical for finding possible treatment targets and evaluating the microbiota's role in a variety of disorders (Mallick et al. 2019) (Fig. 2 and Table 2).

4 Applications of Microbiota Profiling

4.1 *Human Health and Personalized Medicine*

Human health is intrinsically connected to the composition and activity of the gut microbiota, impacting metabolism, immunology, and susceptibility to disease. Personalized medicine in microbiota takes individual microbial communities into account with advanced sequencing and metabolomic technologies and provides individualized therapeutic options. By characterizing unique microbial signatures associated with health or disease conditions, individualized therapies like probiotics, prebiotics, dietary changes, and microbiome-directed drugs can be formulated to re-achieve balance and enhance health (Shukla et al. 2024). This approach is particularly promising in diseases like cancer, obesity, diabetes, and inflammatory diseases, where microbiota-guided medicines may complement conventional treatment and enhance patient outcomes through precision medicine-based care (Kashyap et al. 2017).

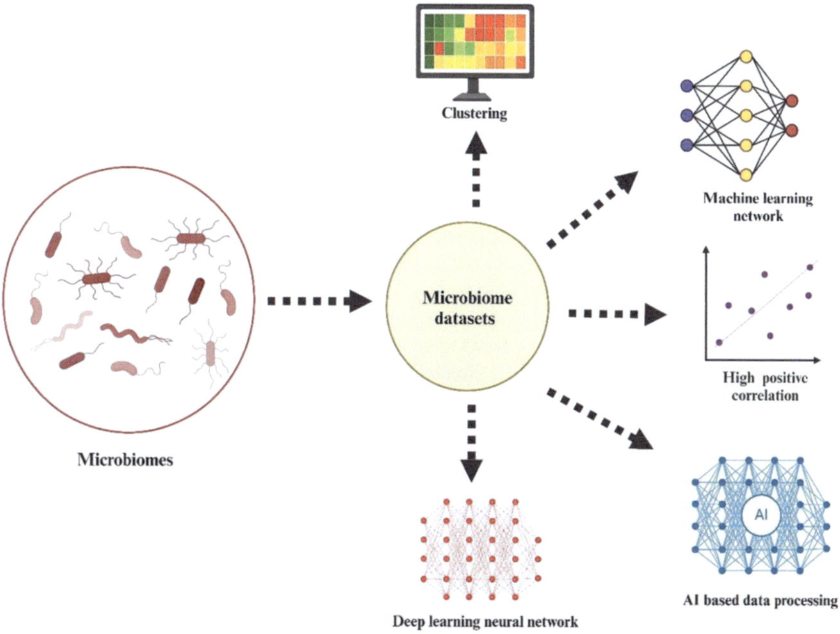

Fig. 2 AI-ML-based analysis approaches employing microbiome datasets

4.2 Microbiome-Based Diagnostics and Therapeutics

Over the last decade, high-throughput technology breakthroughs have advanced study into the trillions of commensal bacteria that live in the gut. The 'omics' technologies used for microbiome study are constantly changing, with large-scale datasets being generated. Even though most of the research is in its initial stages, unique microbial signatures have been associated with the induction of cancer and other diseases like inflammatory bowel disease, neurodegenerative disorders, and so on. The diversity of the gut microbiome has also been seen to affect the efficacy and safety of drugs (Acharjee et al. 2022). The ease of availability and diminishing sequencing prices have rendered RNA-based diagnostics increasingly prevalent in the realm of microbiome, requiring better data-analytical methods to effectively leverage all of the resultant rich biological datasets considering their distinguishing features, that is, compositional nature, heterogeneity, and sparsity. As a result, the gut microbiome is increasingly being recognized as an important component of personalized medicine, as it not only contributes to inter-individual variation in health and disease, but is also a potentially modifiable entity or characteristic that can be targeted by treatments in a personalized manner (Versalovic et al. 2017). Here, artificial intelligence and machine learning-based technologies can potentially open up new areas of biological investigation

Table 2 Summarizing various applications, methodologies, and insights from recent studies in this domain

Applications	Description	Methodologies	Insights	References
Data integration in gut microbiome	Using ML to integrate diverse molecular profiling data (metagenomics, meta transcriptomics, metabolomics) for comprehensive analysis	ML algorithm for identifying key molecular signatures and predicting phenotypes	ML aids in uncovering critical roles of gut microbiota in diseases like obesity, diabetes, liver disease, neurological disorders, and colorectal cancer	Franzosa et al. (2015)
AI in microbiome analysis	Applying AI to explore gut microbiota functionalities and their influence on human health	AI techniques in metabolomics, transcriptomics, proteomics, and genomics	AI facilitates the development of targeted therapies by providing a comprehensive understanding of gut microbiota's role in health and disease	Srivastava and Uddin (2025)
AI-driven microbiome research	Leveraging AI to transform microbiology and microbiome research through ML and deep learning applications	Deep learning architectures and paradigms	AI approaches offer technical advancements and biological insights enhancing the understanding of microbial communities	Bahado-Singh et al. (2019)
ML and deep learning in microbiome studies	Utilizing ML and deep learning to address challenges in microbiome data analysis	Traditional and novel ML methods including deep learning architectures	These approaches help in handling compositional sparse and high-dimensional microbiome data, improving analysis accuracy	Hernández Medina et al. (2022)
AI in gut microbiota analysis	Employing AI in gut microbiota's role in diagnosing and treating disease	Machine learning and deep learning methods	AI aids clinicians in processing and interpreting massive datasets, enhancing understanding of microbiota disease associations	Iadanza et al. (2020)
Supervised ML in microbiomics	Bridging the gap between current practices and optimal methodologies in supervised ML for microbiomics	Supervised machine learning workflows	Emphasis on the need for high-quality, reproducible, and interpretable workflow to meet clinical research standards	Dudek et al. (2024)

(continued)

Table 2 (continued)

Applications	Description	Methodologies	Insights	References
AI and human microbiome	Exploring the role of AI in understanding human microbiome	Machine learning and deep learning	AI assists in interpreting large volumes of microbiome data, contributing to insights to human health and disease	Abavisani et al. (2024)

by developing models that can potentially predict category labels and continuous values. In addition, diagnostic factors would increase the identification of non-invasive markers in key diseases like cancer. One simple but limited diagnostic method is the selective cultivation of faecal samples for the detection of antibiotic-resistant organisms, including enterococci. Nonetheless, this technique is limited to familiar species that can be cultured easily and have been proven to contribute to the identification of gastrointestinal diseases (Damhorst et al. 2021). Faecal microbiota transplantation (FMT) refers to the transfer of faecal material donor solutions into the gastrointestinal tract of a recipient to actively modify the patient's microbial diversity and confer a therapeutic benefit. FMT has its roots in Chinese culture more than 1700 years ago, in the fourth century, when a renowned Chinese physician named Ge Hong was able to treat patients for food poisoning and/or long-term diarrhoea by oral delivery of a humanoid faecal suspension called 'yellow soup'. Dr. Ben Eiseman, the Chief of Surgery at Denver General Hospital, and his colleagues reported their successful treatment of four patients with pseudomembranous colitis using faecal enemas (Wang et al. 2019). Until 1978, researchers were unaware that the illness they were treating, pseudomembranous colitis, was caused by a virulent bacterium known as *Clostridioides difficile*. Patients can receive faecal microbiota through a range of methods, such as enema, nasogastric or nasoduodenal tubes, colonoscopy, or oral capsules (Brandt and Aroniadis 2013). In contrast to other antimicrobial treatments, FMT does not induce an underlying dysbiosis that predisposes the patient to infection. The mechanisms through which FMT can prevent infection are still under research, but they are likely to involve gut ecological restorations, intestinal flora composition and function, and microbe–host signalling transmission. New medical research, especially animal model studies, is increasingly indicating the efficacy of FMT in treating inflammatory bowel disease and metabolic syndrome patients. FMT has also been proposed as a treatment for a range of psychiatric disorders, particularly autism spectrum disorder. While experimental models in animals have shown that FMT is beneficial in preventing energy metabolism issues and other diseases related to the gut microbiome, these results must be thoroughly evaluated (Wang et al. 2019).

4.3 Microbiota Manipulation for Soil Health and Crop Yield

A variety of soil variables have been shown to improve nutrient availability and plant productivity. The bacteria that make up the soil microbial community in the rhizosphere, or the soil around plant roots where complex interactions between roots, soil, and microbes occur, could have the greatest impact. Root exudates serve as substrates and signalling molecules for bacteria, resulting in a complex and interconnected interplay between plants and the microbiome (Chaparro et al. 2012). Single microorganisms such as endophytes, symbionts, pathogens, and plant growth-promoting rhizobacteria (PGPRs) are well studied in the literature, but the greater community of soil microbes, or soil microbiome, may have far-reaching consequences. Every microorganism works in tandem with the soil microbiome as a whole to influence plant health and productivity. Growing data suggest that plants may impact the soil microbiota via secreting root exudates (De Corato 2020). The molecular communication differs depending on the plant's development stage, closeness to adjacent species, management approaches, and a variety of other factors. Plants' interactions with their environs are dynamic, with them observing and reacting to changes as they occur. The root system, which was formerly assumed to merely provide anchoring and nutrition and water uptake, is now recognized as an important component of a plant's interaction with its surroundings. Plants monitor and sense chemical signals from soil microorganisms, and, in response, they release chemical compounds known as root exudates. The release of these compounds varies by plant species, ecotype, and even by root in a plant. Plants emit a variety of substances in the form of root exudates, including sugars, amino acids, flavonoids, aliphatic acids, proteins, and fatty acids, to create a unique habitat in the rhizosphere. All these chemicals have the ability to attract and develop symbiotic and pathogenic relationships in the rhizosphere (Wallenstein 2017). Root exudate concentration and composition change in relation to stimuli received from environment and rhizosphere, plant age, type of soil, and biotic and abiotic factors. De-la-Pena et al., for instance, found the content of root exudates varied when the plant was planted singularly in contrast to interaction with pathogens or symbiotic organisms. Root exudates are utilized as substrates of growth by the soil bacteria and may also be employed as antimicrobials; thus, as the level and concentration of the exudates fluctuate, so also do the microbe inhabitants in the rhizosphere (Schmidt et al. 2016). This chapter's intention is not to summarize all of the positive effects of soil microbes on plant wellness and how they function. Rather, an effort has been made here to underscore some of the recent gains accomplished in this rapidly developing industry, as well as probable useful consequences on integrated and sustainable farm practice (De Corato 2020). For instance, injecting desirable microbes in soil helps improve the plant intake of nutrients, increase growth, impart tolerance for abiotic stress, and mitigate disease. These microorganisms are living and potentially self-sustaining, reducing the requirement for multiple applications and reducing the likelihood of pests and pathogens building resistance to treatment (Schmidt et al. 2016). A potential management technique is applying plant growth-promoting rhizobacteria (PGPRs) as a farming treatment to limit niche vacancy and effectively replace empty niches. It has been shown that PGPRs colonize especially and successfully in low-microbial-biomass soils, so inoculations are more likely to succeed. Helpful bacteria that are well adapted to this environment can rapidly occupy space and resources that otherwise would be accessible to potential invaders of disease, helping to accomplish long-term occupancy of the niche. Besides 'sealing off' open ecological niches and increasing soil resistance to pathogen invasion, PGPRs benefit the plant host by promoting yield, nutrient acquisition, tolerance to stress, and disease resistance. Agricultural practices have emphasized the significance of sustainable farming. There is evidence that shows the close connection between plants and their microbiota. A balanced and even microbiota could be the key to healthier and more productive plants (Neupane et al. 2021). Recent studies have started to suggest the importance of this connection, as well as to examine the system as a whole to better understand the intricacies of the plant–microbiome interaction and how it affects plant health and productivity. There is a complex interaction between plants and soil bacteria, facilitated by root exudates, but still much is to be translated. We know that effective management of soil health and plant productivity needs the coordination of numerous factors and individuals to establish ideal conditions for a healthy plant. PGPRs are known to ensure plant productivity and health, and we should be capable of enhancing the communication among plants and those microbes (Suman et al. 2022). Root exudates that bacteria utilize as signalling compounds and substrates are possibly the solutions to the control of this conversation. Subsequent investigations should identify which essential chemicals and root exudate contents will grow those beneficial microbes that produce healthy and more productive crops (Canarini et al. 2019).

4.4 Applications in Bioremediation and Ecosystem Management

During the past decades, rapid agricultural and industrial growth has polluted the environment with a variety of pollutants, such as heavy metals, polychlorinated biphenyls, plastics, and other agrochemicals. These are found to be non-biodegradable and toxic in nature, whose presence in the environment is very alarming. Their significance and interaction in the ecosystem are enormous, threatening both the ecological environment and human health. Furthermore, the presence of these contaminants reduces soil quality and fertility (Chandran et al. 2020). Physicochemical procedures are used to cure such situations; however, they are ineffective and incur high operational costs. Bioremediation is an effective, widely used, low-cost, and environmentally acceptable clean-up method. Microorganisms have received significant attention as a successful biotechnology method for cleaning the environment. Microbial bioremediation appears to be the most economically friendly and effective method since it inflicts the least damage on the environment.

Bioremediation technology employs microorganisms' metabolic energy to degrade contaminated environments. *Alcaligenes*, *Arthrobacter*, *Aspergillus*, *Bacillus*, *Burkholderia*, *Mucor*, *Penicillium*, *Pseudomonas*, *Stenotrophomonas*, *Talaromyces*, and *Trichoderma* are some of the microbial genera that have been associated with bioremediation. *Archaea* from extreme environments, in particular *Natrialba* and *Haloferax*, have also been recognized as important bioresources for biological remediation. Therefore, employing microorganisms to control environmental pollution is an effective strategy; indeed, bacteria offer a useful platform for a better bioremediation model of numerous environmental pollutants (Kour et al. 2021). Soil pollution on a large scale with oil products is a universal issue, especially in nations that explore, process, sell, and use crude petroleum products, which has intense implications on soil condition. Leakage of oil from tanks, factories, and pipelines as oil sludge can result in a serious environmental and health hazard. It is a thick combination of sediments (aromatics, alkanes, polycyclic, asphaltenes, resin), water, hydrocarbon, oil, and non-hydrocarbon chemicals (sulphur, nitrogen, and trace metals, predominantly nickel, iron, and copper). Oil sludge has the potential to contain 80% oil and 40% solids. Due to their chemical complexity, strong hydrophobicity, and recalcitrant nature, many hydrocarbons and by-products have long residence times in the environment and can easily get into the food chain, creating a continuous risk to public health and the environment. In addition, these substances penetrate into soil and negatively affect ecosystems by modifying soil properties such as pH, moisture, and aeration (Ambaye et al. 2022). Xenobiotics are foreign molecules created by human endeavours like chemical applications in farming and the production of toxic wastes from paint, drugs, plastics, and fabric industries. Environmental xenobiotics encompass chemical mixtures like medicines, oil blends, azo dyes, antibiotics, alkanes, pesticides, polycyclic hydrocarbons, polyaromatics, fuels, chlorinates, and nitro aromatic chemicals. Dichlorodiphenyl trichloroethane (DDT) is one of the most critical xenobiotics that degrade the environment (Kumar and Singh 2025). Xenobiotics in the environment negatively affect the biota. In human beings, such toxic chemicals cause skin disease and, after prolonged exposure, cancer. Environmental xenobiotics can bioaccumulate, leading to their incorporation in the food chain, and this increases as the ecosystem becomes increasingly tropical. The impacts of xenobiotics provide a considerable problem, and their elimination is required due to their widespread use in modern civilization. Nevertheless, because of their refractory nature, these compounds are hard to degrade (Gangola et al. 2018). Compounds containing groups such as halogen, nitro, and sulfonyl are not decomposed into simple inorganic substances by chemical reactions. Consequently, microbial bioremediation is a good technique for eliminating or degrading some pollutants in the environment. The employment of small organisms, which contribute to half of our planet's biomass, is the most effective technique for bioremediation since they can grow and reproduce rapidly on a huge scale within a short period and are cost-effective. Microbes can degrade such toxic compounds with the help of specific enzymes that degrade xenobiotic molecules into harmless end products (Kour et al. 2022).

5 Challenges in Microbiota Profiling

Microbiota profiling raises a number of problems that can affect the accuracy and reproducibility of results. One key difficulty is the complexity and diversity of microbial communities, which differ depending on genetics, diet, environment, and illness status. The sequencing platform, primer selection, and bioinformatics pipelines all have the potential to generate biases in taxonomic resolution and quantification. Furthermore, distinguishing between living and dead microorganisms as well as host contamination, complicates data interpretation (Scherz et al. 2022). Standardization of techniques and databases remains an issue, resulting in inconsistencies between investigations. Overcoming these constraints necessitates advances in multi-omics integration, enhanced computational tools, and standardized procedures for more reliable microbiota profiling (Ericsson and Franklin 2015).

6 Future Directions

Microbiota profiling techniques are quickly developing, thanks to recent advances in multi-omics methodologies, AI, and throughput sequencing technology. Metagenomics, meta transcriptomics, and metabolomics are all being combined to create a more complete picture of microbial populations and their functional roles in health and disease. Single-cell sequencing and spatial transcriptomics are new technologies that provide higher-resolution insights into microbial interactions at the cellular scale (Misheva et al. 2021). Additionally, AI-driven computational techniques improve the accuracy of microbiome data analysis, allowing for better predictions of microbial functions and host–microbiome interactions. The development of lab-on-a-chip technologies and portable sequencing machines make microbiota more accessible and cost-effective. Future research will most likely focus on tailored microbiome-based treatments using these advances to provide targeted interventions for diseases including cancer, metabolic disorders, and immune-related problems (Bäckhed et al. 2012).

7 Conclusion

The fast improvement of microbiota profiling tools has greatly improved our understanding of microbial communities and their roles in human health, disease, and the environment. Traditional methods such as culture-based procedures have given way to high-throughput sequencing methodologies such as 16s rRNA gene sequencing, metagenomics, meta transcriptomics, and metabolomics, all of which provide extensive insights into microbial diversity and functional potential. Third-generation sequencing, single-cell analysis, and artificial intelligence-driven bioinformatics tools have all helped to improve the accuracy, sensitivity, and efficiency of microbiome research. Despite these developments, issues like data interpretation, standardization, and the need for better computational pipelines remain. Future research should focus on combining multi-omics methodologies, and creating low-cost, real-time microbiota profiling tools enable translational applications in personalized medicine, diagnostics, and environmental microbiology. As technology advances, microbiota profiling will become an essential tool for understanding the complicated interactions between microorganisms and their hosts, ultimately paving the path for breakthrough medicinal and biotechnological advancements.

Acknowledgements The authors express heartfelt gratitude towards the leadership and management of JSS College of Pharmacy, JSS Academy of Higher Education & Research (JSS AHER), Mysuru, Karnataka, India.

Competing Interest Statement The authors declare no conflict of interest.

Disclaimer None.

References

Abavisani M, Khoshrou A, Foroushan SK, Ebadpour N, Sahebkar A (2024) Deciphering the gut microbiome: the revolution of artificial intelligence in microbiota analysis and intervention. Curr Res Biotechnol 7:100211. https://doi.org/10.1016/j.crbiot.2024.100211

Aboud O, Liu YA, Fiehn O, Brydges C, Fragoso R, Lee HS, Riess J, Hodeify R, Bloch O (2023) Application of machine learning to metabolomic profile characterization in glioblastoma patients undergoing concurrent chemoradiation. Metabolites 13(2):299. https://doi.org/10.3390/metabo13020299

Acharjee A, Singh U, Choudhury SP, Gkoutos GV (2022) The diagnostic potential and barriers of microbiome based therapeutics. Diagnosis 9(4):411–420. https://doi.org/10.1515/dx-2022-0052

Aguiar-Pulido V, Huang W, Suarez-Ulloa V, Cickovski T, Mathee K, Narasimhan G (2016) Metagenomics, metatranscriptomics, and metabolomics approaches for microbiome analysis. Evol Bioinforma 12s1. https://doi.org/10.4137/EBO.S36436

Akhatou I, González-Domínguez R, Fernández-Recamales Á (2016) Investigation of the effect of genotype and agronomic conditions on metabolomic profiles of selected strawberry cultivars with different sensitivity to environmental stress. Plant Physiol Biochem 101:14–22. https://doi.org/10.1016/j.plaphy.2016.01.016

Amaretti A, Raimondi S, Leonardi A, Quartieri A, Rossi M (2015) Hydrolysis of the rutinose-conjugates flavonoids rutin and hesperidin by the gut microbiota and bifidobacteria. Nutrients 7(4):2788–2800. https://doi.org/10.3390/nu7042788

Ambaye TG, Chebbi A, Formicola F, Prasad S, Gomez FH, Franzetti A, Vaccari M (2022) Remediation of soil polluted with petroleum hydrocarbons and its reuse for agriculture: recent progress, challenges, and perspectives. Chemosphere 293:133572. https://doi.org/10.1016/j.chemosphere.2022.133572

Ardui S, Ameur A, Vermeesch JR, Hestand MS (2018) Single molecule real-time (SMRT) sequencing comes of age: applications and utilities for medical diagnostics. Nucleic Acids Res 46(5):2159–2168. https://doi.org/10.1093/nar/gky066

Bäckhed F, Fraser CM, Ringel Y, Sanders ME, Sartor RB, Sherman PM, Versalovic J, Young V, Finlay BB (2012) Defining a healthy human gut microbiome: current concepts, future directions, and clinical applications. Cell Host Microbe 12(5):611–622. https://doi.org/10.1016/j.chom.2012.10.012

Bahado-Singh RO, Yilmaz A, Bisgin H, Turkoglu O, Kumar P, Sherman E, Mrazik A, Odibo A, Graham SF (2019) Artificial intelligence and the analysis of multi-platform metabolomics data for the detection of intrauterine growth restriction. PLoS One 14(4):e0214121. https://doi.org/10.1371/journal.pone.0214121

Balomenos AD, Tsakanikas P, Aspridou Z, Tampakaki AP, Koutsoumanis KP, Manolakos ES (2017) Image analysis driven single-cell analytics for systems microbiology. BMC Syst Biol 11(1):43. https://doi.org/10.1186/s12918-017-0399-z

Barberis E, Khoso S, Sica A, Falasca M, Gennari A, Dondero F, Afantitis A, Manfredi M (2022) Precision medicine approaches with metabolomics and artificial intelligence. Int J Mol Sci 23(19):11269. https://doi.org/10.3390/ijms231911269

Belkaid Y, Hand TW (2014) Role of the microbiota in immunity and inflammation. Cell 157(1):121–141. https://doi.org/10.1016/j.cell.2014.03.011

Bowden R, Davies RW, Heger A, Pagnamenta AT, de Cesare M, Oikkonen LE, Parkes D, Freeman C, Dhalla F, Patel SY, Popitsch N, Ip CLC, Roberts HE, Salatino S, Lockstone H, Lunter G, Taylor JC, Buck D, Simpson MA, Donnelly P (2019) Sequencing of human genomes with nanopore technology. Nat Commun 10(1):1869. https://doi.org/10.1038/s41467-019-09637-5

Brandt LJ, Aroniadis OC (2013) An overview of fecal microbiota transplantation: techniques, indications, and outcomes. Gastrointest Endosc 78(2):240–249. https://doi.org/10.1016/j.gie.2013.03.1329

Brehm-Stecher BF, Johnson EA (2004) Single-cell microbiology: tools, technologies, and applications. Microbiol Mol Biol Rev 68(3):538–559. https://doi.org/10.1128/MMBR.68.3.538-559.2004

Canarini A, Kaiser C, Merchant A, Richter A, Wanek W (2019) Root exudation of primary metabolites: mechanisms and their roles in plant responses to environmental stimuli. Front Plant Sci 10. https://doi.org/10.3389/fpls.2019.00157

Cavill R, Jennen D, Kleinjans J, Briedé JJ (2016) Transcriptomic and metabolomic data integration. Brief Bioinform 17(5):891–901. https://doi.org/10.1093/bib/bbv090

Chandran H, Meena M, Sharma K (2020) Microbial biodiversity and bioremediation assessment through omics approaches. Front Environ Chem 1. https://doi.org/10.3389/fenvc.2020.570326

Chaparro JM, Sheflin AM, Manter DK, Vivanco JM (2012) Manipulating the soil microbiome to increase soil health and plant fertility. Biol Fertil Soils 48(5):489–499. https://doi.org/10.1007/s00374-012-0691-4

Chen K, Pachter L (2005) Bioinformatics for whole-genome shotgun sequencing of microbial communities. PLoS Comput Biol 1(2):e24. https://doi.org/10.1371/journal.pcbi.0010024

Chen Z, Hui PC, Hui M, Yeoh YK, Wong PY, Chan MCW, Wong MCS, Ng SC, Chan FKL, Chan PKS (2019) Impact of preservation method and 16S rRNA hypervariable region on gut microbiota profiling. Msystems 4(1). https://doi.org/10.1128/msystems.00271-18

Chi J, Shu J, Li M, Mudappathi R, Jin Y, Lewis F, Boon A, Qin X, Liu L, Gu H (2024) Artificial intelligence in metabolomics: a current review. TrAC Trends Anal Chem 178:117852. https://doi.org/10.1016/j.trac.2024.117852

Cuperlovic-Culf M (2018) Machine learning methods for analysis of metabolic data and metabolic pathway modeling. Meta 8(1):4. https://doi.org/10.3390/metabo8010004

Cuscó A, Catozzi C, Viñes J, Sanchez A, Francino O (2019) Microbiota profiling with long amplicons using nanopore sequencing: full-length 16S rRNA gene and the 16S-ITS-23S of the rrn operon. F1000Res 7:1755. https://doi.org/10.12688/f1000research.16817.2

Damhorst GL, Adelman MW, Woodworth MH, Kraft CS (2021) Current capabilities of gut microbiome–based diagnostics and the promise of clinical application. J Infect Dis 223(Supplement_3):S270–S275. https://doi.org/10.1093/infdis/jiaa689

De Corato U (2020) Soil microbiota manipulation and its role in suppressing soil-borne plant pathogens in organic farming systems under the light of microbiome-assisted strategies. Chem Biol Technol Agric 7(1):17. https://doi.org/10.1186/s40538-020-00183-7

del Campo-Moreno R, Alarcón-Cavero T, D'Auria G, Delgado-Palacio S, Ferrer-Martínez M (2018) Microbiota and human health: characterization techniques and transference. Enfermedades Infecciosas y Microbiologia Clinica (English Ed) 36(4):241–245. https://doi.org/10.1016/j.eimce.2018.02.016

Domenzain I, Sánchez B, Anton M, Kerkhoven EJ, Millán-Oropeza A, Henry C, Siewers V, Morrissey JP, Sonnenschein N, Nielsen J (2022) Reconstruction of a catalogue of genome-scale metabolic models with enzymatic constraints using GECKO 2.0. Nat Commun 13(1):3766. https://doi.org/10.1038/s41467-022-31421-1

Donelli G, Vuotto C, Mastromarino P (2013) Phenotyping and genotyping are both essential to identify and classify a probiotic microorganism. Microb Ecol Health Dis 24. https://doi.org/10.3402/mehd.v24i0.20105

Dudek NK, Chakhvadze M, Kobakhidze S, Kantidze O, Gankin Y (2024) Supervised machine learning for microbiomics: bridging the gap between current and best practices. Mach Learn Appl 18:100607. https://doi.org/10.1016/j.mlwa.2024.100607

Ericsson AC, Franklin CL (2015) Manipulating the gut microbiota: methods and challenges: figure 1. ILAR J 56(2):205–217. https://doi.org/10.1093/ilar/ilv021

Fan TW-M (2012) Metabolomics-edited transcriptomics analysis (META). In: The handbook of metabolomics. Humana Press, pp 439–480. https://doi.org/10.1007/978-1-61779-618-0_14

Fraher MH, O'Toole PW, Quigley EMM (2012) Techniques used to characterize the gut microbiota: a guide for the clinician. Nat Rev Gastroenterol Hepatol 9(6):312–322. https://doi.org/10.1038/nrgastro.2012.44

Franzosa EA, Hsu T, Sirota-Madi A, Shafquat A, Abu-Ali G, Morgan XC, Huttenhower C (2015) Sequencing and beyond: integrating molecular "omics" for microbial community profiling. Nat Rev Microbiol 13(6):360–372. https://doi.org/10.1038/nrmicro3451

Gangola S, Bhatt P, Chaudhary P, Khati P, Kumar N, Sharma A (2018) Bioremediation of industrial waste using microbial metabolic diversity, pp 1–27. https://doi.org/10.4018/978-1-5225-3126-5.ch001

Gerasimidis K, Bertz M, Quince C, Brunner K, Bruce A, Combet E, Calus S, Loman N, Ijaz UZ (2016) The effect of DNA extraction methodology on gut microbiota research applications. BMC Res Notes 9(1):365. https://doi.org/10.1186/s13104-016-2171-7

Hernández Medina R, Kutuzova S, Nielsen KN, Johansen J, Hansen LH, Nielsen M, Rasmussen S (2022) Machine learning and deep learning applications in microbiome research. ISME Commun 2(1):98. https://doi.org/10.1038/s43705-022-00182-9

Hu T, Chitnis N, Monos D, Dinh A (2021) Next-generation sequencing technologies: an overview. Hum Immunol 82(11):801–811. https://doi.org/10.1016/j.humimm.2021.02.012

Iadanza E, Fabbri R, Bašić-ČiČak D, Amedei A, Telalovic JH (2020) Gut microbiota and artificial intelligence approaches: a scoping review. Heal Technol 10(6):1343–1358. https://doi.org/10.1007/s12553-020-00486-7

Ibrahim S, Belheouane M (2018) Methods for microbiota analysis: sample collection and laboratory methods. In: The microbiome in rheumatic diseases and infection. Springer International Publishing, pp 13–27. https://doi.org/10.1007/978-3-319-79026-8_2

Ishii S, Tago K, Senoo K (2010) Single-cell analysis and isolation for microbiology and biotechnology: methods and applications. Appl Microbiol Biotechnol 86(5):1281–1292. https://doi.org/10.1007/s00253-010-2524-4

Jervis-Bardy J, Leong LEX, Marri S, Smith RJ, Choo JM, Smith-Vaughan HC, Nosworthy E, Morris PS, O'Leary S, Rogers GB, Marsh RL (2015) Deriving accurate microbiota profiles from human samples with low bacterial content through post-sequencing processing of Illumina MiSeq data. Microbiome 3(1):19. https://doi.org/10.1186/s40168-015-0083-8

Jovel J, Patterson J, Wang W, Hotte N, O'Keefe S, Mitchel T, Perry T, Kao D, Mason AL, Madsen KL, Wong GK-S (2016) Characterization of the gut microbiome using 16S or shotgun metagenomics. Front Microbiol 7:459. https://doi.org/10.3389/fmicb.2016.00459

Karst SM, Ziels RM, Kirkegaard RH, Sørensen EA, McDonald D, Zhu Q, Knight R, Albertsen M (2021) High-accuracy long-read amplicon sequences using unique molecular identifiers with Nanopore or PacBio sequencing. Nat Methods 18(2):165–169. https://doi.org/10.1038/s41592-020-01041-y

Kashyap PC, Chia N, Nelson H, Segal E, Elinav E (2017) Microbiome at the frontier of personalized medicine. Mayo Clin Proc 92(12):1855–1864. https://doi.org/10.1016/j.mayocp.2017.10.004

Kour D, Kaur T, Devi R, Yadav A, Singh M, Joshi D, Singh J, Suyal DC, Kumar A, Rajput VD, Yadav AN, Singh K, Singh J, Sayyed RZ, Arora NK, Saxena AK (2021) Beneficial microbiomes for bioremediation of diverse contaminated environments for environmental sustainability: present status and future challenges. Environ Sci Pollut Res 28(20):24917–24939. https://doi.org/10.1007/s11356-021-13252-7

Kour D, Khan SS, Kour H, Kaur T, Devi R, Judy C, Rai PK, McQuestion C, Spells S, Bianchi A, Mohan R, Rai AK, Yadav AN (2022) Microbe-mediated bioremediation: current research and future challenges. J Appl Biol Biotechnol 10:6–24. https://doi.org/10.7324/JABB.2022.10s202

Kumar P, Singh J (2025) Microbes in xenobiotics biodegradation. In: Development in waste water treatment research and processes. Elsevier, pp 581–612. https://doi.org/10.1016/B978-0-443-13615-3.00012-1

Lin Z, Vicente Gonçalves CM, Dai L, Lu H, Huang J, Ji H, Wang D, Yi L, Liang Y (2014) Exploring metabolic syndrome serum profiling based on gas chromatography mass spectrometry and random forest models. Anal Chim Acta 827:22–27. https://doi.org/10.1016/j.aca.2014.04.008

Logsdon GA, Vollger MR, Eichler EE (2020) Long-read human genome sequencing and its applications. Nat Rev Genet 21(10):597–614. https://doi.org/10.1038/s41576-020-0236-x

Lohani V, Akhiya AR, Kundu S, Akhter MQ, Bag S (2023) Single-cell proteomics with spatial attributes: tools and techniques. ACS Omega 8(20):17499–17510. https://doi.org/10.1021/acsomega.3c00795

Mallick H, Franzosa EA, Mclver LJ, Banerjee S, Sirota-Madi A, Kostic AD, Clish CB, Vlamakis H, Xavier RJ, Huttenhower C (2019) Predictive metabolomic profiling of microbial communities using amplicon or metagenomic sequences. Nat Commun 10(1):3136. https://doi.org/10.1038/s41467-019-10927-1

Misheva M, Ilott NE, McCullagh JSO (2021) Recent advances and future directions in microbiome metabolomics. Curr Opin Endocr Metab Res 20:100283. https://doi.org/10.1016/j.coemr.2021.07.001

Mukhopadhya I, Martin JC, Shaw S, McKinley AJ, Gratz SW, Scott KP (2022) Comparison of microbial signatures between paired faecal and rectal biopsy samples from healthy volunteers using next-generation sequencing and culturomics. Microbiome 10(1):171. https://doi.org/10.1186/s40168-022-01354-4

Müller S, Nebe-von-Caron G (2010) Functional single-cell analyses: flow cytometry and cell sorting of microbial populations and communities. FEMS Microbiol Rev 34(4):554–587. https://doi.org/10.1111/j.1574-6976.2010.00214.x

Neupane A, Bulbul I, Wang Z, Lehman RM, Nafziger E, Marzano S-YL (2021) Long term crop rotation effect on subsequent soybean yield explained by soil and root-associated microbiomes and soil health indicators. Sci Rep 11(1):9200. https://doi.org/10.1038/s41598-021-88784-6

Nowrotek M, Jałowiecki Ł, Harnisz M, Płaza GA (2019) Culturomics and metagenomics: in understanding of environmental resistome. Front Environ Sci Eng 13(3):40. https://doi.org/10.1007/s11783-019-1121-8

Patel JB (2001) 16S rRNA gene sequencing for bacterial pathogen identification in the clinical laboratory. Mol Diagn 6(4):313–321. https://doi.org/10.1007/BF03262067

Petersen LM, Martin IW, Moschetti WE, Kershaw CM, Tsongalis GJ (2019) Third-generation sequencing in the clinical laboratory: exploring the advantages and challenges of nanopore sequencing. J Clin Microbiol 58(1). https://doi.org/10.1128/JCM.01315-19

Prudent E, Raoult D (2019) Fluorescence in situ hybridization, a complementary molecular tool for the clinical diagnosis of infectious diseases by intracellular and fastidious bacteria. FEMS Microbiol Rev 43(1):88–107. https://doi.org/10.1093/femsre/fuy040

Quince C, Walker AW, Simpson JT, Loman NJ, Segata N (2017) Shotgun metagenomics, from sampling to analysis. Nat Biotechnol 35(9):833–844. https://doi.org/10.1038/nbt.3935

Sanches PHG, de Melo NC, Porcari AM, de Carvalho LM (2024) Integrating molecular perspectives: strategies for comprehensive multi-omics integrative data analysis and machine learning applications in transcriptomics, proteomics, and metabolomics. Biology 13(11):848. https://doi.org/10.3390/biology13110848

Santiago-Rodriguez TM, Hollister EB (2021) Multi 'omic data integration: a review of concepts, considerations, and approaches. Semin Perinatol 45(6):151456. https://doi.org/10.1016/j.semperi.2021.151456

Santiago-Rodriguez TM, Le François B, Macklaim JM, Doukhanine E, Hollister EB (2023) The skin microbiome: current techniques, challenges, and future directions. Microorganisms 11(5):1222. https://doi.org/10.3390/microorganisms11051222

Sato MP, Ogura Y, Nakamura K, Nishida R, Gotoh Y, Hayashi M, Hisatsune J, Sugai M, Takehiko I, Hayashi T (2019) Comparison of the sequencing bias of currently available library preparation kits for Illumina sequencing of bacterial genomes and metagenomes. DNA Res 26(5):391–398. https://doi.org/10.1093/dnares/dsz017

Scherz V, Greub G, Bertelli C (2022) Building up a clinical microbiota profiling: a quality framework proposal. Crit Rev Microbiol 48(3):356–375. https://doi.org/10.1080/1040841X.2021.1975642

Schirmer M, D'Amore R, Ijaz UZ, Hall N, Quince C (2016) Illumina error profiles: resolving fine-scale variation in metagenomic sequencing data. BMC Bioinform 17(1):125. https://doi.org/10.1186/s12859-016-0976-y

Schmidt JE, Bowles TM, Gaudin ACM (2016) Using ancient traits to convert soil health into crop yield: impact of selection on maize root and rhizosphere function. Front Plant Sci 7. https://doi.org/10.3389/fpls.2016.00373

Sharpton TJ (2014) An introduction to the analysis of shotgun metagenomic data. Front Plant Sci 5:209. https://doi.org/10.3389/fpls.2014.00209

Sheng Y, Ali RA, Heyland A (2022) Comparative transcriptomics analysis pipeline for the meta-analysis of phylogenetically divergent datasets (CoRMAP). BMC Bioinform 23(1):415. https://doi.org/10.1186/s12859-022-04972-9

Shukla V, Singh S, Verma S, Verma S, Rizvi AA, Abbas M (2024) Targeting the microbiome to improve human health with the approach of personalized medicine: latest aspects and current updates. Clin Nutr ESPEN 63:813–820. https://doi.org/10.1016/j.clnesp.2024.08.005

Srivastava S, Uddin M (2025) Recent trends in metabolomics, machine learning and artificial intelligence. In: Artificial intelligence in biomedical and modern healthcare informatics. Elsevier, pp 91–100. https://doi.org/10.1016/B978-0-443-21870-5.00010-8

Suman J, Rakshit A, Ogireddy SD, Singh S, Gupta C, Chandrakala J (2022) Microbiome as a key player in sustainable agriculture and human health. Front Soil Sci 2:821589. https://doi.org/10.3389/fsoil.2022.821589

Ventimiglia G, Petralia S (2013) Recent advances in DNA microarray technology: an overview on production strategies and detection methods. BioNanoScience 3(4):428–450. https://doi.org/10.1007/s12668-013-0111-8

Versalovic J, Dore J, Guarner F, Luna RA, Ringel Y (2017) Microbiome-based diagnostics: ready for applications in laboratory medicine? Clin Chem 63(11):1674–1679. https://doi.org/10.1373/clinchem.2016.264473

Vitale GA, Geibel C, Minda V, Wang M, Aron AT, Petras D (2024) Connecting metabolome and phenotype: recent advances in functional metabolomics tools for the identification of bioactive natural products. Nat Prod Rep 41(6):885–904. https://doi.org/10.1039/D3NP00050H

Walker AW (2016) Studying the human microbiota. Adv Exp Med Biol 902:5–32. https://doi.org/10.1007/978-3-319-31248-4_2

Walker AW, Martin JC, Scott P, Parkhill J, Flint HJ, Scott KP (2015) 16S rRNA gene-based profiling of the human infant gut microbiota is strongly influenced by sample processing and PCR primer choice. Microbiome 3(1):26. https://doi.org/10.1186/s40168-015-0087-4

Wallenstein MD (2017) Managing and manipulating the rhizosphere microbiome for plant health: a systems approach. Rhizosphere 3:230–232. https://doi.org/10.1016/j.rhisph.2017.04.004

Wang J-W, Kuo C-H, Kuo F-C, Wang Y-K, Hsu W-H, Yu F-J, Hu H-M, Hsu P-I, Wang J-Y, Wu D-C (2019) Fecal microbiota transplantation: review and update. J Formos Med Assoc 118:S23–S31. https://doi.org/10.1016/j.jfma.2018.08.011

Wanichthanarak K, Fahrmann JF, Grapov D (2015) Genomic, proteomic, and metabolomic data integration strategies. Biomark Insights 10s4:BMI.S29511. https://doi.org/10.4137/BMI.S29511

Woelders T, Revell VL, Middleton B, Ackermann K, Kayser M, Raynaud FI, Skene DJ, Hut RA (2023) Machine learning estimation of human body time using metabolomic profiling. Proc Natl Acad Sci 120(18). https://doi.org/10.1073/pnas.2212685120

Yilmaz A, Ustun I, Ugur Z, Akyol S, Hu WT, Fiandaca MS, Mapstone M, Federoff H, Maddens M, Graham SF (2020) A community-based study identifying metabolic biomarkers of mild cognitive impairment and Alzheimer's disease using artificial intelligence and machine learning. J Alzheimer's Dis 78(4):1381–1392. https://doi.org/10.3233/JAD-200305

Zeb Q, Wang C, Shafiq S, Liu L (2019) An overview of single-cell isolation techniques. In: Single-cell omics. Elsevier, pp 101–135. https://doi.org/10.1016/B978-0-12-814919-5.00006-3

Zhang Y, Thompson KN, Branck T, Yan Y, Nguyen LH, Franzosa EA, Huttenhower C (2021) Metatranscriptomics for the human microbiome and microbial community functional profiling. Annu Rev Biomed Data Sci 4(1):279–311. https://doi.org/10.1146/annurev-biodatasci-031121-103035

Microbiota–Drug Interaction and Its Influence on Drug Metabolism and Toxicity

Radheshyam Pal, Pratibha Bhowmick, Jayshree Mahanty, Priyanti Halder, and Mithun Bhowmick

Abstract The human biological system consists of trillions of microscopic living organisms that are connectively responsible for maintaining the homeostasis of the human body; those are called microbiomes. The changes in the microbiome environment of the system may lead to some pathological conditions like diabetes, cancer, allergy, obesity, and gastrointestinal tract (GIT) diseases. Microbiota is responsible for the absorption and metabolism of the drug. Recent clinical trials on microbiomes show some interaction with pharmacological agents that can be found in drug metabolism and safety profiling effects. Microbiota-mediated drug metabolism is essential for drug discovery, which can alter the typical drug interaction with enzymes and change the efficacy level of the drug. Comprehensive investigations of microbiota–drug interactions are necessary for proper drug dosing, efficacious administration, and reduced side effects. This emphasizes the requirements of individual microbiome profiling, which would guide future personalized medicine when involving drug metabolism and toxicity. Moreover, future research should concentrate on understanding these interactions in depth and applying the knowledge to manipulate microbiota profiles so as to enhance drug response along with patient safety.

Keywords Drug metabolism · Drug toxicity · Gut microbiota · Microbiota–drug interactions · Pharmacokinetics · Pharmacodynamics · Probiotics · Adverse drug reactions

R. Pal · P. Bhowmick · J. Mahanty · P. Halder · M. Bhowmick (✉)
Bengal College of Pharmaceutical Sciences and Research, Durgapur, West Bengal, India

S. K. Singh et al. (eds.), *Microbiota Profiling for Precision Medicine*, Interdisciplinary Biotechnological Advances,
https://doi.org/10.1007/978-981-95-2104-3_3

1 Introduction

A number of bodily processes, such as digestion and nutrient absorption, immune system modulation, brain function, detoxification and metabolism, and cancer prevention, depend on the trillions of microorganisms called microbiota that consists of bacteria, viruses, fungi, as well as other microorganisms encountered in our gastrointestinal (GI) tract (GIT), skin, respiratory tract, urogenital tract, mouth, eyes, ears, and other areas (Blum 2017; Ahmed et al. 2022). A microbiota–drug interaction arises when the gut microbiota modifies the pharmacokinetics, toxicity, or potency of the medication (Feng et al. 2020). It is clinically important to target drugs. The microorganisms in our intestines are very important for boosting our immunity and nutrition absorption. The maximum disease is arisen because of the imbalance of microorganisms (Jia et al. 2008). Fecal microbiota transplant (FMT) highlighted the importance of microbiome and probiotics in a disease context including gastrointestinal disorders (Khoruts and Sadowsky 2016). In accordance with a new study, the metabolizing process of important medications is influenced by the vaginal microbiome. Dysbiosis is synonymous with low gut microbiome diversity. Low comparative bacterial diversity and lactobacilli dominance are features of an uncommon example in a normal vaginal microbiome. Lactobacilli can produce lactic acid via fermentation since it is linked to vaginal health, reducing the vagina's pH. This prevents pathogenic organisms from colonizing the vagina. Alteration of vaginal pH occurs due to vaginal dysbiosis. Both the vaginal and gut microbiomes have a site-specific effect on treatments and are implicated in host–microbial interaction (Amabebe and Anumba 2018; Aldunate et al. 2015). Microbiota can affect the host's inflammatory mechanism because of the host cell (Lu et al. 2015). Numerous phylotypes that had an impact on the gut microbiome's long-term diversity declined months after therapy. Throughout life, the human microbiome changes and evolves, even in the absence of antibiotics. The child gut microbiome is particularly susceptible to change due to neonatal exposure to antibiotic (Yassour et al. 2016). Genetic polymorphisms can be influenced by individual patient response to therapeutics and may affect drug target receptor coordination and drug efficacy. Personalized medicine is a new approach to healthcare to reduce toxicity and improve drug efficacy and the study is useful to see the effect of microbiome and drug interaction (Hassan et al. 2021). The pharmacokinetics of widely used immunotherapies, chemotherapeutics, and antibacterial and inflammatory medications are influenced by the human microbiome, which can either increase or decrease the therapeutic efficacy (Weersma et al. 2020).

1.1 Overview of Gut Microbiota

The large number of microorganisms that are available in our GIT are known as gut microbiota. The major microorganism is bacteria (70–80%) including *Firmicutes*, *Bacteroidetes*, and *Proteobacteria* (Srivastava et al. 2022), and there are also present other microorganisms like viruses, fungi, archaea, and protozoa that control many body functions such as digestion, immune system modulation, production of vitamins, detoxification, and influence in brain function (Vemuri et al. 2020). Gut microbiota means the microorganisms present in our gut that maintain our gut health. The dynamic community is mainly composed of the prokaryotic domain. Viruses are also constituents of nature (Anwar et al. 2021). Humans can be classified according to their enterotypes. Each individual inhibits gut microbiota; there are 3.3 million genes. Our biological system contains many bacterial cells as a human cell. According to phylogenetics, bacteria would be classified into distinct groups according to their genetic relatedness (Arumugam et al. 2011).

1.2 Methods to Study Microbiota

The most successful method to characterize the gut microbiota according to the researcher is stool sampling. Fecal samples are frozen at −80 °C. Firstly the sample is purified through centrifugation, allowing the microbes to stick with the other components of the stool (Fu et al. 2016). The other step is incubating the sample to lyse bacterial cells with lysis buffer and vigorous shaking with or without beds. As a result, approaches such as multiple displacement amplification are used to amplify DNA. For gene sequence, a 16S rRNA primer is selected (Ishoey et al. 2008; Shahi et al. 2019). The process of classifying bacteria and normalizing the sequence count is known as operational taxonomic unit (OTU) analysis. The OTU-clustering method is used to determine the genera and species of bacteria (Zalewski et al. 2018). Alpha diversity and beta diversity are the parameters of microbial diversity. A single variation in diversity is called as alpha diversity. A beta diversity means diversity between samples (Kers and Saccenti 2022). The 16 s RNA sequences are determined by a software called QIIME. 16S rRNA gene amplification is a crucial test for contamination because the chemicals used in the steps could be tainted (Kuczynski et al. 2012).

Microcell tissue collection of the distal GIT could provide a more detailed picture of the gut microbiota. The approach is less frequently employed in research and has a number of drawbacks. One limitation of using mucosal tissue biopsy samples to analyze is the presence of human cells and genetic material that can interfere with binding, unlike with fecal collection, where the human cell contamination is negligible (Mottawea et al. 2019). The two commonly used techniques that rarely get used are swab sampling and lavage. It has not yet been thoroughly investigated for protocol consistency. A greater awareness of some aspects of the gut microbiota's impact on health has been rendered possible by the use of animal models in this field of study. In opposition to typical animal models, contamination-free models of animals have been used to evaluate the autonomous functioning of microwaves (Kamareddine et al. 2020). This test additionally employs gnotobiotic models to investigate the influence of microorganisms on animals, usually rats, that have been created and prepared with certain bacterial strains and frequently genetically modified to comprehend the downstream effects of specific bacterial products or other related products. However, the findings of the research with gnotobiotic models must be considered in light of the fact that the animal's immune systems are compromised from an early age (Martín et al. 2016; Lavin et al. 2018).

For further examination, the cell culture system is very important to control the interaction between gut microbes and others. The development of the microorganism in the gastrointestinal system occurs throughout infancy, and microbial colonization is a result of several host-specific variables (Francino 2014). The characteristics of the gut bacterial community that give rise to the observed inter-individual variation are also influenced by the gut environment. An adult human's gut microbiota can contain up to 1 kg of bacteria, the bulk of which are obligatory anaerobes from the genera, along with a variety of yeast and other microorganisms, *Bacteroides*, *Clostridium*, *Lactobacillus*, *Escherichia*, and *Bifidobacteria*. As a result, at least 2000 species of a complex and dynamic ecology are comprised. The advantages that the bacteria offer the host include improved energy recovery from undigested food, protection from infections, and interaction with the immunological and neural systems (Thursby and Juge 2017; Kleessen and Jaana Mättö 2000).

1.3 Importance of Microbiota in Health

There are enormous numbers of microorganisms or symbiotic microorganisms in human beings; more than 100 trillion contribute to human health (Murphy 2018). Especially gut microbiota and other microbiota in humans are considered an "essential organ." The microwave of a human being carries approximately 150 times the variance of genes in a human genome (Zhu et al. 2010). It has been demonstrated in research that humans' gut microbiota is involved in fundamental biological functions. According to age, sex, the diet of the host, and different locations, the microbiota compositions are different. For certain foods that the stomach and intestine cannot digest, here the microbiota helps in digesting the foods (Hasan and Yang 2019). Certain *Bacteroides* break down dietary fibers like xyloglucans, which are frequently present in vegetables. Beneficial microorganisms, including *Lactobacillus* as well as *Bifidobacterium*, use other indigestible fibers, like fructooligosaccharides and oligosaccharides (Larsbrink et al. 2014; Buddington 2001). Short-chain fatty acids (SCFAs) of up to 50–100 mmol of butyric, propionic, and acetic acids are produced daily by a healthy gut flora. Colon can quickly absorb the SCFAs created in regulating gut moduli inflammation, glucose homeostasis, and producing energy. The gut microbiota also supplies the host with a variety of vitamins, including folates, vitamin K, biotin, riboflavin (B2), cobalamin (B12), and other B vitamins (Den Besten et al. 2013). The humoral and cellular mucosal immune system development is stimulated by the gut-colonizing bacteria. Microorganisms' signals and metabolites can be detected by innate immune system cells, both hematopoietic and non-hematopoietic, and converted into physiological reactions. According to a study, the gut microbiota can produce a tolerogenic response that affects gut dendritic cells and prevents the anti-inflammatory pathway's type 17 T-helper cell (TH17) from functioning. Microbiota not only provides health benefits but also sometimes it induces inflammation under certain conditions (Pickard et al. 2017).

1.4 Introduction to Drug Metabolism

The primary location of drug metabolism is the liver. Additionally, the microbiome is essential to medication metabolism. Understanding the characteristics that may cause compounds to be therapeutically inactive or cause toxicity in patients is crucial for the research of drug metabolism and toxicity. With this knowledge, patients can receive better compounds or medications that offer safe and effective treatment (Dhurjad et al. 2022). Numerous metabolic processes involving medications or other xenobiotics are mediated by the gut flora. Heterocyclic reactions and reductive metabolism are the two most significant biotransformations (Wilson and Nicholson 2014). Xenobiotics can reside in fatty tissue for long times causing toxicity and leading to chronic problems like birth defects, stunted growth, respiratory problems, cancer, recurring disease, and behavioral, neurological, and reproductive deficiencies (Kucherenko et al. 2021). However, the human liver and gut have metabolic enzymes that chemically alter xenobiotics to lower their concentration before they enter the bloodstream. This natural detoxification process, often known as "first-pass metabolism," lowers toxicity. The mechanism is further divided into two steps, including phase 1 and phase 2. In phase 1, reactive functional groups are added via hydrolysis, oxidation, reduction, deamination, and demethylation. By conjugating phase 1 metabolite with glucuronic acid, sulfonic acid, glutathione, and certain amino acids, the phase 2 metabolism that follows the phase 1 reaction greatly increases hydrophilicity (Back and Rogers 1987; Grace et al. 2008). Recent research has demonstrated that gut microbes are essential to the metabolism of medicine and convert it into a variety of metabolites, primarily from oral medications. The GI tract is where the microbiome interacts with the drug (Jourova et al. 2016). The gut microbiome is subjected to a range of xenobiotics and is wide and expanding. Several enzymes actively participate in the biotransformation of dietary compounds (complex polysaccharides, proteins, phytochemicals, and lipids) that are derived from the gut microbiota. The altered metabolites act as nutrients and are associated with various health benefits and disease risks (Zhao et al. 2022). When taken orally, xenobiotics may undergo changes in the mouth and digestive tract due to the host's digestive enzymes before being absorbed into tissues and going into the bloodstream. Xenobiotics are easy to absorb; they pass the stomach (epithelial) cells, where host enzymes may further process them before the portal vein carries the xenobiotics through the liver. In the liver, these substances undergo additional metabolic processing (Gad 2018). After entering the circulatory system, the xenobiotics and their metabolites affect distant organs after reaching different tissues. These substances are either further broken down in the circulatory system or eliminated by the kidneys into the urine or the bile duct into the gut lumen. Some metabolites come back to the intestinal lumen, where they may be reabsorbable in the small intestine (through enterohepatic circulation) or pass to the large intestine for elimination through feces. The gut microbiota may alter substances that are poorly absorbed in the small intestine as they pass through the large intestine. These altered metabolites may subsequently interact with the GI tract's lining cells or be reabsorbed by host cells into the bloodstream. These metabolites are ultimately eliminated through feces or urine (Gachon and Firsov 2011). The activities of human xenobiotic metabolizing enzymes and their own bioactivity, bioavailability, and toxicity can directly or indirectly impact the expression and function of these enzymes and their own bioavailability and bioactivity, and thus the final metabolites of ingested drugs can be changed by gut microbial metabolites (Nakov and Velikova 2020). Anaerobic bacteria live in the nutrient-rich, anaerobic environment of the colon and convert complex carbohydrates and other indigestible materials into short-chain fatty acids (SCFA). This procedure aids in digestion and controls host metabolism (Kasubuchi et al. 2015). The effectiveness, quality, and toxicity of medications and xenobiotics are all impacted by the gut microbiota, which has been shown to have a direct and indirect impact on xenobiotic metabolism. Gut microbiota affects drug metabolism through two main mechanisms. The direct biotransformation mechanism involves microbial enzymes that directly metabolize drugs. The indirect biotransformation mechanism, on the other hand, involves microbial metabolites that influence host receptors and signaling pathways (Sun et al. 2019). By directly interacting with a medicine or by generating enzymes that perform major or minor biochemical changes, the gut microbiota can affect a person's reaction to it. This could result in the drug being more or less active or in the production of harmful metabolites. The reactivation of previously inactive drug metabolites is another way that microbiota can indirectly impact medications that are delivered. Drug metabolism by gut bacteria mostly entails hydrolytic and reductive biotransformation activities because the gut milieu is usually anaerobic or just slightly oxygenated (Weersma et al. 2020; Wilson and Nicholson 2017).

1.5 Reductive Metabolism

Sulfanilamide-based azo-antibacterial pro-drugs, such as prontosil (Aldunate et al. 2015; Shahi et al. 2019) and neoprontosil (Ishoey et al. 2008), are the quintessential form of therapeutic medication biotransformation by gut microbes. Gut microbiota has a major role in mediating the reductive biotransformation in the range of 5-aminosalicylic acid (5-ASA) pro-drugs, which are utilized to treat inflammatory bowel disorders and ulcerative colitis (Crouwel et al. 2021). The release of aminosalicylic acid is important for therapeutic action of substances such as sulfasalazine (Zalewski et al. 2018; Kers and Saccenti 2022), olsalazine (Kuczynski et al. 2012), ipsalazide, and balsalazide (Mottawea et al. 2019), as the aminosalicylic acid derives its therapeutic activity from inflammation. Rafii and Cerniglia performed reductive metabolism on azo and nitro polycyclic aromatic hydrocarbons to demonstrate the bacteria belonging to the genera *Clostridia* and *Eubacteria* (Kamareddine et al. 2020; Arvind and Farthing 1988). The study of *Clostridium* and *Eubacteria* on azo and nitro reductase was detected in three strains and discovered that although the enzymes from different bacteria had varied electrophoretic mobilities, the same enzyme was responsible for both azo and nitro reductase activities (Rafii and Cerniglia 1995). The metabolism of benzodiazepines with this functional group, such as nitrazepam, clonazepam, and bromazepam, is significantly influenced by the activity of gut microbial nitro reductase, which converts nitro groups to amines (LinWu et al. 2010). Thus, it has been demonstrated that gut microbiota may create 7-amino nitrazepam, 7-amino clonazepam, and 2-(2-amino-5-bromobenzoyl) pyridine from their parent medications by nitro reduction in both lab and living species. The breakdown of nitro groups might result in negative side effects, such as the teratogenic effects that nitrazepam may cause. Treatment with antibiotics reduced microbiota-driven metabolism in rats, which in turn reduced the teratogenic effects of nitrazepam. When [14C]-clonazepam was administered orally to germ-free rats, only 15% of the radioactive material from the drug was detected in their urine as decreased metabolites. The main metabolite was found to be 7-acetamido-clonazepam, and the introduction of intestinal flora considerably raised nitro reduction to 77% (Takeno and Sakai 1991; Elmer and Remmel 1984). In vitro and in rats, the gut microbiota has been shown to nitro reduce metronidazole, producing the amino metabolite 1-(2-aminoimidazol-1-yl)-3-methoxypropanol-2-ol and acetamide, which is known to cause cancer in rats. The reductive metabolism of the nitro-containing antibiotic chloramphenicol may also be a factor in drug-induced bone marrow destruction, in addition to the toxicity associated with nitrazepam reduction (Koch et al. 1979). P-aminophenyl-2-amino-1,3-propanediol, which is formed in a tiny percentage of patients following oral dosing, has been suggested as the metabolite in question. Patients who generate this metabolite typically have a high percentage of coliform bacteria that can convert the medication into this particular molecule (Dingsdag and Hunter 2018). Studies on people and canines using the pharmaceuticals levosimendan and eltrombopag have shown that the gut bacteria can break down hydrazone connections in drugs. The microbiota-driven reduction of sulfur-containing medications such as sulphinpyrazone and sulindac shows that reductive metabolism also impacts sulfur-containing substances. Sulfides are created during this procedure when these medications are incubated with human or rabbit excrement (Strong et al. 1987). Studies conducted in vitro have demonstrated that intestinal microbiota reduces omeprazole to its sulfide metabolite. Moreover, several strains of gut bacteria have been shown to undergo the reductive metabolism of the benzoxazole ring in the anticonvulsant zonisamide, with *Clostridium*-poor genes showing the maximum activity. Both rats and dogs have shown that gut microbiota reductively metabolizes the Benz isoxazole ring in the antipsychotic medication risperidone (Stiff et al. 1992). Digoxin's reduction by gut bacteria is among the most fascinating instances of how microbiota-driven reductive drug metabolism might affect medication effects. Research by Lindenbaum and others initially brought attention to this biotransformation when they found that each person produced different amounts of reduced metabolites, including dihydrodigoxin.

The fact that less metabolites were eliminated in the urine after intravenous treatment was another sign of microbial participation, indicating that "a variable component of the intestinal flora" was responsible for the reductive metabolism. *Eubacterium lentum* was later found to be the organism in charge of the reductive metabolism of digoxin. It was observed, although, that the bacterium's presence did not always correspond with the lowered metabolites' synthesis. Additionally, it was shown that the production of decreased metabolites was inversely correlated with the amount of arginine present in the growth medium. Interestingly, reduced digoxin metabolites were not detected in infants younger than 8 months, even though *E. lentum* was present. However, by 16 months of age, about one-third of the children studied did produce reduced drug metabolites, although the quantities observed were lower than those in the adult population (around 10%) and were not detected in children under 9 years old (Dobkin et al. 1982, 1983). Digoxin-reducing *E. lentum* was found to be less prevalent in the fecal samples of 77 nursing home residents with *Clostridium difficile* infections who had previously received enteral feeding or antibiotic treatment, according to another study that found an inverse link between the two bacteria. The researchers observed differences in the degree of digoxin decrease within and between populations in addition to age-related effects. Within an Indian community, for example, notable disparities were noted: urban residents produced almost 23% of the decreased metabolites, whereas rural villagers produced roughly 5%. Comparisons between North American respondents, who produced about 36% of reduced metabolites, and a South Indian population, who generated only 13.7%, showed clear demographic differences. Even after Indian subjects relocated to the United States, these differences continued to exist. The question of why *E. lentum* can exist without causing digoxin reduction has been effectively addressed by a recent study by Haiser et al., which has also elucidated the connection between this reduction and arginine when it does occur (Mathan et al. 1989). Even after the relocation of Indian subjects to the United States, these inequities continued. Haiser et al.'s recent study has effectively addressed the question of why *E. lentum* can exist without causing digoxin reduction and has elucidated the connection between this reduction and arginine when it does occur (Haiser et al. 2013). Studies conducted on gnotobiotic mice colonized with digoxin-reducing or non-reducing *E. lentum* showed that mice colonized with the non-reducing strain were unaffected by a high dietary protein consumption, but digoxin's reductive metabolism was reduced, altering serum pharmacokinetics and urine excretion (Haiser et al. 2013). The authors proposed that this effect was "likely due to the inhibitory impact of elevated luminal arginine on cgr operon expression." The findings effectively highlight a complex interaction between the microbiome and drugs, which requires the presence of a specific microbial strain in the host and can also be influenced by dietary factors (Haiser et al. 2013).

2 Direct Microbial Metabolism of Drug

In the intestinal lumen, ingested xenobiotics undergo direct microbial biotransformation, where bacterial enzymes react chemically with the substances. Specifically, bacterial-glucuronidase enzymes in the human intestine act on abundant glucuronides to convert them into glucuronic acids from conjugated compounds (Bik et al. 2018). According to one study, there are 3013 glucuronidases encoded by the microbiome, and each one has distinct functional properties for digesting different glucuronide substrates. Apart from the hydrolytic biotransformations of glucuronide conjugates, the gut microbiota is home to numerous additional bacterial enzymes, such as reductases (nitroreductases and azoreductases), polysaccharide lyases and transferases, lipases, mono- and dioxygenases, sulfatases, endoglycosidases, and glycyl radical enzymes (Pollet et al. 2017). It has been demonstrated that giving probiotics affects the intestinal microbiota's shape and enzyme activity. According to research by Kim and colleagues, probiotic strains change the pharmacokinetics of medications when taken with them. In particular, the addition of *Lactobacillus reuteri* K8 raised the concentrations of *Cyanobacteria*, *Enterococci*, *Clostridium*, and *Bifidobacteria* (Kim et al. 2018). This probiotic strain improved arylsulfatase sulfotransferase activity and the breakdown of acetaminophen, both directly and through the gut microbiota, without influencing intestinal metabolic activity as a whole. On the other hand, acetaminophen degradation was unaffected by the administration of *Lactobacillus rhamnosus* K9, which increased *Deferribacteres* and decreased *Bifidobacteria* (Kim et al. 2018).

According to one study, the glucuronidase enzyme (encoded by the gus gene) induces the production of aglycones, which in turn sets off a chain of events that results in cell destruction when it interacts with enterocytes and produces nonsteroidal anti-inflammatory drug (NSAID)-induced gastrointestinal toxicity (Boelsterli et al. 2013). Nevertheless, mice were shielded from NSAID-induced enteropathy brought on by indomethacin, ketoprofen, and diclofenac when Inh-1[1-((6,8-dimethyl-2-oxo-1,2-dihydroquinolin-3-yl)-3-(4-ethoxyphenyl)-1-(2-hydroxyethyl) thiourea] was administered to block bacterial glucuronidase (Saitta et al. 2014). It was unclear how the gut bacteria contributed to the hepatotoxicity caused by cisplatin. But when antibiotics and cisplatin were taken together, the toxicity decreased, indicating that the gut microbiota may be involved in controlling cisplatin-induced liver damage (Gong et al. 2021).

2.1 *Hydrolysis*

Hydrolases can be made by the gut microbiota or the host enzyme, which cleave compounds ingested into lesser molecules for further metabolism. By introducing a water molecule to a substrate, they cause bond cleavage (Bell and Juge 2021). Proteases, sulfatases, glycosidases, and esterases are examples of common hydrolases found in the gastrointestinal system. Based on their catalytic sites, proteases—which cleave amino acid links in polypeptides by acting on peptide bonds—are divided into four categories: metalloproteases, cysteine proteases, serine proteases, and aspartic proteases. Certain types of proteases are found in different parts of the human gut; for example, the small intestine mostly contains pancreatic serine proteases, but the colon is rich in microbial cysteine and metalloproteases (Giuffrida et al. 2019). The substrate specificities of these enzymes vary, which may result in varying treatment outcomes. Overall, 285 putative serine protease sequences from the intestinal microbiota were found through a study of roughly ten million genes from the human gut metagenomic database. Fifty-six genera from five phyla—Firmicutes, Bacteroidetes, Actinobacteria, Proteobacteria, and Fusobacteria—contributed these sequences (Verberkmoes et al. 2009). Furthermore, the most basic kind of drug conjugates, single amino acid or peptide prodrugs, are made up of a drug joined to a brief linker. The activity of these prodrugs is very pro-for the conjugated form, but they are potent only after release of the drug by the target protease. Legumain is a cysteine protease that exhibits very high selectivity, favoring the recognition of asparagine (Asn) over aspartate (Asp) at the P1 site. As a result, asparagine (Asn) and a cytotoxic medication are coupled with numerous prodrugs intended to be targeted by legumains (Vidmar et al. 2019). Hydrolyzing carbohydrate or glycosyl conjugates is the function of glycosidases. These enzymes use a water molecule and two carboxylic acid residues to dissolve glycosidic bonds, releasing free sugars in the process. The gut microbiota contains large amounts of the three main glycosidases, α-rhamnosidase, β-glucosidase, and β-glucuronidase. Because it hydrolyzes glucuronides released into the gut through biliary secretion, β-glucuronidase is an essential enzyme in the metabolism of xenobiotics (Dabek et al. 2008). The chemotherapy medication irinotecan, which is inactivated in the liver by glucuronidation, is a prominent illustration of this process. However, β-glucuronidase, which is produced by gut bacteria, reactivates irinotecan once it reaches the intestine. The drug's dosage is limited due to the severe gastrointestinal toxicity caused by this reactivation (Chamseddine et al. 2019). The hydrolytic metalloenzymes known as sulfatases degrade the sulfate esters that are created during phase II metabolism. At their active site, they make use of the uncommon amino acid residue formyl glycine. It is thought that this residue's hydrate form transesterifies with the sulfate ester substrate to produce a tetrahedral intermediate (Appel and Bertozzi 2015). After that, this intermediate breaks down to replenish the aldehyde and release sulfate. Ervin and associates used protein structure-guided methods to find 728 different microbiome-encoded sulfatase proteins in Hexose Monophosphate pathway (HMP) stool samples out of a database of 4.8 million unique proteins. They also found that the combined gene catalog database had 9.9 million sequences of 1766 gut microbial sulfatases (Luis et al. 2022). The physical characteristics and activity of xenobiotics and their metabolites are altered by hydrolytic processes. For example, removing a glucuronide from the gut lumen usually makes it less polar, which can help host cells reabsorb it and keep it in the body longer. This is demonstrated by the cancer medicine irinotecan and the glucuronide conjugates of non-steroidal anti-inflammatory medications (Currò 2018). Esterases are a class of hydrolase enzymes that catalyze the hydrolysis of ester groups into alcohol and carboxylic acid. This group comprises arylesterases, carboxylesterases, and acetylesterases (Anspaugh and Roe 2002). Several bacteria, including *Rhodobacter sphaeroides*, *Bacillus subtilis*, *Geobacillus stearothermophilus*, *Pseudomonas fluorescens*, *Arthrobacter globiformis*, *Archaeoglobus fulgidus*, *Alicyclobacillus acidocaldarius*, *Lactobacillus casei*, *Klebsiella* sp., *Sulfolobus solfataricus*, *Serratia* sp. SES-01, and *Sulfolobus tokodaii*, exhibit carboxylic ester hydrolase activities; they can hydrolyze substrates such as phenylacetate, 4-nitrophenylacetate, naphthyl butyrate, and some medications, including ketoprofen, aspirin, and naproxen. Gut bacteria produce arylsulfatases and β-glucuronidases, which are enzymes that can break conjugates made by sulfo- and glucuronosyltransferases, returning them to their parent chemicals and prolonging the half-life of the medicine (Sun et al. 2019; Sousa et al. 2008). Furthermore, hydrolysis frequently comes before other processing steps like fermenting sugars made from indigestible polysaccharides. These reactions' byproducts frequently encourage the development and survival of gut commensals in the digestive system (Cockburn and Koropatkin 2016).

2.2 *Reduction/Oxidation*

By changing a metabolite's charge, electrophilicity, or hybridization, reductive reactions usually cause its polarity to decrease. The activity and durability of these metabolites within the body are typically impacted by this alteration (Guo et al. 2020). Nitro (NO_2), azo (–N=N–), ketone (–C=O), alkene (–C=C–), sulfoxide (–S=O), and N-oxide (–N–O) are the six main categories of reductive transformations. In the human gastrointestinal tract, the bacterial taxa *Clostridium* sp., *Eubacterium* sp., *Clostridium leptum*, *Clostridium paraputrificum*, *Clostridium perfringens*, and *Clostridium clostridioforme* showed nitroreductase and azoreductase activity (Guo et al. 2020; Rafil et al. 1991). They consist of 5-aminosalicylic acid (5-ASA), a non-steroidal anti-inflammatory compound, linked to an inert carrier via an azo bond to prevent rapid absorption. The effectiveness of these drugs depends on the cleavage of the azo bond by azoreductases produced by gut microbiota, which releases 5-ASA. Enzymes called nitroreductases catalyze the reduction of substances that contain nitrogen. There are three types of bacterial nitroreductases: type I (oxygen-insensitive), type II (oxygen-sensitive), or both (Simpson et al. 2023). Based on their choice for electron donors, the extensively studied oxygen-insensitive nitroreductases can be roughly divided into two groups: a major group that employs NADPH as an electron donor, and a minor group that uses both Nicotinamide Adenine Dinucleotide Phosphate (NADPH) and Nicotinamide Adenine Dinucleotide (NADH) (Boddu et al. 2021). *Bacteroides fragilis* produces nitro reductases that convert the naturally occurring nitroaromatic substance chloramphenicol to an amine metabolite. The antibacterial action of this amine metabolite is extremely poor (Smith et al. 2007). Additionally, the gut microbiota performs a number of biotransformation processes that entail oxidation or the transfer of functional groups. This activity is caused by transferase enzymes, which help two substrates transfer functional groups through nucleophilic substitution processes (Krishnan et al. 2015). Furthermore, deaminases encoded by the gut microbiota convert 5-fluorocytosine to 5-fluorouracil, which at least partially accounts for the toxicity of 5-fluorocytosine treatment. The gut bacterium *Blautia* sp. MRG-PMF1 was shown to biotransform 12 structurally different polymethoxy flavones (PMFs) and identify their metabolic intermediates in anaerobic in vitro research (Harris et al. 1986). The identified compounds include 5,7-dimethoxyflavone, 5-hydroxy-7-methoxyflavone, 3,5,7-trimethoxyflavone, 5-hydroxy-3,7-dimethoxyflavone, 5,7,4′-trimethoxyflavone, 5-hydroxy-7,4′-dimethoxyflavone, and 3,5,7,4′-tetramethoxyflavone, along with 5,7,3′,4′-tetramethoxyflavone, 5-hydroxy-3,7,4′-trimethoxyflavone, and pentamethoxyflavone. Following complete demethylation of 5-hydroxy-3,7,3′,4′-tetramethoxyflavone and 5,3′-dihydroxy-3,7,4′-trimethoxyflavone, chrysin, apigenin, galanga, kaempferol, luteolin, and quercetin were produced (Ta and Walle 2007; Nielsen et al. 1998). Chemically activated cosubstrates like adenosine triphosphate (ATP), S-adenosylmethionine (SAM), or acetyl coenzyme A (CoA) are frequently required for further reactions. Hydrolysis is usually required for the enzymatic elimination of acyl groups. Cofactors such as cob(I)alamin and tetrahydrofolate are useful for nucleophilic catalysis and are used by demethylases (Walsh et al. 2018a).

2.3 Deconjugation

The process of deconjugation in microbiota drug metabolism is defined as the ability of gut bacteria to convert non-conjugated forms of a drug into conjugated forms by deconjugating these drugs. Many drugs are initially processed in the liver and excreted into bile as conjugates, such as sulfate or glucuronide conjugates (Currò 2018). Upon reaching the gut, these conjugated drugs can be deconjugated by the gut microbiota through enzymatic action, reverting them to their original forms or transforming them into other metabolites (Currò 2018).

This deconjugation can affect the effectiveness and toxicity of drugs. For instance, deconjugated drugs might be reabsorbed into the bloodstream, potentially modifying their therapeutic effects. Alternatively, some drugs might be inactivated or their side effects adjusted due to microbial deconjugation. Thus, pharmacokinetics and drug metabolism are significantly influenced by the gut flora (Wilson and Nicholson 2017). Various sites and types of drug metabolism via gut microbiota are shown in Fig. 1, and relation between gut microbiota and drug metabolism is shown in Fig. 2.

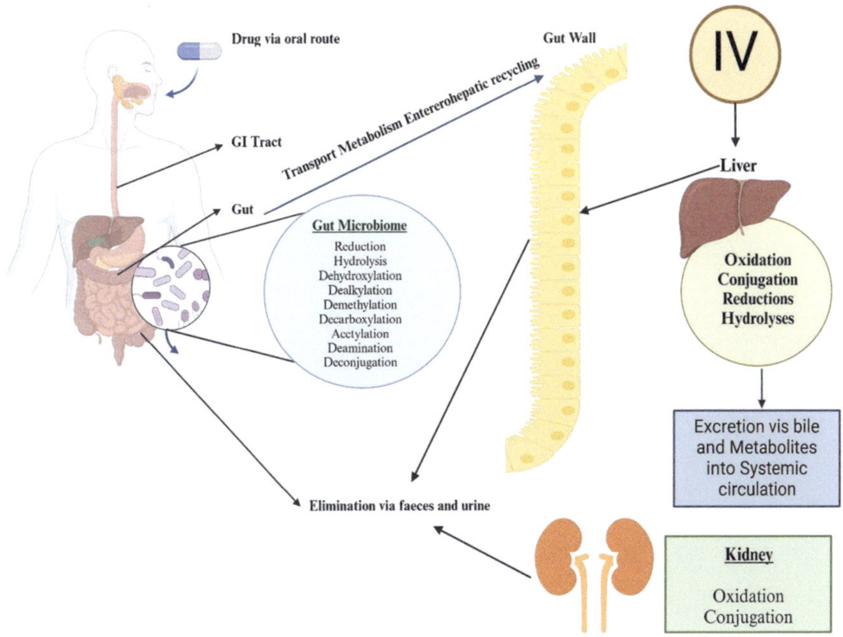

Fig. 1 Overview of different sites and types of drug metabolism facilitated by gut microbiota

2.4 Alteration of Drug Absorption

Drug absorption is another key mechanism of microbiota–drug interactions. Considering the altered absorption of drugs, the drug fraction exposed for microbiota interaction in the gastrointestinal tract varies. The entire human gastrointestinal tract, especially the colon to a larger extent, is colonized by around 100 trillion microorganisms, with more than 700–1000 different fungi, viruses, parasites, and bacteria species constituting what is known as gut microbiota that causes drug–microbiota interaction and influences the pharmacodynamics and pharmacokinetics of drugs, particularly the drug absorption (Qin et al. 2010; Walsh et al. 2018b; Takiishi et al. 2017). The gut microbiome alters drug absorption by different means, discussed as follows.

2.4.1 Gut Barrier Integrity

The gut barrier allows nutrients and drugs to pass into the bloodstream and protects the body from pathogenic bacteria maintaining the balance between the body's immune system and commensal microbes. The gut barrier consists of the outer mucus layer; intestinal epithelial cells; defense proteins including antimicrobial proteins, and secretory immunoglobulin A; cell junctions involving tight junctions

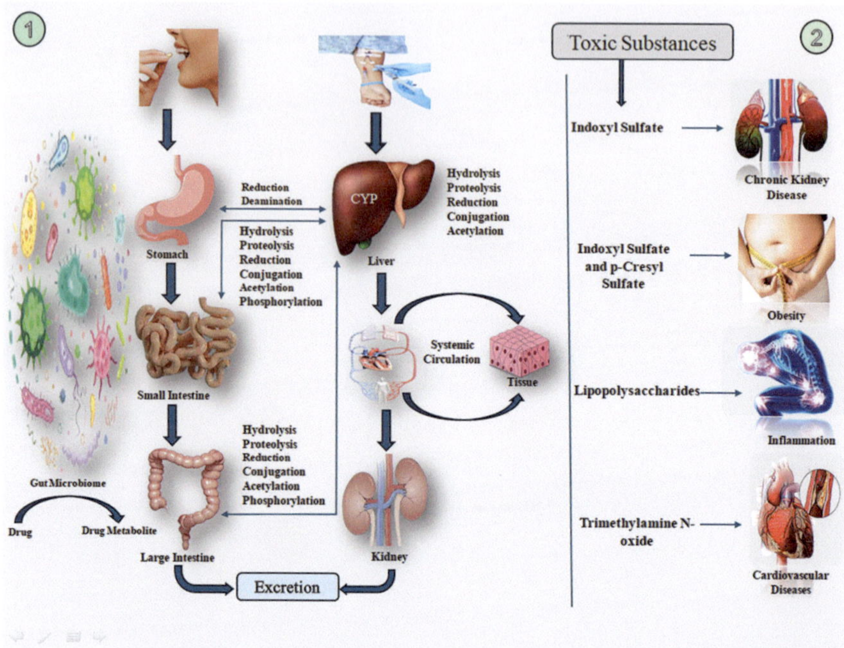

Fig. 2 (1) Highlights the complex relationship between gut microbiota and drug metabolism, focusing on major enzymatic steps such as hydrolysis, proteolysis, reduction, conjugation, acetylation, and phosphorylation that are required for drug transformation and subsequent systemic effects. (2) Illustrates the role of toxic substances produced by the gut microbiota, such as indoxyl sulfate, p-cresyl sulfate, lipopolysaccharides, and trimethylamine N-oxide (TMAO), in the onset of conditions like obesity, inflammation, and cardiovascular diseases

followed by adherens junctions, and desmosomes; and immune cells (Ghosh et al. 2021). The integrity of the gut barrier is crucial for drug absorption, and all of its parts cooperate to control the amount of medication that enters the bloodstream from the intestinal lumen. From food items and other nutrients, the gut microbiota produces a metabolome that can affect several pathophysiological processes and gut health behavior. The integrity of the gut barrier is greatly impacted by gut microbial metabolites, which change the permeability of the barrier because they are produced close to the gut epithelium (Camilleri 2019). The permeability of the epithelium of the small intestine is a crucial factor for the absorption of drugs in the gastrointestinal lumen as the small intestine enriched in gut microbiota is the primary site of absorption (Williams et al. 2013).

2.4.1.1 Increased Gut Permeability (Leaky Gut)

Sometimes the intestinal lining of the GI tract becomes more permeable by loosening and disruption of tight junctions between epithelial cells, called leaky gut (Stillhart et al. 2020; Thaiss et al. 2018; Quigley 2016; Hollander 1999). The gut microbial metabolites such as lipopolysaccharides and short-chain fatty acids trigger the leaky gut. The leaky gut can be induced if the microbial balance inside the gut is compromised by chronic constipation, chronic stress, exposure to environmental pollutants, unhealthy diet, or taking any potent antibiotic. In a leaky gut along with food particles, most of the drugs and harmful bacterial toxins easily pass from the gut lumen into the bloodstream. The more permeable leaky gut increases the absorption and systemic exposure of those drugs that usually get blocked by tight junctions of the gut barrier due to their size and nature, that is, peptide-based drugs. Due to enhanced paracellular transport followed by uncontrolled drug absorption and bypass from first-pass metabolism, the potent drugs may lead to toxicity reaching higher plasma concentrations (Parsons 1977). Drugs that are highly absorbed in the stomach or intestine have limited interaction with gut microbiota present in the colon.

2.4.1.2 Decreased Gut Permeability

Gastrointestinal physiology also varies by enhanced gut barrier integrity because microbial imbalances make it less permeable. It limits the absorption of drugs with large size or high lipophilicity. Since the tighter gut barrier restricts the passive diffusion or paracellular transport of drugs, poorly absorbed drugs retain in small intestine and colon for longer duration increasing their exposure to gut microbes, which proceeds to effective metabolism of the drug (Sugihara et al. 2015).

The gut microbiota controls the integrity of the intestinal barrier in different ways. Commensal gut bacteria, that is, *Faecalibacterium*, *Bifidobacterium*, *Roseburia*, *Bacteroides*, and *Enterobacter* species, produce short-chain fatty acids (SCFAs) like acetate, propionate, and butyrate by fermenting the consumed dietary fibers and carbohydrates. SCFAs maintain gut barrier health, supporting tight junctions' integrity between epithelial cells (Wu et al. 2017). The imbalance in bacterial structural components alternates bacterial metabolic activities and bacterial distribution in the gut epithelium, which is called dysbiosis (DeGruttola et al. 2016). Dysbiosis triggers inflammatory responses that damage the gut epithelial layer and lead to unpredictable drug absorption and increased risk of adverse reactions (Camilleri 2019).

2.4.2 Competition for Binding Sites

The absorption and transportation of drugs are reduced or delayed by competition for binding sites. If certain drugs are co-administered and they are the substrate of one common transporter to be absorbed in the gut lumen, they compete with each other reducing the absorption of one or both drugs. Drugs possessing almost similar chemical structures compete for transporter in the absorption site. Since the availability of carrier molecules is limited, the binding sites on the surface of the carrier get saturated (Alagga et al. 2024). Carrier proteins in the intestine can facilitate or obstruct the rate of drug absorption. Efflux membrane transporters, that is, BCRP, MRP2, and mainly P-glycoprotein (P-gp), hinder drug absorption. P-gp (MDR1) decreases the absorption of most different drugs transporting them back to the gut lumen by the apical membrane (Amin 2013). Drugs such as rifampicin, carbamazepine, doxorubicin, phenobarbital, prazosin, and phenytoin, inducing P-glycoprotein pump, reduce the absorption and bioavailability of other P-gp substrate drugs like fexofenadine, digoxin, and talinolol (Finch and Pillans 2014; Elmeliegy et al. 2020). Vice versa, some common P-gp inhibitor drugs can increase the uptake, absorption, and bioavailability of drugs. A few potent P-glycoprotein inhibitors include verapamil, cyclosporine, amiodarone, clarithromycin, colchicine, and felodipine (Srivalli and Lakshmi 2012). Also, bacterial metabolites like short-chain fatty acids regulate the expression and function of P-gp inhibiting the Nuclear Factor kappa-light-chain-enhancer of activated B cells (NF-κB) pathway (Xie et al. 2021). However, uptake transporters such as organic anion transporting polypeptides (OATPs) and peptide transporters (PEPTs) can improve drug absorption by supporting their movement from the intestinal lumen into the blood circulation (Yu et al. 2017; Wang et al. 2017). The organic cation transporter 1 (OCT1) facilitates the uptake and hepatic metabolism of drugs like metformin (Shu et al. 2007). Gut microbiome causes the formation of various enzymes that can metabolize drugs before they are absorbed. β-glucuronidase produced by certain bacterial species can metabolize drugs that have undergone glucuronidation in the liver. Some gut bacteria can directly degrade drugs, essentially consuming or breaking them down before they get absorbed. The complex behavior of the gut microbiome community manifests such interaction competing for the same niche in the gut with drugs ultimately reducing the amount of active drug available to the host.

3 Impact of Microbiota on Drug Metabolism

The gut microbiota strongly contributes to drug metabolism, influencing the efficacy and toxicity of various drugs. Gut bacteria produce enzymes that can metabolize drugs before their absorption into the systemic circulation. Gut microbiota mediates the conversion of the prodrug into activated drug through reactions like hydrolysis, reduction, or demethylation (Wilson and Nicholson 2017). Microbial products can modulate the expression of drug-metabolizing enzymes and transporters in the host's liver and intestine. Gut microbiota-derived molecules like lipopolysaccharides can influence the expression of enzymes like cytochrome (CYP) 3A4, which is involved in drug metabolism (Liu et al. 2022a).

3.1 Variability in Drug Efficacy Through Drug Activation/Inactivation

Variability in drug efficacy due to drug activation and inactivation signifies drug–microbiota metabolism focusing on the complex interaction between the gut microbiome and drug (Dhurjad et al. 2022). The gut microbiota potentially leads to the regulation of the pharmacokinetics and pharmacodynamics of many drugs. The gut microbiota can activate prodrugs, or inactive compounds to metabolize into their active forms (Koppel et al. 2017). Microbial enzymes can catalyze this transformation leading to the therapeutic efficacy of the drug. Sulfasalazine is a prodrug used to treat inflammatory bowel disorders like ulcerative colitis. It is activated in the colon by bacterial azoreductase, which cleaves the drug into its active form, 5-aminosalicylic acid (5-ASA) (Lima et al. 2024). Conversely, the microbiota can inactivate drugs through metabolic processes, reducing their efficacy or altering their pharmacological profile (Dikeocha et al. 2022). Digoxin, a cardiac glycoside used for heart failure, is partially deactivated by certain strains of *Eggerthella lenta* in the gut (Haiser et al. 2014). Patients with high levels of this bacterium may experience reduced drug efficacy. Numerous factors contribute to the variability in drug efficacy. The human gut microbiota varies widely between individuals, which can lead to significant differences in how drugs are metabolized. Factors including diet, genetics, age, antibiotic use, and lifestyle influence the composition of the gut microbiota (Rinninella et al. 2023). This variability leads to differences in drug metabolism, impacting both therapeutic outcomes and side effects. The gut microbiota is comprised of a broad spectrum of enzymes able to metabolize a diverse set of drugs. Those enzymes can activate, inactivate, or even toxify drugs. Variations in microbial gene content, which determine the production of those enzymes, contribute to inter-individual differences in drug metabolism. The gut microbiome interacts with host drug-metabolizing enzymes found in the liver (i.e., cytochrome P450 enzymes), which further complicates the drug metabolism process (Jourová et al. 2020). A dynamic interplay between microbial metabolism and host metabolism influences drug concentrations and therapeutic outcomes.

3.2 Increased Risk of Adverse Drug Reactions

Drug microbiota metabolism leads to increased adverse drug reactions (ADRs), adding complexity to pharmacotherapy, and suboptimal therapeutic outcomes (Zhao et al. 2023). Understanding these microbiota-mediated processes is vital for minimizing ADRs and improving drug safety. Certain drug–microbiota interactions result in the formation of immune-modulating molecules that may trigger hypersensitivity or inflammatory response (Zheng et al. 2020). Non-steroidal anti-inflammatory drugs like ibuprofen can cause gastrointestinal inflammation and ulcers, partly due to the microbiota. Inflammation can be worsened by microbial metabolites that increase gut permeability, enabling more toxins or drugs to reach the systemic circulation, and causing inflammatory ADRs (Rogers and Aronoff 2016). Antibiotics can disrupt the gut microbiota by depleting beneficial bacterial species. This may lead to the overgrowth of pathogenic bacteria, which can metabolize drugs in harmful ways, increasing the risk of ADRs (Patangia et al. 2022). Oral contraceptives can interfere with the enterohepatic recirculation, reducing their efficacy and potentially leading to unintended pregnancy as an adverse outcome (Li et al. 2023). Similarly, food habits impact the composition of the gut microbiota, and hence drug metabolism. Diets rich in fiber aid the commensal bacteria to produce short-chain fatty acids (SCFAs), which modulate drug metabolism and immune responses, potentially impacting ADR risk. Manipulating the gut microbiota through probiotics or prebiotics may help reduce ADR risk by promoting a more favorable microbial composition for drug metabolism.

3.2.1 Toxic Metabolites

The generation of toxic metabolites by the gut microbiome can potentially increase the risk of adverse drug reactions (ADRs) in drug–microbiota metabolism (Wang et al. 2024). These toxic metabolites result from the biotransformation of drugs by microbial enzymes into harmful forms, either locally (in the gut) or systemically (throughout the body). This can lead to increased toxicity, organ damage, immune reactions, and other negative effects, often complicating the pharmacological outcomes of drugs. Numerous types of enzymes found in the gut microbiota are capable of breaking down a broad range of Active Pharmaceutical Ingredients (APIs). These microbial enzymes can convert drugs into metabolites that may have toxic properties, especially if they accumulate or if the body's detoxification systems are overburdened. Irinotecan, an anti-neoplastic, is metabolized in the liver into its active but detoxified form, SN-38, and excreted as SN-38-glucuronide (Haaz et al. 1997). However, bacterial enzymes such as β-glucuronidases in the gut can reactivate SN-38 from its detoxified form. One significant dose-limiting side effect of irinotecan therapy is severe diarrhea brought on by the local toxicity caused by this reactivation in the intestines. In this context, the formation of toxic metabolite (SN-38) by gut microbiota increases the risk of gastrointestinal damage and systemic toxicity (Klaassen and Cui 2015). Acetaminophen is mainly metabolized in the liver, but it can also be metabolized into potentially hazardous byproducts by the gut microbiota. In patients with disrupted microbiota or predisposing conditions (such as excessive alcohol consumption), the production of toxic metabolites from acetaminophen by gut bacteria can contribute to liver toxicity, especially if the liver's glutathione stores (necessary for detoxification) are depleted. This can lead to hepatotoxicity and liver failure in cases of overdose or impaired metabolism. The gut microbiota can convert dietary components and drug-related compounds into toxic substances, which can then exacerbate adverse drug reactions. The gut microbiota metabolizes certain dietary compounds like choline and carnitine into trimethylamine (TMA), which is further converted into trimethylamine N-oxide (TMAO) by the liver (Klaassen and Cui 2015). An elevated risk of cardiovascular diseases is linked to high levels of TMAO (Zhu et al. 2020). Some drugs that affect the metabolism of choline or interact with the microbiota may lead to elevated TMAO levels, indirectly leading to the higher possibility of heart-related adverse drug reactions.

3.2.2 Unpredictable Drug Responses

Unpredictable drug responses in the context of drug–microbiota metabolism arise due to the complex and highly variable nature of the interactions between drugs, the human body, and the gut microbiota. These unpredictable responses can increase the risk of adverse effects, therapeutic failure, or toxicity, creating challenges for clinicians in ensuring safe and effective drug treatments (Torres-Carrillo et al. 2023). Inter-individual variability in microbiota composition is a key reason for elevated ADR during drug–microbiota metabolism (Zhao et al. 2023). The composition of the gut microbiota differs significantly between individuals, influenced by genetics, diet, environment, age, antibiotic use, and health status (Wen and Duffy 2017). These differences affect how drugs are metabolized by the microbiota, leading to unpredictable drug responses. One drug may remain inactivated in one individual due to high levels of a specific bacterial enzyme, while in another individual, the absence of that enzyme may result in the drug remaining active or being metabolized differently. Due to the diversity and complexity of microbial metabolic pathways the microbiota may activate a prodrug, while on the other hand it may lead to drug inactivation, or the production of toxic metabolites. Prodrugs such as sulfasalazine, which are activated in the gut by microbial azoreductase, may not be fully activated if a patient does not have enough of this enzyme, which would reduce the prodrug's therapeutic efficacy (Long et al. 2024). Conversely, excess microbial activity could lead to overactivation and toxicity. The metabolism of drugs by the microbiota often interacts with host liver enzymes, particularly those involved in the cytochrome P450 (CYP) family. The combined effect of microbial and host metabolism can result in unpredictable drug concentrations in the bloodstream, increasing the risk of adverse drug reactions (ADRs) or therapeutic failure. Oral contraceptives undergo both microbial and hepatic metabolism. Microbiota disruption resulting from antibiotic overuse can alter enterohepatic recirculation, reducing contraceptive efficacy and potentially leading to unintended pregnancies. A change in diet can alter the gut microbial composition and its metabolic activity, making drug responses unpredictable. Antibiotic use can disrupt the balance of the microbiota causing dysbiosis, which may either enhance or reduce drug metabolism, leading to increased risk of toxicity or therapeutic failure. A patient on antibiotics may experience reduced efficacy of oral anticoagulants like warfarin due to microbiota disruption, which can alter the synthesis of vitamin K, a critical cofactor for warfarin's activity. This can lead to either clotting or excessive bleeding due to altered drug response.

Also, changes in the microbiota can impact immune reactions, contributing to unpredictable immune responses to drugs, including hypersensitivity or inflammatory reactions. Unpredictable drug responses can be rectified by some strategies. One promising solution to the unpredictability of drug responses is personalized medicine, which involves tailoring drug treatments based on individual characteristics, including microbiome composition. Advances in microbiome sequencing can help predict how a patient's microbiota will metabolize a specific drug, enabling more accurate dosing and reducing ADR risks. Modifying drug formulations to avoid metabolism by the gut microbiota or using targeted drug delivery methods (e.g., coated capsules or nanoparticles) can help reduce unpredictability by bypassing the microbiota's influence. Targeted drug delivery systems can release a drug at a specific point in the gut where microbiota metabolism is less likely, thereby reducing variability in drug response (Feng et al. 2020). As the gut microbiota is shaped by diet, dietary interventions to modify the gut microbiome can be used to select drug targets and predict responses to drugs.

3.3 *Microbiome Profiling in Personalized Medicine*

Microbiome profiling involves assessing the general structure and functioning ability of an individual's microbiota, particularly on the gut microbiome. It is essential to comprehend the intricate connection between a person's microbiome and their medications, as it significantly impacts drug processing, efficacy, and potential adverse reactions. The human gut has a diverse range of bacteria that create drug-modifying enzymes. These changes can activate, deactivate, or even convert medicines into hazardous molecules. As a result, microbiome profiling gives vital information for predicting how an individual will respond to a specific treatment based on the microbial composition of their gut (Spanogiannopoulos et al. 2016).

3.3.1 Microbiome Profiling in Drug Response

Healthcare professionals can anticipate the activation or deactivation of a drug before it takes effect therapeutically by examining the presence of bacteria linked to drug metabolism. For instance, some individuals lack the required microbes to trigger treatments like sulfasalazine, making it ineffective due to the lack of microbial enzymes. Microbiome analysis can identify persons who are more likely to develop drug toxicity as a result of hazardous byproducts produced by gut bacteria. Individuals with high amounts of β-glucuronidase-producing bacteria may have diarrhea when their gut microbiota converts irinotecan into harmful chemicals (Wilson and Nicholson 2017). Microbiome profiling assists healthcare providers in identifying bacteria that may interact with drugs and cause negative side effects.

This is particularly crucial for drugs with narrow therapeutic effects or the ability to generate toxic compounds during metabolism. Several technologies allow for accurate profiling of an individual's microbiome, including t16S rRNA sequencing that identifies and categorizes bacterial species based on their 16S ribosomal RNA gene (Wallace et al. 2010). It is frequently used because of its low cost and ability to provide a full overview of bacterial composition. Unlike 16S rRNA genes, metagenomics is a comprehensive technique that examines all of the genetic material in a microbiome sample (Suez 2019). This method allows for the identification of metabolic pathways that may have an impact on drug metabolism by providing a comprehensive understanding of the capabilities of the microbiome (Markova et al. 2013).

3.3.2 Clinical Applications of Microbiome Data

In personalized medicine, microbiome profiling can be applied in several ways.

3.3.2.1 Tailored Drug Therapies

Healthcare professionals can customize the choice of drugs and their doses for a patient's microbiome by understanding how the microbiota affects drug metabolism. This is essential for medications like metformin, which is utilized in treating diabetes and needs to engage with the gut microbiota to work effectively (Zimmermann et al. 2019a).

3.3.2.2 Preventing Toxicity

Recognizing patients who are at risk of medication toxicity is critical for preventing harm. Individuals with certain microorganisms in their bodies can metabolize acetaminophen into toxic substances, potentially causing liver damage. In some cases, additional drugs or dose modifications may be suggested for consideration (Maurice et al. 2013).

3.3.2.3 Optimizing Probiotic and Prebiotic Use

Microbiome data can help with drug metabolism and minimize undesirable side effects. Probiotics, for example, could be used to control the microbiota in NSAID-treated patients, reducing gastrointestinal toxicity (Enright et al. 2016).

3.4 Microbiota Modulation

Probiotics are live microorganisms, which in the intestinal tract are used for the therapeutic and prophylactic purposes (supplantation of beneficial microorganisms or stimulation of residual indigenous microbes to restore or make normal microbial balance within the colon). In a personalized medicine context, probiotic therapies are tailored to an individual's microbiota profile. By identifying deficiencies or imbalances in specific microbial species, probiotics can be prescribed to promote health or improve drug metabolism. Certain probiotic strains can help protect the gut from damage and offset the decrease in beneficial bacteria caused by NSAIDs or antibiotics in people (Koppel et al. 2017). Some probiotics can be given to people with gut bacteria that impede the metabolism of drugs like sulfasalazine or digoxin, to enhance the drug's effectiveness or reduce negative side effects. Often prebiotics are indigestible food components that induce the growth of helpful bacteria in the gut (Gopalakrishnan et al. 2018). Customized prebiotic treatments can enhance the proliferation of helpful microbial species that enhance drug metabolism while suppressing harmful species that lead to drug toxicity. For instance, the increase in the growth of *Lactobacillus* and *Bifidobacterium* may assist individuals who have gut inflammation or are experiencing side effects from specific drugs (Wallace et al. 2010). Prebiotics may boost prodrug metabolism. Dietary fibers, for example, work as prebiotics by promoting the development of bacteria that produce SCFAs, which can affect host drug-metabolizing enzymes, influencing drug pharmacokinetics (Sousa et al. 2008). Customized prebiotic treatments can be created by analyzing a person's unique microbiome composition to address deficiencies in important microbial species needed for efficient drug metabolism (Haiser et al. 2013). FMT is the transfer of stool from a healthy donor to a patient in order to grow a wide and favorable microbiome (Compare et al. 2022). Its benefits go beyond just treating recurrent *Clostridium difficile* infections, and also apply to personalized medicine. FMT to correspond to the unique microbiomes of individuals may boost harmony and enhance treatment outcomes (Person and Keefer 2021). This can provide specific advantages to patients who have a microbiome composition that results in adverse reactions to drugs or ineffective drug metabolism (Schaal 2023). Diet significantly influences the microbiome. Personalized dietary interventions can be designed to modify the microbiota in ways that improve medication responses. For example, high-fiber diets encourage the growth of bacteria that produce SCFAs, which alter immunological and metabolic activities and may improve treatment efficacy in patients with inflammatory diseases. Individual dietary recommendations based on microbiome profiling can be provided to promote the growth of microbial species that improve drug metabolism or reduce drug-related negative effects (Mahdy et al. 2023). Patients undergoing antibiotics or chemotherapy, for example, may benefit from dietary modifications that help in microbiota recovery. Phage therapy utilizes bacteriophages, which are viruses that specifically target and eradicate bacteria (Noh et al. 2017). This approach can be tailored to target harmful bacteria in the gut without impacting beneficial microbes, offering a precise tool for manipulating the microbiota (Dhurjad et al. 2022). Phage therapy can be used to specifically target harmful bacteria that result in drug toxicity or interfere with drug metabolism. For instance, bacteriophages might be employed to reduce the quantity of bacteria that produce β-glucuronidase, thereby decreasing the gastrointestinal side effects of medications like irinotecan (Shoubridge et al. 2022).

4 Clinical Approaches to Microbiota–Drug Interaction

4.1 Microbiome-Targeted Drug Development

Due to the intense impact of the microbiota on drug metabolism, researchers have begun to design drugs that account for microbiota–drug interactions. This approach involves developing medications that either avoid harmful microbial interactions or exploit beneficial ones. New drugs are being designed to be less susceptible to microbial metabolism or to be selectively activated by gut bacteria (Winek et al. 2016). One of the challenges in treating infections while maintaining microbiota integrity is that broad-spectrum antibiotics disrupt the microbiome, leading to dysbiosis and adverse effects. Recently, narrow-spectrum antibiotics have been developed that target specific pathogens without affecting beneficial microbiota. These antibiotics minimize the risk of antibiotic-associated diarrhea and reduce the chances of secondary infections such as *Clostridioides difficile* (Song and Shin 2022). *Microbiome-Activated Prodrugs*: Prodrugs are inactive compounds that are metabolized into active drugs by the body. In a microbiome-targeted approach, researchers are designing prodrugs that are activated by specific bacterial enzymes. For instance, experimental drugs for inflammatory bowel disease (IBD) are being developed to be activated by the gut microbiota only in inflamed regions of the gut, reducing systemic side effects (Zimmermann et al. 2019b).

4.2 Modulating the Microbiota to Reduce Drug Toxicity

The manipulation of the gut microbiota to decrease adverse drug reactions (ADRs) has been motivated by the impact of the microbiota on drug toxicity. This method includes either decreasing harmful microbial waste or supporting beneficial ones. Prebiotics and probiotics are utilized to enhance beneficial bacteria to improve drug metabolism or reduce toxicity. For example, probiotics with *Lactobacillus* and *Bifidobacterium* can increase the gut's ability to safely metabolize specific drugs by reducing the production of harmful substances (Amedei and Boem 2018). Researchers have studied how effective it is to use specific antibiotics to reduce the levels of certain bacteria that produce toxic byproducts from cancer drugs. Administering antibiotics before starting irinotecan treatment reduces the formation of harmful bacterial metabolites, decreasing the risk of severe diarrhea without affecting the effectiveness of chemotherapy (Tucker et al. 2015).

4.3 Personalized Microbiome Profiling for Drug Dosing and Selection

Personalizing pharmaceutical treatment according to individual characteristics is crucial for precision medicine, with microbiome evaluation playing a vital role in reaching this objective. Healthcare providers can now analyze a patient's microbiome and forecast how they will respond to certain medications, making it easier to personalize dosages and treatment options (Weersma et al. 2020). Individuals' microbiome compositions differ resulting in different drug metabolism. Microbiome profiling is currently being used in pharmacokinetics to predict how medications will be absorbed, distributed, metabolized, and removed in different people. This helps to avoid under- or over-dosing and lowers the risk of undesirable consequences (Jaswal et al. 2023). Drug choice based on microbiome composition is an emerging strategy in personalized medicine. Clinicians can use microbiome sequencing data to select drugs that are more likely to be metabolized effectively by a patient's microbiota or to avoid drugs that are likely to produce harmful metabolites (Amedei and Boem 2018).

4.4 Microbiota-Related Biomarkers for Drug Safety

Researchers are developing microbe-based indicators to predict potential adverse reactions and ensure treatment safety. Scientists can respond rapidly by looking for early signals of medication toxicity in the stomach, such as bacteria metabolites or enzymes (Zimmermann et al. 2019b). Monitoring levels of β-glucuronidase allows for early detection of possible effects and microbial shifts in order to reduce harm (Le Chatelier et al. 2013). The gut microbiota plays a crucial role in maintaining gut health through the production of short-chain fatty acids (SCFA). Low levels of SCFA in cancer patients are associated with poor results in immunotherapy. The latest developments in biomarker research are centered around utilizing SCFA levels to forecast treatment responses in patients, aiding in the selection of appropriate medications (Rabot et al. 2010).

5 Future Advancement

5.1 Microbiome-Based Diagnostic Tools and Biomarkers

Incorporating microbiome-centered diagnostic tools into healthcare environments would be a major accomplishment in the coming years. These technologies allow for personalized drug treatment by analyzing individual microbiome characteristics to predict how drugs will be processed, their effectiveness, and potential side effects (Song and Shin 2022). Potential upcoming diagnostic methods, like examining samples of feces, breath, or saliva, could offer immediate insights into the makeup of the microbiome. These microbiome images could help healthcare providers predict how a patient will respond to medication and make any necessary treatment adjustments. Progress in high-throughput sequencing and artificial intelligence (AI) will enhance the availability and effectiveness of these diagnostic tools in healthcare facilities (Dhurjad et al. 2022). Discovering microbial indicators allows for the creation of specific pharmacokinetic models that can predict drug absorption, distribution, metabolism, and elimination. These models will take into account both the host's genetic makeup and the composition of the microbiome to enhance drug dosing and minimize side effects (Shoubridge et al. 2022).

5.2 Microbiome-Engineered Therapies

Manipulating the microbiota in microbiome engineering is an exciting frontier for scientists to enhance drug outcomes. This involves altering the microbiome to enhance drug processing and creating microbial treatments through synthetic biology that can either produce or adjust drugs in the human gut (Wallace et al. 2010). One possible future therapy technique is to use genetically modified intestinal bacteria to produce drugs in the body. These bacteria can be programmed to generate drugs, eliminate harmful substances, or release therapeutic molecules in response to signals such as inflammation or disease. This revolutionary method enables precise drug distribution, enhancing efficacy while minimizing the risk of unwanted effects (Noh et al. 2017). In the future, probiotics will be designed to boost drug metabolism rather than overall gut health. When intelligent probiotics are used with specific medications, they can enhance the growth of beneficial bacteria, improve drug absorption, and prevent negative side effects (Song and Shin 2022).

5.3 Artificial Intelligence (AI) and Machine Learning in Drug–Microbiome Interaction Prediction

AI and machine learning will play a pivotal role in predicting microbiota–drug interactions in the future. These technologies can analyze vast datasets of microbiome sequences, drug interactions, and clinical outcomes to predict how an individual's microbiome will interact with a specific drug, thereby optimizing personalized treatment strategies (Weersma et al. 2020). By combining microbiome data and pharmacokinetics, AI can generate predictive models that anticipate how medicine will be digested by different microbiota compositions. This allows healthcare providers to change dosages or choose alternate therapies based on specific predictions (Tucker et al. 2015).

5.4 Machine Learning for Drug Discovery

Machine learning can help find new microbiome-based therapeutic targets and accelerate the development of microbiome-modulating treatments. Machine learning algorithms may identify essential bacterial species or enzymes that influence medication metabolism by evaluating large-scale microbiome data, leading drug development to either exploit or avoid these microbial interactions (Weersma et al. 2020).

5.5 Fecal Microbiota Transplantation (FMT) as a Tool for Modulating Drug Metabolism

Fecal microbiota transplantation (FMT) has proven to be successful in the treatment of conditions like *Clostridium difficile* infections. In the future, microbiota modulation using FMT may enhance drug metabolism and decrease toxicity (Wilson and Nicholson 2017). If a patient's microbiome is leading to drug side effects or lack of effectiveness, FMT might be used to swap out the troublesome microbiota with a more beneficial balance. This may be particularly advantageous for patients with stubborn bacteria that hinder the body's ability to process medication (Le Chatelier et al. 2013). The future advancement on microbiota and drug metabolism is shown in Table 1.

Table 1 Future advancements in the field of microbiota–drug interactions and their influence on drug metabolism and toxicity

S. no.	Advancement	Description	References
1.	Personalized medicine	Development of microbiome-based biomarkers for predicting individual responses to drugs, leading to personalized treatment plans	Rabot et al. (2010)
2.	Microbiome profiling	Advances in sequencing technologies will enable comprehensive profiling of the microbiota, facilitating better understanding of its role in drug metabolism	Macpherson and Harris (2004)
3.	Targeted probiotics	Development of probiotics that can modulate specific microbiota functions, enhancing drug efficacy and reducing toxicity	Lavecchia (2015)
4.	Microbiota modulation therapies	Research into drugs that can specifically alter the gut microbiota composition to improve drug metabolism and minimize adverse effects	Liu et al. (2022b)
5.	Artificial intelligence and machine learning	Utilizing AI and machine learning to predict microbiota–drug interactions based on large datasets, leading to better drug design and safety evaluations	Nuzzo and Brown (2019)
6.	Gut–brain axis research	Exploring the impact of microbiota on the pharmacokinetics of neuroactive drugs, providing insights into drug interactions affecting central nervous system (CNS) medications	Wilson and Nicholson (2017)
7.	Development of microbiota-targeted drugs	Creating drugs designed to interact with gut microbiota to enhance therapeutic outcomes and reduce side effects	Qi et al. (2022)
8.	Clinical trials focusing on microbiota	Increased number of clinical trials evaluating the effects of microbiota on drug metabolism, aiding in regulatory approvals and clinical guidelines	Zhang et al. (2017)

6 Challenges and Limitation

6.1 Inter-Individual Variability of the Gut Microbiota

Studying the relationship between microbiota and drugs is difficult because of the differences in microbiome composition across individuals. Genetics, dietary choices, individual habits, and environmental conditions collectively impact the unique gut microbiota of each person. The variety can result in important differences in how drugs are processed in various patients (Tucker et al. 2015). Even among healthy people, the overall composition of gut microbiota can vary significantly, challenging predictions about how a certain microbiome will react to a specific medicine. This diversity makes it difficult to design uniform treatment methods that account for the microbiome's role in drug metabolism (Shoubridge et al. 2022).

6.2 Complexity of Microbial Enzyme Functions

The complex microbial enzyme activity poses challenges in understanding how microbiota and drugs interact with each other. Lots of enzymes in the gut microbiome responsible for regulating drug metabolism are not well comprehended. Identifying the particular strains or enzymes responsible for drug metabolism is difficult due to the functional overlap exhibited by certain bacteria (Markova et al. 2013). Even within the same bacterial species, metabolic activity can fluctuate due to changes in gene expression or environmental factors. Furthermore, some microbial enzymes may produce unexpected or harmful metabolites that were not anticipated during medication development, confusing therapeutic safety and efficacy estimations (Compare et al. 2022).

6.3 Dynamic Nature of the Microbiome

The gut microbiome is dynamic and changes based on diet, sickness, drugs, and environmental shifts. Longitudinal predictions of drug metabolism and toxicity are complicated by the dynamic nature of the microbial community, which can change significantly during treatment (David et al. 2014). Temporal shifts in microbiome composition mean that predictions of drug metabolism based on microbiome profiling may be limited to a specific window in time, reducing the reliability of long-term treatment plans. Such shifts are particularly problematic in chronic conditions requiring long-term medication use.

6.4 Lack of Standardized Research Methodologies

The lack of definite techniques for sequencing, studying, and understanding microbiomes hinders the investigation of links between microbiota and medications. Variations in sampling methods, sequencing technology (e.g., 16S rRNA versus whole-genome sequencing), and data interpretation make it challenging to compare results between studies and implement them in clinical treatments. Discrepancies in microbiome data collection and analysis provide contradictory results. Differences between in vitro, in vivo, and human investigations add to the lack of translational clarity (Swanson 2015).

6.5 Ethical and Regulatory Challenges

The use of the microbiota for therapeutic purposes, such as fecal microbiota transplantation (FMT), tailored probiotics, or targeted antibiotics, creates ethical and regulatory issues. The long-term ramifications of modifying the gut microbiota are unknown, and unexpected effects such as microbial resistance, dysbiosis, or disease transmission may cause risk. Safety continues to be a primary concern due to the possibility of microbiome changes leading to dysbiosis or the proliferation of harmful microorganisms. Moreover, extensive regulatory review was necessary to ensure patient safety when utilizing modified microorganisms or FMT (Zhernakova et al. 2016).

7 Conclusion

Drug metabolism is significantly influenced by the human gut microbiota, a complex ecosystem of bacteria that affects both toxicity and efficacy. The complex relationship known as "drug–microbiota interactions" has important ramifications for both drug development and customized treatment. The gut microbiota has an amazing ability to break down drugs into a wide range of different enzymes. These microbial changes can affect how well drugs work as medicines by changing how they are absorbed, distributed, metabolized, and excreted. Interactions between drugs and bacteria can take many different forms. A number of medications are activated by changes in microbial metabolism, and these can have a substantial effect on the effectiveness of the medication. On the other hand, certain medications may produce hazardous metabolites due to microbial metabolism, which may result in unfavorable drug responses. For example, certain gut microbes have the ability to metabolize certain medicines into substances that cause inflammation in the intestines or disrupt the liver's drug-metabolizing enzymes. It is essential to comprehend these interactions in order to maximize medication therapy and reduce side effects. By predicting drug response and customizing treatment techniques, a person's gut microbiota profile can be used to enable customized medical approaches. In order to modify the gut microbiota and improve medication efficacy while reducing toxicity, future research should concentrate on creating targeted therapies, such as microbiota transplantation or dietary changes.

References

Ahmed MM, Farhan MS, Hamad AA, Edan AI (2022) Role of microbiota in health and disease. Res J Pharm Technol 15(10):4825–4828

Alagga AA, Pellegrini MV, Gupta V (2024) Drug absorption. In: StatPearls [Internet]. StatPearls Publishing

Aldunate M, Srbinovski D, Hearps AC, Latham CF, Ramsland PA, Gugasyan R et al (2015) Antimicrobial and immune modulatory effects of lactic acid and short chain fatty acids produced by vaginal microbiota associated with eubiosis and bacterial vaginosis. Front Physiol 6:164

Amabebe E, Anumba DO (2018) The vaginal microenvironment: the physiologic role of lactobacilli. Front Med 5:181

Amedei A, Boem F (2018) I've gut a feeling: microbiota impacting the conceptual and experimental perspectives of personalized medicine. Int J Mol Sci 19(12):3756

Amin ML (2013) P-glycoprotein inhibition for optimal drug delivery. Drug Target Insights 7:DTI-S12519

Anspaugh DD, Roe RM (2002) Assays for the classification of two types of esterases: carboxylic ester hydrolases and phosphoric triester hydrolases. Curr Protoc Toxicol 13(1):4–10

Anwar H, Iftikhar A, Muzaffar H, Almatroudi A, Allemailem KS, Navaid S et al (2021) Biodiversity of gut microbiota: impact of various host and environmental factors. Biomed Res Int 2021(1):5575245

Appel MJ, Bertozzi CR (2015) Formylglycine, a post-translationally generated residue with unique catalytic capabilities and biotechnology applications. ACS Chem Biol 10(1):72–84

Arumugam M, Raes J, Pelletier E, Le Paslier D, Yamada T, Mende DR et al (2011) Enterotypes of the human gut microbiome. Nature 473(7346):174–180

Arvind AS, Farthing MJG (1988) New aminosalicylic acid derivatives for the treatment of inflammatory bowel disease. Aliment Pharmacol Ther 2(4):281–289

Back DJ, Rogers SM (1987) First-pass metabolism by the gastrointestinal mucosa. Aliment Pharmacol Ther 1(5):339–357

Bell A, Juge N (2021) Mucosal glycan degradation of the host by the gut microbiota. Glycobiology 31:691–696

Bik EM, Ugalde JA, Cousins J, Goddard AD, Richman J, Apte ZS (2018) Microbial biotransformations in the human distal gut. Br J Pharmacol 175(24):4404–4414

Blum HE (2017) The microbiome: key player in human health and disease. Review article. Вестник Адыгейского государственного университета. Серия 4: Естественно-математические и технические науки 2(201):40–48

Boddu RS, Perumal O, Divakar K (2021) Microbial nitroreductases: a versatile tool for biomedical and environmental applications. Biotechnol Appl Biochem 68(6):1518–1530

Boelsterli UA, Redinbo MR, Saitta KS (2013) Multiple NSAID-induced hits injure the small intestine: underlying mechanisms and novel strategies. Toxicol Sci 131(2):654–667

Buddington RK (2001) The use of nondigestible oligosaccharides to manage the gastrointestinal ecosystem. Microb Ecol Health Dis 13(1):9–15

Camilleri M (2019) Leaky gut: mechanisms, measurement and clinical implications in humans. Gut 68(8):1516–1526

Chamseddine AN, Ducreux M, Armand JP, Paoletti X, Satar T, Paci A, Mir O (2019) Intestinal bacterial β-glucuronidase as a possible predictive biomarker of irinotecan-induced diarrhea severity. Pharmacol Ther 199:1–15

Cockburn DW, Koropatkin NM (2016) Polysaccharide degradation by the intestinal microbiota and its influence on human health and disease. J Mol Biol 428(16):3230–3252

Compare D, Sgamato C, Nardone OM, Rocco A, Coccoli P, Laurenza C, Nardone G (2022) Probiotics in gastrointestinal diseases: all that glitters is not gold. Dig Dis 40(1):123–132

Crouwel F, Buiter HJ, de Boer NK (2021) Gut microbiota-driven drug metabolism in inflammatory bowel disease. J Crohns Colitis 15(2):307–315

Currò D (2018) The role of gut microbiota in the modulation of drug action: a focus on some clinically significant issues. Expert Rev Clin Pharmacol 11(2):171–183

Dabek M, McCrae SI, Stevens VJ, Duncan SH, Louis P (2008) Distribution of β-glucosidase and β-glucuronidase activity and of β-glucuronidase gene gus in human colonic bacteria. FEMS Microbiol Ecol 66(3):487–495

David LA, Maurice CF, Carmody RN, Gootenberg DB, Button JE, Wolfe BE et al (2014) Diet rapidly and reproducibly alters the human gut microbiome. Nature 505(7484):559–563

DeGruttola AK, Low D, Mizoguchi A, Mizoguchi E (2016) Current understanding of dysbiosis in disease in human and animal models. Inflamm Bowel Dis 22(5):1137–1150

Den Besten G, Van Eunen K, Groen AK, Venema K, Reijngoud DJ, Bakker BM (2013) The role of short-chain fatty acids in the interplay between diet, gut microbiota, and host energy metabolism. J Lipid Res 54(9):2325–2340

Dhurjad P, Dhavaliker C, Gupta K, Sonti R (2022) Exploring drug metabolism by the gut microbiota: modes of metabolism and experimental approaches. Drug Metab Dispos 50(3):224–234

Dikeocha IJ, Al-Kabsi AM, Miftahussurur M, Alshawsh MA (2022) Pharmacomicrobiomics: influence of gut microbiota on drug and xenobiotic metabolism. FASEB J 36(6):e22350

Dingsdag SA, Hunter N (2018) Metronidazole: an update on metabolism, structure–cytotoxicity and resistance mechanisms. J Antimicrob Chemother 73(2):265–279

Dobkin JF, Saha JR, Butler VP Jr, Neu HC, Lindenbaum J (1982) Inactivation of digoxin by Eubacterium lentum, an anaerobe of the human gut flora. Trans Assoc Am Physicians 95:22–29

Dobkin JF, Saha JR, Butler VP Jr, Neu HC, Lindenbaum J (1983) Digoxin-inactivating bacteria: identification in human gut flora. Science 220(4594):325–327

Elmeliegy M, Vourvahis M, Guo C, Wang DD (2020) Effect of P-glycoprotein (P-gp) inducers on exposure of P-gp substrates: review of clinical drug–drug interaction studies. Clin Pharmacokinet 59:699–714

Elmer GW, Remmel RP (1984) Role of the intestinal microflora in clonazepam metabolism in the rat. Xenobiotica 14(11):829–840

Enright EF, Gahan CG, Joyce SA (2016) The impact of the gut microbiota on drug metabolism and clinical outcome. Yale J Biol Med 89(3):375–382

Feng W, Liu J, Ao H, Yue S, Peng C (2020) Targeting gut microbiota for precision medicine: focusing on the efficacy and toxicity of drugs. Theranostics 10(24):11278

Finch A, Pillans P (2014) P-glycoprotein and its role in drug-drug interactions. Aust Prescr 37(4)

Francino MP (2014) Early development of the gut microbiota and immune health. Pathogens 3(3):769–790

Fu BC, Randolph TW, Lim U, Monroe KR, Cheng I, Wilkens LR et al (2016) Characterization of the gut microbiome in epidemiologic studies: the multiethnic cohort experience. Ann Epidemiol 26(5):373–379

Gachon F, Firsov D (2011) The role of circadian timing system on drug metabolism and detoxification. Expert Opin Drug Metab Toxicol 7(2):147–158

Gad SC (2018) The role of the microbiome on metabolism, absorption, and toxicity of xenobiotics in the gastrointestinal tract. In: Toxicology of the gastrointestinal tract, 2nd edn. CRC Press, pp 329–339

Ghosh S, Whitley CS, Haribabu B, Jala VR (2021) Regulation of intestinal barrier function by microbial metabolites. Cell Mol Gastroenterol Hepatol 11(5):1463–1482

Giuffrida P, Caprioli F, Facciotti F, Di Sabatino A (2019) The role of interleukin-13 in chronic inflammatory intestinal disorders. Autoimmun Rev 18(5):549–555

Gong S, Feng Y, Zeng Y, Zhang H, Pan M, He F et al (2021) Gut microbiota accelerates cisplatin-induced acute liver injury associated with robust inflammation and oxidative stress in mice. J Transl Med 19:1–13

Gopalakrishnan V, Spencer CN, Nezi L, Reuben A, Andrews MC, Karpinets TV et al (2018) Gut microbiome modulates response to anti–PD-1 immunotherapy in melanoma patients. Science 359(6371):97–103

Grace D, Abraham S, Varghese A, Sathianarayanan S (2008) Absorption and metabolism of xenobiotics: an overview. Int J Nutr Wellness 7:1–5

Guo Y, Lee H, Jeong H (2020) Gut microbiota in reductive drug metabolism. Prog Mol Biol Transl Sci 171:61–93

Haaz MC, Rivory L, Jantet S, Ratanasavanh D, Robert J (1997) Glucuronidation of SN-38, the active metabolite of irinotecan, by human hepatic microsomes. Pharmacol Toxicol 80(2):91–96
Haiser HJ, Gootenberg DB, Chatman K, Sirasani G, Balskus EP, Turnbaugh PJ (2013) Predicting and manipulating cardiac drug inactivation by the human gut bacterium Eggerthella lenta. Science 341(6143):295–298
Haiser HJ, Seim KL, Balskus EP, Turnbaugh PJ (2014) Mechanistic insight into digoxin inactivation by Eggerthella lenta augments our understanding of its pharmacokinetics. Gut Microbes 5(2):233–238
Harris BE, Manning BW, Federle TW, Diasio RB (1986) Conversion of 5-fluorocytosine to 5-fluorouracil by human intestinal microflora. Antimicrob Agents Chemother 29(1):44–48
Hasan N, Yang H (2019) Factors affecting the composition of the gut microbiota, and its modulation. PeerJ 7:e7502
Hassan R, Allali I, Agamah FE, Elsheikh SS, Thomford NE, Dandara C, Chimusa ER (2021) Drug response in association with pharmacogenomics and pharmacomicrobiomics: towards a better personalized medicine. Brief Bioinform 22(4):bbaa292
Hollander D (1999) Intestinal permeability, leaky gut, and intestinal disorders. Curr Gastroenterol Rep 1(5):410–416
Ishoey T, Woyke T, Stepanauskas R, Novotny M, Lasken RS (2008) Genomic sequencing of single microbial cells from environmental samples. Curr Opin Microbiol 11(3):198–204
Jaswal K, Todd OA, Behnsen J (2023) Neglected gut microbiome: interactions of the non-bacterial gut microbiota with enteric pathogens. Gut Microbes 15(1):2226916
Jia W, Li H, Zhao L, Nicholson JK (2008) Gut microbiota: a potential new territory for drug targeting. Nat Rev Drug Discov 7(2):123–129
Jourova L, Anzenbacher P, Anzenbacherova E (2016) Human gut microbiota plays a role in the metabolism of drugs. Biomed Pap Med Fac Univ Palacky Olomouc Czech Repub 160(3):317
Jourová L, Vavreckova M, Zemanova N, Anzenbacher P, Langova K, Hermanova P et al (2020) Gut microbiome alters the activity of liver cytochromes P450 in mice with sex-dependent differences. Front Pharmacol 11:01303
Kamareddine L, Najjar H, Sohail MU, Abdulkader H, Al-Asmakh M (2020) The microbiota and gut-related disorders: insights from animal models. Cells 9(11):2401
Kasubuchi M, Hasegawa S, Hiramatsu T, Ichimura A, Kimura I (2015) Dietary gut microbial metabolites, short-chain fatty acids, and host metabolic regulation. Nutrients 7(4):2839–2849
Kers JG, Saccenti E (2022) The power of microbiome studies: some considerations on which alpha and beta metrics to use and how to report results. Front Microbiol 12:796025
Khoruts A, Sadowsky MJ (2016) Understanding the mechanisms of faecal microbiota transplantation. Nat Rev Gastroenterol Hepatol 13(9):508–516
Kim JK, Choi MS, Jeong JJ, Lim SM, Kim IS, Yoo HH, Kim DH (2018) Effect of probiotics on pharmacokinetics of orally administered acetaminophen in mice. Drug Metab Dispos 46(2):122–130
Klaassen CD, Cui JY (2015) Mechanisms of how the intestinal microbiota alters the effects of drugs and bile acids. Drug Metab Dispos 43(10):1505–1521
Kleessen EB, Jaana Mättö B (2000) Culture-based knowledge on biodiversity, development and stability of human gastrointestinal microflora. Microb Ecol Health Dis 12(2):53–63
Koch RL, Chrystal EJ, Beaulieu BB Jr, Goldman P (1979) Acetamide—a metabolite of metronidazole formed by the intestinal flora. Biochem Pharmacol 28(24):3611–3615
Koppel N, Maini Rekdal V, Balskus EP (2017) Chemical transformation of xenobiotics by the human gut microbiota. Science 356(6344):eaag2770
Krishnan S, Alden N, Lee K (2015) Pathways and functions of gut microbiota metabolism impacting host physiology. Curr Opin Biotechnol 36:137–145
Kucherenko SV, Ovcharenko AM, Pushenko SL (2021) Xenobiotics: a threat to the health of living organisms. In: E3S web of conferences, vol 285. EDP Sciences, p 03006

Kuczynski J, Stombaugh J, Walters WA, González A, Caporaso JG, Knight R (2012) Using QIIME to analyze 16S rRNA gene sequences from microbial communities. Curr Protoc Microbiol 27(1):1E–5

Larsbrink J, Rogers TE, Hemsworth GR, McKee LS, Tauzin AS, Spadiut O et al (2014) A discrete genetic locus confers xyloglucan metabolism in select human gut Bacteroidetes. Nature 506(7489):498–502

Lavecchia A (2015) Machine-learning approaches in drug discovery: methods and applications. Drug Discov Today 20(3):318–331

Lavin R, DiBenedetto N, Yeliseyev V, Delaney M, Bry L (2018) Gnotobiotic and conventional mouse systems to support microbiota based studies. Curr Protoc Immunol 121(1):e48

Le Chatelier E, Nielsen T, Qin J, Prifti E, Hildebrand F, Falony G, Almeida M, Arumugam M, Batto JM, Kennedy S, Leonard P (2013) Richness of human gut microbiome correlates with metabolic markers. Nature 500(7464):541–546

Li L, Yang X, Tran D, Seo SK, Lu Y (2023) Combined oral contraceptives as victims of drug interactions. Drug Metab Dispos 51(6):718–732

Lima SF, Pires S, Rupert A, Oguntunmibi S, Jin WB, Marderstein A et al (2024) The gut microbiome regulates the clinical efficacy of sulfasalazine therapy for IBD-associated spondyloarthritis. Cell Rep Med 5(3)

LinWu SW, Wang AH, Peng FC (2010) Flavin-containing reductase: new perspective on the detoxification of nitrobenzodiazepine. Expert Opin Drug Metab Toxicol 6(8):967–981

Liu J, Tan Y, Cheng H, Zhang D, Feng W, Peng C (2022a) Functions of gut microbiota metabolites, current status and future perspectives. Aging Dis 13(4):1106

Liu J, Gao Z, Liu C, Liu T, Gao J, Cai Y, Fan X (2022b) Alteration of gut microbiota: new strategy for treating autism spectrum disorder. Front Cell Dev Biol (10):792490

Long AR, Mortara EL, Mendoza BN, Fink EC, Sacco FX, Ciesla MJ, Stack TM (2024) Sequence similarity network analysis of drug-and dye-modifying azoreductase enzymes found in the human gut microbiome. Arch Biochem Biophys 757:110025

Lu K, Mahbub R, Fox JG (2015) Xenobiotics: interaction with the intestinal microflora. ILAR J 56(2):218–227

Luis AS, Baslé A, Byrne DP, Wright GS, London JA, Jin C et al (2022) Sulfated glycan recognition by carbohydrate sulfatases of the human gut microbiota. Nat Chem Biol 18(8):841–849

Macpherson AJ, Harris NL (2004) Interactions between commensal intestinal bacteria and the immune system. Nat Rev Immunol 4(6):478–485

Mahdy MS, Azmy AF, Dishisha T, Mohamed WR, Ahmed KA, Hassan A et al (2023) Irinotecan-gut microbiota interactions and the capability of probiotics to mitigate Irinotecan-associated toxicity. BMC Microbiol 23(1):53

Markova SM, De Marco T, Bendjilali N, Kobashigawa EA, Mefford J, Sodhi J et al (2013) Association of CYP2C9* 2 with bosentan-induced liver injury. Clin Pharmacol Ther 94(6):678–686

Martín R, Bermúdez-Humarán LG, Langella P (2016) Gnotobiotic rodents: an in vivo model for the study of microbe–microbe interactions. Front Microbiol 7:409

Mathan VI, Wiederman J, Dobkin JF, Lindenbaum J (1989) Geographic differences in digoxin inactivation, a metabolic activity of the human anaerobic gut flora. Gut 30(7):971–977

Maurice CF, Haiser HJ, Turnbaugh PJ (2013) Xenobiotics shape the physiology and gene expression of the active human gut microbiome. Cell 152(1):39–50

Mottawea W, Butcher J, Li J, Abujamel T, Manoogian J, Mack D, Stintzi A (2019) The mucosal–luminal interface: an ideal sample to study the mucosa-associated microbiota and the intestinal microbial biogeography. Pediatr Res 85(6):895–903

Murphy JF (2018) The human microbiome: an emerging paradigm for better health. MOJ Immunol 6(2):54–55

Nakov R, Velikova T (2020) Chemical metabolism of xenobiotics by gut microbiota. Curr Drug Metab 21(4):260–269

Nielsen SE, Breinholt V, Justesen U, Cornett C, Dragsted LO (1998) In vitro biotransformation of flavonoids by rat liver microsomes. Xenobiotica 28(4):389–401

Noh K, Kang YR, Nepal MR, Shakya R, Kang MJ, Kang W et al (2017) Impact of gut microbiota on drug metabolism: an update for safe and effective use of drugs. Arch Pharm Res 40:1345–1355

Nuzzo A, Brown JR (2019) The microbiome factor in drug discovery and development. Chem Res Toxicol 33(1):119–124

Parsons RL (1977) Drug absorption in gastrointestinal disease with particular reference to malabsorption syndromes. Clin Pharmacokinet 2:45–60

Patangia DV, Anthony Ryan C, Dempsey E, Paul Ross R, Stanton C (2022) Impact of antibiotics on the human microbiome and consequences for host health. Microbiology 11(1):e1260

Person H, Keefer L (2021) Psychological comorbidity in gastrointestinal diseases: update on the brain-gut-microbiome axis. Prog Neuropsychopharmacol Biol Psychiatry 107:110209

Pickard JM, Zeng MY, Caruso R, Núñez G (2017) Gut microbiota: role in pathogen colonization, immune responses, and inflammatory disease. Immunol Rev 279(1):70–89

Pollet RM, D'Agostino EH, Walton WG, Xu Y, Little MS, Biernat KA et al (2017) An atlas of β-glucuronidases in the human intestinal microbiome. Structure 25(7):967–977

Qi Q, Li J, Yu B, Moon JY, Chai JC, Merino J et al (2022) Host and gut microbial tryptophan metabolism and type 2 diabetes: an integrative analysis of host genetics, diet, gut microbiome and circulating metabolites in cohort studies. Gut 71(6):1095–1105

Qin J, Li R, Raes J, Arumugam M, Burgdorf KS, Manichanh C et al (2010) A human gut microbial gene catalogue established by metagenomic sequencing. Nature 464(7285):59–65

Quigley EM (2016) Leaky gut–concept or clinical entity? Curr Opin Gastroenterol 32(2):74–79

Rabot S, Rafter J, Rijkers GT, Watzl B, Antoine JM (2010) Guidance for substantiating the evidence for beneficial effects of probiotics: impact of probiotics on digestive system metabolism. J Nutr 140(3):677S–689S

Rafii F, Cerniglia CE (1995) Reduction of azo dyes and nitroaromatic compounds by bacterial enzymes from the human intestinal tract. Environ Health Perspect 103(suppl 5):17–19

Rafil F, Franklin WIRT, Heflich RH, Cerniglia CE (1991) Reduction of nitroaromatic compounds by anaerobic bacteria isolated from the human gastrointestinal tract. Appl Environ Microbiol 57(4):962–968

Rinninella E, Tohumcu E, Raoul P, Fiorani M, Cintoni M, Mele MC et al (2023) The role of diet in shaping human gut microbiota. Best Pract Res Clin Gastroenterol 62:101828

Rogers MA, Aronoff DM (2016) The influence of non-steroidal anti-inflammatory drugs on the gut microbiome. Clin Microbiol Infect 22(2):178–1e1

Saitta KS, Zhang C, Lee KK, Fujimoto K, Redinbo MR, Boelsterli UA (2014) Bacterial β-glucuronidase inhibition protects mice against enteropathy induced by indomethacin, ketoprofen or diclofenac: mode of action and pharmacokinetics. Xenobiotica 44(1):28–35

Schaal PD (2023) Exploring the microbial ecology and energetics of wild and domesticated Atlantic salmon (Salmo salar) [PhD thesis]. Cork: University College Cork. Available from: https://hdl.handle.net/10468/16371

Shahi SK, Zarei K, Guseva NV, Mangalam AK (2019) Microbiota analysis using two-step PCR and next-generation 16S rRNA gene sequencing. J Vis Exp 152:10–3791

Shoubridge AP, Choo JM, Martin AM, Keating DJ, Wong ML, Licinio J, Rogers GB (2022) The gut microbiome and mental health: advances in research and emerging priorities. Mol Psychiatry 27(4):1908–1919

Shu Y, Sheardown SA, Brown C, Owen RP, Zhang S, Castro RA et al (2007) Effect of genetic variation in the organic cation transporter 1 (OCT1) on metformin action. J Clin Invest 117(5):1422–1431

Simpson JB, Sekela JJ, Carry BS, Beaty V, Patel S, Redinbo MR (2023) Diverse but desolate landscape of gut microbial azoreductases: a rationale for idiopathic IBD drug response. Gut Microbes 15(1):2203963

Smith AL, Erwin AL, Kline T, Unrath WC, Nelson K, Weber A, Howald WN (2007) Chloramphenicol is a substrate for a novel nitroreductase pathway in Haemophilus influenzae. Antimicrob Agents Chemother 51(8):2820–2829

Song EJ, Shin JH (2022) Personalized diets based on the gut microbiome as a target for health maintenance: from current evidence to future possibilities. J Microbiol Biotechnol 32(12):1497

Sousa T, Paterson R, Moore V, Carlsson A, Abrahamsson B, Basit AW (2008) The gastrointestinal microbiota as a site for the biotransformation of drugs. Int J Pharm 363(1–2):1–25

Spanogiannopoulos P, Bess EN, Carmody RN, Turnbaugh PJ (2016) The microbial pharmacists within us: a metagenomic view of xenobiotic metabolism. Nat Rev Microbiol 14(5):273–287

Srivalli KMR, Lakshmi PK (2012) Overview of P-glycoprotein inhibitors: a rational outlook. Braz J Pharm Sci 48:353–367

Srivastava A, Prabhakar MR, Mohanty A, Meena SS (2022) Influence of gut microbiome on the human physiology. Syst Microbiol Biomanufacturing 2:1–15

Stiff DD, Robicheau JT, Zemaitis MA (1992) Reductive metabolism of the anticonvulsant agent zonisamide, a 1, 2-benzisoxazole derivative. Xenobiotica 22(1):1–11

Stillhart C, Vučićević K, Augustijns P, Basit AW, Batchelor H, Flanagan TR et al (2020) Impact of gastrointestinal physiology on drug absorption in special populations—an UNGAP review. Eur J Pharm Sci 147:105280

Strong HA, Renwick AG, George CF, Liu YF, Hill MJ (1987) The reduction of sulphinpyrazone and sulindac by intestinal bacteria. Xenobiotica 17(6):685–696

Suez J (2019) Microbiome-based rational design of therapeutics. Doctoral dissertation, The Weizmann Institute of Science (Israel)

Sugihara M, Takeuchi S, Sugita M, Higaki K, Kataoka M, Yamashita S (2015) Analysis of intra- and intersubject variability in oral drug absorption in human bioequivalence studies of 113 generic products. Mol Pharm 12(12):4405–4413

Sun C, Chen L, Shen Z (2019) Mechanisms of gastrointestinal microflora on drug metabolism in clinical practice. Saudi Pharm J 27(8):1146–1156

Swanson HI (2015) Drug metabolism by the host and gut microbiota: a partnership or rivalry? Drug Metab Dispos 43(10):1499–1504

Ta N, Walle T (2007) Aromatase inhibition by bioavailable methylated flavones. J Steroid Biochem Mol Biol 107(1–2):127–129

Takeno S, Sakai T (1991) Involvement of the intestinal microflora in nitrazepam-induced teratogenicity in rats and its relationship to nitroreduction. Teratology 44(2):209–214

Takiishi T, Fenero CIM, Câmara NOS (2017) Intestinal barrier and gut microbiota: shaping our immune responses throughout life. Tissue Barriers 5(4):e1373208

Thaiss CA, Levy M, Grosheva I, Zheng D, Soffer E, Blacher E et al (2018) Hyperglycemia drives intestinal barrier dysfunction and risk for enteric infection. Science 359(6382):1376–1383

Thursby E, Juge N (2017) Introduction to the human gut microbiota. Biochem J 474:1823–1836

Torres-Carrillo N, Martínez-López E, Torres-Carrillo NM, López-Quintero A, Moreno-Ortiz JM, González-Mercado A, Gutiérrez-Hurtado IA (2023) Pharmacomicrobiomics and drug–infection interactions: the impact of commensal, symbiotic and pathogenic microorganisms on a host response to drug therapy. Int J Mol Sci 24(23):17100

Tucker AC, Mimee MK (2015) Programming a Human Commensal Bacterium, Bacteroides Thetaiotaomicron, to Sense and Respond to Stimuli in the Murine Gut Microbiota. Cell Systems 1(1):62–71

Vemuri R, Shankar EM, Chieppa M, Eri R, Kavanagh K (2020) Beyond just bacteria: functional biomes in the gut ecosystem including virome, mycobiome, archaeome and helminths. Microorganisms 8(4):483

Verberkmoes NC, Russell AL, Shah M, Godzik A, Rosenquist M, Halfvarson J et al (2009) Shotgun metaproteomics of the human distal gut microbiota. ISME J 3(2):179–189

Vidmar R, Vizovišek M, Turk D, Turk B, Fonović M (2019) Characterization of legumain degradome confirms narrow cleavage specificity. Acta Chim Slov 66(1):50

Wallace BD, Wang H, Lane KT, Scott JE, Orans J, Koo JS et al (2010) Alleviating cancer drug toxicity by inhibiting a bacterial enzyme. Science 330(6005):831–835
Walsh CT, Tu BP, Tang Y (2018a) Eight kinetically stable but thermodynamically activated molecules that power cell metabolism. Chem Rev 118(4):1460–1494
Walsh J, Griffin BT, Clarke G, Hyland NP (2018b) Drug–gut microbiota interactions: implications for neuropharmacology. Br J Pharmacol 175(24):4415–4429
Wang CY, Liu S, Xie XN, Tan ZR (2017) Regulation profile of the intestinal peptide transporter 1 (PepT1). Drug Des Devel Ther 11:3511–3517
Wang S, Ju D, Zeng X (2024) Mechanisms and clinical implications of human gut microbiota-drug interactions in the precision medicine era. Biomedicine 12(1):194
Weersma RK, Zhernakova A, Fu J (2020) Interaction between drugs and the gut microbiome. Gut 69(8):1510–1519
Wen L, Duffy A (2017) Factors influencing the gut microbiota, inflammation, and type 2 diabetes. J Nutr 147(7):1468S–1475S
Williams HD, Trevaskis NL, Charman SA, Shanker RM, Charman WN, Pouton CW, Porter CJ (2013) Strategies to address low drug solubility in discovery and development. Pharmacol Rev 65(1):315–499
Wilson ID, Nicholson JK (2014) The modulation of drug efficacy and toxicity by the gut microbiome. In: Metabonomics and gut microbiota in nutrition and disease. Springer, London, pp 323–341
Wilson ID, Nicholson JK (2017) Gut microbiome interactions with drug metabolism, efficacy, and toxicity. Transl Res 179:204–222
Winek K, Dirnagl U, Meisel A (2016) The gut microbiome as therapeutic target in central nervous system diseases: implications for stroke. Neurotherapeutics 13(4):762–774
Wu W, Sun M, Chen F, Cao AT, Liu H, Zhao Y et al (2017) Microbiota metabolite short-chain fatty acid acetate promotes intestinal IgA response to microbiota which is mediated by GPR43. Mucosal Immunol 10(4):946–956
Xie QS, Zhang JX, Liu M, Liu PH, Wang ZJ, Zhu L et al (2021) Short-chain fatty acids exert opposite effects on the expression and function of p-glycoprotein and breast cancer resistance protein in rat intestine. Acta Pharmacol Sin 42(3):470–481
Yassour M, Vatanen T, Siljander H, Hämäläinen AM, Härkönen T, Ryhänen SJ et al (2016) Natural history of the infant gut microbiome and impact of antibiotic treatment on bacterial strain diversity and stability. Sci Transl Med 8(343):343ra81
Yu J, Zhou Z, Tay-Sontheimer J, Levy RH, Ragueneau-Majlessi I (2017) Intestinal drug interactions mediated by OATPs: a systematic review of preclinical and clinical findings. J Pharm Sci 106(9):2312–2325
Zalewski S, Stewart CJ, Embleton ND, Berrington JE (2018) Brief guide to the analysis, interpretation and presentation of microbiota data. Arch Dis Child Educ Pract 103(6):327–330
Zhang M, Sun K, Wu Y, Yang Y, Tso P, Wu Z (2017) Interactions between intestinal microbiota and host immune response in inflammatory bowel disease. Front Immunol 8:942
Zhao Y, Zhong X, Yan J, Sun C, Zhao X, Wang X (2022) Potential roles of gut microbes in biotransformation of natural products: an overview. Front Microbiol 13:956378
Zhao Q, Chen Y, Huang W, Zhou H, Zhang W (2023) Drug-microbiota interactions: an emerging priority for precision medicine. Signal Transduct Target Ther 8(1):1–27
Zheng D, Liwinski T, Elinav E (2020) Interaction between microbiota and immunity in health and disease. Cell Res 30(6):492–506
Zhernakova A, Kurilshikov A, Bonder MJ, Tigchelaar EF, Schirmer M, Vatanen T et al (2016) Population-based metagenomics analysis reveals markers for gut microbiome composition and diversity. Science 352(6285):565–569
Zhu B, Wang X, Li L (2010) Human gut microbiome: the second genome of human body. Protein Cell 1(8):718–725

Zhu Y, Li Q, Jiang H (2020) Gut microbiota in atherosclerosis: focus on trimethylamine N-oxide. APMIS 128(5):353–366

Zimmermann M, Zimmermann-Kogadeeva M, Wegmann R, Goodman AL (2019a) Mapping human microbiome drug metabolism by gut bacteria and their genes. Nature 570(7762):462–467

Zimmermann M, Zimmermann-Kogadeeva M, Wegmann R, Goodman AL (2019b) Separating host and microbiome contributions to drug pharmacokinetics and toxicity. Science 363(6427):eaat9931

Immunomodulatory Effects of Microbiome-Derived Therapies

Krishnendu Ghosh and Daniel Sands

Abstract The human microbiome plays a crucial role in modulating the immune system, influencing both health and disease states. Recent studies have highlighted the potential of microbiome-derived therapies as an innovative approache to harness various immunomodulatory effects. Microbiome-derived metabolites such as short-chain fatty acids (SCFAs) can shape arrays of immune responses. These therapeutic factors also enhance the integrity of the gut barrier, promote anti-inflammatory cytokine production, and modulate T cell differentiation, ultimately leading to improved outcomes in conditions such as autoimmune diseases, neurodegenerative disorders, allergies, infections, and cancer. Clinical trials investigating the use of probiotics, prebiotics, fecal microbiota transplantation (FMT), and phage therapies have demonstrated promising results, indicating a reduction in disease severity and an increase in the quality of life for patients. Challenges pertaining to the need for personalized treatment protocols and a deeper understanding of individual microbiome compositions need to be dealt with more scientific outlook. Future studies should aim to elucidate the specific interactions between microbiota and host immune systems, paving the way for targeted microbiome-based therapies. This chapter underscores the potential of microbiome-derived therapies as a novel strategy for immune modulation, emphasizing the importance of further research to optimize these interventions and explore their application in a broader range of diseases.

Keywords Microbiome · Immune modulation · Disease · Therapies · Fecal microbiota transplantation (FMT) · Prebiotics · Probiotics

K. Ghosh (✉)
Department of Pathology, Carver College of Medicine, University of Iowa, Iowa City, IA, USA
e-mail: krishnendu-ghosh@uiowa.edu

D. Sands
Department of Microbiology & Immunology, University of Iowa, Iowa City, IA, USA

S. K. Singh et al. (eds.), *Microbiota Profiling for Precision Medicine*, Interdisciplinary Biotechnological Advances,
https://doi.org/10.1007/978-981-95-2104-3_4

1 Introduction

Microbes are the most abundant and diverse forms of life on Earth. These microorganisms inhabit a wide range of environments, from extreme conditions such as icy polar regions and deep-sea hydrothermal vents to the human gut, where they play pivotal roles in ecosystem dynamics and host health. The word microbiome refers to the clustered diversity of microorganisms—primarily bacteria, but also archaea, viruses, and fungi, an approximate hundred trillion in number (Wang et al. 2017; Rappuoli et al. 2023)—that inhabit various environments within and on the human body and contribute to essential functions, including digestion, nutrient absorption, and the synthesis of vitamins, as well as being a critical modulator of body's immune system. They interact with host cells through complex signaling pathways, influencing inflammation and other immunological interactions. There are some excellent articles by Gilbert et al. (2018), Mousa et al. (2022), and Sharon et al. (2022) that have encompassed a nice overall view of human microbiome.

Human skin and lungs were once thought to be mostly interactive with the external world in terms of sensitivity and respiration, arguably replaced by the gastrointestinal (GI) tract/gut. The cumulative gene of human gut microbiome is almost 150 times that of the entire human genome. Gut microbiome plays essential roles in number of biophysiological aspects of homeostasis and weight gain and loss (Piccioni et al. 2022). Thus, gut maintains number of interacting axes with other vital organs and system of the body like brain, heart, lungs, skin, bone, reproductive systems, kidney, and bladder (Afzaal et al. 2022). Apart from eubiotic (beneficial) microbiota of gut, body parts like skin, lung, oral cavity, upper respiratory tract, and urethro-genital system also harbor host of microbial species.

The composition of microbiome diversity varies greatly among individuals and is influenced by number of factors like diet, lifestyle, genetics, and environmental cues. Dysbiosis, the imbalance in microbial populations or harmful interaction of microbiome, has been implicated in various pathological conditions (Hrncir, 2022) ranging from cardiovascular diseases, allergies, and neuropsychiatric conditions to chronic kidney diseases, obesity, inflammatory bowel disease (IBD) (Carding et al. 2015; DeGruttola et al. 2016), and even personality traits (Johnson, 2020). Cheng et al. (2020) exhaustively summarized "*gutMDisorder*," an exhaustive database on dysbiosis of human gut microbiome related to a holistic multimodal associative pathobiological approach.

The research and understanding of human microbiome over the last few decades have yielded the emergence of microbiome-derived therapeutics as a promising frontier in medicine, as the elusive microbe–host physiology axis has developed innovative treatments for a variety of health conditions (Hill and Round, 2021). This understanding has driven research into microbiome-based interventions, notably *probiotics* (live beneficial microorganisms like lactic acid bacteria in milk products), *prebiotics* (fiber-rich compounds like berries fostering growth of beneficial microbiome), *postbiotics* (non-living byproducts like short-chain fatty acids [SCFAs] derived from prebiotics by probiotic bacteria), *paraprobiotics* (inactive insoluble microbial cells or cell fractions), *synbiotics* (probiotics + prebiotics), and *fecal microbiota transplantation* (FMT) (medical procedure involving healthy microbiota from donor transplanted in the colon of recipient). Recent advancements in high-throughput sequencing and computational biology in course of "*Human Microbiome Project*" have facilitated a comprehensive understanding of interaction between functional diversity of microbiome and the host, paving the way for microbiome-targeted therapies (Creasy et al. 2021). This has resulted into the advent of more cutting-edge microbiome-derived therapeutics like *phage therapy* (treatment using phage virus to kill pathogenic bacteria), *live biotherapeutics* (biological product that contain beneficial microorganisms for treating disease; this should not be confused with vaccine), and *microbial mimetics* (models of bacterial membranes and antimicrobial peptides [AMPs] used as antibiotics and therapeutics) with enhanced efficacy, effective sustainability, and decreased side effects (Gulliver et al. 2022; Strathdee et al. 2023). Microbiome-derived therapeutics have gained interest for their prospective in immune modulation, leveraging the intricate relationship between microbial communities and host immune responses. Gut microbiota plays an essential role in modeling the immune system by altering the functional and developmental aspects of immune cells (Graham and Xavier, 2023). Probiotics are thought to escalate the production of anti-inflammatory cytokines while suppressing pro-inflammatory pathways, thereby promoting immune tolerance and homeostasis (see Fig. 1). Additionally, microbial metabolites like short-chain fatty acids

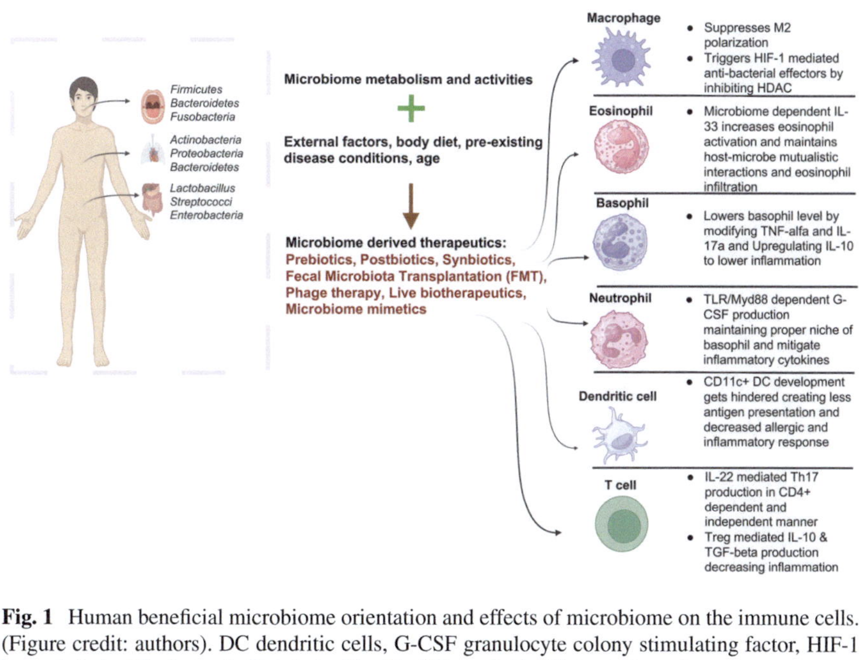

Fig. 1 Human beneficial microbiome orientation and effects of microbiome on the immune cells. (Figure credit: authors). DC dendritic cells, G-CSF granulocyte colony stimulating factor, HIF-1 hypoxia inducible factor-1, IL interleukin, Myd88 myeloid differentiation primary response protein 88, TGF-beta transforming growth factor-beta, TLR toll-like receptor, TNF-alfa tumor necrosis factor-alfa

(SCFAs) have been reported to modulate immune responses by acting on G-protein-coupled receptors (GPCRs) and promoting the differentiation of regulatory T cells (Venegas et al. 2019). FMT has also demonstrated the ability to restore immune balance in conditions like IBD, where dysbiosis is associated with chronic inflammation (Boicean et al. 2023). Overall, the therapeutic application of microbiome-derived interventions holds promise for managing immune-related disorders, offering a novel approach to restore immune homeostasis and mitigate inflammation. In the successive sections, we will address the various strategies of microbiome-derived therapies with special reference to the interaction and modulation of body immune response.

2 Historical Perspective of Microbiome-Derived Therapeutics

The use of microbiome-derived therapeutics has a rich history, evolving from ancient practices to modern biomedicine. In ancient civilizations, the concept of using microorganisms for health can be traced back to practices like fermentation. Ancient Egyptians, Greeks, and Romans used fermented foods, such as yogurt, to improve digestion and overall health due to the presence of certain beneficial species of *Lactobacillus* that are known to promote gut health by modulating the microbiome and inhibiting pathogenic microbes (Leeuwendaal et al. 2022). While the beneficial microbiome in probiotic fermented foods was not fully understood at that time, yet those ancient civilizations intuitively recognized their therapeutic properties (Cuamatzin-Garcia et al. 2022). FMT also has an ancient root dating back to the fourth century in China, where it was used as a remedy for food poisoning and diarrhea. The treatment was referred to as "*yellow soup*," which was a mixture of fecal matter from a healthy person and water and it was administered to patients suffering from gastrointestinal disorders (Stripling and Rodriguez, 2018). Similarly, in the sixteenth century, the physician Girolamo Fracastoro from Italy, who first proposed transmission of pathogen from human to human ("*seminaria*") as the cause of epidemic infections (Bottaccioli et al. 2021), also recommended the use of "*healthy excrement*" to treat dysentery. Although these practices were based on empirical knowledge rather than scientific understanding, they were some of the earliest forms of microbiome-based therapeutics.

In the late nineteenth century, the inception of *germ theory of disease* by Koch and Pasteur revolutionized medicine acting as a foundational framework for understanding the relationship between pathogens and infectious diseases (Carlsson and Råberg 2024). Koch's postulates consist of four key criteria: (1) the basic difference between healthy and unhealthy individual relies on absence and presence of microbes; (2) after isolating the causative microbes from the diseased host, they could be cultured outside; (3) the cultured microbes, when passed to another healthy but susceptible individual, will create the same type of disease; and (4) isolation of microbe from this experimentally infective individual should be of same entity to that of the original one. Koch's theory revolutionized the field of microbiology by linking specific microorganisms to particular diseases, providing the basis for the study of pathogens, diagnostics, and the development of vaccines. Although modern microbiology has advanced beyond the rigid application of Koch's postulates, especially with the advent of molecular techniques and the discovery of asymptomatic carriers, Koch's work remains central to the understanding of infectious diseases (Singh et al. 2016). Metchnikoff, the "*father of innate immunity*," could be considered as one of the first to describe the effects of the microbiome. He performed radical but relevant arrays of experiments to embed the idea of effect of beneficial microbiome on health (Podolsky, 2012). Penicillin, the first antibiotic, discovered by Fleming in 1928, resulted into the successive usage of antibiotics to treat bacterial infections. While antibiotics saved millions of lives, their overuse also led to the disruption of the gut microbiome and the emergence of antibiotic-resistant bacteria. This has led to renewed interest in the therapeutic potential of the microbiome in recent decades.

By the 1950s, doctors began using fecal transplants to treat severe gastrointestinal infections, particularly *Clostridium difficile* infections, though the practice did not gain widespread acceptance until the late twentieth century. The first clinical success of FMT was reported in 1983 by Dr. Thomas Borody in Australia. Since then, FMT has been accepted as a successful treatment for *C. difficile* infections and ongoing clinical trials are exploring its potential to treat immune-related diseases, GI disorders, and even neurological problems (Aroniadis and Brandt 2013).

The twenty-first century sees microbiome-derived therapeutics in a more minute and analytical way. In 2007, the U.S. National Institutes of Health (NIH) launched the Human Microbiome Project (HMP) for mapping the microbial communities of human body for better understanding of their role in health and disease conditions. The project, completed in 2016, revealed the vast complexity and diversity of the microbiome and provided new insights into how alterations in the microbiome can contribute to diseases like altered neurological issues, diabetes mellitus (Crudele et al. 2023), IBD, and obesity (Turnbaugh et al. 2007). The HMP and subsequent studies have provided the foundation for developing microbiome-based therapeutics. By identifying specific microbial species and their roles in disease, researchers have been able to develop targeted interventions, including probiotics, prebiotics, and even engineered microbes (Voth and Khanna, 2020).

3 Transmission of Microbiota in Human Babies

In her ground-breaking book named "*10% Human*," Dr. Alanna Collen described the fascinating way of initial microbial transfer from mother koala (*Phascolarctos cinereus*) to her baby in a form of soft fecal paste (made up within her body with predigested eucalyptus leaves and an inoculum of her gut bacteria) that mother koala forces her baby to eat, which would disseminate baby's gut with formidable inception of beneficial microbes (William Collins Publication, UK, 2015, chap. 7, p. 204). Trillions of microorganisms coexist with human host in a symbiotic relationship. The inception of this bond begins as early as during the gestation period, which continues to get developed post parturition and thus predominantly deluged from mother to baby (Browne et al. 2022). There are several factors that can influence the extent and composition of microbiota passed from mother to infants.

Broadly, the routes can be categorized into prenatal, perinatal, and postnatal transmission.

3.1 Prenatal Transmission

Historically, the womb was thought to be a sterile environment, but recent research has challenged this notion, suggesting that the prenatal microbiota may influence early infant microbial development. Although the fetus is largely isolated from external microbial exposure, some studies have detected microorganisms in the amniotic fluid, placenta, and umbilical cord like *Fusobacterium*, *Lactobacillus*, and *Bifidobacterium*. These bacteria are thought to play potential role at early stages of immune development in the fetus. The maternal immune system might be an important factor that selectively prevents and/or allows passing of microbes across the placenta (Xiao and Zhao 2023). However, the extent and functional significance of these microbial populations in the fetus remain unclear.

3.2 Perinatal Transmission

The perinatal period (encompasses birth and the immediate postpartum period) is critical for the transfer of microbiota. During this phase, the infant is first exposed to the mother's vaginal and gastrointestinal microbiota (in case of normal birth through vaginal canal) and this exposure has effects on the infant's microbiome. The vaginal canal is a major source of beneficial microbes, such as *Lactobacillus*, *Bifidobacterium*, and *Prevotella*, that play important role to develop a balanced gut microbiota in the infant. Research has demonstrated that the microbiome of vaginally born babies nearly resembles that of the mothers' vaginal microbial diversity, particularly in the first many days and weeks of life (Dunn et al. 2017). This early microbial exposure is thought to prime the infant's immune system and support the settlement of beneficial microbes. Infants delivered by cesarean section (C-section) are exposed to a different host of microbial species by bypassing the vaginal canal. Instead, they are increasingly exposed to skin microbes from the mother although these infants inclined toward having less diverse beneficial microbes like *Bifidobacterium* and *Lactobacillus* with a more prevalence of potentially pathogenic bacteria (namely, *Clostridia* and *Staphylococcus*). The lack of exposure to vaginal and fecal microbiota during C-section delivery has been associated with an advanced risk of immune-related circumstances (asthma, allergies, and autoimmune conditions), suggesting that early microbial exposure is critical for proper development of immunity (Arora et al. 2017).

3.3 Postnatal Transmission

Post birth, the baby's microbiota continues to develop by interacting with the surrounding environment. Breast milk, a finest source of natural prebiotic with beneficial microorganisms, is mostly responsible for beneficial microbes in the infant's gut. Beneficial bacteria, such as *Bifidobacterium* and *Lactobacillus*, are often transmitted through breast milk. Direct skin-to-skin contact with the mother also contributes to microbiota transmission. The skin microbiota, which consists of bacteria like *Staphylococcus epidermidis* and *Propionibacterium acnes*, plays an important role in the early settlement of good microbes in the infant's skin and gut (Serghiou et al. 2023).

Maternal factors include the association of healthy fiber-rich diet of mother with increased number of beneficial bacteria and at the same time usage of antibiotics with reduced microbiome load, while mode of delivery (vaginal vs. cesarean), gestational age (preterm infants tend to have a less diverse microbiota at birth), and types of milk-feeding practices (whether breast milk, formulated milk, or a combination) contribute as infant factors (Bogaert et al. 2023).

Box 1 Colostrum: An Ambrosia from Mother to Infant

Colostrum, the concentrated yellowish milk produced by mammals during the initial days postpartum, plays a central role in shaping the early microbiota of infants. Rich in nutrients, immunoglobulins (high IgA), cytokines, and a diverse array of microbial components, colostrum serves not only as a nutritional source but also as a medium through which beneficial microbes are passed from mother to baby. The bacterial composition of colostrum is characterized by the presence of several core genera that are typically involved in early gut colonization. *Lactobacillus* species are often the most abundant bacteria in colostrum and breast milk and significantly contributes to the enhancement of the infant's gut microbial diversity. *Bifidobacterium*, another commonly found bacteria, enhances gut health by producing SCFAs and maintaining mucosal integrity. Recent studies have also identified *Bacteroides*, *Enterococcus*, and *Firmicutes* as key bacterial families present in colostrum, which contribute to the infant's gut microbiome and immune maturation. While much less is known about the fungal microbiota in colostrum, studies suggest that yeasts such as *Candida* spp. and *Saccharomyces* spp. are found in both colostrum and breast milk. The significance of these fungi remains unclear but their presence may have implications for the early immune system, as they can influence the gut's inflammatory responses. Viral components, such as bacteriophages, have also been detected in colostrum, indicating their prospective role in modulating the microbiota and protecting against pathogenic microbes (Francese et al. 2023; Boix-Amorós et al. 2017; Hurley and Theil 2011; Thapa 2005).

4 Microbiome and Immune System

Multifaceted interactions prevail between the microbiota and immunobiological system, explaining the action of microbial communities shaping immune responses, modulating inflammation, and contributing to the development of immune-related diseases. The immune system, a robust elusive system on its own with diverse network of cells, tissues, and signaling molecules, is a key player that defends the body against pathogens and maintains homeostasis. It is categorically divided into the innate and adaptive, both of which are influenced by microbial signals (Jordan and Clarke 2024). Dysbiosis, an imbalance in microbial communities, is thus one of the major contributing factors for diseases like autoimmune disorders and IBD.

4.1 Regulation

The process of immune system development begins early in life, with the microbiome catering an important part in the establishment of immune tolerance and the growth of immune memory (Jain 2020). Microbial exposure during early childhood helps prime the immune system, especially the gut-associated lymphoid tissue (GALT), the largest immune-related structure in the body. Studies have shown that mice lacking microbiota result into inadequately developed immune conditions with deficits in both the innate and adaptive immune responses (Fiebiger et al. 2016). The microbiome helps maintain immune regulation through the induction of regulatory T cells (T_{regs}), which are critical for preventing autoimmunity and chronic inflammation. These T_{regs} are influenced by various microbial metabolites. A lack of specific microbial communities can lead to an imbalance in T_{regs} and increased sensitivity to autoimmune diseases (Liu et al. 2021). The microbiome modulates functionalities of different immune cell types including dendritic cells (DCs), macrophages, and T cells. Metabolites from certain gut microbes (acetate, propionate, butyrate, etc.) can directly influence the activity of DCs, enhancing their ability to induce immune tolerance (Shvets et al. 2024). By binding to G-protein-coupled receptors (GPCRs) such as GPR43 and GPR109A, SCFAs promote anti-inflammatory cytokine production and improve intestinal barrier integrity (Sun et al. 2017). These metabolites can also regulate the differentiation of T-helper (Th) cells, like Th17 subset, which plays a critical role in mucosal immunity and inflammation. Th17 cells are controlled by gut microbiota and alterations in microbial composition can lead to shifts in Th17 responses (Brockmann et al. 2023), contributing to autoimmune diseases like IBD and multiple sclerosis (MS). The skin microbiome, while less studied than the gut microbiome, is also crucial in modulating immune responses. Skin-resident bacteria influence different immune cells, including Langerhans cells, T cells, and mast cells. For example, the skin microbiota can influence the expression of antimicrobial peptides (AMPs) and shape the local immune response (Nakatsuji et al. 2021). Figure 1 briefly documents the basic role of human microbiome in the immune system.

4.2 Dysregulation (Dysbiosis)

In dysbiosis, an overgrowth of pathogenic microorganisms is directly related to arrays of immune-related diseases. In autoimmune conditions like rheumatoid arthritis (RA) and type 1 diabetes mellitus (T1DM), microbiota imbalances have been shown to trigger inflammation. In the case of RA, for example, certain gut bacteria (e.g., *Prevotella copri*) have been linked to disease onset (Horta-Baas et al. 2017). Similarly, in T1DM, the early-life microbiome composition, including reduced diversity and altered *Bacteroides/Firmicutes* ratio, has been correlated with an increased outcome of disease (Zheng et al. 2018; Yuan et al. 2022). IBD, which includes Crohn's disease and ulcerative colitis (UC), is strongly linked with altered microbiota as studies have shown patients with IBD often exhibit exacerbated loads of pro-inflammatory bacteria such as *Escherichia coli* and *Fusobacterium nucleatum*, which can induce epithelial barrier dysfunction and exacerbate mucosal inflammation (Gu et al. 2020). Recent research has also shown that restoring a healthy microbiome via microbiota-targeted therapies can improve clinical outcomes in IBD patients (Shan et al. 2022).

4.2.1 In Allergies

There is insistent evidence of relationship between microbiome and allergic diseases like asthma, allergic rhinitis, and atopic dermatitis. The "hygiene hypothesis" suggests that reduced exposure to diverse microbes early in life may weaken developmental aspects of immune system, leading to an increased susceptibility to allergies and asthma. Studies have shown that children with higher microbial diversity in their early life have a lower risk of developing allergic conditions (Okada et al. 2010). Moreover, specific bacteria, such as *Bifidobacterium* and *Lactobacillus*, are thought to promote immune tolerance and protect against allergic responses by regulating Th1/Th2 polarization (Ren et al. 2019).

4.2.2 In Skin Diseases

Dysbiosis in the skin has also been implicated in the pathogenesis of several inflammatory skin conditions, such as atopic dermatitis, psoriasis, and acne vulgaris. For example, in atopic dermatitis, there is often an overabundance of *Staphylococcus aureus*, which exacerbates skin inflammation and disrupts the epidermal barrier. Conversely, reduced levels of *Staphylococcus epidermidis*, a commensal bacterium that produces antimicrobial peptides, can impair skin defense mechanisms (Landemaine et al. 2023). Skin dysbiosis can also affect the immune response by differential production of cytokines and antimicrobial peptides, which can trigger or exacerbate inflammation (Koh et al. 2022).

4.2.3 In Cancer

Microbiome dysbiosis has appeared as a substantial factor in the development and progression of various types of cancer, influencing both tumor initiation and the response to treatment. Dysbiosis can affect multiple aspects of tumor biology, including immune modulation, inflammation, and metabolic pathways (Jain et al. 2021). For instance, in colorectal cancer (CRC), dysbiosis has been associated with an overgrowth of specific bacteria such as *Fusobacterium nucleatum* and *Bacteroides fragilis*, which can promote tumorigenesis through mechanisms such as the activation of pro-inflammatory signaling pathways (e.g., NF-κB [Nuclear factor kappa-light-chain-enhancer of activated B cells]) and the modulation of immune responses, including the suppression of antitumor T cell activity (Garvey 2024). Dysbiosis can also alter the metabolism of certain carcinogens, thereby enhancing their mutagenic potential (Sheflin et al. 2014). Beyond gastrointestinal cancers, alterations in the skin and oral microbiomes have been linked to skin cancer and head and neck cancers, respectively.

4.2.4 In Neurodegenerative Diseases

Microbiome dysbiosis has arguably been identified as a contributor to the pathogenesis of neurodegenerative diseases. The gut–brain axis, a looped communication between the gastrointestinal system and the central nervous system, provides a crucial link through which microbial imbalances can influence neuroinflammation, neuronal function, and disease progression (Ullah et al. 2023). In neurodegenerative diseases, dysbiosis is thought to be a positive contributor to the disease condition. In Alzheimer's disease, for instance, a decrease in beneficial bacteria like *Bifidobacterium* and *Lactobacillus* and an increase in bacteria such as *Fusobacterium* and *Enterococcus* in the gut have been linked to exacerbated neuroinflammation and amyloid plaque deposition (Liu et al. 2020). Similarly, in Parkinson's disease, an overrepresentation of *Firmicutes* and a depletion of *Bacteroidetes* in the gut have thought to been associated with the progression of motor dysfunction and neurodegeneration (Dutta et al. 2019; Loh et al. 2024).

4.2.5 In Obesity

Obesity is also thought to be associated with alteration of microbial communities in the gut, exemplified by an increase in *Firmicutes* and a decrease in *Bacteroidetes*, linked to enhanced energy extraction from the diet. Dysbiosis may also affect the balance of pro-inflammatory and anti-inflammatory cytokines, contributing to an initial chronic inflammation, an iconic indicator of obesity and metabolic syndrome (Vetrani et al. 2022). The dysbiotic condition leads to increased intestinal permeability ("leaky gut"), increasing translocation of bacterial endotoxins (like lipopolysaccharide [LPS]) into the bloodstream, which in turn exacerbates systemic inflammation and insulin resistance. Moreover, gut microbiota influence host metabolism through the modulation of bile acids (BAs) and regulation of appetite-controlling hormones like ghrelin and leptin, suggesting that microbiome-driven shifts in microbial communities can affect both food intake and energy expenditure (Breton et al. 2022). Studies indicate that restoring a balanced gut microbiome through probiotics, prebiotics, and/or dietary interventions may mitigate some of the obesity-associated metabolic instabilities, presenting potential therapeutic strategies for weight management (Puljiz et al. 2023).

4.2.6 In Sepsis

Microbiome dysbiosis plays a significant role in the pathobiology of sepsis, a life-threatening condition due to systemic inflammatory reaction to infection. Sepsis is often associated with either disruptions or a complete loss of microbial diversity in the gut, which is further believed to impair the gut barrier function, leading to increased intestinal permeability of various microbial components. These microbial products can trigger widespread inflammation and immune system dysregulation, development of a condition known as systemic inflammatory response syndrome (SIRS), and multiple organ dysfunction (Piccioni et al. 2024). In sepsis, a reduction in beneficial commensals, such as *Lactobacillus* and *Bifidobacterium*, and an overgrowth of potentially harmful bacteria, like *Enterococcus* and *Clostridium*, have been reported. The imbalance between these microbial populations results in an altered immune response, where the normally protective immune mechanisms are impaired and pro-inflammatory pathways (e.g., NF-κB, TNF-α [tumor necrosis factor-alpha]) are activated, further exacerbating the sepsis cascade (Shang et al. 2024).

5 Microbiome-Derived Therapeutics and Their Functions

Microbiome-derived therapeutics are currently an emerging regime of treatments that primarily depends upon the intricate connections between the human microbiota and host health. Microbiome-derived therapeutics (probiotics, prebiotics, postbiotics, and fecal microbiota transplantation) aim to restore a balanced microbial ecosystem. These strategies can modulate body's microbial diversity, enhance beneficial microbial functions, or suppress harmful pathogens, ultimately improving health outcomes.

5.1 Prebiotics

Prebiotics are defined as non-digestible food ingredients rendering benefits to the host by promoting selective growth and/or activity of beneficial gut microbes. The most well-known prebiotics are certain types of nutritional fibers, oligosaccharides, and polyphenolic compounds that usually do not get digested in the upper portion of GI tract (Rodríguez-Daza et al. 2021). However, after reaching the colon, they are fermented by localized microbial communities, leading to the production of SCFAs (category includes butyrate, acetate, and propionate). These SCFAs are having many health benefits, including promoting barrier function of intestine, modulating immune responses, and influencing metabolic processes. Prebiotics can further be subdivided into two types based on their structural entities, origin, and the microbiota they target (Davani-Davari et al. 2019). Key categories of microbiome-derived prebiotics include the following.

5.1.1 Oligosaccharides

5.1.1.1 Fructo-Oligosaccharides (FOS) and Inulin

These oligosaccharides, found in foods like chicory root, onions, and garlic, are among the most studied prebiotics. They are selectively fermented by *Bifidobacteria* and *Lactobacilli*, leading to SCFA production and enhanced gut health.

5.1.1.2 Galacto-Oligosaccharides (GOS)

Found in dairy products and legumes, GOS can increase *Bifidobacterial* populations and improve gut health. They are particularly beneficial for infants and have been shown to influence immune system development.

5.1.2 Polyphenols

Polyphenols, found in foods like fruits, vegetables, tea, and cocoa, are bioactive compounds that have been shown to act as prebiotics. These compounds are said to promote the growth of some of the members of beneficial microbiota, such as *Bifidobacterium* spp. (Dias et al. 2021), which are thought to have antioxidant and anti-inflammatory effects. Flavonoids, a subclass of polyphenols (like quercetin and catechins, found in apples, citrus fruits, and tea), have been reported to show prebiotic effects by modulating gut microbial communities and boosting SCFA production.

5.1.3 Dietary Fibers

Dietary fibers like cellulose and pectin serve as important substrates of gut-related fermentation process. Foods high in fiber, such as whole grains, legumes, and fruits, are thus a natural source of prebiotics.

5.1.4 Resistant Starch

Found in foods like unripe bananas, legumes, and cooked-then-cooled potatoes, fermentation of resistant starch by gut microbes into SCFAs plays a significant role in gut health and metabolic regulation (Portincasa et al. 2022).

5.1.5 Other Complex Carbohydrates

Certain plant-derived complex carbohydrates, such as arabinoxylan (from wheat and rye), beta-glucan (from oats), and mannan oligosaccharides (from palm kernel and soybeans), also exhibit prebiotic effects. These compounds boost the growth of beneficial microbes resulting in improved gut health.

Prebiotics act as substrates for beneficial gut microbes, particularly *Bifidobacteria* and *Lactobacilli*, leading to their selective proliferation that enhances microbial diversity and promotes the dominance of beneficial taxa, which can outcompete pathogenic or harmful microorganisms (Virk et al. 2024). SCFAs, as fermentation end products, help to acidify lumen of gut to restrict the growth of pathogenic microbes as well as supply energy to colon epithelial cells helping increased gut health and barrier integrity. SCFAs also modulate functions of immune system with their influence over dendritic cells, macrophages (Duan et al. 2023), and regulatory T cells (T_{regs}), prevailing an anti-inflammatory condition in gut. Butyrate upscales development of T_{regs} and production of anti-inflammatory cytokines (e.g., interleukin-10 [IL-10]), maintains immune tolerance, and prevents inflammatory responses by inhibiting pro-inflammatory cytokines (e.g., TNF-α, IL-6) (Lyu et al. 2022). Additionally, prebiotics can influence the activation of pattern recognition receptors (PRRs), such as toll-like receptors (TLRs), on immune cells, facilitating pathogen recognition and enhancing the immune response without inducing chronic inflammation. This immune regulation is particularly relevant for managing chronic inflammatory diseases like IBD and metabolic syndrome (see Table 1). SCFAs enhance tight junction protein expression, preventing leaky gut and thus reducing intestinal permeability (Harbison et al. 2019). Emerging evidences opine that prebiotics may also influence the gut–brain axis concerning to the modulation of the central nervous system, mood, cognition, and behavior.

5.2 *Probiotics*

Probiotics are defined as the live microorganisms that in ideal extent benefit host's health. The therapeutic aspects of probiotics are related to their ability to influence gut microbiota composition, enhance intestinal barrier function, modify immune responses, and ease systemic inflammation. The efficacy of probiotics depends heavily on the specific strain used related to the provision of health benefits.

Table 1 Effective immune modulations of microbiome-derived therapeutics in certain diseases

Diseases	Microbiome-derived therapeutics used	Effects of microbiome-derived therapeutics	Immune effector modulations involved	References
Ulcerative colitis (UC) and inflammatory bowel disease (IBD)	FMT	Decreases ulcerative colitis and IBD both in isolative and in combination with other drugs	↑CD4⁺ T cells; ↑IL-10; ↓Macrophage; ↑TGF-β	Burrello et al. (2018); Perry et al. (2015)
Immune checkpoint inhibitors (ICI)-related colitis	Oral administration of *Bifidobacterium*	Maintains activity of anti-CTLA-4 immunotherapy without the risk of colitis	↑IL-10 activity of intestinal T_{reg}	Wang et al. (2018)
	Oral administration of *Lactobacillus reuteri*	Maintains activity of anti-CTLA-4 and anti-PD-1 immunotherapy without the risk of colitis	↓Leukocytes infiltration into intestinal wall; ↓TNF-α; ↓IFN-γ; ↓IL-6	Renga et al. (2022)
Various cancers	SCFAs	Modulate tumor immune microenvironment (TME)	↑IL-17; ↑T_{reg}; ↑γδ T cells	Lee et al. (2020)
Rheumatoid arthritis (RA)	SCFAs; probiotics (*Lactobacillus rhamnosus*, *Lactobacillus reuteri*)	Enhance the activities of disease-modifying antirheumatic drugs (DMARDs)	↓IL-1α; ↓IL-6; ↓IL-10; ↓IL-12	Bodkhe et al. (2019)
Type 1 diabetes mellitus (T1DM)	SCFAs	SCFAs, mainly acetate, lowers autoimmune cells and decreases proinflammatory cytokines	↓autoreactive T cells in the pancreatic lymph node; ↓IL-21; ↑IL-10RA	Marino et al. (2017); Zheng et al. (2017)

5.2.1 *Lactobacillus*

Lactobacillus spp. is among the most well-researched probiotics. These are natural dwellers in human gut, as well as in fermented foods like yogurt. *Lactobacillus* strains, such as *Lactobacillus rhamnosus*, *Lactobacillus acidophilus*, and *Lactobacillus reuteri*, have demonstrated arrays of benefits, including the modulation of immune responses, inhibition of pathogenic bacteria, and support of gut barrier physiology.

5.2.2 *Bifidobacterium*

Bifidobacterium is a genus of bacteria that plays a crucial role in dietary fibers fermentation and the production of beneficial SCFAs. Strains like *Bifidobacterium bifidum*, *Bifidobacterium longum*, and *Bifidobacterium breve* are widely used as probiotics and have been shown to promote gut health, improve intestinal motility, and reduce inflammation by modulating T_{reg} (Sun et al. 2020).

5.2.3 *Saccharomyces boulardii*

Saccharomyces boulardii is a non-pathogenic yeast that shows beneficial effects on gastrointestinal health, particularly as therapeutic and preventive measure of diarrhea. It also plays role in the restoration of good gut microbiota promoting integrity of the intestinal mucosal layer (Ansari et al. 2023).

5.2.4 *Escherichia coli* Nissle 1917

This strain of *E. coli*, originally isolated from healthy human intestines, has been used as a probiotic for the management of inflammatory bowel diseases, particularly ulcerative colitis. It has been reported to show an anti-inflammatory effect and also restore gut microbiota balance in patients with IBD (Behrouzi et al. 2020).

5.2.5 *Enterococcus faecium*

Known for enhancing gut barrier function and modulation of immune responses (Pi et al. 2022).

5.2.6 *Streptococcus thermophilus*

A strain commonly found in yogurt, which helps with lactose digestion and supports immune function by showing anti-inflammatory attributes (Allouche et al. 2024).

5.2.7 *Lactococcus lactis*

Used in dairy fermentation, this bacterium has immune-modulating properties and can produce antimicrobial peptides. It has also been reported to show preventive efficacies to intestinal tissue injuries (Darby et al. 2019).

Probiotics directly impact the microbial composition of the gut, frequently by promoting the growth of salutary bacteria while suppressing pathogenic or opportunistic microbes. They can compete for nutrients, cling to intestinal epithelial cells, and alter the microbial terrain by producing antimicrobial substances like bacteriocins and lactic acid. One of the key mechanisms through which probiotics support health is by improving the structural entity of the gut epithelial barrier (Yan and Polk 2020). Probiotics increase the expression of junctional proteins (like occludin, claudins) that close the openings between enterocytes, preventing the translocation of harmful bacteria and toxins across the intestinal wall. Probiotics, particularly those producing SCFAs like butyrate, may influence the brain through neural, endocrine, and immune-signaling pathways. Studies have shown that certain probiotics can affect mood, cognition, and behavior and may provide therapeutic benefits for neuropsychiatric conditions like depression, anxiety, and autism spectrum disorder (ASD) (He et al. 2023). See Table 1 for the functional significances of prebiotics on various diseases. Probiotics have been shown to modulate gut motility by influencing the contractility of smooth muscle in the gastrointestinal tract. They can also enhance digestion by producing enzymes that break down complex carbohydrates, proteins, and lipids, improving nutrient absorption and overall digestive health. Probiotics modulate both the innate and adaptive immune systems. They can interact with GALT to stimulate the production of anti-inflammatory cytokines (e.g., interleukin-10, TGF-β [Transforming Growth Factor-beta]) and regulatory T cells (T_{regs}), which help regulate immune tolerance (Cristofori et al. 2021). Probiotics also enhance the production of IgA (immunoglobulin A), an important antibody that protects mucosal surfaces. Additionally, some probiotics can stimulate the production of antimicrobial peptides (AMPs) that aid in pathogen clearance. Through their modulation of the gut microbiome and immune system, probiotics have been shown to reduce systemic and local inflammation. They can attenuate the activation of the NPLR3-mediated inflammasome and the production of pro-inflammatory cytokines (e.g., TNF-α, IL-6), thus helping to mitigate chronic inflammation (Yang, Wang, Chen, Guo, Dong 2023; Yang, Chen, Cai 2023) that is characteristic of conditions like IBD and metabolic syndrome.

5.3 *Postbiotics*

Postbiotics are soluble factors or products that are produced by microorganisms during fermentation processes or cell lysis and provide health benefits to the host without the need to introduce live microorganisms. These bioactive compounds include a wide range of metabolites, namely SCFAs, antimicrobial peptides, cell wall fragments, enzymes, and other microbial products, that can modulate various aspects of human health (Scott et al. 2022).

The growing interest in postbiotics stems from their potential to offer therapeutic benefits similar to probiotics, but without some of the associated risks of live organisms. Additionally, postbiotics have more stable formulations, fewer regulatory concerns, and are easier to incorporate into food or pharmaceutical applications.

5.3.1 Short-Chain Fatty Acids (SCFAs)

SCFAs are the terminal products of the fermentation of fiber-enriched diet by gut microbiota, especially those from the *Firmicutes* and *Bacteroidetes* phyla. Predominantly SCFAs include acetate, propionate, and butyrate, of which butyrate is most well characterized. These metabolites are key postbiotics that influence host physiology by promoting gut health, reducing inflammation, enhancing immune function, and constructively maintaining intestinal epithelial barrier (see Fig. 2). SCFAs control the immune system by modulating G-protein-coupled receptors (GPCRs) or act as histone deacetylase inhibitors (HDACi) (Chen et al. 2003). These molecules interact with some of the specific types of GPCRs like GPR41, GPR43, and GPR109A and also interface with the aryl hydrocarbon (AhR) receptor (Kim 2023), actively interacting with both immune and non-immune cells within the intestines. For non-immune intestinal epithelial cells, SCFAs induce the production of antimicrobial peptides (such as β-Defensins) promoting barrier integrity. They also help in the induction of mucin production through prostaglandin stimulation and thus tighten barrier junction by upregulating junctional proteins (Deleu et al. 2021). Regarding the immune functions, SCFAs can block LPS-driven maturation of DCs and can attenuate macrophage responses to LPS. Similar to epithelial cells, SCFAs, particularly butyrate, can drive antimicrobial peptide production in macrophages. Finally, in neutrophils, SCFAs similarly attenuate their responses, decreasing neutrophil extracellular traps (NETs) and migration. In lymphoid cells, SCFAs drive broadly immunoregulatory responses via acetylating the Foxp3-associated (Smith et al. 2013) histones and promoting fatty acid oxidation, both of which induce T_{reg} formation. SCFAs can also trigger increased retinoic acid production, which allows DCs to generate more peripheral T_{regs} (Facchin et al. 2024). SCFAs can promote a Th1 phenotype, biasing away from a Th17 phenotype through T-bet expression, and can provoke more IgA class shifting and increase CXCR5 for more B cells with upregulated AhR and hypoxia inducible factor (HIF) in innate lymphoid cells (ILCs), driving IL-22 production leading to increased epithelial barrier support (Yang et al. 2020). Emerging data show that SCFA components escalate the activity of cytotoxic T-lymphocytes (CTLs) and chimeric antigen receptor T (CAR-T) in course of showing antitumor effects (González-Brito and Uribe-Herranz 2023) governed by epigenetically reprogrammed function of mammalian target of rapamycin (mTOR) that inhibits class I histone deacetylase activity and elevates production of effector molecules like CD25, interferon gamma (IFN-γ), and tumor necrosis factor-alpha (TNF-α). Thus, altogether they play crucial role in antitumor activity of antigen-specific CTLs and ROR1-targeting CAR-T cells, as observed in different syngeneic murine melanoma and pancreatic cancer models proving the role of SCFAs in cancer immunotherapy (Luu et al. 2021).

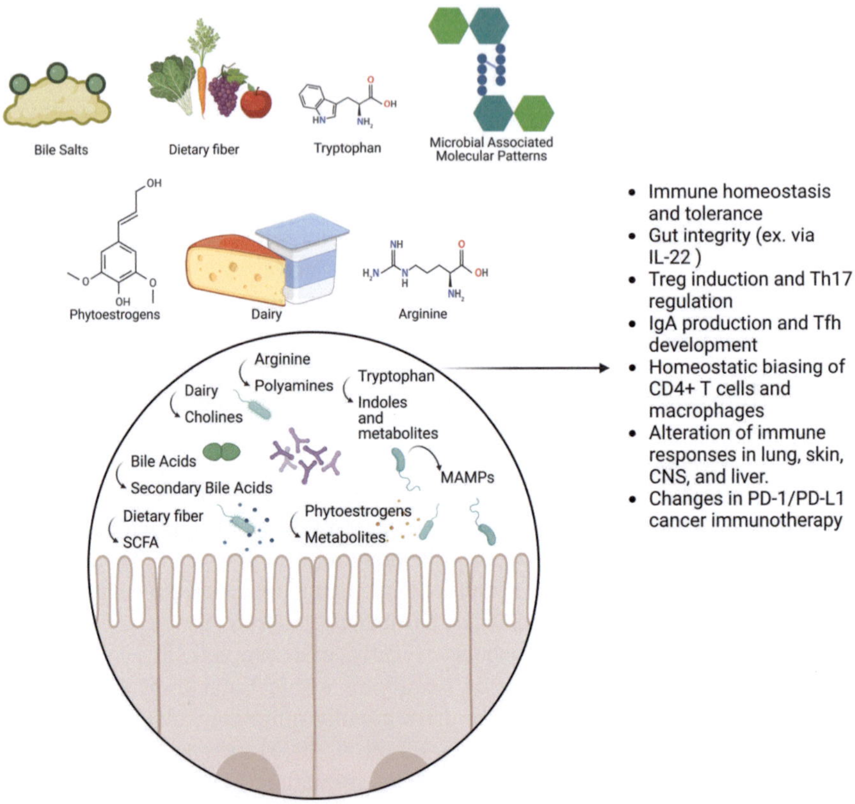

Fig. 2 Efficacy and immunological modulations catered by microbiome-derived therapeutics. (Figure credit: authors). MAMP microbe-associated molecular pattern, PD-1 programmed cell death protein-1, PD-L1 programmed cell death ligand-1, SCFA short-chain fatty acid

5.3.2 Microbial Cell Wall Components

These include peptidoglycans, lipopolysaccharides (LPS), and teichoic acids, which are byproducts of bacterial metabolism and cell wall synthesis. While LPS has been traditionally associated with endotoxin activity, certain LPS variants from beneficial microbes can have immunomodulatory effects and promote anti-inflammatory responses (Mohr et al. 2022).

5.3.3 Bacteriocins

These are antimicrobial peptides produced by certain bacterial species that restrain the growth of competing microorganisms. Postbiotic bacteriocins can have therapeutic applications in controlling harmful pathogens and enhancing the gut microbial balance (Cheruvari and Kammara 2024).

5.3.4 Exopolysaccharides (EPS)

EPS are complex polysaccharides produced by some probiotic bacteria and contribute to their biofilm formation, gut adherence, and interaction with the host. These polymers have prebiotic, anti-inflammatory, and immune-regulating properties (Angelin and Kavitha 2020).

5.3.5 Microbial Enzymes

Certain microbial enzymes such as amylases, proteases, and lipases can be considered postbiotics if they are secreted during microbial growth or fermentation. These enzymes break down complex dietary molecules and may support gut digestion and health (Da et al. 2024).

5.3.6 Bioactive Metabolites

Other bioactive molecules such as vitamins (e.g., B-vitamins, vitamin K), polyamines, and aromatic amino acid derivatives can be generated by gut bacteria having number of physiological effects on the host, including regulating the immune system, modulating gut motility, and influencing mood and cognition (Thorakkattu et al. 2022).

5.3.6.1 Bile Acids (BAs)

Bile acids (BAs), another important bioactive component, are an amphipathic cholesterol derivative that help emulsify lipids from food. Bile acids are synthesized in the liver and stored in the gallbladder. The primary bile acids are either cholic acid (CA) or chenodeoxycholic acid (CDCA). These enter the enterohepatic circulation, in which ~95% of the primary BAs are reabsorbed, but 5% descend further into the gut. These primary BAs are then processed by commensal gut microbes into secondary BAs (majorly deoxycholic acid [DCA] and lithocholic acid [LCA]) through the hydrolysis of glycine or taurine via bile salt hydrolases (BSHs) (Prete et al. 2020). BAs help in the sustainability of beneficial microbiota through numbers of molecular signaling cascades of nuclear receptors including Farnesoid X, Vitamin D receptor, retinoid-related orphan receptor gamma t (RORγt), Pregnane X receptors (PXR), and some GPCRs, including Takeda-G-protein-coupled receptor 5 (TGR5) (Collins et al. 2023).

5.3.6.2 Tryptophan Catabolites

Tryptophan catabolites are another group of important metabolites that interact with aryl hydrocarbon receptor (AhR). Tryptophan is the largest and least abundant canonical amino acids in cells. Tryptophan is only gained from the diet, which can be derived from sources including dairy, fish, grains, and some fruits. It is metabolized mainly by the kynurenine and minorly through the serotonin pathway (Gupta et al. 2023). Tryptophan when digested by the microbial community (e.g., *Lactobacillus* spp.) can be processed into several indoles—including indole-3-aldehyde (I3A), indoleacetic acid (IAA), and 3-indoleproprionate (IPA)—all of which are AhR ligands. AhR signaling is necessary for gut homeostasis and gut barrier integrity and can be important for immune signaling for immune effector cells (IECs), Th17, innate lymphoid cells (ILCs), DCs, macrophages, and neutrophils (Li 2023).

5.3.6.3 Phytoestrogens

Phytoestrogens, another important constituent, are plant-derived compounds that structurally resemble to body estrogen. These compounds can mimic or modulate the effects of estrogen in the body and are found in a variety of foods, particularly in legumes, seeds, and grains. The bioactive metabolites of phytoestrogens include isoflavones (in soy products), lignans (in flaxseeds), and coumestans (in clover) (Canivenc-Lavier and Bennetau-Pelissero 2023). Recent studies suggest phytoestrogen isoflavone metabolites may induce regulatory $CD8^{+}$ T cells via engagement of estrogen receptors (ERs), both ERα and Erβ; help in the production of low levels of IFN-γ, perforin, and IL-10; and may be related to autoimmune encephalomyelitis (Peterson et al. 2023). Recent evidence also suggests role of phytoestrogens in treating cancers (Farhat et al. 2023).

Postbiotics employ their effects via a variety of mechanisms, many of which overlap with the actions of probiotics (see Fig. 2). However, because postbiotics are often non-living compounds, their mechanisms can be more directly predictable and less prone to the variability seen with live microorganisms. Postbiotics are produced by an array of microorganisms in the human microbiome. *Firmicutes* (e.g., *Faecalibacterium prausnitzii*, *Roseburia* spp., *Clostridia*) are primarily responsible for the fermentation of fiber to produce butyrate, while *Bacteroidetes* (e.g., *Bacteroides fragilis*) contribute to the production of acetate and propionate (Stojanov et al. 2020). *Lactobacillus* species are implicated to the fermentation of lactose and production of bioactive metabolites that influence gut health and immune responses. *Akkermansia muciniphila*, a gut bacterium, is involved in mucin degradation and has been associated with the production of metabolites that help maintain gut integrity and modulate inflammation (Glover et al. 2022). It is also related to the production of biologically active components such as SCFAs and other mucin-degrading enzymes.

5.4 *Paraprobiotics*

Paraprobiotics are inactivated microbial cells or microbial components that retain their beneficial health effects without the need to be live. Paraprobiotics represent an emerging area of microbiome-based therapies with potential applications in modulating gut health, immune function, and metabolic regulation. Paraprobiotics are distinct from probiotics in that they do not involve the live administration of microorganisms but rather focus on the health-promoting properties of microbial cells or their components once they are inactivated or broken down and interact with the host and modulating biological processes (Teame et al. 2020). The growing interest in paraprobiotics is driven by their ability to elicit beneficial responses like live probiotics, but without some of the challenges associated with probiotic therapies—such as survival in the gastrointestinal tract, potential risk of infections in immunocompromised individuals, and variability in microbial effects (Siliciano et al. 2021). Paraprobiotics encompass a variety of microbial-derived products including heat-inactivated microbes (entire bacterial cells that have been killed through heat or other means but retain their bioactive properties), microbial cell wall fragments (components of bacterial cell walls, such as peptidoglycans, lipoteichoic acid, and lipopolysaccharides), bacterial lysates (made by breaking down bacterial cells, typically by mechanical, enzymatic, or physical processes, releasing intracellular components and proteins that retain activity even after the bacteria are no longer viable), microbial exopolysaccharides (high molecular weight carbohydrates produced by bacteria during fermentation that play roles in microbial adhesion, biofilm formation, and gut modulation), and microbial metabolites and secreted proteins (these include secreted antimicrobial peptides, enzymes, and signaling molecules) (Mudaliar et al. 2024).

5.5 *Synbiotics*

A synbiotic formulation is designed to offer health benefits by combining probiotics and prebiotics, both of which are obtained from the microbiome or its interaction with the host. The combination of probiotics and prebiotics aims to create synergistic effects. Synbiotics are primarily classified into two types—Class I (combination of a specific prebiotic and probiotic strain that has demonstrated efficacy when used together) and Class II (combination of multiple pro- and prebiotics that support each other through a more generalized symbiotic relationship) (Lee et al. 2024).

5.6 *Fecal Microbiota Transplantation (FMT)*

FMT refers to the collection of fecal material from a healthy donor into the recipient's gastrointestinal tract to restore the recipient's beneficial microbial community. The procedure involves the collection, processing, and infusion of donor stool into the recipient, typically via an enema, colonoscopy, or oral capsule (Wynn et al. 2023). The brief steps of FMT include:

5.6.1 Screening of Donor

The first and one of the most critical steps in FMT is the selection and screening of a healthy donor. Donors are typically selected based on strict criteria to minimize the risk of transmitting infectious diseases and harmful pathogens. This involves arrays of blood test (for detecting disease like HIV, hepatitis, and syphilis) and stool test (for pathogenic factors like *C. difficile*, *Salmonella*, *Shigella*, *Giardia*, and *Entamoeba histolytica*, as well as confirming donor's healthy microbiota) (Ng et al. 2024).

5.6.2 Stool Collection and Processing

After donor selection and screening, stool is collected, typically in a sterile container. The stool is then processed and prepared for transplantation. This involves homogenization (stool is blended with sterile saline or a preservation solution to create a suspension), filtration (stool suspension is filtered to remove large particles like undigested food), and storage or banking (prepared stool may be frozen for later use for future transplantations preserving the viability of the microbiota) (Nicco et al. 2020).

5.6.3 Preparation of the Recipient

Before undergoing FMT, the recipient may need to prepare their gastrointestinal system to optimize the success of the transplantation. Recipient is usually instructed to fast for a certain period before the procedure to clear the digestive tract and minimize the risk of rejection. Recipient is also monitored for the potential pre-existence of any other infection that, if found, would be treated prior to transplantation (Tian et al. 2024).

5.6.4 Administration Routes and Methods

Usually, stool suspension is administered directly into the colon through a colonoscope, allowing for targeted distribution of microbes throughout the large intestine. In certain instances, stool suspension is delivered via the rectum using an enema. In a less common method, FMT can be administered through a tube that is inserted through the nose and into the stomach or upper part of the small intestine. Recent advancements have led to the development of encapsulated FMT preparations, which consist of freeze-dried stool in capsules. These capsules are ingested orally, providing a non-invasive alternative to traditional methods (Gulati et al. 2020).

5.6.5 Post-Transplantation Follow-Up

After FMT, the recipient is typically monitored for any immediate adverse effects, which are generally rare but may include gastrointestinal discomfort, bloating, and diarrhea. In some cases, recipients may experience transient changes in bowel movements as the new microbiota establishes itself. Regular stool tests are done to ensure that no harmful pathogens are introduced from the donor. In some cases, microbiota from the donor may not persist long-term in the recipient, and additional interventions (like second or third repeats of transplantation) may be required to maintain microbiome health sustainability (van Lingen et al. 2023).

The key principle behind FMT is the reintroduction of a diverse and functional microbiome that can outcompete pathogenic microorganisms, re-establish microbial balance and immune system modulation, and restore normal host functions. There is a good number of evidences regarding the utility of FMT in recurrent *Clostridioides difficile* infection (rCDI) (Shao et al. 2023). The immune-modulatory effects of FMT are particularly beneficial in conditions such as IBD, where immune-mediated inflammation of the gut is a central feature and in cases of recurring *C. difficile* infection, where restoring a healthy microbiome significantly reduces inflammation and infection recurrence. FMT has been shown to reverse dysbiosis by replenishing beneficial microbes, which in turn influences immune system regulation. Furthermore, FMT has been reported to restore GI epithelial barrier integrity, preventing rapid translocation of pathogens into the bloodstream, reducing systemic inflammation and immune activation (Soveral et al. 2022). Microbiota-derived signals that activate pattern recognition receptors (PRRs) on immune cells may favor production of anti-inflammatory cytokines (e.g., IL-10) over pro-inflammatory ones (e.g., TNF-α, IL-6). Recent study suggested that therapeutic FMT simultaneously activates $CD4^+$ T cells and Natural Killer (NK) cells-mediated IL-10 production reducing colonic inflammation, especially in colitis (Burrello et al. 2018; Yang, Wang, Chen, Guo, Dong 2023; Yang, Chen, Cai 2023).

6 Challenges of Microbiome-Derived Therapeutics

Over the past decade, significant advances in microbiome research have opened newer windows in the development of microbiome-derived therapeutics. These therapeutics have made radical transformation in healthcare. Although having significant results in various preclinical and clinical studies, the field of microbiome therapeutics faces several significant challenges. These challenges span scientific, clinical, regulatory, and technical domains.

6.1 Complexity of Microbial Diversity and Composition

A major challenge of microbiome-derived therapeutics is the inherent complexity and individual variability of the human microbiome. The microbial composition varies widely across individuals due to factors like diet, genetics, age, geography, and lifestyle. These things complicate the development of universally effective therapies, as a therapeutic intervention that benefits one individual may not work for another. Studies have shown that the microbiomes derived from healthy individuals can vary greatly in terms of both species composition and functional capacity (Yadav and Chauhan 2022).

6.2 Gap of Knowledge in Disease Specificity of Microbes

The vast role of microbiome in health and disease conditions is not yet fully understood. For some diseases, specific microbial communities or dysbiosis are linked to disease states. However, it often poses a tough challenge to understand if the changes in the microbiome are a cause or a consequence of other disease states. The challenge of identifying precise microbial signatures that predict or correlate with specific diseases continues to be a key hurdle for developing diverse therapies.

6.3 Complexities of Host–Microbe Interaction

Host–microbiota interaction is intricate and dynamic, making it difficult to predict how microbiome-based therapeutics will influence human physiology. Microbial metabolites (SCFAs, bile acids, and vitamins) can interact with host metabolic pathways, immune responses, and even brain function. Disrupting these interactions, either intentionally or unintentionally, could have unforeseen consequences (Badal et al. 2019).

6.4 Risk of Unintended Consequences

The introduction of exogenous microbial strains into the human body through probiotics, FMT, or engineered microbiome therapeutics presents safety concerns, particularly with respect to unintended side effects. Some microbial strains may harbor pathogenic traits or could potentially transfer antibiotic resistance genes. For example, there have been reports of infections caused by the introduction of non-native microbial strains like *Escherichia coli* and *Clostridium difficile*, following FMT (Park and Seo 2021). The long-term safety of microbiome-based interventions remains a significant concern, as the full consequences of modifying the microbiome are still largely unknown.

6.5 Dearth of Information Regarding Efficacies

Despite having an upscaling in the usage of microbiome-derived therapeutics for various specific disease conditions, many of these therapies still lack robust, large-scale clinical trials demonstrating their efficacy. For instance, although FMT has been a promising treatment strategy for recurrent *C. difficile* infection, its success in other diseases, such as metabolic disorders or autoimmune conditions, specially across different age groups, has been less clear (Hourigan et al. 2021). The clinical outcomes of microbiome-based interventions can also vary significantly across studies owing to the factors like patient population, delivery method, and the microbiome profile of participants. The lack of standardized protocols for therapeutic administrations, such as the precise microbial composition, dosage, and delivery mechanisms, further complicates the determination of therapeutic efficacy.

6.6 Lack of Standardized Approaches

There is no consensus on the best methods for characterizing and modifying the microbiome for therapeutic purposes. Different studies often use distinct methodologies for microbiome sequencing, data analysis, and reporting, which makes it difficult to compare results across trials. Standardization of microbiome analysis methods, such as sequencing depth, bioinformatics pipelines, and taxonomic classification, is needed to enhance reproducibility and facilitate the development of regulatory guidelines (Allegretti et al. 2024).

6.7 Ambiguity in the Regulatory Usage

The approaches to microbiota-based therapies are evolving continuously. Microbial therapies, particularly those involving live microorganisms, are crucially regulated by international agencies like the U.S. Food and Drug Administration (FDA) and the European Medicines Agency (EMA). However, the lack of a well-defined regulatory framework specifically tailored to microbiome-based therapeutics poses a challenge to the development of these therapies. For instance, FMT is classified as biologic in some jurisdictions, while others may consider it a drug or even a medical device (Carlson 2020). This ambiguity complicates clinical trials, approval processes, and post-market surveillance.

6.8 Problem in Clinical Trial Design

The design of clinical trials for microbiome therapies presents unique challenges. Microbiome-related outcomes are difficult to measure, and many microbiome-based interventions do not produce immediate effects. Traditional endpoints, such as clinical symptom resolution, may not adequately capture the broader effects of microbiome modulation. Moreover, microbiome therapies may not always yield predictable or linear outcomes, necessitating novel trial designs, such as adaptive trials or long-term follow-up, to assess their safety and efficacy (Young 2021).

6.9 Manufacturing Challenge

Scaling up the production of microbiome-derived therapeutics presents significant logistical and technological challenges. The production of live microbial products must abide by strict Good Manufacturing Practices (GMP) ensuring consistency, potency, and safety. Moreover, the storage and transportation of probiotics and other live microbial products can be challenging due to their sensitivity to temperature, pH, and other environmental factors (Thanush et al. 2023). For engineered microbiomes or FMT, ensuring the quality and reproducibility of the product can be even more complicated. The preparation of FMT, for instance, involves isolating microbial communities from donor fecal matter, which is then processed and transplanted into recipients. Maintaining the viability of specific beneficial microbes throughout the process is critical, as is preventing contamination with pathogens.

6.10 *Ethical Issues*

Microbiome-based therapies, especially FMT, raise ethical concerns related to donor screening, microbiome ownership, and potential misuse of microbial products. Donor screening for pathogens is essential for ensuring the safety of FMT, but the ethical implications of using human-derived material in treatments are still debated. Furthermore, as the commercialization of microbiome therapies grows, there are concerns about equitable access and the potential for exploitation of vulnerable populations (Metselaar and Widdershoven 2017).

7 Future Directions of Microbiome-Derived Therapeutics

As research progresses, numerous exciting opportunities are emerging in microbiome-based interventions. However, several key areas require further investigation to enhance our understanding and unlock the full therapeutic potential of microbiome modulation.

7.1 *Personalized Microbiome Medicine*

Personalization of microbiome-derived therapies based on individual microbiome profiles is gradually becoming promising. As microbiome sequencing becomes more affordable and widespread, it would be easier to identify disease- and therapy-specific microbial signatures. By leveraging these insights, microbiome therapeutics (e.g., probiotics, prebiotics, FMT) can be tailored to individuals, maximizing therapeutic efficacy while minimizing adverse effects. More research-driven throughput is essential to explain how different microbiome compositions interact with individual host factors, like genetics, diet, lifestyle, and immune-health. By integrating microbiome data with host genomics, metabolomics, and other omics data, scientists can develop more accurate predictive models for treatment outcomes. Personalized therapies could involve not only microbial supplementation (e.g., probiotics) but also dietary interventions that modify the microbiome to promote health.

7.2 *Newly Emerged Microbial Therapeutics*

Advances in synthetic biology hold great promise for engineering microbes or microbial consortia with specific therapeutic functions. For example, synthetic biology can be used to design microbes capable of producing therapeutic molecules (e.g., proteins, vitamins, or metabolic enzymes) within the gut. Researchers are also exploring engineered probiotics that can deliver drugs or modulate immune responses in a highly targeted manner. Technologies like CRISPR (Clustered Regularly Interspaced Short Palindromic Repeats)-Cas9 (CRISPR-associated protein 9) have opened the door to editing the genomes of microbes, allowing researchers to engineer microbial strains that can specifically target or treat disease (Careaga 2024). This could be particularly useful in conditions like IBD, where specific microbial interventions could be designed to modulate immune responses or reduce inflammation. Research is also focusing on identifying not only the specific bioactive metabolites produced by the microbiome but also the usage of more cutting-edge microbial components that could be harnessed for therapeutic use.

7.2.1 Phage Therapy

This refers to the use of bacteriophages to target and eliminate specific bacterial pathogens, which is gaining renewed interest as an alternative or adjunct to traditional antibiotic treatments, coping positively up with antibiotic resistance. Bacteriophages, viruses that specifically infect bacteria, offer a highly targeted approach, capable of lysing pathogenic bacteria while sparing beneficial microbiota, thus reducing the collateral damage associated with broad-spectrum antibiotics. Phage therapy can be particularly effective in treating multidrug-resistant organisms (MDROs), such as *Pseudomonas aeruginosa*, *Staphylococcus aureus*, and *Clostridium difficile*, for which treatment options are limited (Hibstu et al. 2022). Moreover, phages can be engineered or selected for their ability to target bacterial strains resistant to multiple antibiotics, offering a precision medicine approach to infection control. Currently, phage therapy has also been reported to treat both localized (e.g., wound infections) and systemic infections (e.g., sepsis) (Sawa et al. 2024). However, challenges remain, including the need for personalized phage cocktails to account for bacterial diversity, issues surrounding phage resistance, regulatory hurdles, and the optimization of delivery methods, especially for complex or deep-seated infections. Despite these challenges, ongoing research into phage biology, phage engineering, and regulatory frameworks holds promise for integrating phage therapy into clinical practice, offering a powerful tool in the fight against antibiotic-resistant infections (Lin et al. 2017).

7.2.2 Live Biotherapeutics

Live biotherapeutics, also known as live microbial products, are a class of therapeutics that utilize living microorganisms, such as bacteria, fungi, or yeast, to exert a therapeutic effect on the host. Unlike traditional probiotics, which typically aim to restore or maintain a balanced microbiome, live biotherapeutics are specifically designed, manufactured, and administered to treat or prevent diseases through a defined mechanism of action. These microorganisms can be engineered to produce bioactive metabolites, modulate the immune system, restore microbiome homeostasis, or inhibit pathogenic microbes (Tarantino et al. 2019). Live biotherapeutics have shown promising result in treating a range of conditions including gastrointestinal disorders like IBD, metabolic diseases like obesity and type 2 diabetes, autoimmune diseases, and some cancers. One of the key advantages of live biotherapeutics is their potential for precision medicine, as they can be personalized to target specific pathways, environmental stressors, or host–microbe interactions. A new study indicates treating ulcerative colitis (UC) using *Firmicutes* spores obtained from ethanol-washed human stool, termed *SER-287*. It is thought to be more superior to regular FMT having less risk of transfer of harmful bacteria into patients (Henn et al. 2021). However, the clinical development of live biotherapeutics faces challenges, such as ensuring product consistency, safety, and stability, particularly when these products are manufactured on a large scale. Additionally, the regulatory landscape for live biotherapeutics is still evolving, requiring rigorous testing to demonstrate safety, efficacy, and long-term outcomes (Cordaillat-Simmons et al. 2020). Despite these hurdles, live biotherapeutics represent an exciting and rapidly growing field, with probabilities to revolutionize treatments for different chronic and complex diseases.

7.2.3 Microbiome Mimetics

Microbiome mimetics are synthetic or engineered formulations designed to replicate or modulate the functions of the human microbiome, particularly its beneficial microbial communities, to restore or maintain health. These mimetics can be composed of a variety of components, including microbial-derived metabolites, engineered bacteria, or chemical compounds mimicking activities of specific microbiome-derived molecules. Unlike direct administration of traditional probiotics, microbiome mimetics mimic the beneficial actions of the microbiome through more controlled, standardized, and often non-living interventions. For example, microbiome mimetics can include SCFAs, which are known to support improved gut health and immunity (Nastasi et al. 2015). Similarly, synthetic peptides or small molecules that mimic microbial signaling pathways could be used to influence host metabolism, immune responses, or even brain–gut interactions. The advantage of microbiome mimetics is that they offer a more stable, scalable, and targeted alternative to live microbial therapies, as they are less susceptible to issues related to microbial viability, strain-specific variations, or the complexity of microbial-engineering processes. However, despite their potential, the field of microbiome mimetics is still naïve that requires further research to identify key molecular signatures and pathways in health and disease, optimizing delivery methods and establishment of safety and efficacy for human use (Bajaj et al. 2022; Shintani et al. 2023).

7.3 *Microbiome and Futuristic Drug Development Strategies*

One of the most critical areas for future research is understanding how the microbiome interacts with drugs, in terms of both drug metabolism and efficacy. Many drugs, including chemotherapeutics, antibiotics, and immunosuppressants, are metabolized by gut microbes, which can affect their pharmacokinetics, toxicity, and efficacy. Research in this area will help identify microbiome-mediated drug–drug interactions and optimize drug therapies by taking microbiome composition into account. In addition to using microbiome-derived therapies, future research will explore whether the microbiome itself can be targeted by drugs to treat diseases. For example, drugs could be developed that selectively modulate microbial composition or function by inducing the growth of beneficial microbes or restricting harmful ones (Manrique et al. 2024). This could be particularly relevant in conditions like IBD or obesity, where dysbiosis plays a key role.

Box 2 Microbiome and Space Medicine

Space exploration presents unique challenges to astronaut's health, primarily due to microgravity, high exposure of radiation, isolation, and prolonged confinement leading to both physiological imbalance as well as dysbiosis of the gut microbiota, a phenomenon often referred to as "spaceflight-associated dysbiosis," having significant adverse implications for astronaut health. Understanding the role of microbiome and microbiome-derived therapeutics in maintaining health aspects of space travelers, is a promising area of research. Studies suggest that specific microbiome is necessary to maintain astronaut health while in space. Using in vivo and in vitro models for microgravity research, it would be worth to do exhaustive study in order to retain proper ratios of beneficial bacterial species in the gut of the space traveler. Metabolomics and proteomics investigations of microbiota (e.g., *Serratia marcescens*, *Escherichia coli*, *Klebsiella* sp., *Bacillus* sp., etc.) found in the gut of cockroach and tardigrades might indicate toward their high-level resistance against radiation, which could be an orchestrated mode of study of human astronaut's gut flora for better protection against space radiation. A recent multi-omics study by Tierney et al. (2024) suggests a high shift of plaque-associated bacteria (like *Fusobacteriota*) with space flight condition, which further correlates with immune cell gene expression. This study indicates link between microbiome and shifts of immune responses during spaceflight that can be pivotal in designing future space missions, spacecraft design, and space habitat planning with better sustainable health condition of astronauts (Siddiqui et al. 2023; Tierney et al. 2024).

8 Conclusion

Microbiome-derived therapeutics represent a promising frontier in immunomodulation, leveraging the intricate relationship between the microbial communities and host's immune system, enhancing immune function, restoring immune balance, and treating immune-pathologies. The gut microbiome, in particular, plays a central role in shaping immune responses, influencing both systemic and mucosal immunity modulating regulation of T cell differentiation, cytokine production, and immune tolerance. With potential ability to treat various life-threatening diseases, these microbiome-derived therapeutics also boost immune responses in immunocompromised individuals and prevent infections. However, while the immunomodulatory effects of microbiome-based therapies are promising, further exhaustive research is necessary for complete understanding of microbiome-influenced immune functions to cater optimal therapeutic regimens for long-term safety and efficacy. As the field advances, microbiome-derived therapeutics could offer personalized, non-invasive alternatives or adjuncts to conventional treatments, revolutionizing the management of immune-related diseases and reshaping the landscape of immunotherapy.

References

Afzaal M, Saeed F, Shah YA, Hussain M, Rabail R, Socol CT, Hassoun A, Pateiro M, Lorenzo JM, Rusu AV, Aadil RM (2022) Human gut microbiota in health and disease: unveiling the relationship. Front Microbiol 13:999001

Allegretti JR, Khanna S, Mullish BH, Feuerstadt P (2024) The progression of microbiome therapeutics for the management of gastrointestinal diseases and beyond. Gastroenterology 167:885

Allouche R, Hafeez Z, Dary-Mourot A, Genay M, Miclo L (2024) Streptococcus thermophilus: a source of postbiotics displaying anti-inflammatory effects in THP 1 macrophages. Molecules 29(7):1552

Angelin J, Kavitha M (2020) Exopolysaccharides from probiotic bacteria and their health potential. Int J Biol Macromol 162:853–865

Ansari F, Alian Samakkhah S, Bahadori A, Jafari SM, Ziaee M, Khodayari MT, Pourjafar H (2023) Health-promoting properties of Saccharomyces cerevisiae var. boulardii as a probiotic; characteristics, isolation, and applications in dairy products. Crit Rev Food Sci Nutr 63(4):457–485

Aroniadis OC, Brandt LJ (2013) Fecal microbiota transplantation: past, present and future. Curr Opin Gastroenterol 29(1):79–84

Arora N, Sadovsky Y, Dermody TS, Coyne CB (2017) Microbial vertical transmission during human pregnancy. Cell Host Microbe 21(5):561–567

Badal VD, Wright D, Katsis Y, Kim HC, Swafford AD, Knight R, Hsu CN (2019) Challenges in the construction of knowledge bases for human microbiome-disease associations. Microbiome 7:1–15

Bajaj JS, Ng SC, Schnabl B (2022) Promises of microbiome-based therapies. J Hepatol 76(6):1379–1391

Behrouzi A, Mazaheri H, Falsafi S, Tavassol ZH, Moshiri A, Siadat SD (2020) Intestinal effect of the probiotic Escherichia coli strain Nissle 1917 and its OMV. J Diabetes Metab Disord 19:597–604

Bodkhe R, Balakrishnan B, Taneja V (2019) The role of microbiome in rheumatoid arthritis treatment. Ther Adv Musculoskelet Dis 11:1759720X19844632

Bogaert D, Van Beveren GJ, de Koff EM, Parga PL, Lopez CEB, Koppensteiner L et al. (2023) Mother-to-infant microbiota transmission and infant microbiota development across multiple body sites. Cell Host Microbe 31(3):447–460
Boicean A, Birlutiu V, Ichim C, Anderco P, Birsan S (2023) Fecal microbiota transplantation in inflammatory bowel disease. Biomedicine 11(4):1016
Boix-Amorós A, Martinez-Costa C, Querol A, Collado MC, Mira A (2017) Multiple approaches detect the presence of fungi in human breastmilk samples from healthy mothers. Sci Rep 7(1):13016
Bottaccioli AG, Lazzari D, Bottaccioli F (2021) Promoting the resilience of the Italian population against SARS-CoV-2. Front Psych 11:560017
Breton J, Galmiche M, Déchelotte P (2022) Dysbiotic gut bacteria in obesity: an overview of the metabolic mechanisms and therapeutic perspectives of next-generation probiotics. Microorganisms 10(2):452
Brockmann L, Tran A, Huang Y, Edwards M, Ronda C, Wang HH, Ivanov II (2023) Intestinal microbiota-specific Th17 cells possess regulatory properties and suppress effector T cells via c-MAF and IL-10. Immunity 56(12):2719–2735
Browne HP, Shao Y, Lawley TD (2022) Mother–infant transmission of human microbiota. Curr Opin Microbiol 69:102173
Burrello C, Garavaglia F, Cribiù FM, Ercoli G, Lopez G, Troisi J et al. (2018) Therapeutic faecal microbiota transplantation controls intestinal inflammation through IL10 secretion by immune cells. Nat Commun 9(1):5184
Canivenc-Lavier MC, Bennetau-Pelissero C (2023) Phytoestrogens and health effects. Nutrients 15(2):317
Carding S, Verbeke K, Vipond DT, Corfe BM, Owen LJ (2015) Dysbiosis of the gut microbiota in disease. Microb Ecol Health Dis 26(1):26191
Careaga MBL (2024) Engineering the microbiome: CRISPR leads the way. Scientist 38(1):30–38
Carlson PE (2020) Regulatory considerations for fecal microbiota transplantation products. Cell Host Microbe 27(2):173–175
Carlsson F, Råberg L (2024) The germ theory revisited: a noncentric view on infection outcome. Proc Natl Acad Sci 121(17):e2319605121
Chen JS, Faller DV, Spanjaard RA (2003) Short-chain fatty acid inhibitors of histone deacetylases: promising anticancer therapeutics? Curr Cancer Drug Targets 3(3):219–236
Cheng L, Qi C, Zhuang H, Fu T, Zhang X (2020) gutMDisorder: a comprehensive database for dysbiosis of the gut microbiota in disorders and interventions. Nucleic Acids Res 48(D1):D554–D560
Cheruvari A, Kammara R (2024) Bacteriocins future perspectives: substitutes to antibiotics. Food Control 110834
Collins SL, Stine JG, Bisanz JE, Okafor CD, Patterson AD (2023) Bile acids and the gut microbiota: metabolic interactions and impacts on disease. Nat Rev Microbiol 21(4):236–247
Cordaillat-Simmons M, Rouanet A, Pot B (2020) Live biotherapeutic products: the importance of a defined regulatory framework. Exp Mol Med 52(9):1397–1406
Creasy HH, Felix V, Aluvathingal J, Crabtree J, Ifeonu O, Matsumura J et al. (2021) HMPDACC: a Human Microbiome Project Multi-omic data resource. Nucleic Acids Res 49(D1):D734–D742
Cristofori F, Dargenio VN, Dargenio C, Miniello VL, Barone M, Francavilla R (2021) Anti-inflammatory and immunomodulatory effects of probiotics in gut inflammation: a door to the body. Front Immunol 12:578386
Crudele L, Gadaleta RM, Cariello M, Moschetta A (2023) Gut microbiota in the pathogenesis and therapeutic approaches of diabetes. EBioMedicine 97:104821
Cuamatzin-Garcia L, Rodriguez-Rugarcia P, El-Kassis EG, Galicia G, Meza-Jiménez ML, Baños-Lara MDR et al. (2022) Traditional fermented foods and beverages from around the world and their health benefits. Microorganisms 10(6):1151
Da M, Sun J, Ma C, Li D, Dong L, Wang LS, Chen F (2024) Postbiotics: enhancing human health with a novel concept. eFood 5(5):e180

Darby TM, Owens JA, Saeedi BJ, Luo L, Matthews JD, Robinson BS et al. (2019) Lactococcus lactis subsp. cremoris is an efficacious beneficial bacterium that limits tissue injury in the intestine. Iscience 12:356–367

Davani-Davari D, Negahdaripour M, Karimzadeh I, Seifan M, Mohkam M, Masoumi SJ et al. (2019) Prebiotics: definition, types, sources, mechanisms, and clinical applications. Foods 8(3):92

DeGruttola AK, Low D, Mizoguchi A, Mizoguchi E (2016) Current understanding of dysbiosis in disease in human and animal models. Inflamm Bowel Dis 22(5):1137–1150

Deleu S, Machiels K, Raes J, Verbeke K, Vermeire S (2021) Short chain fatty acids and its producing organisms: an overlooked therapy for IBD? EBioMedicine 66:103293

Dias R, Pereira CB, Pérez-Gregorio R, Mateus N, Freitas V (2021) Recent advances on dietary polyphenol's potential roles in celiac disease. Trends Food Sci Technol 107:213–225

Duan H, Wang L, Huangfu M, Li H (2023) The impact of microbiota-derived short-chain fatty acids on macrophage activities in disease: mechanisms and therapeutic potentials. Biomed Pharmacother 165:115276

Dunn AB, Jordan S, Baker BJ, Carlson NS (2017) The maternal infant microbiome: considerations for labor and birth. MCN: Am J Matern Child Nurs 42(6):318–325

Dutta SK, Verma S, Jain V, Surapaneni BK, Vinayek R, Phillips L, Nair PP (2019) Parkinson's disease: the emerging role of gut dysbiosis, antibiotics, probiotics, and fecal microbiota transplantation. J Neurogastroentero Motil 25(3):363

Facchin S, Bertin L, Bonazzi E, Lorenzon G, De Barba C, Barberio B et al. (2024) Short-chain fatty acids and human health: from metabolic pathways to current therapeutic implications. Life 14(5):559

Farhat EK, Sher EK, Džidić-Krivić A, Banjari I, Sher F (2023) Functional biotransformation of phytoestrogens by gut microbiota with impact on cancer treatment. J Nutr Biochem 118:109368

Fiebiger U, Bereswill S, Heimesaat MM (2016) Dissecting the interplay between intestinal microbiota and host immunity in health and disease: lessons learned from germfree and gnotobiotic animal models. Eur J Microbiol Immunol 6(4):253–271

Francese R, Peila C, Donalisio M, Lamberti C, Cirrincione S, Colombi N et al. (2023) Viruses and human milk: transmission or protection? Adv Nutr 14

Garvey M (2024) Intestinal dysbiosis: microbial imbalance impacts on colorectal cancer initiation, progression and disease mitigation. Biomedicines 12(4):740

Gilbert JA, Blaser MJ, Caporaso JG, Jansson JK, Lynch SV, Knight R (2018) Current understanding of the human microbiome. Nat Med 24(4):392–400

Glover JS, Ticer TD, Engevik MA (2022) Characterizing the mucin-degrading capacity of the human gut microbiota. Sci Rep 12(1):8456

González-Brito A, Uribe-Herranz M (2023) The potential role of short chain fatty acids improving ex vivo T and CAR-T cell fitness and expansion for cancer immunotherapies. Front Immunol 14:1083303

Graham DB, Xavier RJ (2023) Conditioning of the immune system by the microbiome. Trends Immunol 44(7):499–511

Gu X, Song LJ, Li LX, Liu T, Zhang MM, Li Z et al. (2020) Fusobacterium nucleatum causes microbial dysbiosis and exacerbates visceral hypersensitivity in a colonization-independent manner. Front Microbiol 11:1281

Gulati M, Singh SK, Corrie L, Kaur IP, Chandwani L (2020) Delivery routes for faecal microbiota transplants: available, anticipated and aspired. Pharmacol Res 159:104954

Gulliver EL, Young RB, Chonwerawong M, D'Adamo GL, Thomason T, Widdop JT, Rutten EL, Rossetto Marcelino V, Bryant RV, Costello SP, O'Brien CL (2022) The future of microbiome-based therapeutics. Aliment Pharmacol Ther 56(2):192–208

Gupta SK, Vyavahare S, Blanes ILD, Berger F, Isales C, Fulzele S (2023) Microbiota-derived tryptophan metabolism: impacts on health, aging, and disease. Exp Gerontol 183:112319

Harbison JE, Roth-Schulze AJ, Giles LC, Tran CD, Ngui KM, Penno MA et al. (2019) Gut microbiome dysbiosis and increased intestinal permeability in children with islet autoimmunity and type 1 diabetes: a prospective cohort study. Pediatr Diabetes 20(5):574–583

He J, Gong X, Hu B, Lin L, Lin X, Gong W et al. (2023) Altered gut microbiota and short-chain fatty acids in Chinese children with constipated autism Spectrum disorder. Sci Rep 13(1):19103

Henn MR, O'Brien EJ, Diao L, Feagan BG, Sandborn WJ, Huttenhower C et al. (2021) A phase 1b safety study of SER-287, a spore-based microbiome therapeutic, for active mild to moderate ulcerative colitis. Gastroenterology 160(1):115–127

Hibstu Z, Belew H, Akelew Y, Mengist HM (2022) Phage therapy: a different approach to fight bacterial infections. Biologics 16:173–186

Hill JH, Round JL (2021) SnapShot: microbiota effects on host physiology. Cell 184(10):2796–2796

Horta-Baas G, Romero-Figueroa MDS, Montiel-Jarquín AJ, Pizano-Zárate ML, García-Mena J, Ramírez-Durán N (2017) Intestinal dysbiosis and rheumatoid arthritis: a link between gut microbiota and the pathogenesis of rheumatoid arthritis. J Immunol Res 2017(1):4835189

Hourigan SK, Nicholson MR, Kahn SA, Kellermayer R (2021) Updates and challenges in fecal microbiota transplantation for Clostridioides difficile infection in children. J Pediatr Gastroenterol Nutr 73(4):430–432

Hrncir T (2022) Gut microbiota dysbiosis: triggers, consequences, diagnostic and therapeutic options. Microorganisms 10(3):578

Hurley WL, Theil PK (2011) Perspectives on immunoglobulins in colostrum and milk. Nutrients 3(4):442–474

Jain N (2020) The early life education of the immune system: moms, microbes and (missed) opportunities. Gut Microbes 12(1):1824564

Jain T, Sharma P, Are AC, Vickers SM, Dudeja V (2021) New insights into the cancer–microbiome–immune axis: decrypting a decade of discoveries. Front Immunol 12:622064

Johnson KVA (2020) Gut microbiome composition and diversity are related to human personality traits. Hum Microb J 15:100069

Jordan CK, Clarke TB (2024) How does the microbiota control systemic innate immunity? Trends Immunol 45(2):94–102

Kim CH (2023) Complex regulatory effects of gut microbial short-chain fatty acids on immune tolerance and autoimmunity. Cell Mol Immunol 20(4):341–350

Koh LF, Ong RY, Common JE (2022) Skin microbiome of atopic dermatitis. Allergol Int 71(1):31–39

Landemaine L, Da Costa G, Fissier E, Francis C, Morand S, Verbeke J et al. (2023) Staphylococcus epidermidis isolates from atopic or healthy skin have opposite effect on skin cells: potential implication of the AHR pathway modulation. Front Immunol 14:1098160

Lee J, d'Aigle J, Atadja L, Quaicoe V, Honarpisheh P, Ganesh BP et al. (2020) Gut microbiota–derived short-chain fatty acids promote poststroke recovery in aged mice. Circ Res 127(4):453–465

Lee S, Choi SP, Choi HJ, Jeong H, Park YS (2024) A comprehensive review of synbiotics: an emerging paradigm in health promotion and disease management. World J Microbiol Biotechnol 40(9):280

Leeuwendaal NK, Stanton C, O'toole PW, Beresford TP (2022) Fermented foods, health and the gut microbiome. Nutrients 14(7):1527

Li S (2023) Modulation of immunity by tryptophan microbial metabolites. Front Nutr 10:1209613

Lin DM, Koskella B, Lin HC (2017) Phage therapy: an alternative to antibiotics in the age of multi-drug resistance. World J Gastrointest Pharmacol Ther 8(3):162

Liu S, Gao J, Zhu M, Liu K, Zhang HL (2020) Gut microbiota and dysbiosis in Alzheimer's disease: implications for pathogenesis and treatment. Mol Neurobiol 57:5026–5043

Liu Y, Tran DQ, Lindsey JW, Rhoads JM (2021) The association of gut microbiota and Treg dysfunction in autoimmune diseases. Adv Exp Med Biol 1278:191–203

Loh JS, Mak WQ, Tan LKS, Ng CX, Chan HH, Yeow SH et al. (2024) Microbiota–gut–brain axis and its therapeutic applications in neurodegenerative diseases. Signal Transduct Target Ther 9(1):37

Luu M, Riester Z, Baldrich A, Reichardt N, Yuille S, Busetti A et al. (2021) Microbial short-chain fatty acids modulate CD8+ T cell responses and improve adoptive immunotherapy for cancer. Nat Commun 12(1):4077

Lyu M, Suzuki H, Kang L, Gaspal F, Zhou W, Goc J et al. (2022) ILC3s select microbiota-specific regulatory T cells to establish tolerance in the gut. Nature 610(7933):744–751

Manrique P, Montero I, Fernandez-Gosende M, Martinez N, Cantabrana CH, Rios-Covian D (2024) Past, present, and future of microbiome-based therapies. Microbiome Res Rep 3(2):23

Mariño E, Richards JL, McLeod KH, Stanley D, Yap YA, Knight J et al. (2017) Gut microbial metabolites limit the frequency of autoimmune T cells and protect against type 1 diabetes. Nat Immunol 18(5):552–562

Metselaar S, Widdershoven G (2017) Ethical issues in fecal microbiota transplantation: taking into account identity and family relations. Am J Bioeth 17(5):53–55

Mohr AE, Crawford MS, Jasbi P, Fessler S, Sweazea KL (2022) Lipopolysaccharide and the gut microbiota: considering structural variation. FEBS Lett 596(7):849–875

Mousa WK, Chehadeh F, Husband S (2022) Recent advances in understanding the structure and function of the human microbiome. Front Microbiol 13:825338

Mudaliar SB, Poojary SS, Bharath Prasad AS, Mazumder N (2024) Probiotics and paraprobiotics: effects on microbiota-gut-brain Axis and their consequent potential in neuropsychiatric therapy. Probiotics Antimicrob Proteins 16:1–25

Nakatsuji T, Cheng JY, Gallo RL (2021) Mechanisms for control of skin immune function by the microbiome. Curr Opin Immunol 72:324–330

Nastasi C, Candela M, Bonefeld CM, Geisler C, Hansen M, Krejsgaard T et al. (2015) The effect of short-chain fatty acids on human monocyte-derived dendritic cells. Sci Rep 5(1):16148

Ng RW, Dharmaratne P, Wong S, Hawkey P, Chan P, Ip M (2024) Revisiting the donor screening protocol of Faecal microbiota transplantation (FMT): a systematic review. Gut 73(6):1029–1031

Nicco C, Paule A, Konturek P, Edeas M (2020) From donor to patient: collection, preparation and cryopreservation of fecal samples for fecal microbiota transplantation. Diseases 8(2):9

Okada H, Kuhn C, Feillet H, Bach JF (2010) The 'hygiene hypothesis' for autoimmune and allergic diseases: an update. Clin Exp Immunol 160(1):1–9

Park SY, Seo GS (2021) Fecal microbiota transplantation: is it safe? Clin Endosc 54(2):157–160

Perry T, Jovel J, Patterson J, Wong G, Fedorak RN, Thiesen A et al. (2015) Fecal microbial transplant after ileocolic resection reduces ileitis but restores colitis in IL-10−/− mice. Inflamm Bowel Dis 21(7):1479–1490

Peterson SR, Jensen S, Mangalam AK (2023) Gut microbiome-derived isoflavone metabolites may ameliorate experimental autoimmune encephalomyelitis (EAE) through estrogen receptor signaling in CD8 T-cells. J Immunol 210(1_Supplement):77–14

Pi X, Teng W, Fei D, Zhao G, Liu W (2022) Effects of live combined Bacillus subtilis and enterococcus faecium on gut microbiota composition in C57BL/6 mice and in humans. Front Cell Infect Microbiol 12:821662

Piccioni A, Cicchinelli S, Valletta F, De Luca G, Longhitano Y, Candelli M, Ojetti V, Sardeo F, Navarra S, Covino M, Franceschi F (2022) Gut microbiota and autoimmune diseases: a charming real world together with probiotics. Curr Med Chem 29(18):3147–3159

Piccioni A, Spagnuolo F, Candelli M, Voza A, Covino M, Gasbarrini A, Franceschi F (2024) The gut microbiome in sepsis: from dysbiosis to personalized therapy. J Clin Med 13(20):6082

Podolsky SH (2012) Metchnikoff and the microbiome. Lancet 380(9856):1810–1811

Portincasa P, Bonfrate L, Vacca M, De Angelis M, Farella I, Lanza E et al. (2022) Gut microbiota and short chain fatty acids: implications in glucose homeostasis. Int J Mol Sci 23(3):1105

Prete R, Long SL, Gallardo AL, Gahan CG, Corsetti A, Joyce SA (2020) Beneficial bile acid metabolism from Lactobacillus plantarum of food origin. Sci Rep 10(1):1165

Puljiz Z, Kumric M, Vrdoljak J, Martinovic D, Ticinovic Kurir T, Krnic MO et al. (2023) Obesity, gut microbiota, and metabolome: from pathophysiology to nutritional interventions. Nutrients 15(10):2236

Rappuoli R, Young P, Ron E, Pecetta S, Pizza M (2023) Save the microbes to save the planet. A call to action of the International Union of the Microbiological Societies (IUMS). One Health Outlook 5(1):5

Ren D, Wang D, Liu H, Shen M, Yu H (2019) Two strains of probiotic Lactobacillus enhance immune response and promote naive T cell polarization to Th1. Food Agric Immunol 30(1):281–295
Renga G, Nunzi E, Pariano M, Puccetti M, Bellet MM, Pieraccini G et al. (2022) Optimizing therapeutic outcomes of immune checkpoint blockade by a microbial tryptophan metabolite. J Immunother Cancer 10(3):e003725
Rodríguez-Daza MC, Pulido-Mateos EC, Lupien-Meilleur J, Guyonnet D, Desjardins Y, Roy D (2021) Polyphenol-mediated gut microbiota modulation: toward prebiotics and further. Front Nutr 8:689456
Sawa T, Moriyama K, Kinoshita M (2024) Current status of bacteriophage therapy for severe bacterial infections. J Intensive Care 12(1):44
Scott E, De Paepe K, Van de Wiele T (2022) Postbiotics and their health modulatory biomolecules. Biomolecules 12(11):1640
Serghiou IR, Webber MA, Hall LJ (2023) An update on the current understanding of the infant skin microbiome and research challenges. Curr Opin Microbiol 75:102364
Shan Y, Lee M, Chang EB (2022) The gut microbiome and inflammatory bowel diseases. Annu Rev Med 73(1):455–468
Shang W, Zhang S, Qian H, Huang S, Li H, Liu J, Chen D (2024) Gut microbiota and sepsis and sepsis-related death: a Mendelian randomization investigation. Front Immunol 15:1266230
Shao T, Hsu R, Hacein-Bey C, Zhang W, Gao L, Kurth MJ et al. (2023) The evolving landscape of fecal microbial transplantation. Clin Rev Allergy Immunol 65(2):101–120
Sharon I, Quijada NM, Pasolli E, Fabbrini M, Vitali F, Agamennone V, Dötsch A, Selberherr E, Grau JH, Meixner M, Liere K (2022) The core human microbiome: does it exist and how can we find it? A critical review of the concept. Nutrients 14(14):2872
Sheflin AM, Whitney AK, Weir TL (2014) Cancer-promoting effects of microbial dysbiosis. Curr Oncol Rep 16:1–9
Shintani T, Shintani H, Sato M, Ashida H (2023) Calorie restriction mimetic drugs could favorably influence gut microbiota leading to lifespan extension. GeroScience 45(6):3475–3490
Shvets Y, Khranovska N, Senchylo N, Ostapchenko D, Tymoshenko I, Onysenko S et al. (2024) Microbiota substances modulate dendritic cells activity: a critical view. Heliyon 10(5):e27125
Siciliano RA, Reale A, Mazzeo MF, Morandi S, Silvetti T, Brasca M (2021) Paraprobiotics: a new perspective for functional foods and nutraceuticals. Nutrients 13(4):1225
Siddiqui R, Elmoselhi AB, Khan NA (2023) Space medicine: gut microbiome of hardy species is a potential source to counter disorders during space travel. Future Sci OA 9(7):FSO868
Singh VP, Proctor SD, Willing BP (2016) Koch's postulates, microbial dysbiosis and inflammatory bowel disease. Clin Microbiol Infect 22(7):594–599
Smith PM, Howitt MR, Panikov N, Michaud M, Gallini CA, Bohlooly-y M et al. (2013) The microbial metabolites, short-chain fatty acids, regulate colonic Treg cell homeostasis. Science 341(6145):569–573
Soveral LF, Korczaguin GG, Schmidt PS, Nunes IS, Fernandes C, Zárate-Bladés CR (2022) Immunological mechanisms of fecal microbiota transplantation in recurrent Clostridioides difficile infection. World J Gastroenterol 28(33):4762
Stojanov S, Berlec A, Štrukelj B (2020) The influence of probiotics on the firmicutes/bacteroidetes ratio in the treatment of obesity and inflammatory bowel disease. Microorganisms 8(11):1715
Strathdee SA, Hatfull GF, Mutalik VK, Schooley RT (2023) Phage therapy: from biological mechanisms to future directions. Cell 186(1):17–31
Stripling J, Rodriguez M (2018) Current evidence in delivery and therapeutic uses of fecal microbiota transplantation in human diseases—Clostridium difficile disease and beyond. Am J Med Sci 356(5):424–432
Sun M, Wu W, Liu Z, Cong Y (2017) Microbiota metabolite short chain fatty acids, GPCR, and inflammatory bowel diseases. J Gastroenterol 52:1–8
Sun S, Luo L, Liang W, Yin Q, Guo J, Rush AM et al. (2020) Bifidobacterium alters the gut microbiota and modulates the functional metabolism of T regulatory cells in the context of immune checkpoint blockade. Proc Natl Acad Sci 117(44):27509–27515

Tarantino V, Savaia V, D'Agostino R, Silvestri M, Passali F, Di Girolamo S, Ciprandi G (2019) Bacteriotherapy in children with recurrent upper respiratory tract infections. Eur Rev Med Pharmacol Sci 23:39–43

Teame T, Wang A, Xie M, Zhang Z, Yang Y, Ding Q et al. (2020) Paraprobiotics and postbiotics of probiotic Lactobacilli, their positive effects on the host and action mechanisms: a review. Front Nutr 7:570344

Thanush D, Basavaraj HC, Gowrav MP (2023) Current regulation and initial considerations for successful development and commercialization of microbiome therapies. Adv Gut Microbiome Res 2023(1):6657515

Thapa BR (2005) Health factors in colostrum. Indian J Pediatr 72:579–581

Thorakkattu P, Khanashyam AC, Shah K, Babu KS, Mundanat AS, Deliephan A et al. (2022) Postbiotics: current trends in food and pharmaceutical industry. Foods 11(19):3094

Tian H, Wang X, Fang Z, Li L, Wu C, Bi D et al. (2024) Fecal microbiota transplantation in clinical practice: Present controversies and future prospects. hLife 2(6):269–228

Tierney BT, Kim J, Overbey EG, Ryon KA, Foox J, Sierra MA et al. (2024) Longitudinal multiomics analysis of host microbiome architecture and immune responses during short-term spaceflight. Nature. Microbiology 9:1–15

Turnbaugh PJ, Ley RE, Hamady M, Fraser-Liggett CM, Knight R, Gordon JI (2007) The human microbiome project. Nature 449(7164):804–810

Ullah H, Arbab S, Tian Y, Liu CQ, Chen Y, Qijie L et al. (2023) The gut microbiota–brain axis in neurological disorder. Front Neurosci 17:1225875

van Lingen E, Baunwall S, Lieberknecht S, Benech N, Ianiro G, Sokol H et al. (2023) Short- and long-term follow-up after fecal microbiota transplantation as treatment for recurrent Clostridioides difficile infection in patients with inflammatory bowel disease. Therap Adv Gastroenterol 16:17562848231156285

Venegas DP, Marjorie K, Landskron G, González MJ, Quera R, Dijkstra G, Harmsen HJ, Faber KN, Hermoso MA (2019) Corrigendum: short chain fatty acids (SCFAs)-mediated gut epithelial and immune regulation and its relevance for inflammatory bowel diseases. Front Immunol 10

Vetrani C, Di Nisio A, Paschou SA, Barrea L, Muscogiuri G, Graziadio C et al. (2022) From gut microbiota through low-grade inflammation to obesity: key players and potential targets. Nutrients 14(10):2103

Virk MS, Virk MA, He Y, Tufail T, Gul M, Qayum A et al. (2024) The anti-inflammatory and curative exponent of probiotics: a comprehensive and authentic ingredient for the sustained functioning of major human organs. Nutrients 16(4):546

Voth E, Khanna S (2020) The Integrative Human microbiome project: a mile stone in the understanding of the gut microbiome. Expert Rev Gastroenterol Hepatol 14(8):639–642

Wang B, Yao M, Lv L, Ling Z, Li L (2017) The human microbiota in health and disease. Engineering 3(1):71–82

Wang F, Yin Q, Chen L, Davis MM (2018) Bifidobacterium can mitigate intestinal immunopathology in the context of CTLA-4 blockade. Proc Natl Acad Sci 115(1):157–161

Wynn AB, Beyer G, Richards M, Ennis LA (2023) Procedure, screening, and cost of fecal microbiota transplantation. Cureus 15(2)

Xiao L, Zhao F (2023) Microbial transmission, colonisation and succession: from pregnancy to infancy. Gut 72(4):772–786

Yadav M, Chauhan NS (2022) Microbiome therapeutics: exploring the present scenario and challenges. Gastroenterol Rep 10:goab046

Yan F, Polk DB (2020) Probiotics and probiotic-derived functional factors—mechanistic insights into applications for intestinal homeostasis. Front Immunol 11:1428

Yang W, Yu T, Huang X, Bilotta AJ, Xu L, Lu Y et al. (2020) Intestinal microbiota-derived short-chain fatty acids regulation of immune cell IL-22 production and gut immunity. Nat Commun 11(1):4457

Yang D, Wang Z, Chen Y, Guo Q, Dong Y (2023) Interactions between gut microbes and NLRP3 inflammasome in the gut-brain axis. Comput Struct Biotechnol J 21:2215–2227

Yang R, Chen Z, Cai J (2023) Fecal microbiota transplantation: emerging applications in autoimmune diseases. J Autoimmun 141:103038

Young VB (2021) Unexpected results from a phase 2 trial of a microbiome therapeutic for Clostridioides difficile infection: lessons for the future. Clin Infect Dis 72(12):2141–2143

Yuan X, Wang R, Han B, Sun C, Chen R, Wei H et al. (2022) Functional and metabolic alterations of gut microbiota in children with new-onset type 1 diabetes. Nat Commun 13(1):6356

Zheng L, Kelly CJ, Battista KD, Schaefer R, Lanis JM, Alexeev EE et al. (2017) Microbial-derived butyrate promotes epithelial barrier function through IL-10 receptor–dependent repression of claudin-2. J Immunol 199(8):2976–2984

Zheng P, Li Z, Zhou Z (2018) Gut microbiome in type 1 diabetes: a comprehensive review. Diabetes Metab Res Rev 34(7):e3043

Alzheimer's Disease and Gut Microbiota

Shashwat Pandey and Preeti Yadav

Abstract Alzheimer's disease (AD) is a progressive neurodegenerative disorder that manifests as cognitive decline and memory impairment, significantly affecting the elderly population. Despite extensive research, the exact pathogenesis of AD remains elusive. Emerging evidence has highlighted the role of the gut–brain axis, suggesting a potential link between gut microbiota and AD. The gut microbiota, a complex community of microorganisms residing in the gastrointestinal tract, is known to influence various physiological processes, including immune modulation, metabolic functions, and the maintenance of the blood–brain barrier integrity. Dysbiosis, an imbalance in the gut microbiota composition, has been associated with systemic inflammation and metabolic disturbances, both of which are critical factors in AD development. Altered gut microbiota can influence neuroinflammation, amyloid-beta plaque formation, and tau protein pathology, which are hallmarks of AD. Several studies have demonstrated that individuals with AD exhibit distinct gut microbiota profiles compared to healthy individuals, indicating a possible contributory role of microbial dysbiosis in AD pathogenesis.

This abstract reviews current findings on the interplay between gut microbiota and AD, focusing on how microbial metabolites, such as short-chain fatty acids and lipopolysaccharides, impact neuro.

Keywords Alzheimer's disease · Dysbiosis · Tau hyperphosphorylation · Gut microbiota · Neurofibrillary tangles

S. Pandey (✉)
Department of Pharmaceutical Sciences, Babasaheb Bhimrao Ambedkar University, Lucknow, Uttar Pradesh, India

Institute of Pharmaceutical Sciences, University of Lucknow, (LUIPS), Lucknow, India
e-mail: Shashwatpandey0510@gmail.com

P. Yadav
Department of Pharmaceutical Sciences, Babasaheb Bhimrao Ambedkar University, Lucknow, Uttar Pradesh, India

S. K. Singh et al. (eds.), *Microbiota Profiling for Precision Medicine*, Interdisciplinary Biotechnological Advances,
https://doi.org/10.1007/978-981-95-2104-3_5

1 Introduction

An essential function of the gut microbiota is to regulate the bidirectional signaling that underlies the axis between gut and brain (Bistoletti et al. 2020). Modifications in the makeup of the gut microbiome as well as dysbiosis are being demonstrated to have a role in the pathogenesis of several illnesses in humans, including allergies, type 2 diabetes, being overweight, *i*rritable *b*owel *s*yndrome (IBS), metabolic syndrome, obesity, colorectal cancer including Alzheimer's disease (AD) (West et al. 2015). Development of protein-based misfolded amyloid-beta (Aβ) fibrils along with oligomers, combined with tangled neurofibrillary fibers composed of hyperphosphorylated tau protein, in the central nervous system (CNS) along with additional brain areas characterizes Alzheimer's disease, an irreversible neurodegenerative illness (Grüning 2014). Diabetes type 2 with Alzheimer's disease have been proposed to be intricately linked. Specifically, changes in the gut microbiota can stimulate proinflammatory cytokines and enhance intestinal permeability, resulting in the generation of insulin resistance, a condition that has also been linked to Alzheimer's disease (AD) (Sochocka et al. 2019). Moreover, it is well-documented that bacteria that inhabit the gut microbiome release immunogenic combinations of amyloid particles, lipopolysaccharides (LPSs), and additional metabolic exudates into their immediate surroundings (Eckel and Ametaj 2020). Microbial amyloids may stimulate signaling pathways implicated in neurotoxicity and Alzheimer's disease development, whereas the gut microbiome may augment inflammatory reactions to the buildup of the amyloid beta in the brain (Van Giau et al. 2018).

It was proposed that the makeup of the gut's microbiota as well as the synthesis or accumulation of amyloid-containing proteins could be influenced by both nutrients and foods that are consumed (Malik and Yadav 2023). According to this, modifying the gut microbiota through certain dietary treatments and the consumption of probiotics as well as prebiotics may be a useful tactic to lower the amount of Aβ in persistent inflammation linked to AD, thereby avoiding or lessening AD symptoms (Scott et al. 2015).

Alzheimer's disease (AD) is a progressive neurodegenerative disorder, most common cause of dementia among older adults, and its prevalence continues to rise with the aging global population. Despite significant research, the exact mechanisms that lead to Alzheimer's disease remain elusive, although factors like amyloid-beta (Aβ) plaques, neurofibrillary tangles of tau protein, oxidative stress, and neuroinflammation are known to play key roles in its pathophysiology (Matej et al. 2019). In recent years, there has been growing interest in the role of the gut microbiota—the diverse community of microorganisms residing in the gastrointestinal tract—in neurodegenerative diseases like Alzheimer's disease. The gut–brain axis, a bidirectional communication system involving the central nervous system (CNS) and the enteric nervous system (ENS), suggests that the gut microbiota may influence brain function and, potentially, the development of neurodegenerative diseases. Evidence shows that imbalance in the gut microbiota can lead to increased intestinal permeability, inflammation, and the release of microbial metabolites that may affect brain health. These microbial alterations may contribute to the accumulation of amyloid plaques, and other hallmark features of Alzheimer's disease. Furthermore, the gut microbiota influences immune system regulation, which may impact the inflammatory processes observed in Alzheimer's pathology. Alzheimer's disease (AD) involves the gradual deterioration and death of neurons in the brain and nervous system. This loss of neurons leads to cognitive, motor, and behavioral impairments. In AD, misfolded amyloid-beta (Aβ) proteins aggregate to form plaques outside neurons, while tau proteins accumulate inside neurons, forming neurofibrillary tangles (Bancher et al. 1989). These aggregates disrupt neuronal communication, triggering a cascade of neurotoxicity. Neurons have high energy demands, making them particularly sensitive to mitochondrial dysfunction. When mitochondria fail to produce sufficient energy or become damaged, they generate reactive oxygen species (ROS), leading to oxidative stress. Oxidative stress results in damage to DNA, proteins, and lipids, further exacerbating neuronal dysfunction and promoting cell death. This oxidative damage plays a critical role in the progression of AD. In Alzheimer's disease, amyloid-beta plaques interfere with synaptic transmission, leading to the loss of communication between neurons. Tau pathology further disrupts axonal transport, exacerbating synaptic failure. The investigation into the relationship between Alzheimer's disease and the gut microbiota offers promising new avenues for understanding disease mechanisms and developing novel therapeutic approaches, such as probiotics, prebiotics, and dietary interventions, aimed at modulating the gut microbiome to delay or prevent the onset of neurodegenerative diseases like Alzheimer's disease (Moloney et al. 2021). The chemistry of the gut microbiota in Alzheimer's disease (AD) involves complex interactions between microbial metabolites, neuroinflammation, and neurodegenerative processes. Gut bacteria produce short-chain fatty acids (SCFAs) like butyrate, acetate, and propionate through the fermentation of dietary fibers. These SCFAs play an essential role in maintaining the gut–brain axis by reducing inflammation, preserving the integrity of the blood–brain barrier (BBB), and modulating immune responses. However, in AD, dysbiosis leads to reduced SCFA production and increased levels of pro-inflammatory microbial metabolites such as lipopolysaccharides (LPS). LPS, a component of the outer membrane of gram-negative bacteria, can enter the bloodstream when the intestinal barrier is compromised, promoting systemic inflammation and crossing the blood–brain barrier to induce neuroinflammation. This heightened inflammatory state contributes to amyloid-beta plaque formation and tau hyperphosphorylation, hallmarks of AD pathology (Liu et al. 2015). Additionally, the gut microbiota influences neurotransmitter production (e.g., serotonin, GABA), affecting cognitive function. As a result, gut dysbiosis and altered microbial metabolite profiles are emerging as critical factors in the progression of Alzheimer's disease.

This chapter reviews recent research on the probable roles of the intestinal in the spread of prion-like amino acids along with amyloid, the function of the microbiota in the gut in controlling the relationship between the gut and the brain, the possibility of amyloidogenic consequences of gut microbes, along with the potential influence of specific meals with plant-related infections on the makeup of the intestinal bacteria while the development of amyloid particles (Pistollato et al. 2016). It also highlights a few of the drawbacks of these investigations and the advantages of specific nutrients derived from plants for the avoidance of Alzheimer's disease (Singh et al. 2016).

2 Tau Hyperphosphorylation in Alzheimer's Disease (AD)

Tau protein is a microtubule-associated protein that plays a critical role in stabilizing microtubules, which are essential for maintaining the structure and function of neurons, as well as for intracellular transport. In Alzheimer's disease (AD), tau undergoes a pathological process called hyperphosphorylation, which is a key feature in the disease's neurodegenerative mechanism. Under normal conditions, tau binds to microtubules in neurons to maintain their stability and support axonal transport, allowing the movement of nutrients, organelles, and signaling molecules along the neuron. In AD, tau becomes excessively phosphorylated at multiple sites, leading to hyperphosphorylation. Phosphorylation refers to the addition of phosphate groups to proteins, a regulatory process that normally controls protein activity (Hu et al. 2016). However, when tau becomes hyperphosphorylated, it loses its ability to bind to microtubules effectively, causing microtubule destabilization. Hyperphosphorylated tau detaches from microtubules and starts to aggregate into insoluble, twisted fibers known as neurofibrillary tangles (NFTs). These tangles accumulate inside neurons, disrupting cellular function and contributing to neuronal death. As hyperphosphorylated tau detaches from microtubules, the loss of microtubule stability leads to impaired axonal transport. This disrupts the flow of essential materials within neurons, leading to synaptic dysfunction and, ultimately, neuronal degeneration. The presence of NFTs in regions of the brain crucial for memory and cognition, such as the hippocampus and cortex, is strongly associated with the cognitive decline seen in AD patients. Tau pathology correlates more closely with disease progression and cognitive impairment than amyloid plaques, another hallmark of AD. Hyperphosphorylated tau and NFTs trigger inflammatory responses in the brain. Microglia, the brain's resident immune cells, become activated and release pro-inflammatory cytokines, further contributing to neuronal damage (Metaxas and Kempf 2016).

3 Triggers of Tau Hyperphosphorylation in AD

The precise triggers for tau hyperphosphorylation in AD are not fully understood, but potential factors include, amyloid-beta (Aβ) pathology; Aβ plaques may initiate tau hyperphosphorylation by activating kinases (enzymes that add phosphate groups) or inhibiting phosphatases (enzymes that remove phosphate groups).

Increased levels of oxidative stress in AD can modify tau and promote its hyperphosphorylation. Mutations in the tau gene (MAPT) or other genetic factors may influence tau hyperphosphorylation (abnormal activity of enzymes such as glycogen synthase kinase-3β (GSK-3β) and cyclin-dependent kinase 5 (CDK5), which regulate tau phosphorylation, are implicated in the hyperphosphorylation process in AD (Rademakers et al. 2004). Tau hyperphosphorylation and the formation of NFTs are considered critical targets for AD therapy. Efforts to develop tau-targeted treatments include drugs that inhibit kinases responsible for tau phosphorylation (e.g., GSK-3β inhibitors) are being explored to prevent or reduce tau hyperphosphorylation. These aim to block the formation of tau aggregates, thereby preventing or reversing the formation of NFT. Approaches that use antibodies to target pathological tau and clear it from the brain are also under investigation. Tau hyperphosphorylation in AD leads to the formation of neurofibrillary tangles, disruption of neuronal transport, synaptic dysfunction, and neuroinflammation, all contributing to cognitive decline. Understanding and targeting tau pathology are critical for developing effective treatments for Alzheimer's disease (Leveille et al. 2021).

4 Gut Microbiota Mechanism in Alzheimer's Disease

The gut microbiota refers to the diverse community of microorganisms, including bacteria, viruses, fungi, and archaea, that reside in the human gastrointestinal tract, primarily in the colon. This complex ecosystem plays a crucial role in maintaining human health, influencing everything from digestion and immune function to metabolism and even brain health. The most abundant microorganisms in the gut are primarily from the phyla Firmicutes, Bacteroidetes, Actinobacteria, and Proteobacteria (Shin et al. 2015). These bacteria are involved in, digesting indigestible polysaccharides (e.g., fiber) to produce short-chain fatty acids (SCFAs) like butyrate, acetate, and propionate, which serve as energy sources and have anti-inflammatory properties, also synthesizing essential vitamins such as vitamin K and some vitamins B. Although less abundant than bacteria, gut fungi like Candida and Saccharomyces play a role in nutrient metabolism and immune response. Viruses, especially bacteriophages (viruses that infect bacteria), influence the composition of the bacterial population by targeting specific bacterial strains. Methanogens, a type of archaea, are involved in the digestion process, particularly in methane production, which affects gas formation and energy metabolism in the gut. The gut microbiota has recently been implicated in the pathophysiology of neurodegenerative diseases, including Alzheimer's disease (AD) (Dieterich et al. 2018). Emerging research suggests that the gut microbiota can influence brain function through a bidirectional communication pathway known as the gut–brain axis. This axis involves multiple mechanisms that may contribute to the development and progression of Alzheimer's disease (Rizzatti et al. 2017). The gut–brain axis connects the central nervous system (CNS) with the enteric nervous system (ENS) and involves Vagus nerve signaling, direct communication between the gut and brain through the vagus nerve. The gut microbiota influences immune responses, which can affect neuroinflammation in the brain. Gut bacteria produce neurotransmitters (e.g., serotonin, dopamine) and neuroactive compounds that can affect brain function. Hormones like cortisol, which can be influenced by the gut microbiota, affect the stress response and brain function. Altered gut microbiota can trigger systemic inflammation through the release of pro-inflammatory cytokines like interleukin-6 (IL-6), tumor necrosis factor alpha (TNF-α), and IL-1β. These cytokines can cross the blood–brain barrier (BBB) and induce neuroinflammation, which plays a significant role in Alzheimer's disease progression. Dysbiosis (imbalance) in the gut microbiota can lead to the activation of microglia (immune cells in the brain). Chronic microglial activation is associated with the development of amyloid-beta plaques and tau tangles, two hallmarks of Alzheimer's pathology. Some gut bacteria produce lipopolysaccharides (LPS), endotoxins that can increase intestinal permeability (leaky gut) and enter the bloodstream. LPS has been shown to promote neuroinflammation by crossing the BBB and activating immune responses in the brain. Certain gut bacteria can produce amyloid-like proteins. These bacterial amyloids may trigger a misfolding cascade in host amyloid proteins, leading to increased production and deposition of amyloid-beta (Aβ) in the brain. Gut dysbiosis can impair the brain's ability to clear amyloid-beta, exacerbating its accumulation (Liu et al. 2020).

The mechanisms include disrupted autophagy, impaired glymphatic clearance, and neuroinflammation-driven dysfunction in clearance pathways. SCFAs like butyrate, acetate, and propionate, produced by beneficial gut bacteria, have anti-inflammatory and neuroprotective effects. They can modulate microglial function, reduce oxidative stress, and promote neurotrophic factors like brain-derived neurotrophic factor (BDNF), which supports neuronal health. In Alzheimer's disease, dysbiosis often results in reduced SCFA production, which diminishes their neuroprotective and anti-inflammatory benefits, contributing to disease progression. Gut dysbiosis can weaken the BBB, allowing harmful substances (e.g., toxins, inflammatory cytokines) from the gut to enter the brain. Increased BBB permeability is linked to neuroinflammation and neuronal damage in Alzheimer's disease (Marizzoni et al. 2020). The gut microbiota can influence the metabolism of tryptophan, an amino acid precursor to serotonin. In Alzheimer's disease, gut dysbiosis may shift tryptophan metabolism toward the kynurenine pathway (KP), leading to the production of neurotoxic metabolites (e.g., quinolinic acid) that contribute to neurodegeneration. Diets rich in fats and sugars can alter gut microbiota composition, increasing the abundance of pro-inflammatory bacteria and decreasing beneficial bacteria that produce SCFAs. Diets rich in fiber, antioxidants, and polyphenols (e.g., Mediterranean diet) have been shown to promote beneficial gut microbiota, reduce inflammation, and potentially reduce the risk of Alzheimer's disease by maintaining a healthier gut–brain axis.

5 The Gut–Brain Axis: Exploring the Role of the Gut in Amyloid Protein Dissemination

The link between the gut as well as the brain establishes a close connection between the gut with the cerebral cortex (Levinthal and Strick 2020). Factors associated with surroundings stimulate the CNS specifically hypothalamic and pituitary, Furthermore, the CNS modulates permeation, mucus production, movement, including resistance via bilateral autonomic connections involving the enteric nerves, the gut muscular sections, including the intestinal membrane (Fried et al. 2021). While individual symptoms can be modulated by signals from the stomach to the central nervous system, the CNS itself can alter how the gut operates. Modifications to these reciprocal exchanges might potentially be involved in the development of CNS diseases and neurological inflammation (Polazzi and Contestabile 2002).

Specifically, it has been demonstrated that the stomach contributes to the spread of prionic amino acids, which cause spongiform encephalopathies with transmission in mammals (Gallardo and Delgado 2021). A growing body of research indicates that prion-like peptide clumping may be a common feature among several of the main CNS disorders, including ALS, Parkinson's disease (PD), along with Huntington's disease (Moda et al. 2023). This theory suggests that AD might be seen as a prion-like illness and that Aβ associated with AD ought to be reconstructed as prion-like protein-based nucleating entities (Arar et al. 2023). Small levels of prions can often be broken down by protein degradation processes; nevertheless, prions have the ability to self-proliferate and compromise central nervous system functioning when they build up and beyond a threshold (Dey et al. 2024).

Prions have been shown to be resistant to digestion when taken orally, and they may be absorbed by microfold layers or villous columnar epithelial cells that exist in the gastrointestinal tract (Drobny et al. 2021). After penetrating the epithelium of the gut, prions may gather in Peyer's lesions (significant locations for inducing colonic resistance) along with other lymphocyte follicles where they communicate with cells called dendritic cells or monocytes (Wilson et al. 2020). Prion-like peptides may then propagate to the intestinal neurological system, which controls gastrointestinal processes, through interactions with follicle dendritic cells, and ultimately to the CNS. The immunological, humorous, including endocrine connections that define the relationship between the gut and the brain may be the mechanism for this (Murray 2020).

Through the use of electronic microscopy as well as immunohistochemistry labelling, systemically decrepit amyloid particles were found in Peyer's spots within a senescence-accelerated mouse strain (SAM-P/1) (Doifode et al. 2021). Throughout a different research, it was discovered that if Aβ was given to mice prior to being weaned it bonded to the increased green-florescent protein including collected 3 h following delivery in the nucleus of column epithelial cells, the vault, Peyer's areas, including the lymph nodes (Chaney 2013). It is still unclear, nevertheless, if Peyer's spots directly contribute to the spread of Aβ as well as the start of AD in people.

Additionally, either similar or dissimilar amyloidogenic ancestors have the potential to function as amyloid-enhancing agents which can be administered parenterally or orally (Sattianayagam 2012). Compared with regular brain specimens, cerebral samples from individuals with AD had greater amounts of all ubiquitin as well as ubiquitin–protein combinations as well as higher activity in amyloid-enhancing peptide (Chan 1997). These proteins are produced either continually at various levels throughout various regular human cells or in reaction to triggers of inflammation. Additionally, this was discovered that treating glial cells grown in vitro using amyloid A from the blood increased their responsiveness, as evidenced by the increased levels of inducible NO synthase along with a number of inflammatory mediator genes, including IL-6, TNF-α, IL-12 p40, IL-23 p19, along with IL-10 (Sochocka et al. 2017), especially in microglial cell lines in addition to some degree, in the astrocytes. This implies that blood levels of amyloid A may contribute to the development of AD (Singh 2022).

6 Functions of Gut Microbiota

The gut microbiota ferments dietary fibers that are otherwise indigestible by human enzymes, producing SCFAs, which have various beneficial effects on gut health and energy balance. The microbiota helps metabolize proteins, fats, and polyphenols, facilitating the absorption of nutrients. The gut microbiota plays a critical role in educating the immune system by helping to distinguish between harmful pathogens and benign antigens (Qian et al. 2022).

A balanced microbiota promotes immune tolerance while maintaining defenses against infections. Microbial metabolites such as SCFAs can help regulate inflammation and maintain the gut's immune balance by influencing immune cell function, particularly in the gut-associated lymphoid tissue (GALT). The microbiota strengthens the intestinal epithelial barrier by promoting the production of mucus and tight junction proteins, preventing the invasion of pathogens and toxins. A healthy microbiome also protects against harmful pathogens through competitive exclusion—outcompeting potential invaders for space and resources—and by producing antimicrobial substances. Gut bacteria influence host metabolism by regulating energy extraction from food, storing fat, and modulating glucose metabolism. Dysbiosis (microbial imbalance) can contribute to metabolic diseases such as obesity, diabetes, and non-alcoholic fatty liver disease (Verhaar et al. 2022). The gut microbiota plays a critical role in producing neurotransmitters like serotonin, dopamine, and GABA, which can affect brain function and mood. Through the gut–brain axis, there is bidirectional communication between the gut microbiota and the central nervous system (CNS), which influences stress response, cognitive function, and emotional behavior. Some gut microbes help in the detoxification of harmful compounds and metabolizing certain drugs, affecting the efficacy and toxicity of medications. The microbiota can activate or inactivate certain drug compounds, thus impacting pharmacokinetics (Bostanciklioğlu 2019).

7 The Role of Gut Microbiota in Regulating the Gut–Brain Axis and the Generation of Amyloid Proteins

The gut microbiota, which is made up of up to 10 distinct types of microbes, primarily bacteria from around 1000 distinct species, which additionally include bacteriophage fragments, molds, viruses, as well as archaea, regulates the relationship between the gut and the brain (Lagier et al. 2012). The microbiome of the gut has a vital function in keeping a normal functioning gut in addition to influencing communication via the gut–brain axis, therefore affecting the well-being of a person (Liang et al. 2018).

It is suggested that people with IBS may also experience associated mental problems along with breathing, neurological, sleep-related, as well as mental illnesses. This is consistent with the theory that ongoing imbalance can impair the biological bidirectional signaling that characterizes the gut–brain axis (Mitrea et al. 2022). Similar to that, nonceliac allergy to gluten, believed to promote intestinal dysbiosis, is additionally related to neurological inflammation with a greater risk of Alzheimer (Mohan et al. 2020). A group of Taiwanese individuals with recently identified IBS also revealed that those with the condition appeared to have a higher chance of getting PD (Tai et al. 2023). Furthermore, the complicated and immunological combinations of functioning LPSs, amyloid particles, and additional bacterial exudates are released from the outermost cells of microbiome-popping species—such as fungus and bacteria—into their surroundings. Specifically, LPSs have been demonstrated to alter intestinal permeability, irritation, including balance in sepsis, necrotizing enter colitis, including IBS (Doney et al. 2022). Furthermore, bacterial amyloids are connected to physiological and structural adaptability, adherence excitement, collection, biofilm creation, penetration of tissue by microbes, and disease infectivity (Bleem 2019). Numerous bacterial species, including Staphylococcus aureus, Salmonella enterica, Mycobacterium tuberculosis (TB), along with *Escherichia coli*, are capable of producing active external amyloid fibers (Van Gerven et al. 2018). There are several ways in which these microbes could communicate with their host surroundings. Specifically, it was discovered that the *E. coli* toxin enhanced the production of Aβ strands in vitro; thus as a result, it may have a role in the etiology of AD (Ghosh et al. 2023).

While amyloids as well as monomers LPSs are usually dissolving, they have the potential to polymerize into intractable protein fiber clumps with time (Alberti and Hyman 2021). These clumps may be the cause of cross-seeding events, oxidative damage induction ones, along with proteostasis change (Ahanger et al. 2024). For instance, amyloid proteins have been demonstrated to seed themselves through polymerization via in vitro methods as well as in vivo by functioning as aggregating nucleus (Ivanova et al. 2021). It has also been proposed in conjunction with the idea of structural imitation that was created in the last few years to explain why structures in biology evolved in response to selective forces (Veit 2022). It is possible that microbial amyloid particles stimulate the natural immune system to increase the body's reaction to brain Aβ.

8 Gut Microbiota and Gut-Derived Amyloids' Part in the Startups of Alzheimer's Disease

Although not proven, it is plausible whether human intestinal microbiota contributes to the etiology of neurological illnesses with amyloid genic traits, including AD, given the unknown quantities of LPSs with amyloid particles in the gut of people (Wang 2023). More specifically, changes in the gut's microbiota as well as increased intestinal permeability may contribute to the onset of diabetes, which is linked to the pathophysiology of AD, as well as a general rise in chronic neurological inflammation, along with malfunction of certain areas of the brain, including the brain's hippocampus along with cerebellum (Snelson et al. 2021). With ageing comes a further significance for the microbes in the gut in the production and spread of amyloid, as the barrier between the BBB along the gut epithelial gets more susceptible to small-molecule penetration (Narengaowa et al. 2021). Lower-density lipoprotein receptors-related protein 1 regulates amyloid elimination, whereas amyloid cerebral inflow across the BBB is believed to be regulated through its receptor of accelerated glycosylation processes (RAGE) while relying on amyloidosis guides through a protein called E and J. Individuals with AD tend to have altered versions of those transportation systems (Chen et al. 2024). Furthermore, by creating a different junctional stickiness molecule, prion-like proteins—PrP C in particular—may also aid in macrophage transendothelial penetration, potentially causing or exacerbating inflammation.

Furthermore, the makeup of the gut microbes alters with age, favoring bacterial species over Firmicutes as well as Bifidobacteria, despite the fact that there are significant interindividual differences among seniors (Salazar et al. 2023). Various dietary plans including residential arrangements (such as belonging, future residential treatment, daily healthcare facility, or therapy) may have an impact on this heterogeneity (Markert et al. 2021).

It was proposed that amyloids produced by bacteria may seep out of the intestinal tract and build up in the cerebral cortex and systemic circulation to clarify how the gastrointestinal microbiota may play a role in the pathophysiology of AD (Choi et al. 2022). ROS may rise as a result, and nuclear factor kappa B (NF-κB) signaling may be activated, upregulating microRNA-34a (miRNA-34a). Consequently, miRNA-34a would suppress TREM2 expression, impairing phagocytosis and causing Aβ 42 amino acids buildup (Bazrgar et al. 2021) (Fig. 1).

Furthermore, amyloid particles as well as LPSs produced by bacteria have the potential to worsen the leaky state within the intestine and raise the amount of mediators along with other minor inflammatory substances that are strongly linked to AD, like IL-17A as well as IL-22 (Choi et al. 2022). These messenger molecules may cross the BBB while through the gut to enter the cerebral cortex. Once inside, they may cause immunogenic responses, the discharge of oxygen radicals, and the signaling of proteins such as toll-like receptor 2/1 (TLR2/1), CD14, with NF-κB, all of which have been linked to neuronal degeneration.

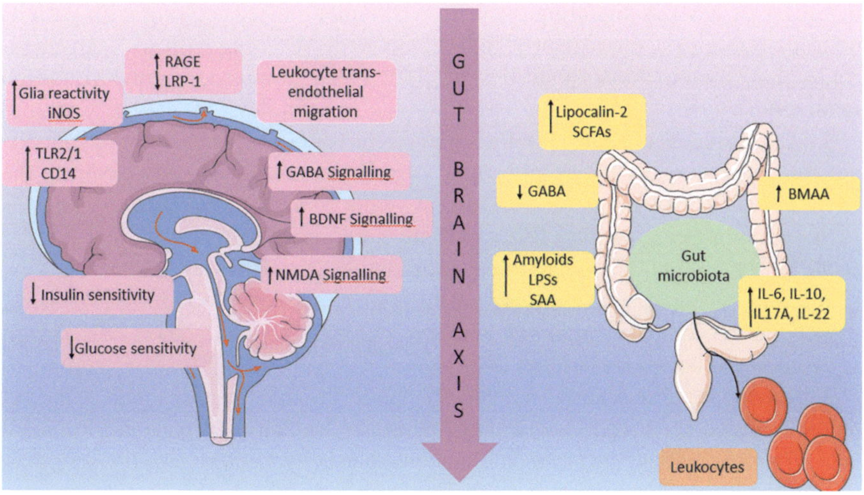

Fig. 1 Alzheimer's disease begins with disruptions in the gut–brain axis and amyloidogenesis caused by microorganisms

The roles played by particular gut microbes in the development or development of Alzheimer's disease including memory loss remain unclear to this day. But certain bacterial species, like gram-positive voluntary anaerobic or microaerophilic microbes lactic acid as well as Bifidobacterium, can metabolize glutamate and generate the main inhibitory neurotransmitter GABA (Saeed et al. 2022). Changes in the bacteria in the gut may jeopardize the body's ability to generate GABA as well as short-chain lipids. In contrast, reduced amounts of short-chain lipids may have detrimental effects on immunity, epithelial cells development, cerebral insulin as well as energy balance, and Alzheimer's disease, mood disorders, anxiety, along with synaptic development impairments (Dalile et al. 2019). These effects may also affect the operation of the central and peripheral nervous systems, respectively.

Surprisingly, the level of BDNF was downregulated in the memory region of a free of germs mice model, as well as BDNF signaling was downregulated in the cerebral cortex along with blood of patients with AD (such as animals that weren't organically colonized by microbes) (Tang et al. 2021). It might be beneficial to evaluate the potential consequences of gastrointestinal dysbiosis on individuals both in vitro and vivo, given that gnotobiotic animal models might have inherent drawbacks (Papadimitriou et al., 2015), might lack certain human-specific intestinal bacteria, or may possess beneficial distinct dimensions of intestinal bacteria, while immune systems compared to those observed in humans.

Additionally, it has been discovered that gut cyanobacteria generate the non-proteinogenic amino acids β-N-methylamino-L-alanine, which is a neurotoxin that interacts via the glutamate receptors of N-methyl-D-aspartate (NMDA) to cause excitotoxicity (Hill et al. 2014). Additionally, it is referred to that disorders of NMDA signaling occur in Alzheimer's disease along with various neuropathologies. Consequently, individuals with AD along with ALS have neurones that contain high quantities of β-N -methylamino-L-alanine (Ludolph et al. 1998).

Lastly, a process called atopobiosis may occur when gut germs move beyond the gut into bloodstreams along with other organs via microfold cells covering the Peyer's lesions. This might have an impact on the course of inflammatory illnesses such necrotizing enterocolitis, IBS, especially neuroinflammatory disorders (Niesler et al. 2021). Comparably, electron microscopy by transmission has revealed the presence of coccus-shaped microorganisms in specimens of blood from AD patients, namely on the cellular exteriors of leukocytes along with white blood cells as well as inside erythrocytes (Henriques 2021).

Furthermore, these findings point to the possibility that the microbiota of the gut along with goods derived from it may operate as regulators of gene–environment conversations, influencing brain chromosomal plasticity, neuron gene transcription, as well as ultimately neuronal functioning (Dash et al. 2022). These rationales support the emerging hypothesis that the microbiota of the intestinal tract itself may be an imprinted entity.

9 Factors Influencing Gut Microbiota Composition

A diet rich in plant-based fibers promotes a healthy, diverse microbiota, while a high-fat, high-sugar diet can lead to dysbiosis. Overuse of antibiotics can disrupt the gut microbiota by killing beneficial bacteria, leading to long-term changes and increasing susceptibility to infections. Probiotics (live beneficial bacteria) and prebiotics (non-digestible fibers that promote the growth of beneficial bacteria) can help maintain or restore healthy microbiota. The composition of the microbiota evolves with age, from infancy to old age, influencing health at different stages of life. Host genetics can influence the composition and functionality of the gut microbiota. Psychological stress, lack of sleep, and sedentary lifestyles can negatively impact the microbiota and contribute to dysbiosis (Hasan and Yang 2019). An imbalance in the gut microbiota, termed dysbiosis, has been linked to various health conditions, including irritable bowel syndrome (IBS), inflammatory bowel disease (IBD), metabolic disorders, obesity, type 2 diabetes, cardiovascular disease, neurodegenerative diseases, Alzheimer's disease, Parkinson's disease, autoimmune diseases, multiple sclerosis, rheumatoid arthritis, mental health conditions, depression, anxiety, and autism spectrum disorders (ASD).

10 Role of Microbiota in Metabolism and Immune Modulation

The gut microbiota is a key player in maintaining metabolic health and regulating the immune system. Disruptions in the gut flora, often caused by poor diet, antibiotics, or diseases, can lead to metabolic disorders (like obesity, diabetes) and immune dysfunction (like autoimmunity, allergies). The relationship between the gut microbiota, metabolism, and immunity is a dynamic process (Vandenplas et al. 2020). Gut bacteria help in breaking down complex carbohydrates, fiber, and proteins that the human body cannot digest on its own. The fermentation of these compounds produces short-chain fatty acids (SCFAs), such as butyrate, acetate, and propionate, which are vital for colon health and energy regulation. Certain gut bacteria promote energy extraction from food, which can influence body weight. For example, Firmicutes have been associated with increased calorie absorption, while Bacteroidetes are linked to leaner body mass. Dysbiosis, or an imbalance in gut flora, is often observed in obesity and metabolic disorders. The gut microbiota interact with the host's glucose metabolism. For instance, SCFAs can enhance insulin sensitivity, reduce inflammation, and regulate glucose homeostasis. Dysregulation in this interaction can lead to conditions like type 2 diabetes. Gut bacteria transform bile acids into secondary bile acids, which are essential for fat digestion and cholesterol regulation. Disruption in this process can affect lipid metabolism and contribute to liver diseases and metabolic syndromes. The gut microbiota is crucial in shaping both the innate and adaptive immune systems. In early life, they help educate immune cells, promoting the development of immune tolerance and ensuring a balance between pro- and anti-inflammatory responses (Kato et al. 2014). The gut microbiota reinforces the intestinal barrier by promoting the production of mucus and antimicrobial peptides, limiting the invasion of harmful pathogens. They also compete with pathogens for nutrients and space, thus preventing infections. SCFAs produced by gut microbes help regulate immune cells like T-regulatory cells, which suppress inappropriate immune responses and reduce inflammation. This prevents excessive immune reactions and conditions like inflammatory bowel disease (IBD). The gut microbiota can affect immune cells throughout the body, not just in the gut. This systemic influence plays a role in autoimmune diseases, allergies, and even cancer immunotherapy responses. For instance, specific bacteria have been shown to enhance the efficacy of immune checkpoint inhibitors used in cancer treatment (Palomares et al. 2017).

11 Neuroinflammation and Gut-Microbiota

The development and operation of the body's defenses are significantly influenced by the gut flora. It is thought because the immune response comes into touch with the gut microbiota for the initial time during delivery, which means this relationship has a significant impact on its growth (Butel et al. 2018). But the elements of breast milk—live bacteria, compounds, IgA, antibodies, including cytokines—are linked to the development of the gut bacteria alongside the host's reactions to this initial colonization (Boudry et al. 2021).

Research suggests that changes in the gut bacteria impact CNS activities, which in turn trigger the onset of depression and depressed symptoms (Carlessi et al. 2021). Mice exposed to various stimuli for a period of 5 weeks, for instance, exhibited behavior resembling depression and had higher amounts of IL-1 in their hippocampal regions (Zhao et al. 2019). Probiotics B. infantis 35,624 taken orally by humans is also linked to higher levels of IL-10 transcription in the bloodstream. In the blood of depressed people, elevated levels of TNF-α, IL-6, with IL-1β were found by Valkanova, Ebmeier, with Allan (2013). Thus, the immune response has the ability to regulate the equilibrium among the cerebral cortex and the stomach (Levinthal and Strick 2020).

A disturbance with the continuous interplay connecting host with microbiota within the immunity system can result in the emergence of central nervous system diseases (Doroszkiewicz et al. 2021). The receptors that recognize motifs are known as pathogen-associated molecular patterns (PAMPs), and they include LPS. Group of these receptors includes TLR family, that is expressed by both native immune cells along with CNS persistent cell types, such as neurones along with glial cell types (Jerala 2007). These molecules are crucial in identifying bacterial targets for illness. TLRs trigger both inflammatory and antibacterial natural immune responses after identifying elements of microbes, such as viruses, microbes, along with fungi (Xia et al. 2021).

These TLRs are categorized into subdivisions which bind to one another and are triggered by certain ligands (Patra et al. 2022). For instance, TLR4 is capable of recognizing LPS, TLR2 may recognize various PAMPs in conjunction with TLR1 or TLR6, as well as TLR5 can recognize bacterial flagellin (Wicherska-Pawłowska et al. 2021). Important internal transcription variables including NF-κB along with interferon I are activated upon the stimulation of a TLR that triggers a signaling cascade (Cornut et al. 2020). These stem cells release a variety of proinflammatory substances, including TNF-α, IL-6, IL-1α, as well as IL-1β, after they are triggered (Dabrowska et al. 2021). These chemicals promote the growth of TH17 & CD4+T helper 1 effector cells (TH1). Inconsequently, TH17 generates IL-17A, IL-17F, and IL-22, which lead to persistent inflammation (Pedersen 2021).

By stimulating the immune system and triggering signaling processes, the gastrointestinal bacteria can support distinct subsets including CD4 + T cells (Kim 2021). The number of Treg cells along with CD4 + T cells were much less in animals receiving a long-lasting antibiotic treatment (Zhang et al. 2021b). Through a polysaccharide A process, Bacteroides fragilis stimulates the growth of TH1 cells. Additionally, research has shown that GF mice lack Foxp3+ Treg lymphoid cells which regulate the growth and operation of effector T cell populations (Zhang et al. 2021b). It has also been shown that the gut microbiota is crucial for the development of Treg cells. Through altering histone 6, SFCAs also cause Foxp3+ Tregs to proliferate.

Overproduction of inflammatory chemicals (TNF-α, IL-1β, IL-1α, & IL-6) within the peripheral may penetrate the blood–brain barrier and enter brain-resident neurones (Salas-Venegas et al. 2022). They interact with receptors that are produced by glial tissue and neurones in the cerebral cortex. Transmembrane dispersion constitutes one of the prospective pathways whereby cytokines might penetrate the blood–brain barrier; this process necessitates a small size with elevated liposolubility (Teixeira et al. 2020). Because of this property, the intestinal nutrients can pass through the blood–brain barrier. The receptors that recognise motifs are known as PAMPs, and they include LPS, these molecules are crucial in identifying bacterial targets for illness. TLRs trigger both inflammatory as well as antibacterial natural immune responses after identifying elements of microbes, such as viruses, microbes, along with fungi (Xia et al., 2021; Galea 2021). Conversely, when GF mature mice are exposed to the free of pathogenic organism intestinal microbiota, their intestinal barrier permeability decreases and their expression of intersection amino acids is positively regulated (Lu et al. 2021).

12 The Inflammasome, the Microbiota of the Gut, Along with Stimulation of Microglia

According to research, inflammasome is an important driver of the neuroinflammatory reactions brought on by anxiety, thus deregulation of its activity may play a role in the development of major depressive disorder (MDD) (Réus et al. 2023). In addition to IL-1β as well as IL-18, oxidative damage along with stimulation of caspase-1 can also trigger inflammatory processes, particularly nucleic acid attaching and oligomerisation domain-like receptors NLRP3 (nucleotide-binding domain, leucine-rich–containing family, pyrin domain–containing-3) (Ramachandran et al., 2024). A research investigation using rats lacking a genetic mutation or pharmacological suppression of caspase-1 showed that blocking of caspase-1 could lessen stress- alongside depression-like behavior (Li et al. 2023). Additionally, these benefits were linked to a modification within the microbiota–gut–inflammasome–brain pathway that provided protection against a stressful reaction. Rats under prolonged stress showed increased levels of NF-κB, NLRP3, with IL-1β within the prefrontal cortex (Rutsch et al. 2020). Microglial activity as well as astrocyte damage were additionally seen. However, therapy using antidepressant medication fluoxetine succeeded to restore these alterations. According to research by Westfall along with Pasinetti (2019), combining a nutritional polyphenolic using *Lactiplantibacillus plantarum* along with *Bifidobacterium longum* can lessen the symptoms of persistent stress, including anxiousness and stress-like behavior (Shanmugam et al. 2022). Furthermore, it has been proposed that *B. longum* and *L. plantarum* prevent NLRP3-mediated IL-1β production in microglia neurones. This was shown how age-induced shifts, primarily monocyte penetration into the cerebral cortex along with activated microglial cells, might be reversed by focusing on the microbes in the gut with

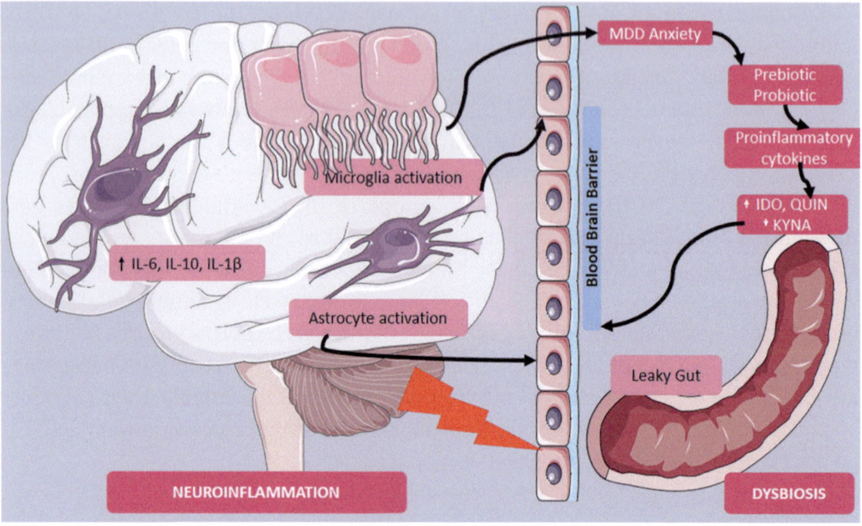

Fig. 2 Connection between MDD and the gut microbiota–brain pathway. Anxiety can cause dysbiosis, a drop in SCFAs, as well as an increase in proinflammatory cytokines, primarily IL-6 and IFN-γ, via dysregulating the gut microbiota. This state of inflammation can cause bacterial migration and intestinal permeability (leaky gut). A decrease in QUIN/KYNA synthesis may result from a rise in IDO (Indoleamine 2,3-dioxygenase) brought on by inflammatory cytokines. Hazardous metabolites from the kynurenine pathway particularly inflammatory cytokines may cause disruption to the blood–brain barrier, which would raise IL-6, IL-1β, including the inflammasome NLRP3 in brain-resident cells. Disorders of the mood including anxiety with MDD might result from the inflammatory procedure in the mind, which also causes astrocyte atrophy along with microglial activation

FOS-Inulin (Cruz-Pereira 2022). This suggests how prebiotic-driven modifications to the makeup of the gut microbiota might be a novel approach for the amelioration of age-associated neuroinflammatory disorders as well as cognitive function (Boehme et al. 2020). Another study found that microbes such as *Lactobacillus reuteri* NK33 along with B. adolescentis NK98 protected mice from developing colitis and depressive or anxiety disorders brought on by immobilization stress (Mindus et al. 2021) (Fig. 2).

Zhang et al. (2017) found that administering an anti-mouse IL-6 MR16-1 enhanced the Bacteroidetes proportion in prone mice and produced antidepressant properties in an interpersonal defeat anxiety model (Yu et al. 2024). These findings indicate that IL-6 suppression produces antidepressants impacts by restoring the changed structure within the intestinal microbiota. IL-6 is essential and has a significant impact on MDD (Ting et al. 2020). Recently, blood IL-6 levels have been shown to be lower in refractory to individuals with MDD who used ketamine, an inhibitor of the NMDA receptor, for their depression symptoms (Yang et al. 2015). Furthermore, Getachew et al. (2018) found that Wistar rats that were treated with ketamine for 24 h saw changes in microbial diversity. Further research on the effects of ketamine on the microbiome in MDD patients and in depression-stricken animals may be fruitful.

13 Biomarkers, in Alzheimer's Disease

Biomarkers are crucial for diagnosing and monitoring the progression of Alzheimer's disease (AD), as they provide measurable indicators of the biological processes associated with the disease (Mantzavinos and Alexiou 2017). In AD, biomarkers are often related to the pathological hallmarks of the disease, including amyloid-beta (Aβ) plaques, tau protein tangles, neurodegeneration, and neuroinflammation. Aβ42 in cerebrospinal fluid (CSF), the deposition of amyloid plaques in the brain, is a hallmark of AD (Palmqvist et al. 2019). Aβ42 levels in CSF are typically reduced in Alzheimer's disease, as this peptide gets sequestered into plaques within the brain. Positron emission tomography (PET) scans using tracers like [18F]-florbetapir allow visualization of amyloid plaques in the brain. Elevated amyloid deposition is a core feature of AD and is detectable long before clinical symptoms appear. Recently, blood-based biomarkers have gained interest. The Aβ42/Aβ40 ratio in plasma shows promise in reflecting amyloid deposition in the brain (Wilczyńska and Waszkiewicz 2020). Tau protein becomes abnormally phosphorylated and forms neurofibrillary tangles in AD. Elevated t-tau levels in CSF reflect the degree of neuronal damage or neurodegeneration. p-tau levels, particularly at positions p-tau181 and p-tau217, are specific markers of tau pathology in AD. Elevated p-tau levels in CSF correlate with tau tangles and amyloid deposition. Like Aβ, tau tangles can be visualized in the brain using PET imaging with tau-specific tracers, such as [18F] flortaucipir. This imaging method allows for the assessment of tau spread in the brain, which closely correlates with cognitive decline. Neurofilament light chain (NfL) is a structural protein in neurons, and elevated levels in CSF or blood indicate neurodegeneration (Gaiottino et al. 2013). NfL is not specific to AD but is a general marker of neuronal damage seen in various neurodegenerative diseases, including Alzheimer's disease. Magnetic resonance imaging (MRI) is used to assess brain atrophy, particularly in regions such as the hippocampus, which is involved in memory and is significantly affected in AD. MRI is a key tool in tracking disease progression over time. Proteins associated with synaptic function, such as neurogranin and SNAP-25, are elevated in the CSF of AD patients. These proteins reflect synaptic loss and dysfunction, which are central to the cognitive decline seen in AD. Neuroinflammation plays a critical role in AD pathology. Elevated levels of pro-inflammatory cytokines such as IL-6, TNF-α, and IL-1β in the CSF or plasma are often observed in AD patients (Wang et al. 2015). Imaging of activated microglia, the brain's immune cells, can be performed using PET tracers like [11C]-PK11195, which binds to the translocator protein (TSPO). This technique helps to monitor neuroinflammation in AD. Recent advances have made blood-based biomarkers a promising area for non-invasive AD diagnosis and monitoring. Plasma Aβ42/Aβ40 ratio, reflecting amyloid plaque deposition in the brain, correlates with

AD pathology. Plasma p-tau181 and p-tau217, emerging as strong indicators of tau pathology in AD, biomarkers can differentiate between AD and other neurodegenerative conditions, potentially offering a cost-effective screening tool. Plasma NfL levels, as a marker of neuronal damage, are elevated in AD and correlate with disease progression (Mattsson et al. 2017). The Apo lipoprotein E (APOE) gene is the strongest genetic risk factor for late-onset Alzheimer's disease. Individuals carrying one or two copies of the ε4 allele have a significantly increased risk of developing AD. While not diagnostic, APOE genotyping can help assess genetic predisposition to the disease. Extracellular vesicles, or exosomes, derived from neurons and carrying tau or amyloid markers, are being investigated for their potential in early detection (Rastogi et al. 2021). Changes in brain glucose metabolism, observed through FDG-PET imaging, are linked to neuronal dysfunction in AD. Additionally, gut microbiota-derived metabolites are being explored for their potential role in influencing neurodegeneration. Biomarkers play a vital role in early detection, diagnosis, and monitoring of Alzheimer's disease, providing insights into the underlying pathology and disease progression. Advances in CSF, imaging, and blood-based biomarkers are improving the ability to diagnose AD early, track disease progression, and even assess responses to emerging therapies (Hu et al. 2022).

14 Microbiota Biomarkers in Alzheimer's Disease (AD)

The emerging link between gut microbiota and Alzheimer's disease (AD) has prompted research into specific microbial biomarkers that could help in early diagnosis, monitor disease progression, and even guide therapeutic interventions. The gut microbiota can influence neuroinflammation, amyloid-beta accumulation, and tau pathology through the gut–brain axis, making microbiota-based biomarkers a promising area of investigation (Schächtle and Rosshart 2021). SCFAs, particularly butyrate, acetate, and propionate, are produced by the fermentation of dietary fibers by gut bacteria. SCFAs help regulate inflammation, maintain gut barrier integrity, and modulate brain function. In AD patients, levels of butyrate and other SCFAs are often reduced, correlating with increased neuroinflammation and cognitive decline (Marizzoni et al. 2020). Measuring SCFA concentrations in fecal samples or blood could serve as an indicator of gut health and systemic inflammation linked to AD. LPS is a component of the outer membrane of gram-negative bacteria and a potent pro-inflammatory molecule. In AD, a compromised gut barrier can allow LPS to enter the bloodstream, leading to systemic inflammation and neuroinflammation when it crosses the blood–brain barrier. Elevated levels of circulating LPS in the blood or CSF are associated with increased neuroinflammation, amyloid-beta accumulation, and cognitive decline. LPS levels could thus serve as a biomarker for gut permeability and inflammation in AD patients. Dysbiosis refers to an imbalance in the composition of gut microbiota, often characterized by a decrease in beneficial bacteria and an increase in harmful ones (Brüssow 2020). In AD, there is often a reduction in beneficial bacterial genera such as Bifidobacterium, Lactobacillus, and

Faecalibacterium (butyrate producers). These bacteria are linked to anti-inflammatory effects and gut–brain axis regulation. Elevated levels of Proteobacteria (including *E. coli*) and Firmicutes, associated with pro-inflammatory properties, have been observed in AD patients (Chen et al. 2022). Fecal microbiota analysis can identify changes in bacterial diversity and abundance, offering a potential biomarker for early AD diagnosis or disease monitoring. Gut bacteria play a crucial role in the metabolism of tryptophan, an essential amino acid, into various metabolites, including serotonin (a neurotransmitter) and kynurenine (a modulator of neuroinflammation) (Gao et al. 2020). In AD, alterations in tryptophan metabolism, particularly increased kynurenine levels, have been linked to neuroinflammation and cognitive decline. Measuring tryptophan and kynurenine levels in the gut or blood could provide insights into microbiota-driven neuroinflammatory processes in AD. Bile acids, produced by the liver and metabolized by gut bacteria, have been implicated in AD. They influence neuroinflammation and neuronal function. Altered bile acid profiles have been observed in AD patients, with decreased levels of primary bile acids (produced in the liver) and increased secondary bile acids (produced by gut bacteria). Monitoring bile acid levels in blood or feces may serve as a biomarker for gut microbiota alterations in AD. Indoles are metabolites of tryptophan produced by gut bacteria. Indole derivatives like indole-3-acetic acid (IAA) and indole-3-propionic acid (IPA) have neuroprotective effects (Konopelski and Mogilnicka 2022). Reduced levels of neuroprotective indole derivatives, such as IPA, have been linked to neurodegeneration and cognitive impairment in AD. Measuring these indoles in the blood or feces may provide biomarkers for gut–brain communication dysfunction in AD. The gut microbiota influences the production of neurotransmitters such as serotonin, gamma-aminobutyric acid (GABA), and dopamine, which play key roles in brain function and cognition. Alterations in gut-derived neurotransmitter levels, particularly reductions in GABA and serotonin, may be associated with cognitive decline and mood disorders in AD. Monitoring these neurotransmitter levels in the gut or blood could serve as a biomarker for AD-related changes in the gut–brain axis (Strandwitz 2018). Fragments of bacterial DNA (microbial DNA) can be detected in blood due to increased gut permeability ("leaky gut") in AD patients. Elevated levels of microbial DNA in blood samples could indicate systemic inflammation and compromised gut barrier function, both of which are linked to AD progression. Analyzing microbial DNA in blood can thus serve as a non-invasive biomarker for gut health in AD. MicroRNAs (miRNAs) are small non-coding RNA molecules that regulate gene expression and are influenced by the gut microbiota. Specific changes in fecal microRNA expression profiles have been observed in AD patients. These miRNAs can regulate inflammation, neurodegeneration, and gut permeability, making them potential biomarkers for AD diagnosis and progression (Doroszkiewicz et al. 2022).

15 The Cysturenine Pathway and Gut Microbiota

The kynurenine pathway (KP) is involved in the progression of psychosis and major depressive disorder (MDD), among other mental diseases (Savitz 2020). Furthermore, tryptophan is an initial form of KP, which appears to be regulated in a way that is connected to the gut flora. The intestinal microbiota's initial colonization and subsequent changes within richness as well as makeup during life affect tryptophan supply and are crucial for serotonergic signaling in the central nervous system (Choudhuri et al. 2022). The gut bacteria can influence host behavior via recruiting the brain–gut system's serotonergic signaling pathway as well as tryptophan synthesis (O'Mahony et al. 2015). Tryptophan can be converted to serotonin by bacterial gut molecules including short-chain fats, which can control this pathway and stop tryptophan from being converted to KP (Agus et al. 2018).

A necessary source of the chemical serotonin related KP compounds is tryptophan. Every tissue has an enzyme called IDO, while the liver has the molecule TDO, which metabolizes tryptophan partially to serotonin with the remaining portion to phenylalanine (KP) (Badawy and Guillemin 2022). The hormone interferon-γ (IFN-γ) as well as IL-6 in particular, which are stimulated by inflammatory mediators in intestinal inflammation, cause the formation of IDO as well as alter tryptophan digestion, which shifts the process of the hormone serotonin to that of kynurenine along with its derivatives (Haq et al. 2021). Inflammation of the gut has been linked to alterations in brain function as well as MDD; mediators including TNF-α, IL-2, and IFN-γ may be linked to an increased chance of developing MDD (Carlessi et al. 2021).

B. infantis-treated rats have lower IDO activation as well as higher plasma amounts of tryptophan. Furthermore, B. infantis therapy encouraged actions similar to those of antidepressants. KP function is reduced in GF mice; nevertheless, the KP can return to its usual level in an identical animal with an adequate microbiota repositioning (Pedroza Matute and Iyavoo 2023). The two distinct routes that produce KYNA are a NMDA receptor inhibitor as well as neuroprotector, potentially QUIN. By stimulating NMDA receptors within the cerebral cortex, QUIN produces excitotoxicity; KYNA counteracts these side effects. MDD is linked to both a decrease in KYNA and an overproduction of QUIN (Hughes et al. 2022). Individuals with MDD had a reduced amount of KYNA, whereas patients who had attempted suicide had higher amounts of QUIN in their CSF. In addition, mice exposed to prolonged stress showed changes in the makeup of their gut microbiota, a reduction in Lactobacillus, including an elevation in kynurenine concentrations (Marin et al. 2017). However, metabolic alterations and behavioral abnormalities were improved even when gut Lactobacillus concentrations were restored to normal (Soto et al. 2018). Catabolism progresses to full destruction at the conclusion of KP, yielding NAD or ATP. The tryptophan inside KYN is broken down either TDO or IDO to initiate this full oxidation, after that kynurenine is broken down by KMO to OHK or KATs to KYNA. Kynureninase keeps breaking down OHK to create HAA. Last but not least, HAA can be broken down in NAD, ATP, or trace quantities of QUIN (Xu et al. 2017).

Different neurones in the cerebral cortex, such as macrophages as well as astrocytes, produce IDO. Elevated kynurenine synthesis throughout irritation and elevated tryptophan breakdown brought on by pro-inflammatory cytokines can cross the blood–brain barrier and enter the brain (Tanaka et al. 2021). Neuronal protective and/or neurotoxic alterations are influenced by the waste products of this system. It appears that QUIN as well as KYNA are produced by macrophages along with microglia within the neurones (Biernacki et al. 2020). Enhanced macrophage invasion as well as stimulation of microglia throughout the cerebral cortex following infections may contribute to astrocyte-mediated cytotoxicity (Rodríguez et al. 2022). An inflammatory condition linked to MDD might encourage the neuropathic route of KP synthesis, maintaining the glial cell population in an unbalanced state which might lead to the chronic and recurring character of MDD (Athira et al. 2020).

16 Therapeutic Approaches Targeting Gut Microbiota in AD

The microbiota comprises a population of symbiotic bacteria that play crucial regulatory roles in both health and illness (Hou et al. 2022). They may be neutral, helpful, or harmful for the human. Trillions of microbes, including bacteria, fungus, viruses, as well as archaea, inhibit the tissues of humans (Hassan et al. 2024). These germs settle on the layer of skin terrain, within the respiratory system, genitourinary systems, along with most significantly, the gut. The intestinal tract is home to around 95% of all living things that make up the body's microbiome (Fuloria et al. 2022). The gut microbial community is mostly made up of anaerobic microbes (primarily bacterial), fungi, along with viruses (Jaswal et al. 2023). The biosynthesis of vitamin B alongside K, the breakdown of essential nutrients (liver chemicals sterols along with medications), the decomposition of unprocessed sugars, the formation of SCFAs along with defense against external pathogens are just a few of the numerous functions that these different kinds of microbes perform in human health (Rastall et al. 2022). In adulthood, all four primary microbial phyla are Bacteroidetes (~48%), Firmicutes (~51%), Proteobacteria along with Actinobacteria (1%). Through symbiotic relationships, the gastrointestinal microbiota controls human pathology and wellness state as individuals evolve together. Mitochondria, derived from ancestral microbes, share a close functional link with gut microbiota. Microbial metabolites regulate mitochondrial energy production and oxidative stress, while mitochondria influence host immune and metabolic balance. In Alzheimer's, gut dysbiosis disrupts this crosstalk, leading to mitochondrial dysfunction, neuroinflammation, and disease progression (Hampl and Roger 2024). Furthermore, these structures are especially susceptible to compounds generated by different microbes linked to the intestinal and its surrounding mucosa. Via the bilateral gastrointestinal–brain axis, which involves several pathways including the neurological system both parasympathetic as well as sympathetic parts from the ANS including t neuroimmune structure, the intestinal bacteria interacts directly with the CNS (Gershon and Margolis 2021).

The axis between the gut including the brain is in charge of the healthy operation of the gastrointestinal tract as well as the sympathetic nerve system, hormonal glands, along with certain brain areas, including the frontal lobe along with brain (Reutov and Sorokina 2022). This is a molecular communication system. Additionally, the link between the gut and the brain affects both typical and unusual behavioral outcomes as well as CNS growth. Numerous metabolites from bacteria possess physiological importance beyond their requirement for anabolic as well as catabolic processes (Weissbrodt et al. 2020). The intestinal microbiota produces bioactive compounds such as short-chain fatty acids (SCFAs), metabolizes essential amino acids like tryptophan, modulates systemic immune responses, and synthesizes key neurotransmitters including γ-aminobutyric acid (GABA), noradrenaline, and dopamine. In addition, it promotes the release of neurotrophic factors such as nerve growth factor (NGF), glial cell-derived neurotrophic factor (GDNF), and brain-derived neurotrophic factor (BDNF), which are critical in the context of neurological disorders (Morris et al., 2017). Specifically, within both the peripheral and central nervous systems of adults, BDNF plays an essential role in neuronal survival, synaptic plasticity, and proper functional regulation.

It's important to note that mice without germs displayed a reduced amount of BDNF production within their hippocampus, both in the level of protein and mRNA stages, which is linked to cognitive impairment (Maqsood and Stone 2016). SCFAs, compounds also linked to neural communication, are created when microbes degrade indigestible sugars within the intestine. They do this by modifying the production of many neurotransmitters that control behavior as well as memory (Myhrer 2003). The primary compounds produced by intestinal microbes are propionate, acetate, along with butyrate, which can function as signals or as precursors for human physiology (Hosseini et al. 2011). Butyrate and butyric acid, found in different parts of the gut, support health by helping good bacteria grow, improving food digestion and nutrient absorption, and limiting harmful microbes (Verma et al., 2018). In particular, systemic treatment of butyrate induces a behavioral reaction similar to that of a depression medication, as it has been demonstrated that butyric acid and valeric acid, along with propionic acid, can effectively prevent Aβ peptides from converting into hazardous clusters in vitro (Sun et al. 2016).

The breakdown and ingestion of proteins, which are necessary for the creation of proteins as well as the formation of bioactive chemicals that control vital signaling as well as metabolic processes within the support, are mostly mediated by the microbiota of the gastrointestinal tract (Abdallah et al. 2020). The pathophysiology of AD may be aided by disruption of protein balance. Actually, GABA levels are altered as a result of alterations in glutamate breakdown in the AD brain, which has an impact on the functioning of neurones (Corona et al. 2011). GABA is generated by several bacterial species, primarily lactic acid microbes, through glutamate breakdown (Cui et al. 2020). As the primary inhibitory neurotransmitter throughout the nervous system, it controls mature brain activity, plasticity of synapses, along with cortical adaptability and reorganization (Turrigiano and Nelson 2004). As such, it serves as a crucial mediator for the modulation of neurological chemistry by bacteria.

This chemical messenger has been shown to be modified in a number of ailments, such as seizures while psychosis and malfunctions in the pathway of GABA have been linked to the etiology of a number of long-term neurological illnesses (Teleanu et al. 2022). Both AD and moderate cognitive decline individuals showed alterations in the metabolic pathways for methionine, tyrosine, tryptophan, along with purines (Connell et al. 2022). Fascinatingly, tryptophan serves as a necessary protein that is mostly present in fruits, nuts, dairy goods, including meats. The substance is additionally taken up by the epithelium of the intestine and circulated throughout the bloodstream, yet it is additionally broken down in both directions by the microbes of the gut producing a number of molecules that are involved in the relationship between the gut and the brain (Nightingale and Spiller 2023). It serves as the initial form for multiple compounds, including serotonin as well as kynurenine. Once within the cerebral cortex, kynurenine may pass through the BBB as well as it is the beginning of neuroactive glutamatergic chemicals, which include the beneficial kynurenic acids (Kearns 2024). Emotion as well as cognition is significantly influenced by serotonin, a chemical messenger that is present in the stomach and the brain. Deregulation of tryptophan seems connected to harmful circumstances in the cerebral cortex as well as gastric system, and changes in the amount of serotonin may be related to the beginning of psychological and digestive disorders (Parolisi et al. 2023). The BBB permeable is regulated by microbes and associated compounds. These include SCFAs, which can enter the circulation and use close junction molecule overexpression to control the functioning within the BBB (Fock and Parnova 2023).

It's important to note that modified levels of proteins involved in tight junctions as well as enhanced BBB permeability are linked to an absence in intestinal microbes (Panwar et al. 2021). The close connection that begins throughout pregnancy and continues across life among the gastrointestinal flora as well as BBB was illustrated by Braniste et al. Researchers demonstrated the BBB's porosity was reduced as well as the production of proteins involved in tight junctions was increased when mice without germs were treated with microorganisms that generate SCFAs or when feces from animals with free of pathogens gut flora were transferred into free of microbes rats (Aburto and Cryan 2024). As we age, the variety in our microbiota's makeup declines as well as the quantity of helpful bacteria drops. By disrupting the intestinal junctions and increasing the passage of microbial products such as lipopolysaccharide (LPS), these changes impair cytochrome activity and hinder energy production in gut cells (Allam-Ndoul et al., 2020).

This state encourages "inflammaging" a phenomenon closely linked to a number of illnesses related to age, while a robust microbiome is essential for long life (Santoro et al. 2021). Furthermore, alterations in the composition of the gut microbiota can impair the normal function of the gastrointestinal barrier, increasing intestinal permeability. Such dysbiosis may result from dietary changes, medication use, or diseases that disturb intestinal balance (Takiishi et al., 2017). Therefore, regions in the CNS which regulate memory can be reached by neuroactive substances including chemicals produced by gut microbes. Neurodevelopmental disorders like autism as well as neurodegenerative disorders like AD, PD, as well as Huntington's disease are known to exhibit dysbiosis. The goal of current study is to learn more about the underlying processes (Bicknell et al. 2023).

Because of its effects upon society as a whole, Alzheimer's disease condition is among the most frequent neurodegenerative ailment in seniors and presents a significant challenge to medical care (Winblad et al. 2016). This is the primary cause of memory and is characterized by gradual decrease in mental abilities due to shrinking of the brain's cortex, swelling of the airways, and death of neurones in the brain's hippocampus as well as the cerebellum (Mukherjee et al. 2023). Progression of AD has been linked with deficits in memory, ocular spatial disorder, including lack of sleep. External Aβ plaques along with NFT made of tau protein that has been hyperphosphorylated are two important molecular markers of the illness (Zhang et al. 2021a). Amyloid peptides created by a class I membrane protein known as APP accumulate extracellularly to form plaques.

When treating AD immediately, APP was thought to be an achievable goal in therapeutic approaches. α- as well as γ-secretases can catalyze a nonamyloidogenic fragmentation procedure for APP that results in the production of an attached to the membrane αCTF along with a sAPPα (Gay et al. 2018). In addition, β- with γ-secretases may break down APP to release two main fragments: sAPPβ with a membrane-located βCTF. Subsequent breakdown of βCTF yields Aβ peptides by lengths ranging between 37 and 43 amino acid residues, the most prevalent and harmful comprising Aβ(1-42) along with Aβ(1-40) (Siegel et al. 2017). All oligomeric aggregates formed by Aβ monomers have the ability to combine into consistent fibrils.

17 Delivery Approaches of Microbiota in AD

There are several promising delivery approaches aimed at modifying gut microbiota as potential therapies for Alzheimer's disease (AD). These approaches target the gut–brain axis with the goal of restoring a healthy microbial balance, reducing neuroinflammation, and protecting against neurodegeneration. The primary delivery methods include, Probiotics, live microorganisms that confer health benefits when consumed in adequate amounts, typically found in foods or supplements. Probiotics can restore beneficial gut bacteria, enhance the production of short-chain fatty acids (SCFAs), and reduce systemic inflammation (Zhan et al. 2023). Certain strains like Lactobacillus and Bifidobacterium have been shown to modulate immune responses and improve cognitive function in animal models of Alzheimer's disease. Probiotics are administered orally in the form of capsules, tablets, or functional foods like yogurt and fermented products. They aim to replenish beneficial microbes and enhance gut–brain communication. Prebiotics are non-digestible fibers that selectively stimulate the growth and activity of beneficial gut bacteria. Prebiotics promote the production of SCFAs, which have neuroprotective effects by modulating inflammation, oxidative stress, and enhancing neuronal health. They also help maintain the integrity of the blood–brain barrier (BBB) and reduce neuroinflammation. Prebiotics can be consumed through dietary fibers (e.g., inulin, oligosaccharides) found in certain foods or as supplements (Rezende et al. 2021). Their role is to fuel the growth of healthy microbiota in the gut. Synbiotics are a combination of both probiotics and prebiotics designed to work synergistically to enhance the survival and activity of probiotics. Improving gut health and modulating the immune system can strengthen the body's ability to reduce neuroinflammation and support better cognitive function. Synbiotics, available as supplements or functional foods, combine beneficial bacteria with prebiotic fibers to enhance these effects on gut health (Tadesse 2012). Fecal microbiota transplant (FMT) involves the transfer of fecal matter from a healthy donor to a recipient, with the goal of restoring a balanced microbiota. FMT aims to restore a healthy gut microbial ecosystem, reduce inflammation, and improve the gut–brain axis (Sadowski et al. 2024). Early research in animal models has shown that FMT from healthy donors can improve cognitive function and reduce amyloid-beta pathology in AD models. FMT is typically delivered through colonoscopy, enema, or oral capsules containing fecal microbiota. Although still experimental for Alzheimer's disease, this approach has shown promise in treating other conditions like Clostridium difficile infections. Postbiotics are bioactive compounds produced by probiotics or beneficial gut bacteria during fermentation, such as SCFAs, enzymes, or peptides. Postbiotics, especially butyrate, have anti-inflammatory and neuroprotective properties (Maiuolo et al. 2024). They can modulate immune responses, enhance neuronal health, and reduce oxidative stress. Postbiotics can be administered orally in supplement form or through specific functional foods designed to deliver these active compounds directly to the gut. Modifying the diet to support a healthy gut microbiome by increasing fiber, antioxidants, and polyphenols while reducing processed foods and sugars can have significant benefits. Fiber rich diets, such as the Mediterranean diet, encourage the growth of beneficial bacteria and the production of short-chain fatty acids (SCFAs), which reduce inflammation and support brain health. Polyphenols found in fruits, vegetables, and tea also have antioxidant and anti-inflammatory effects that protect neurons. Implementing sustained dietary changes can naturally modulate gut microbiota and provide long-term benefits for both gut and brain health (Bashir et al. 2024). Bacteriophages are viruses that specifically target and destroy certain bacteria without affecting others. Phage therapy can be used to selectively eliminate pathogenic bacteria that contribute to dysbiosis and inflammation in Alzheimer's disease. By reducing harmful bacteria, this approach can help rebalance the microbiome and reduce neuroinflammation. Phage therapy is typically delivered orally, either in liquid or capsule form, targeting specific bacterial strains in the gut. Administering key metabolites produced by gut bacteria, such as short-chain fatty acids (SCFAs, e.g., butyrate) and other neuroprotective compounds, may help regulate inflammation, protect neurons, and improve overall brain health (Ragavan & Hemalatha 2024). These metabolites can help modulate neuroinflammation, support neuronal health, and restore the gut–brain axis. Butyrate, in particular, has shown promise in reducing inflammation and promoting cognitive function in Alzheimer's disease models.

These metabolites can be delivered as supplements or through dietary modifications that promote their natural production in the gut. These various delivery approaches, whether through direct supplementation with beneficial bacteria (probiotics), enhancing their growth with prebiotics, or restoring the microbiome with more advanced techniques like FMT, offer promising avenues for potentially mitigating Alzheimer's disease. Each approach aims to target different aspects of gut–brain communication and reduce neuroinflammation, which is central to the pathogenesis of Alzheimer's disease (Cerovic et al. 2019).

18 New Treatment Modalities: Fecal Transplants as Well as Diet-Based Methods

Retirement, infections, poor eating patterns, as well as lifestyle choices can change the makeup and variety of the gut bacteria, which can encourage the development as well as advancement of neurological conditions like AD (MoosWalter et al. 2016). As dysbiosis is closely linked to changes in gastrointestinal permeability, disruptions in the blood–brain barrier, including neuroinflammatory operations, all of which play a significant role in the onset of AD, the microbes in the gut may also be crucial in combating AD (Solanki et al. 2023). Particularly, recent research efforts have focused upon the function of the microbiome of humans in controlling numerous neurochemical routes via the gut–brain direction in addition to searching for new treatments for microbes control in the lack of a conclusive therapy for Alzheimer's disease, in the majority of treatments just putting off the decline of memory and thinking skills (Khan et al. 2020).

A diet heavy in carbs, saturated oils, as well as processed food can be harmful to one's health, reducing microbiome diversity, causing neurological inflammation as well as impairing cognitive function (Kurowska et al. 2023). Healthy eating habits, on the contrary together, have neuroprotective qualities and may support cognitive function. In this sense, a high consumption of unsaturated oils rich in many polyphenols, as well as veggies, beans, fruits, grains, and other foods, make up the numerous components comprising the Mediterranean way of life that are thought to be beneficial for AD patients (Hoffman and Gerber 2013). Furthermore, it was just recently determined that oral bacteriotherapy is a recognized therapeutic and preventative strategy for inflammatory diseases and gastrointestinal infections (d'Ettorre et al. 2020). There have been reports of the beneficial benefits of the bacteria as well as lactic acid-producing bacteria treating CNS-related illnesses (Suganya and Koo 2020).

Probiotics can be administered by certain supplements or meals fortified with microorganisms and other diet-based treatment approaches including foods high in prebiotics that are restricting calories, supplementing using polyphenols as well as consuming digestion-resistant materials (Wernimont et al. 2020). FMT and this involve transferring feces coming from an adult donor towards an individual's intestines in order to reclaim the variety, while functionality of the community of microbes, has become a different offering beneficial possibility towards AD in addition to nutritional therapies (Bibbò et al. 2020). This method is presently being researched over its potential use in extraintestinal circumstances like metabolism along with neuropathological disorders. It has been effective in digestive circumstances like IBD, vomiting, and IBS alongside constipation. Subsequently it is currently regarded as an effective therapy for frequent Clostridium difficile diseases. The injection of living microorganisms along with their related metabolites makes FTM an effective therapeutic technique with mild and temporary adverse effects (Bonfili et al. 2021). Nevertheless, so as to prevent the patient from becoming contaminated with dangerous bacteria that might cause serious illnesses, it is crucial to thoroughly screen both the donor and the fecal matter.

19 Challenges and Future Directions

The relationship between Alzheimer's disease (AD) and gut microbiota represents an emerging frontier in neuroscience and microbiology, where gut–brain axis research has suggested a significant influence of the microbiota on neurodegenerative processes. The gut microbiota affects brain function and health through immune modulation, production of neuroactive compounds, and impact on systemic inflammation. Here are some of the primary challenges and potential future directions in understanding and leveraging this relationship for AD treatment. The exact mechanisms by which gut microbiota influence Alzheimer's pathology are not yet fully understood. While the microbiota can modulate neuroinflammation, amyloid-beta aggregation, and neurodegeneration, the pathways and molecular mechanisms remain complex, often involving immune, endocrine, and nervous systems interactions (Goyal et al. 2021). Dissecting these pathways is crucial for therapeutic applications. The human microbiome is highly individual-specific, influenced by genetics, diet, lifestyle, and environment. This variability complicates the establishment of a clear link between certain microbial species and AD. More standardized studies are needed to identify universally relevant microbial biomarkers. Exploring the use of probiotics, prebiotics, synbiotics, and fecal microbiota transplants (FMT) may offer potential for altering microbiota composition to slow AD progression (Kaźmierczak-Siedlecka et al. 2020). These approaches require further refinement to ensure safety, consistency, and efficacy in modifying the microbiome for therapeutic benefit in AD. Identifying specific gut microbiota profiles or metabolites as biomarkers for early diagnosis of AD could revolutionize the field. By profiling the microbiota of individuals at risk, it may be possible to predict disease onset and progression, allowing for early interventions.

20 Conclusion

Therapeutic approaches targeting the gut microbiota hold promise in addressing the complex pathology of Alzheimer's disease (AD) by modulating the gut–brain axis. Mounting evidence suggests that gut dysbiosis contributes to neuroinflammation, amyloid-beta accumulation, tau hyperphosphorylation, and cognitive decline in AD. Strategies that aim to restore a healthy microbiota—such as probiotics, prebiotics, dietary interventions, and fecal microbiota transplants—have demonstrated potential to reduce inflammation, enhance gut barrier integrity, and improve neurocognitive function. While these interventions show promise, clinical translation requires rigorous testing to better understand the optimal treatment parameters, such as specific microbial strains, dosing, and long-term safety. Future research should also explore personalized approaches that consider individual variations in gut microbiota composition and genetic risk factors. By modulating the gut microbiota, these therapeutic strategies could complement existing AD treatments, offering a novel and potentially transformative approach to managing Alzheimer's disease. While promising, the role of the gut microbiota in Alzheimer's disease is still an evolving field with many challenges. Future research focused on understanding underlying mechanisms, identifying specific biomarkers, and creating personalized and microbiota-targeted therapies will be crucial to advancing Alzheimer's treatments. Continued exploration may yield novel therapeutic strategies, ultimately improving the quality of life and prognosis for AD patients.

References

Abdallah A et al (2020) Gastrointestinal interaction between dietary amino acids and gut microbiota: with special emphasis on host nutrition. Curr Protein Pept Sci 21(8):785–798

Aburto MR, Cryan JF (2024) Gastrointestinal and brain barriers: unlocking gates of communication across the microbiota–gut–brain axis. Nat Rev Gastroenterol Hepatol 21(4):222–247

Agus A, Planchais J, Sokol H (2018) Gut microbiota regulation of tryptophan metabolism in health and disease. Cell Host Microbe 23(6):716–724

Ahanger IA et al (2024) Comprehensive perspective towards the management of proteinopathies by elucidating protein misfolding and aggregation. CNS Neurol Disord Drug Targets 23(2):153–180

Alberti S, Hyman AA (2021) Biomolecular condensates at the nexus of cellular stress, protein aggregation disease and ageing. Nat Rev Mol Cell Biol 22(3):196–213

Allam-Ndoul B, Castonguay-Paradis S, Veilleux A (2020) Gut microbiota and intestinal trans-epithelial permeability. Int J Mol Sci 21(17):6402

Arar S, Haque MA, Kayed R (2023) Protein aggregation and neurodegenerative disease: structural outlook for the novel therapeutics. Proteins: Struct Funct Bioinf 93:1314

Athira KV et al (2020) An overview of the heterogeneity of major depressive disorder: current knowledge and future prospective. Curr Neuropharmacol 18(3):168–187

Badawy AAB, Guillemin GJ (2022) Species differences in tryptophan metabolism and disposition. Int J Tryptophan Res 15:11786469221122512

Bancher C et al (1989) Accumulation of abnormally phosphorylated τ precedes the formation of neurofibrillary tangles in Alzheimer's disease. Brain Res 477(1–2):90–99

Bashir B et al (2024) Opening avenues for treatment of neurodegenerative disease using postbiotics: breakthroughs and bottlenecks in clinical translation. Ageing Res Rev 95:102236
Bazrgar M et al (2021) The role of microRNA-34 family in Alzheimer's disease: a potential molecular link between neurodegeneration and metabolic disorders. Pharmacol Res 172:105805
Bibbò S et al (2020) Fecal microbiota transplantation: screening and selection to choose the optimal donor. J Clin Med 9(6):1757
Bicknell B et al (2023) Neurodegenerative and neurodevelopmental diseases and the gut-brain axis: the potential of therapeutic targeting of the microbiome. Int J Mol Sci 24(11):9577
Biernacki T et al (2020) Kynurenines in the pathogenesis of multiple sclerosis: therapeutic perspectives. Cells 9(6):1564
Bistoletti M et al (2020) The microbiota-gut-brain axis: focus on the fundamental communication pathways. Prog Mol Biol Transl Sci 176:43–110
Bleem A (2019) Biophysical strategies to inhibit bacterial amyloid formation and undermine biofilm infections. Diss.
Boehme M et al (2020) Mid-life microbiota crises: middle age is associated with pervasive neuroimmune alterations that are reversed by targeting the gut microbiome. Mol Psychiatry 25(10):2567–2583
Bonfili L et al (2021) Microbiota modulation as preventative and therapeutic approach in Alzheimer's disease. FEBS J 288(9):2836–2855
Bostanciklioğlu M (2019) The role of gut microbiota in pathogenesis of Alzheimer's disease. J Appl Microbiol 127(4):954–967
Boudry G et al (2021) The relationship between breast milk components and the infant gut microbiota. Front Nutr 8:629740
Brüssow H (2020) Problems with the concept of gut microbiota dysbiosis. Microb Biotechnol 13(2):423–434
Butel M-J, Waligora-Dupriet A-J, Wydau-Dematteis S (2018) The developing gut microbiota and its consequences for health. J Dev Orig Health Dis 9(6):590–597
Carlessi AS et al (2021) Gut microbiota–brain axis in depression: the role of neuroinflammation. Eur J Neurosci 53(1):222–235
Cerovic M, Forloni G, Balducci C (2019) Neuroinflammation and the gut microbiota: possible alternative therapeutic targets to counteract Alzheimer's disease? Front Aging Neurosci 11:284
Chan SL (1997) Lysosome-related generation of potentially amyloidogenic serum amyloid A polypeptides in activated murine macrophages and the potential role of ubiquitin in this process
Chaney SY (2013) Retinal development and age related degeneration following gestational lead exposure
Chen L et al (2022) Different oral and gut microbial profiles in those with Alzheimer's disease consuming anti-inflammatory diets. Front Nutr 9:974694
Chen F et al (2024) The vascular contribution of apolipoprotein E to Alzheimer's disease. Brain 147:awae156
Choi H, Lee D, Mook-Jung I (2022) Gut microbiota as a hidden player in the pathogenesis of Alzheimer's disease. J Alzheimer's Dis 86(4):1501–1526
Choudhuri S, Panda J, Maitra S (2022) Influence of gut microbial flora in body's serotonin turnover and associated diseases. In: Microbiome, immunity, digestive health and nutrition. Elsevier, pp 245–264
Connell E et al (2022) Microbial-derived metabolites as a risk factor of age-related cognitive decline and dementia. Mol Neurodegener 17(1):43
Cornut M, Bourdonnay E, Henry T (2020) Transcriptional regulation of inflammasomes. Int J Mol Sci 21(21):8087
Corona C et al (2011) New therapeutic targets in Alzheimer's disease: brain deregulation of calcium and zinc. Cell Death Dis 2(6):e176–e176
Cruz-Pereira JS (2022) Microbiota-gut-immune-brain communication across the lifespan. PhD Thesis, University College Cork
Cui Y et al (2020) Production of gamma-aminobutyric acid from lactic acid bacteria: a systematic review. Int J Mol Sci 21(3):995

d'Ettorre G et al (2020) Challenges in the management of SARS-CoV2 infection: the role of oral bacteriotherapy as complementary therapeutic strategy to avoid the progression of COVID-19. Front Med 7:389
Dabrowska S et al (2021) Immunomodulatory and regenerative effects of mesenchymal stem cells and extracellular vesicles: therapeutic outlook for inflammatory and degenerative diseases. Front Immunol 11:591065
Dalile B et al (2019) The role of short-chain fatty acids in microbiota–gut–brain communication. Nat Rev Gastroenterol Hepatol 16(8):461–478
Dash S, Syed YA, Khan MR (2022) Understanding the role of the gut microbiome in brain development and its association with neurodevelopmental psychiatric disorders. Front Cell Dev Biol 10:880544
Dey A et al (2024) Alzheimer's disease pathology and assistive nanotheranostic approaches for its therapeutic interventions. Int J Mol Sci 25(17):9690
Dieterich W, Schink M, Zopf Y (2018) Microbiota in the gastrointestinal tract. Med Sci 6(4):116
Doifode T et al (2021) The impact of the microbiota-gut-brain axis on Alzheimer's disease pathophysiology. Pharmacol Res 164:105314
Doney E et al (2022) Inflammation-driven brain and gut barrier dysfunction in stress and mood disorders. Eur J Neurosci 55(9–10):2851–2894
Doroszkiewicz J, Groblewska M, Mroczko B (2021) The role of gut microbiota and gut–brain interplay in selected diseases of the central nervous system. Int J Mol Sci 22(18):10028
Doroszkiewicz J, Groblewska M, Mroczko B (2022) Molecular biomarkers and their implications for the early diagnosis of selected neurodegenerative diseases. Int J Mol Sci 23(9):4610
Drobny A et al (2021) Molecular communication between neuronal networks and intestinal epithelial cells in gut inflammation and Parkinson's disease. Front Med 8:655123
Eckel EF, Ametaj BN (2020) Bacterial endotoxins and their role in periparturient diseases of dairy cows: mucosal vaccine perspectives. Dairy 1(1):61–90
Fock E, Parnova R (2023) Mechanisms of blood–brain barrier protection by microbiota-derived short-chain fatty acids. Cells 12(4):657
Fried S et al (2021) Interactions between the microbiota and enteric nervous system during gut-brain disorders. Neuropharmacology 197:108721
Fuloria S et al (2022) Introduction to microbiome. In: Microbiome in inflammatory lung diseases. Springer, pp 13–28
Gaiottino J et al (2013) Increased neurofilament light chain blood levels in neurodegenerative neurological diseases. PLoS One 8(9):e75091
Galea I (2021) The blood–brain barrier in systemic infection and inflammation. Cell Mol Immunol 18(11):2489–2501
Gallardo MJ, Delgado FO (2021) Animal prion diseases: a review of intraspecies transmission. Open Vet J 11(4):707–723
Gao K et al (2020) Tryptophan metabolism: a link between the gut microbiota and brain. Adv Nutr 11(3):709–723
Gay M et al (2018) A phenotypic approach to the discovery of compounds that promote non-amyloidogenic processing of the amyloid precursor protein: toward a new profile of indirect β-secretase inhibitors. Eur J Med Chem 159:104–125
Gershon MD, Margolis KG (2021) The gut, its microbiome, and the brain: connections and communications. J Clin Invest 131(18):e143768
Ghosh S, Ali R, Verma S (2023) Aβ-oligomers: a potential therapeutic target for Alzheimer's disease. Int J Biol Macromol 239:124231
Goyal D, Ali SA, Singh RK (2021) Emerging role of gut microbiota in modulation of neuroinflammation and neurodegeneration with emphasis on Alzheimer's disease. Prog Neuro-Psychopharmacol Biol Psychiatry 106:110112
Grüning CSR (2014) Selection and characterization of binding proteins specific for amyloidogenic proteins of Alzheimer disease

Hampl V, Roger AJ (2024) The evolutionary origin of mitochondria and mitochondrion-related organelles. In: Endosymbiotic organelle acquisition: solutions to the problem of protein localization and membrane passage. Springer, pp 89–121
Haq S, Grondin JA, Khan WI (2021) Tryptophan-derived serotonin-kynurenine balance in immune activation and intestinal inflammation. FASEB J 35(10):e21888
Hasan N, Yang H (2019) Factors affecting the composition of the gut microbiota, and its modulation. PeerJ 7:e7502
Hassan SS et al (2024) The role of bacteria in human health and disease: friend or foe. Eur J Mod Med Pract 4(9):280–297
Henriques APC (2021) Exploring antibacterial graphene-based biomaterials for hemodialysis catheters. Universidade do Porto (Portugal)
Hill TA et al (2014) Constraining cyclic peptides to mimic protein structure motifs. Angew Chem Int Ed 53(48):13020–13041
Hoffman R, Gerber M (2013) The Mediterranean diet: health and science. John Wiley & Sons
Hosseini E et al (2011) Propionate as a health-promoting microbial metabolite in the human gut. Nutr Rev 69(5):245–258
Hou K et al (2022) Microbiota in health and diseases. Signal Transduct Target Ther 7(1):1–28
Hu W et al (2016) Hyperphosphorylation determines both the spread and the morphology of tau pathology. Alzheimers Dement 12(10):1066–1077
Hu S, Yang C, Luo H (2022) Current trends in blood biomarker detection and imaging for Alzheimer's disease. Biosens Bioelectron 210:114278
Hughes TD et al (2022) The kynurenine pathway and kynurenine 3-monooxygenase inhibitors. Molecules 27(1):273
Ivanova MI et al (2021) Biophysical processes underlying cross-seeding in amyloid aggregation and implications in amyloid pathology. Biophys Chem 269:106507
Jaswal K, Todd OA, Behnsen J (2023) Neglected gut microbiome: interactions of the non-bacterial gut microbiota with enteric pathogens. Gut Microbes 15(1):2226916
Jerala R (2007) Structural biology of the LPS recognition. Int J Med Microbiol 297(5):353–363
Kato LM et al (2014) The role of the adaptive immune system in regulation of gut microbiota. Immunol Rev 260(1):67–75
Kaźmierczak-Siedlecka K et al (2020) Therapeutic methods of gut microbiota modification in colorectal cancer management–fecal microbiota transplantation, prebiotics, probiotics, and synbiotics. Gut Microbes 11(6):1518–1530
Kearns R (2024) The kynurenine pathway in gut permeability and inflammation. Inflammation 48:1–15
Khan S, Barve KH, Kumar MS (2020) Recent advancements in pathogenesis, diagnostics and treatment of Alzheimer's disease. Curr Neuropharmacol 18(11):1106–1125
Kim CH (2021) Control of lymphocyte functions by gut microbiota-derived short-chain fatty acids. Cell Mol Immunol 18(5):1161–1171
Konopelski P, Mogilnicka I (2022) Biological effects of indole-3-propionic acid, a gut microbiota-derived metabolite, and its precursor tryptophan in mammals' health and disease. Int J Mol Sci 23(3):1222
Kurowska A et al (2023) The role of diet as a modulator of the inflammatory process in the neurological diseases. Nutrients 15(6):1436
Lagier J-C et al (2012) Human gut microbiota: repertoire and variations. Front Cell Infect Microbiol 2:136
Leveille E, Ross OA, Gan-Or Z (2021) Tau and MAPT genetics in tauopathies and synucleinopathies. Parkinsonism Relat Disord 90:142–154
Levinthal DJ, Strick PL (2020) Multiple areas of the cerebral cortex influence the stomach. Proc Natl Acad Sci 117(23):13078–13083
Li M et al (2023) Caspase-1 affects chronic restraint stress-induced depression-like behaviors by modifying GABAergic dysfunction in the hippocampus. Transl Psychiatry 13(1):229

Liang S, Wu X, Jin F (2018) Gut-brain psychology: rethinking psychology from the microbiota–gut–brain axis. Front Integr Neurosci 12:33
Liu Z et al (2015) The ambiguous relationship of oxidative stress, tau hyperphosphorylation, and autophagy dysfunction in Alzheimer's disease. Oxidative Med Cell Longev 2015(1):352723
Liu S et al (2020) Gut microbiota and dysbiosis in Alzheimer's disease: implications for pathogenesis and treatment. Mol Neurobiol 57:5026–5043
Lu L et al (2021) Gut microbiota and bone metabolism. FASEB J 35(7):e21740
Ludolph AC et al (1998) Amyotrophic lateral sclerosis and glutamate. Restor Neurol Neurosci 13(1–2):59–67
Maiuolo J et al (2024) The postbiotic properties of butyrate in the modulation of the gut microbiota: the potential of its combination with polyphenols and dietary fibers. Int J Mol Sci 25(13):6971
Malik S, Yadav JK (2023) Amyloids and amyloid-like protein aggregates in foods: challenges and new perspectives. Curr Protein Pept Sci 24(5):393–403
Mantzavinos V, Alexiou A (2017) Biomarkers for Alzheimer's disease diagnosis. Curr Alzheimer Res 14(11):1149–1154
Maqsood R, Stone TW (2016) The gut-brain axis, BDNF, NMDA and CNS disorders. Neurochem Res 41:2819–2835
Marin IA et al (2017) Microbiota alteration is associated with the development of stress-induced despair behavior. Sci Rep 7(1):43859
Marizzoni M et al (2020) Short-chain fatty acids and lipopolysaccharide as mediators between gut dysbiosis and amyloid pathology in Alzheimer's disease. J Alzheimer's Dis 78(2):683–697
Markert J et al (2021) Food practices and nutrition of children and adolescents in residential care: a scoping review. Appetite 167:105640
Matej R, Tesar A, Rusina R (2019) Alzheimer's disease and other neurodegenerative dementias in comorbidity: a clinical and neuropathological overview. Clin Biochem 73:26–31
Mattsson N et al (2017) Association of plasma neurofilament light with neurodegeneration in patients with Alzheimer disease. JAMA Neurol 74(5):557–566
Metaxas A, Kempf SJ (2016) Neurofibrillary tangles in Alzheimer's disease: elucidation of the molecular mechanism by immunohistochemistry and tau protein phospho-proteomics. Neural Regen Res 11(10):1579–1581
Mindus C et al (2021) Lactobacillus-based probiotics reduce the adverse effects of stress in rodents: a meta-analysis. Front Behav Neurosci 15:642757
Mitrea L et al (2022) Guts imbalance imbalances the brain: a review of gut microbiota association with neurological and psychiatric disorders. Front Med 9:813204
Moda F et al (2023) Secondary protein aggregates in neurodegenerative diseases: almost the rule rather than the exception. Front Biosci (Landmark Ed) 28(10):255
Mohan M, Okeoma CM, Sestak K (2020) Dietary gluten and neurodegeneration: a case for preclinical studies. Int J Mol Sci 21(15):5407
Moloney CM, Lowe VJ, Murray ME (2021) Visualization of neurofibrillary tangle maturity in Alzheimer's disease: a clinicopathologic perspective for biomarker research. Alzheimers Dement 17(9):1554–1574
MoosWalter H et al (2016) Microbiota and neurological disorders: a gut feeling. BioRes Open Access 5:137
Morris G et al (2017) The role of the microbial metabolites including tryptophan catabolites and short chain fatty acids in the pathophysiology of immune-inflammatory and neuroimmune disease. Mol Neurobiol 54:4432–4451
Mukherjee S et al (2023) History, origin and types of neurological disorders. In: Applications of stem cells and derived exosomes in neurodegenerative disorders. Springer, pp 1–32
Murray E (2020) Immune challenge during puberty: role of the gut microbiota and neurobehavioural outcomes. Université d'Ottawa/University of Ottawa
Myhrer T (2003) Neurotransmitter systems involved in learning and memory in the rat: a meta-analysis based on studies of four behavioral tasks. Brain Res Rev 41(2–3):268–287
Narengaowa et al (2021) The oral-gut-brain AXIS: the influence of microbes in Alzheimer's disease. Front Cell Neurosci 15:633735

Niesler B et al (2021) Disorders of the enteric nervous system—a holistic view. Nat Rev Gastroenterol Hepatol 18(6):393–410
Nightingale JMD, Spiller R (2023) Normal intestinal anatomy and physiology. In: Intestinal failure. Springer, pp 13–33
O'Mahony SM et al (2015) Serotonin, tryptophan metabolism and the brain-gut-microbiome axis. Behav Brain Res 277:32–48
Palmqvist S et al (2019) Cerebrospinal fluid and plasma biomarker trajectories with increasing amyloid deposition in Alzheimer's disease. EMBO Mol Med 11(12):e11170
Palomares O et al (2017) Mechanisms of immune regulation in allergic diseases: the role of regulatory T and B cells. Immunol Rev 278(1):219–236
Panwar S, Sharma S, Tripathi P (2021) Role of barrier integrity and dysfunctions in maintaining the healthy gut and their health outcomes. Front Physiol 12:715611
Papadimitriou K et al (2015) Discovering probiotic microorganisms: in vitro, in vivo, genetic and omics approaches. Front Microbiol 6:58
Parolisi S et al (2023) Possible role of tryptophan metabolism along the microbiota-gut-brain axis on cognitive & behavioral aspects in phenylketonuria. Pharmacol Res 197:106952
Patra R, Das NC, Mukherjee S (2022) Toll-like receptors (TLRs) as therapeutic targets for treating SARS-CoV-2: an immunobiological perspective. In: Coronavirus therapeutics–volume I: basic science and therapy development. Springer, pp 87–109
Pedersen SHF (2021) Reviews of physiology, biochemistry and pharmacology. Springer Nature
Pedroza Matute S, Iyavoo S (2023) Exploring the gut microbiota: lifestyle choices, disease associations, and personal genomics. Front Nutr 10:1225120
Pistollato F et al (2016) Role of gut microbiota and nutrients in amyloid formation and pathogenesis of Alzheimer disease. Nutr Rev 74(10):624–634
Polazzi E, Contestabile A (2002) Reciprocal interactions between microglia and neurons: from survival to neuropathology. Rev Neurosci 13(3):221–242
Qian X et al (2022) Mechanisms of short-chain fatty acids derived from gut microbiota in Alzheimer's disease. Aging Dis 13(4):1252
Rademakers R, Cruts M, Van Broeckhoven C (2004) The role of tau (MAPT) in frontotemporal dementia and related tauopathies. Hum Mutat 24(4):277–295
Ragavan ML, Hemalatha S (2024) The functional roles of short chain fatty acids as postbiotics in human gut: future perspectives. Food Sci Biotechnol 33(2):275–285
Ramachandran R et al (2024) NLRP3 inflammasome: a key player in the pathogenesis of life-style disorders. Exp Mol Med 56(7):1488–1500
Rastall RA et al (2022) Structure and function of non-digestible carbohydrates in the gut microbiome. Benefic Microbes 13(2):95–168
Rastogi S et al (2021) The evolving landscape of exosomes in neurodegenerative diseases: exosomes characteristics and a promising role in early diagnosis. Int J Mol Sci 22(1):440
Réus GZ et al (2023) Major depressive disorder as a neuro-immune disorder: origin, mechanisms, and therapeutic opportunities. Neurosci Biobehav Rev 155:105425
Reutov VP, Sorokina EG (2022) Causal relationship between physiological and pathological processes in the brain and in the gastrointestinal tract: the brain–intestine axis. Biophysics 67(6):972–986
Rezende ESV, Lima GC, Naves MMV (2021) Dietary fibers as beneficial microbiota modulators: a proposed classification by prebiotic categories. Nutrition 89:111217
Rizzatti G et al (2017) Proteobacteria: a common factor in human diseases. Biomed Res Int 2017(1):9351507
Rodríguez AM, Rodríguez J, Giambartolomei GH (2022) Microglia at the crossroads of pathogen-induced neuroinflammation. ASN Neuro 14:17590914221104566
Rutsch A, Kantsjö JB, Ronchi F (2020) The gut-brain axis: how microbiota and host inflammasome influence brain physiology and pathology. Front Immunol 11:604179
Sadowski K et al (2024) Exploring fecal microbiota transplantation for modulating inflammation in parkinson's disease: a review of inflammatory markers and potential effects. Int J Mol Sci 25(14):7741

Saeed NK et al (2022) Gut microbiota in various childhood disorders: implication and indications. World J Gastroenterol 28(18):1875
Salas-Venegas V et al (2022) The obese brain: mechanisms of systemic and local inflammation, and interventions to reverse the cognitive deficit. Front Integr Neurosci 16:798995
Salazar J et al (2023) Exploring the relationship between the gut microbiota and ageing: a possible age modulator. Int J Environ Res Public Health 20(10):5845
Santoro A, Bientinesi E, Monti D (2021) Immunosenescence and inflammaging in the aging process: age-related diseases or longevity? Ageing Res Rev 71:101422
Sattianayagam PT (2012) The pathogenesis, investigation and management of systemic amyloidosis. UCL (University College London)
Savitz J (2020) The kynurenine pathway: a finger in every pie. Mol Psychiatry 25(1):131–147
Schächtle MA, Rosshart SP (2021) The microbiota-gut-brain axis in health and disease and its implications for translational research. Front Cell Neurosci 15:698172
Scott KP et al (2015) Manipulating the gut microbiota to maintain health and treat disease. Microb Ecol Health Dis 26(1):25877
Shanmugam H, Ganguly S, Priya B (2022) Plant food bioactives and its effects on gut microbiota profile modulation for better brain health and functioning in autism spectrum disorder individuals: a review. Food Front 3(1):124–141
Shin N-R, Whon TW, Bae J-W (2015) Proteobacteria: microbial signature of dysbiosis in gut microbiota. Trends Biotechnol 33(9):496–503
Siegel G et al (2017) The Alzheimer's disease γ-secretase generates higher 42: 40 ratios for β-amyloid than for p3 peptides. Cell Rep 19(10):1967–1976
Singh D (2022) Astrocytic and microglial cells as the modulators of neuroinflammation in Alzheimer's disease. J Neuroinflammation 19(1):206
Singh SK et al (2016) Overview of Alzheimer's disease and some therapeutic approaches targeting Aβ by using several synthetic and herbal compounds. Oxidative Med Cell Longev 2016(1):7361613
Snelson M et al (2021) Gut microbiome, prebiotics, intestinal permeability and diabetes complications. Best Pract Res Clin Endocrinol Metab 35(3):101507
Sochocka M, Diniz BS, Leszek J (2017) Inflammatory response in the CNS: friend or foe? Mol Neurobiol 54:8071–8089
Sochocka M et al (2019) The gut microbiome alterations and inflammation-driven pathogenesis of Alzheimer's disease—a critical review. Mol Neurobiol 56:1841–1851
Solanki R, Karande A, Ranganathan P (2023) Emerging role of gut microbiota dysbiosis in neuroinflammation and neurodegeneration. Front Neurol 14:1149618
Soto M et al (2018) Gut microbiota modulate neurobehavior through changes in brain insulin sensitivity and metabolism. Mol Psychiatry 23(12):2287–2301
Strandwitz P (2018) Neurotransmitter modulation by the gut microbiota. Brain Res 1693:128–133
Suganya K, Koo B-S (2020) Gut–brain axis: role of gut microbiota on neurological disorders and how probiotics/prebiotics beneficially modulate microbial and immune pathways to improve brain functions. Int J Mol Sci 21(20):7551
Sun J et al (2016) Antidepressant-like effects of sodium butyrate and its possible mechanisms of action in mice exposed to chronic unpredictable mild stress. Neurosci Lett 618:159–166
Tadesse S (2012) Probiotics, prebiotics and synbiotics as functional food ingredients: production, health benefits and safety. J Biol Act Prod Nat 2(3):124–134
Tai Y-C et al (2023) Irritable bowel syndrome based on Rome IV diagnostic criteria associates with non-motor symptoms of Parkinson's disease. Parkinsonism Relat Disord 113:105496
Takiishi T, Fenero CIM, Câmara NOS (2017) Intestinal barrier and gut microbiota: shaping our immune responses throughout life. Tissue Barriers 5(4):e1373208
Tanaka M et al (2021) Immune influencers in action: metabolites and enzymes of the tryptophan-kynurenine metabolic pathway. Biomedicine 9(7):734
Tang W et al (2021) Roles of gut microbiota in the regulation of hippocampal plasticity, inflammation, and hippocampus-dependent behaviors. Front Cell Infect Microbiol 10:611014
Teixeira MI et al (2020) Current insights on lipid nanocarrier-assisted drug delivery in the treatment of neurodegenerative diseases. Eur J Pharm Biopharm 149:192–217

Teleanu RI et al (2022) Neurotransmitters—key factors in neurological and neurodegenerative disorders of the central nervous system. Int J Mol Sci 23(11):5954
Ting EY-C, Yang AC, Tsai S-J (2020) Role of interleukin-6 in depressive disorder. Int J Mol Sci 21(6):2194
Turrigiano GG, Nelson SB (2004) Homeostatic plasticity in the developing nervous system. Nat Rev Neurosci 5(2):97–107
Valkanova V, Ebmeier KP, Allan CL (2013) CRP, IL-6 and depression: a systematic review and metaanalysis of longitudinal studies. Journal of affective disorders 150(3):736–44
Van Gerven N et al (2018) The role of functional amyloids in bacterial virulence. J Mol Biol 430(20):3657–3684
Van Giau V et al (2018) Gut microbiota and their neuroinflammatory implications in Alzheimer's disease. Nutrients 10(11):1765
Vandenplas Y et al (2020) Factors affecting early-life intestinal microbiota development. Nutrition 78:110812
Veit W (2022) Scaffolding natural selection. Biol Theory 17(2):163–180
Verhaar BJH et al (2022) Gut microbiota composition is related to AD pathology. Front Immunol 12:794519
Verma MS et al (2018) A common mechanism links activities of butyrate in the colon. ACS Chem Biol 13(5):1291–1298
Wang Y (2023) The role of the gut microbiota and microbial metabolites in the pathogenesis of Alzheimer's disease. CNS Neurol Disord Drug Targets 22(4):577–598
Wang W-Y et al (2015) Role of pro-inflammatory cytokines released from microglia in Alzheimer's disease. Ann Transl Med 3(10):136
Weissbrodt DG et al (2020) Basic microbiology and metabolism. In: Biological wastewater treatment: principles, modelling and design, 2nd edn. IWA Publishing, London, pp 11–76
Wernimont SM et al (2020) The effects of nutrition on the gastrointestinal microbiome of cats and dogs: impact on health and disease. Front Microbiol 11:1266
West CE, Jenmalm MC, Prescott SL (2015) The gut microbiota and its role in the development of allergic disease: a wider perspective. Clin Exp Allergy 45(1):43–53
Wicherska-Pawłowska K, Wróbel T, Rybka J (2021) Toll-like receptors (TLRs), NOD-like receptors (NLRs), and RIG-I-like receptors (RLRs) in innate immunity. TLRs, NLRs, and RLRs ligands as immunotherapeutic agents for hematopoietic diseases. Int J Mol Sci 22(24):13397
Wilczyńska K, Waszkiewicz N (2020) Diagnostic utility of selected serum dementia biomarkers: amyloid β-40, amyloid β-42, tau protein, and YKL-40: a review. J Clin Med 9(11):3452
Wilson BA, Winkler M, Ho BT (2020) Bacterial pathogenesis: a molecular approach. John Wiley & Sons
Winblad B et al (2016) Defeating Alzheimer's disease and other dementias: a priority for European science and society. Lancet Neurol 15(5):455–532
Xia P et al (2021) Research progress on toll-like receptor signal transduction and its roles in antimicrobial immune responses. Appl Microbiol Biotechnol 105(13):5341–5355
Xu K et al (2017) Redox properties of tryptophan metabolism and the concept of tryptophan use in pregnancy. Int J Mol Sci 18(7):1595
Yang J-J et al (2015) Serum interleukin-6 is a predictive biomarker for ketamine's antidepressant effect in treatment-resistant patients with major depression. Biol Psychiatry 77(3):e19–e20
Yu H et al (2024) The gut microbiome from middle-aged women with depression modulates depressive-like behaviors and plasma fatty acid metabolism in female middle-aged mice. J Psychiatr Res 173:139–150
Zhan Y et al (2023) The potential of the gut microbiome for identifying Alzheimer's disease diagnostic biomarkers and future therapies. Front Neurosci 17:1130730
Zhang H et al (2021a) Interaction between Aβ and tau in the pathogenesis of Alzheimer's disease. Int J Biol Sci 17(9):2181
Zhang X et al (2021b) An antibiotic-impacted microbiota compromises the development of colonic regulatory T cells and predisposes to dysregulated immune responses. MBio 12(1):10–1128
Zhao X et al (2019) Behavioral, inflammatory and neurochemical disturbances in LPS and UCMS-induced mouse models of depression. Behav Brain Res 364:494–502

Parkinson's Disease and Gut Microbiota

Swathi NL, Tungki Pratama Umar, Wesam Ibrahim Abo-Elenien, Shrikrishna Bhagat, Ahmed Sermed Al Sakini, and Sachchida Nand Rai

Abstract Parkinson's disease (PD) is a progressive neurodegenerative disorder characterized by motor symptoms, such as tremors, bradykinesia, rigidity, and postural instability. The underlying pathophysiology involves degeneration of dopaminergic neurons in the substantia nigra, leading to dopamine deficiency in the striatum. In addition, alpha-synuclein aggregation, mitochondrial dysfunction, and neuroinflammation contribute to neuronal death. Traditional treatment modalities primarily focus on symptomatic relief through pharmacological and surgical intervention. However, emerging research highlights the significant role of the gut microbiota in PD pathogenesis, suggesting that gut dysbiosis, alpha-synuclein propagation from the gut to the brain, and systemic inflammation may influence disease progression. This chapter aims to explore the synergy between the pathophysiology of Parkinsonism and treatment modalities, emphasizing the role of the gut microbiota and inflammation in PD. Furthermore, it compares traditional treatments with

Swathi NL (✉)
Pharm D (Ph.D), Department of Nursing, Manipal College Of Nursing, Manipal Academy of Higher Education, Manipal, Karnataka, India
e-mail: swathi.mconmpl2025@learner.manipal.edu

T. P. Umar
Division of Surgery and Interventional Science, Faculty of Medical Sciences, University College London, London, UK

W. I. Abo-Elenien
Faculty of Pharmacy, Alexandria University, Alexandria, Egypt

S. Bhagat
Life Sciences, Institute of Life Sciences, Bhubaneswar, Odisha, India

A. S. Al Sakini
College of Medicine, University of Baghdad, Baghdad, Iraq

S. N. Rai (✉)
Centre of Experimental Medicine and Surgery (CEMS), Institute of Medical Sciences (IMS), Banaras Hindu University (BHU), Varanasi, Uttar Pradesh, India
e-mail: sn.rai@bhu.ac.in

S. K. Singh et al. (eds.), *Microbiota Profiling for Precision Medicine*,
Interdisciplinary Biotechnological Advances,
https://doi.org/10.1007/978-981-95-2104-3_6

gut microbiota-based approaches and discusses their mechanisms, outcomes, and processes. Understanding these interactions is crucial for developing comprehensive therapeutic strategies that address both symptoms and underlying disease mechanisms.

Keywords Parkinson's disease · Gut microbiota · Microbiome · Pathophysiology · Inflammation · Alpha-synuclein · Treatment modalities · Traditional therapy · Microbiota-based therapy

1 Introduction

Parkinson's disease (PD) is a degenerative neurological disorder that not only predominantly affects motor function but also includes many non-motor symptoms. The present therapeutic landscape for Parkinson's disease is complex, with a focus on symptom management, enhancement of quality of life, and deceleration of disease progression.

Pharmacological Interventions: Pharmacological treatments constitute the foundation for Parkinson's disease management. Levodopa is the most often recommended medicine, typically administered in conjunction with carbidopa to augment its efficacy and mitigate adverse effects (Hayes 2019). Additional drugs include dopamine agonists, MAO-B inhibitors, and COMT inhibitors, which assist in managing motor symptoms and prolonging the effectiveness of levodopa (Reich and Savitt 2019; Bloem et al. 2021).

Physical and Occupational Therapies: Physiotherapy is essential for enhancing motor functions, including gait, balance, and mobility. Research has indicated that physiotherapy can yield considerable short-term improvements in speed, balance, and total motor function (Tomlinson et al. 2012). Occupational therapy aims to enhance activities of daily living and perceived occupational performance, ultimately improving patients' quality of life. Methods, including dual-task gait and balance training, have been advantageous for individuals with mild-to-moderate Parkinson's disease (De Freitas et al. 2020).

Complementary therapies such as music therapy and exercise are widely acknowledged for their beneficial effects on both motor and non-motor symptoms. Music therapy has demonstrated efficacy in enhancing motor skills, communication, swallowing, and mental health in patients with Parkinson's disease (Machado Sotomayor et al. 2021). Consistent exercise encompassing motor cognitive training and virtual reality rehabilitation has shown advantages in improving gait, balance, and overall physical health (Xu et al. 2019; Truijen et al. 2022; Johansson et al. 2023).

Advanced Surgical Treatments: For patients who exhibit an insufficient response to pharmacotherapy, surgical treatments such as deep brain stimulation (DBS) present an alternative therapeutic option. Deep brain stimulation (DBS) has been useful in ameliorating motor symptoms and augmenting quality of life, particularly when medication-related side effects pose significant challenges (Raza et al. 2019; Reich and Savitt 2019).

The incorporation of technology in the treatment of Parkinson's disease has introduced virtual reality (VR) and telerehabilitation as novel methodologies. Virtual reality rehabilitation training has demonstrated efficacy in enhancing gait, balance, and mobility, functioning as a viable alternative or adjunct to conventional rehabilitation techniques (Lei et al. 2019; Xu et al. 2019; Truijen et al. 2022; Monaghan et al. 2023). Telerehabilitation facilitates remote therapy sessions and enhances accessibility for individuals with mobility impairments (Truijen et al. 2022).

Care of Specific Symptoms: Addressing non-motor symptoms is crucial for holistic Parkinson's disease care. Treatments for dysphagia include interdisciplinary strategies incorporating speech-language pathology and nutritional adjustments. Pain management techniques are essential because pain is a common yet sometimes unacknowledged condition that profoundly affects patients' quality of life (Cattaneo and Jost 2023). Furthermore, the management of orthostatic hypotension, a disorder marked by a rapid decrease in blood pressure upon standing, necessitates both pharmacological and nonpharmacological interventions to stabilize blood pressure (Fanciulli et al. 2020).

1.1 Limitations of Traditional Therapies in Addressing Disease Progression

The symptoms of Parkinson's disease (PD), a degenerative neurological condition that mostly affects movement, include bradykinesia (slowness of movement), stiffness, and tremors. Traditional therapies for PD have largely focused on managing motor symptoms to improve patients' quality of life. However, these conventional treatments have notable limitations in terms of addressing disease progression.

One of the cornerstone treatments for PD is levodopa–carbidopa therapy, which replenishes dopamine levels in the brain to alleviate motor symptoms (Wirdefeldt et al. 2016). While effective in the early stages, the long-term use of levodopa–carbidopa often leads to diminishing returns. Patients may experience motor fluctuations, such as the "on-off" phenomenon, where the effectiveness of the medication varies throughout the day, and dyskinesias, which are involuntary erratic movements (Sivanandy et al. 2021). These side effects can significantly affect daily functioning and reduce the overall efficacy of treatment over time.

Recent advancements have introduced formulations, such as continuous subcutaneous foslevodopa/foscarbidopa infusions, to provide more stable dopamine levels. Clinical trials have shown that while these continuous infusions can reduce "off" time and improve motor function, they encounter challenges such as infusion site complications and the need for specialized delivery systems. Furthermore, these therapies do not address the underlying neurodegenerative processes in PD, indicating that disease progression continues unabated (Aldred et al. 2023).

Systematic reviews have highlighted that traditional pharmacological treatments primarily offer symptomatic relief without modifying the disease course (Sivanandy et al. 2021; Bougea 2024). Despite various drug therapies targeting different pathways involved in PD, none has been conclusively proven to halt or reverse the neurodegeneration characteristic of the disorder. This limitation underscores the necessity for novel therapeutic approaches that not only manage symptoms but also slow disease progression.

Additionally, traditional therapies often require frequent dosing and adjustments, which can burden both patients and health care providers. The variability in patient responses to these treatments further complicates management, making it difficult to maintain consistent long-term symptom control (Sivanandy et al. 2021; Bougea 2024).

While traditional therapies, such as levodopa–carbidopa and continuous infusion systems, play a crucial role in managing Parkinson's disease symptoms, they are limited in their ability to slow or stop disease progression. The side effects and diminishing efficacy over time highlight the need for innovative treatments that can address both the symptoms and the underlying pathological mechanisms of PD.

2 Parkinson's Disease Pathophysiology

Parkinson's disease (PD) is characterized by the accumulation of the protein alpha-synuclein (α-syn), also known as Lewy bodies, and cell death in the basal ganglia. Up to 70% of dopamine-secreting neurons in the substantia nigra pars compacta are affected by the end of life (Wolters and Braak 2006; Sulzer 2007). Parkinson's disease-related vagal nerve dysfunction and specific movement abnormalities are caused by dopaminergic neuronal loss (Wolters and Braak 2006; Yang et al. 2019). Prior to 1980, there was little research on the pathogenesis of Parkinson's disease. Parkinson's disease's cause was unknown when Lewy bodies (LBs) were first described. Neurites in the GI tract of those who have Parkinson's disease have been identified as one of the disease's clinical and pathological characteristics (Oyanagi et al. 1990; Edwards et al. 1992). However, little is known about LBs within the movement disorder community (Wakabayashi et al. 1988).

Gastrointestinal (GI) malfunction is an essential early non-motor symptom of PD. Approximately 80% of patients experience gastrointestinal issues (Kim and Sung 2015). Constipation is especially common in Parkinson's disease which can occur years prior motor symptoms appear (Cersosimo et al. 2013; Chen et al. 2015). Various epidemiological studies have found that constipation increases the risk of Parkinson's disease to some degree (Adams-Carr et al. 2016). Furthermore, it predicts a worse disease course in Parkinson's patients (Camacho et al. 2021). Constipation has been included as a clinical biomarker of Parkinson's disease (PD) in the International Movement Disorders Society's (MDS) diagnostic criteria for the prodromal phase of the disease (Berg et al. 2015), based on substantial evidence from previous studies.

In the early 1980s, when researchers discovered α-synuclein (α-syn) accumulation in enteric tissues of patients with Parkinson's disease, raising the possibility that the GI system serves as a significant contributor to the pathogenesis (Wakabayashi et al. 1988; Edwards et al. 1992; Wolters and Braak 2006; Minguez-Castellanos et al. 2007). Braak et al. (2003) suggested that pathogenic α-syn inclusions could originate in the gastrointestinal system (Hawkes et al. 2007). Epidemiological studies have found that vagotomy and appendectomy may reduce the risk of developing Parkinson's disease (Svensson et al. 2015; Liu et al. 2017). Animal studies have supported this hypothesis. Kim et al. (2019) found that truncal vagotomy prevents α-syn transmission from the gut to the central nervous system (CNS) in rats.

Moreover, α-syn has previously been shown to be transmitted to the CNS and higher cortical regions via the transsynaptic cell-to-cell route (Klingelhoefer and Reichmann 2015). However, there is insufficient evidence to support the origin of α-syn in the gut. In addition, some studies have produced contradictory results. Some studies have suggested that α-syn transmission along the gut–brain axis (GBA) may be bidirectional (Ulusoy et al. 2017; Arotcarena et al. 2020).

The findings regarding changes in the gut microbiota in patients with PD are not the same, but some have been reinforced with strong evidence (Ulusoy et al. 2017; Arotcarena et al. 2020; Prajjwal et al. 2023). In particular, patients with PD have lower *Lachnospiraceae* and *Prevotellaceae* abundances but higher *Verrucomicrobiaceae* and *Lactobacillaceae* abundances (Zhang et al. 2023). It is noteworthy that variations in gut microbiota may exist among different racial/ethnic groups and that these variations may be associated with elements such as genetic heritage, dietary habits, environmental conditions, and the type of testing employed. Moreover, microbiota modifications (such as increased abundance of Christensenellaceae or Oscillospira) are linked to an increased risk of developing PD, implying that specific microbiome changes can be used to diagnose the disease early (Zhang et al. 2023).

2.1 Gut–Brain Axis in PD Development

The gut microbiota plays a crucial role in the regulation of gastrointestinal and immunological functions. Numerous human problems, including gastrointestinal, neurological, metabolic, pulmonary, and cardiovascular diseases, have been identified to be associated with the disruption of gut microbial equilibrium (Lynch and Pedersen 2016). The body's second brain, gut bacteria, is the essential center of the gut–brain connection (Rhee et al. 2009). This intricate process is characterized by the term "microbiota-gut-brain" (Rhee et al. 2009). PD is known to affect not only the CNS but also the digestive system and enteric nervous system (ENS). Pathological changes in the ENS appear early in PD even before pathological changes in the CNS (Montanari et al. 2023). Neuronal and glial injuries within the ENS have been observed in patients with Parkinson's disease and animal models (Chalazonitis and Rao 2018). Several research groups have reported Lewy cell pathology in biopsied enteric neurons from patients with Parkinson's disease. Parkinson's disease patients have been found to have atrophic degeneration of neurons in the myenteric and submucosal plexuses, as well as colocalized α-syn deposits (Chalazonitis et al. 2022). Constipation symptoms in patients may be caused by dysfunction of vasoactive intestinal peptide secretomotor neurons in the submucosal plexus (Giancola et al. 2017). Furthermore, glial markers, such as GFAP and Sox-10, are elevated and linked to pro-inflammatory cytokines in the gut (Natale et al. 2021). Animal studies have shown that aged rats experience neuronal loss and changes in neurochemical phenotypes of the ENS, including dystrophic enteric neurons with α-syn aggregates. Pathologic α-syn can also affect the cytoskeleton of ENS neurons (Zhang et al. 2023).

As Parkinson's disease (PD) is linked to the gut–brain axis, dysbiosis related to the gut and changes in microbe-derived components are significant risk factors for Parkinson's disease (Zhang et al. 2023). The etiology of PD can be clarified by researching the mechanism through which the microbiota–gut–brain axis influences the neurological system.

3 Traditional Treatment Modalities for Parkinson's Disease

Parkinson's disease (PD) is a progressive neurodegenerative disorder characterized by the loss of dopamine-producing neurons in the substantia nigra leading to both motor and non-motor symptoms. The management of PD primarily focuses on symptom relief and improving quality of life. Traditional treatment modalities include pharmacological and surgical interventions that target different stages of the disease. This section discusses the different classes of medications used in PD, their mechanisms of action, the limitations of traditional therapies, and treatments tailored for the early, mid, and advanced stages of the disease.

3.1 Classes of Medications Used in PD and Their Mechanisms of Action

The management of Parkinson's disease (PD) relies heavily on a range of traditional pharmacological therapies aimed at alleviating motor symptoms and improving the quality of life. Each class of medication targets specific aspects of disease pathology or symptomatology, with varying mechanisms of action, efficacy, and side effect profiles. Figure 1 provides a visual summary of the therapeutic options.

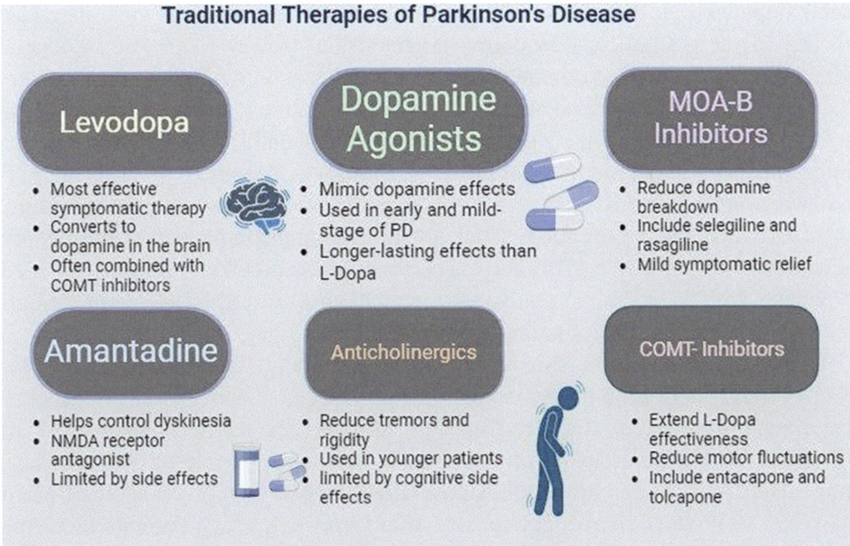

Fig. 1 Synopsis of conventional pharmacological treatments for Parkinson's disease

3.1.1 Levodopa

Levodopa remains the cornerstone of Parkinson's disease (PD) treatment due to its effectiveness in alleviating motor symptoms such as bradykinesia, rigidity, and tremor. It is a precursor of dopamine, which crosses the blood–brain barrier (BBB) and is converted into dopamine by the enzyme aromatic L-amino acid decarboxylase (AADC) within the brain. To enhance its efficacy and reduce peripheral side effects such as nausea and vomiting, levodopa is often combined with carbidopa or benserazide, which inhibits peripheral conversion (Olanow et al. 2001). Although levodopa is highly effective, its long-term use is associated with motor complications including motor fluctuations and dyskinesias. These challenges necessitate careful dose management and potential introduction of adjunct therapies. Additionally, optimizing dopaminergic therapies is crucial in managing PD-related pain, particularly fluctuation-related pain, as clinical observations have shown that patients frequently experience pain during their "off" states. Adjustments in levodopa dosing can improve non-motor fluctuations, including pain, which are integral to the overall symptomatology of PD (Rukavina et al. 2019).

3.1.2 Dopamine Agonists

The brain's dopamine receptors are directly stimulated by dopamine agonists such as rotigotine, pramipexole, and ropinirole. Unlike levodopa, these medications do not require conversion to dopamine and have a longer half-life, thus providing more stable dopaminergic stimulation. Dopamine agonists can be used as monotherapies in early PD or as adjuncts to levodopa in later stages. However, they are associated with side effects, such as somnolence, hallucinations, and impulse control disorders (Jankovic and Poewe 2012). The decision to use dopamine agonists versus levodopa in early PD involves balancing the need for symptom control with the risk of developing side effects and motor complications. Moreover, dopamine agonists are particularly associated with impulse control disorders, such as pathological gambling, hypersexuality, and binge eating, which are more common in patients with a history of addiction or impulsive personality. Therefore, these drugs are typically avoided in patients with these risk factors, and are generally not prescribed to elderly patients because of their high risk of hallucinations (Kalia and Lang 2015).

Dopamine agonists can be used as both monotherapy and adjuvant therapy for PD by directly activating postsynaptic dopamine receptors and avoiding degenerating dopamine neurons. Additionally, they have longer half-lives than those of levodopa. While pramipexole and ropinirole are extended-release formulations, immediate formulations are administered three times a day. A transdermal patch was used to produce the rotigotine. But after three–five years, patients receiving dopamine agonist monotherapy frequently require levodopa as an adjuvant (Stocchi et al. 2015; Zesiewicz et al. 2020).

3.1.3 Monoamine Oxidase-B (MAO-B) Inhibitors

Monoamine oxidase-B (MAO-B) inhibitors, such as selegiline and rasagiline, increase dopaminergic activity by inhibiting the MAO-B enzyme, which breaks down dopamine in the brain. They are used in early Parkinson's disease (PD) to delay the need for levodopa and to enhance its effects. Although there are potential neuroprotective effects, conclusive evidence is lacking. MAO-B inhibitors, such as safinamide, alleviate the pain associated with motor fluctuations (Jankovic and Poewe 2012; Rukavina et al. 2019). While MAO-B inhibitors are generally well tolerated, serotonin syndrome is a rare but serious side effect, particularly when combined with selective serotonin reuptake inhibitors (SSRIs). However, this risk is significantly lower than that of non-selective MAO inhibitors (Riederer and Laux 2011).

3.1.4 Catechol-O-Methyltransferase (COMT) Inhibitors

COMT inhibitors such as entacapone and tolcapone extend the half-life of levodopa by inhibiting its metabolism in the periphery. This results in more stable plasma levels of levodopa and reduced motor fluctuations in patients with advanced PD. COMT inhibitors have been used as adjuncts to levodopa therapy (Olanow et al. 2001). Tolcapone, although more potent, is less frequently used because of its association with hepatotoxicity, necessitating regular monitoring of liver function.

3.1.5 Anticholinergics

By inhibiting the action of acetylcholine, anticholinergic drugs like benztropine and trihexyphenidyl can lessen stiffness and tremors. However, owing to their side effects, including cognitive impairment, dry mouth, and urinary retention, their use is generally limited to younger patients or those with predominant tremors (Olanow et al. 2001). Anticholinergics are generally avoided in older adults because of their heightened risk of cognitive side effects.

3.1.6 Amantadine

Amantadine is an N-methyl-D-aspartate (NMDA) receptor antagonist that provides modest symptomatic relief in PD and is particularly effective against levodopa-induced dyskinesia. Although the exact mechanism of action is not fully understood, it is believed to involve increased dopamine release and inhibition of dopamine reuptake. In addition, amantadine has antiviral properties and has historically been used to treat influenza. The adverse effects of amantadine include confusion, hallucinations, and livedo reticularis. Its potential neuroprotective benefits remain the subject of ongoing research, although clinical trials confirming its efficacy in slowing PD progression are lacking (Yuan et al. 2010; Giugni and Okun 2014).

3.2 Limitations of Traditional Therapies

Traditional Parkinson's therapies, such as levodopa and dopaminergic medications, have improved symptom management and quality of life for patients. However, as the disease progresses, challenges arise, including motor complications, side effects of chronic medication use, and inadequate control of non-motor symptoms. Motor complications such as fluctuating symptom control and dyskinesias become more common with prolonged levodopa use, necessitating complex medication regimens and adjustments. Additionally, traditional therapies often fail to address non-motor symptoms, such as cognitive decline, mood disorders, and autonomic dysfunction, necessitating a more holistic approach involving multidisciplinary teams.

Table 1 summarizes some of the key limitations of traditional Parkinson's disease therapies, along with relevant references.

Table 1 Key limitations of traditional Parkinson's disease therapies

Limitation	Description	Reference
Motor fluctuations	Long-term use of levodopa can lead to fluctuations in motor control, characterized by "on" periods (good symptom control) and "off" periods (return of symptoms)	Ahlskog and Muenter (2001) and Stocchi (2006)
Dyskinesias	Levodopa-induced dyskinesias are involuntary, erratic movements caused by excessive dopaminergic stimulation, often occurring after prolonged use of the medication	Ahlskog and Muenter (2001) and Olanow et al. (2001)
Non-motor symptoms	Traditional therapies primarily target motor symptoms and are less effective against non-motor symptoms like cognitive decline, mood disorders, and autonomic dysfunction	Poewe and Mahlknecht (2009)
Side effects	Dopaminergic treatments are associated with side effects such as hallucinations, orthostatic hypotension, and impulse control disorders, limiting their use, especially in the elderly	Jankovic (2008) and Zesiewicz et al. (2020)
Cognitive impairment and psychiatric effects	Dopamine agonists and anticholinergics are inappropriate for elderly people due to the possibility of cognitive impairment, hallucinations, and other mental side effects	Olanow et al. (2001) and Jankovic (2008)
Complicated dosing regimens	The need to adjust dosages and incorporate adjunct therapies to manage motor complications and side effects can complicate treatment regimens and reduce patient compliance	Olanow et al. (2001) and Zesiewicz et al. (2020)

3.3 *Pharmacological Treatment for Parkinson's Disease*

3.3.1 Levodopa Therapy

Levodopa remains the most effective treatment for Parkinson's disease (PD). It works by replenishing dopamine levels and significantly improving motor symptoms and quality of life (QoL). Concerns regarding levodopa-induced neurotoxicity have been widely investigated, with studies showing no conclusive evidence that levodopa accelerates disease progression (Jankovic and Poewe 2012). Early use of levodopa can restore basal ganglia function, but it may also lead to earlier onset of motor complications, such as dyskinesia, particularly in younger patients.

Advantages: Provide robust symptom relief and improve QoL.

Disadvantages: Early initiation may result in motor complications, such as dyskinesia.

3.3.2 Dopamine Agonists

Dopamine agonists, such as pramipexole and ropinirole, can be used as monotherapies in the early stages of PD to delay levodopa initiation. These drugs directly stimulate dopamine receptors and offer effective symptom control with a lower risk of motor complications than levodopa (Rascol et al. 2000). They are especially advantageous in younger patients but can cause neuropsychiatric side effects, such as impulse control disorders, hallucinations, and cognitive issues.

Advantages: Delays levodopa therapy and reduces the risk of motor complications.

Disadvantages: Requires monitoring of neuropsychiatric side effects.

3.3.3 MAO-B Inhibitors

Monoamine oxidase B (MAO-B) inhibitors such as rasagiline and selegiline reduce dopamine breakdown in the brain and provide mild symptomatic relief in PD. These inhibitors are particularly useful in patients with mild symptoms who wish to delay the initiation of levodopa therapy (Jankovic and Poewe 2012). While there is ongoing research into their neuroprotective effects, clinical benefits in slowing disease progression have not yet been conclusively proven.

Advantages: Mild symptomatic relief and potential delay in levodopa initiation.

Disadvantages: Uncertain neuroprotective effects.

3.3.4 COMT Inhibitors

Catechol-O-methyltransferase (COMT) inhibitors such as entacapone and tolcapone are typically added to levodopa therapy to prolong its effects by inhibiting its breakdown. This helps in managing motor fluctuations and reducing the number and severity of "off" periods by providing more consistent dopaminergic stimulation (Olanow et al. 2001).

Advantages: Extending levodopa duration and reducing motor fluctuations.

Disadvantages: Can increase the risk of levodopa-related side effects.

3.3.5 Deep Brain Stimulation (DBS)

Patients with advanced Parkinson's disease (PD) who have severe motor fluctuations and dyskinesias even after taking the best medication can benefit from DBS, a surgical procedure. Electrodes are implanted in areas such as the subthalamic nucleus or globus pallidus interna to modulate brain activity, improve motor control, and reduce medication requirements (Jankovic 2008). DBS can greatly enhance QoL, but carries risks such as surgical complications and neuropsychiatric side effects.

Advantages: Reduce motor complications, improve motor function, and reduce medication needs.

Disadvantages: Surgical risks, careful patient selection, and postoperative management.

3.3.6 Levodopa–Carbidopa Intestinal Gel (LCIG)

Levodopa–carbidopa intestinal gel (LCIG) therapy provides continuous delivery of levodopa and carbidopa via a percutaneous endoscopic gastrostomy (PEG) tube directly into the small intestine. This method ensures stable plasma levels and reduces motor fluctuations and "off" periods in patients with advanced PD (Antonini et al. 2017). LCIG can significantly improve motor control and QoL, although it is invasive and carries risks related to the procedure, such as infection or tube dislocation.

Advantages: Continuous dopaminergic stimulation reduces motor complications.

Disadvantages: Invasiveness and procedural risks.

3.3.7 Continuous Dopaminergic Stimulation (CDS)

Continuous dopaminergic stimulation (CDS) aims to provide a stable dopaminergic environment in the brain to reduce motor fluctuations and dyskinesia. LCIG or a continuous subcutaneous apomorphine infusion can accomplish this. By eliminating the peaks and troughs associated with oral drugs, CDS helps people with advanced Parkinson's disease (PD) manage their motor symptoms and quality of life (Stocchi et al. 2014). However, CDS methods require close monitoring to manage potential complications such as device-related infections or site reactions.

Advantages: Reduces motor fluctuations and enhances QoL.

Disadvantages: Invasive, with potential complications at infusion sites.

3.3.8 Neuroprotective Agents

Despite extensive research, no neuroprotective therapy has been reported to alter the course of PD. MAO-B inhibitors such as rasagiline have demonstrated neuroprotective potential in preclinical models; however, their clinical impact remains uncertain

(Olanow et al. 2000). Other agents such as antioxidants (e.g., coenzyme Q10) and anti-inflammatory drugs have been explored without conclusive benefits in large clinical trials.

Advantages: Ongoing research is ongoing to explore potential treatments.

Disadvantages: There are no proven neuroprotective agents to date.

4 Gut Microbiota and Parkinson's Disease

4.1 The Role of Gut Microbiota in Brain Health

Gut microbiota, often referred to as the "second brain," plays an essential role in human health and influences multiple physiological processes. This ecosystem comprises approximately 100 trillion microorganisms, including bacteria, fungi, and viruses, and its gut microbiome genome contains approximately three million genes, vastly exceeding the human genome (Franzosa et al. 2015). Through a coevolutionary process, humans and their gut microbiota have established a symbiotic relationship; while the host provides a nutrient-rich environment, the microbiota supports digestion, immunity, and other physiological functions (Caballero-Flores et al. 2023).

Central to the influence of gut microbiota on brain health is the microbiota–gut–brain axis (GBA), a bidirectional communication network connecting the gut and the central nervous system (CNS). This axis operates through the immune, endocrine, metabolic, and neural pathways (Carabotti et al. 2015; Mart'yanov et al. 2021). Advances in omics technologies have revealed that the gut microbiota impacts brain functions through metabolites such as short-chain fatty acids (SCFAs), amino acids, and polyunsaturated fatty acids (PUFAs). These metabolites contribute to brain processes, such as myelination, microglial activation, and synaptic plasticity, thereby shaping complex behaviors (Hou et al. 2021; Damiani et al. 2023).

In turn, the brain affects gut health via the hypothalamic–pituitary–adrenal (HPA) axis and autonomic nervous system. For example, stress-induced norepinephrine secretion can promote proliferation of gut pathogens (Mart'yanov et al. 2021). Experimental findings, such as reduced anxiety-like behavior in germ-free mice and the anxiolytic effects of probiotics such as *Lactobacillus rhamnosus* (JB-1), further underscore the interdependence between the gut and brain (Heijtz et al. 2011).

4.2 Evidence Linking Dysbiosis to Parkinson's Disease

Gut dysbiosis, an imbalance in gut microbiota composition, has emerged as a potential contributor to neurodegenerative diseases, including Parkinson's disease (PD). Dysbiosis can result from factors such as dietary changes, antibiotic use, infections, aging, and diseases that disrupt gut homeostasis (Ma et al. 2019).

Dysbiosis has been linked to hallmark pathophysiological processes in PD, including alterations in neurotransmitter systems (dopaminergic, serotonergic, and glutamatergic), increased neuroinflammation, and motor dysfunction. Studies have shown that gut microbial changes often precede the onset of motor symptoms in PD, suggesting a preclinical role of dysbiosis (Ferreiro et al. 2023; Palacios et al. 2023).

Notably, reduced levels of *Prevotellaceae* and increased levels of *Enterobacteriaceae* have been observed in patients with PD, with the latter correlating with motor symptom severity (Scheperjans et al. 2015). Other findings include overgrowth of genera such as *Bifidobacterium*, *Lactobacillus*, and *Akkermansia* and decreased levels of SCFA-producing bacteria, such as *Lachnospiraceae* and *Faecalibacterium* (Nishiwaki et al. 2020; Ji et al. 2021). Meta-analyses have consistently linked dysbiosis with a pro-inflammatory gut environment in PD patients, contributing to gastrointestinal symptoms and systemic inflammation (Romano et al. 2021; Wallen et al. 2022).

Interestingly, the gut microbiome in individuals with rapid eye movement (REM) sleep behavior disorder (RBD), a prodromal phase of PD, also exhibits dysbiotic patterns, including reduced butyrate-producing bacteria and increased pro-inflammatory genera, such as *Collinsella*. These findings suggest that gut dysbiosis may be an early biomarker for neurodegeneration (Huang et al. 2023).

4.3 *Mechanisms of Gut Microbiota's Contribution to Parkinson's Disease*

The link between gut dysbiosis and Parkinson's disease extends to several mechanisms, including inflammation, α-synuclein aggregation, and altered gut permeability.

1. Alpha-Synuclein Aggregation and Gut–Brain Communication.

Misfolded α-synuclein, a hallmark of PD, has been detected in the enteric nervous system (ENS), where it may originate and propagate into the brain via the vagus nerve (Braak et al. 2003; Breen et al. 2019). This prion-like transmission is hypothesized to be triggered by microbial dysbiosis, which promotes aggregation of α-synuclein fibrils. For instance, bacteria involved in secondary bile acid synthesis may elevate deoxycholic acid levels, facilitating pathological α-synuclein formation (Wlodarska et al. 2017).

2. Increased Gut Permeability ("Leaky Gut").

Gut dysbiosis often disrupts the intestinal barrier integrity, leading to increased gut permeability. This "leaky gut" condition allows bacterial endotoxins, such as lipopolysaccharides (LPS), to translocate into the systemic circulation and CNS, triggering neuroinflammatory cascades (Keshavarzian et al. 2015).

3. Neuroinflammation.

Inflammation is central to PD pathogenesis, with elevated levels of cytokines (interleukin-6 (IL-6), tumor necrosis factor alpha (TNF-α), and IL-1β) observed in the brains and cerebrospinal fluid of patients (Lin et al. 2019). Dysbiosis-induced gut inflammation exacerbates systemic immune activation, creating a feedback loop that fuels neuroinflammation. Specific gut pathogens, such as *Escherichia coli* and *Ralstonia*, have been implicated in intestinal inflammation and endotoxin-mediated α-synuclein aggregation (Forsyth et al. 2011).

4. Microbial Metabolites and Oxidative Stress.

The metabolic output of the gut microbiota influences oxidative stress and inflammation in PD. For example, reduced SCFA production may impair microglial function, whereas elevated indole metabolite levels are linked to anti-inflammatory properties in PD models (Kishimoto et al. 2019).

4.4 *Impact of Gut Microbiota on Systemic and Neuroinflammation*

Neuroinflammation mediated by activated microglia is a critical process in PD progression. Gut dysbiosis amplifies systemic inflammation, which can cross the blood–brain barrier and activate microglia in the CNS. Dysbiotic microbiota may also alter dopamine and serotonin metabolism, further exacerbating neuroinflammatory responses (Sampson et al. 2016).

Under normal conditions, SCFAs produced by gut bacteria play an anti-inflammatory role. However, SCFA deficiencies in patients with PD may predispose them to microglial overactivation, contributing to sustained neuroinflammation and neurodegeneration (Sanchez-Guajardo et al. 2013). Similarly, LPS from gut pathogens activates toll-like receptor 4 (TLR4) signaling, reducing autophagy in microglia and impairing their ability to clear α-synuclein aggregates (Tu et al. 2021).

Furthermore, interventions targeting gut dysbiosis, such as probiotic supplementation, dietary modifications, or even acupuncture, have shown the potential to modulate gut–brain interactions, reduce inflammation, and slow neurodegenerative processes in preclinical PD models (Jang et al. 2020) (Fig. 2).

5 Inflammation in Parkinsonism

Neuroinflammation is an important feature of neurodegenerative diseases including Parkinson's disease (PD). This multifactorial process involves distinct genetic, environmental, and molecular factors (Vos et al. 2016). Several genetic mutations and polymorphisms are linked to elevated neuroinflammation in PD, including

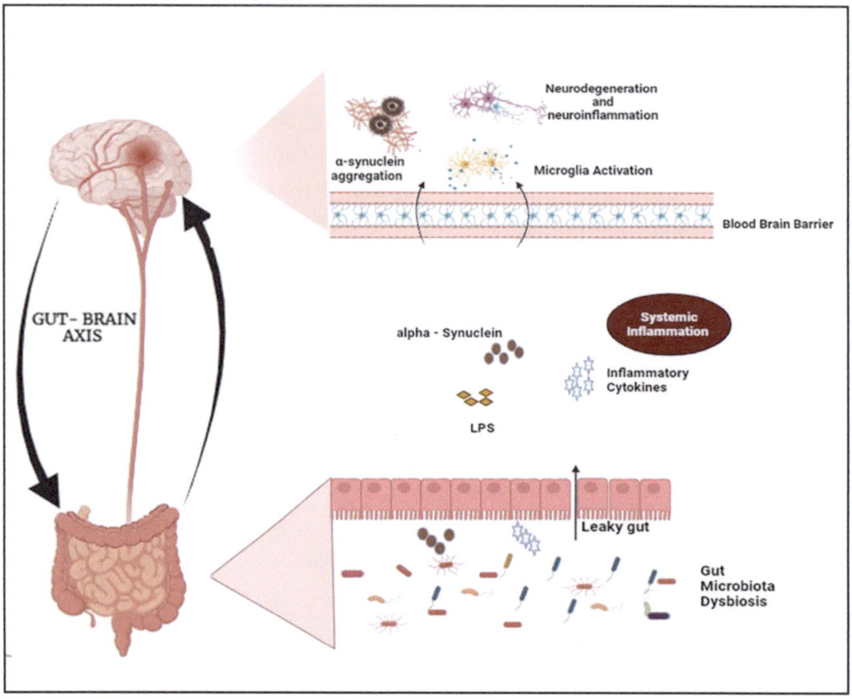

Fig. 2 Gut microbiota dysbiosis induces pathogenesis of PD through gut–brain axis

mutations in the alpha-synuclein-encoding (SNCA), LRRK2, GBA, and PARK7 (deglycase DJ-1-encoding) genes (Tansey et al. 2022). These mutations are associated with abnormal microglial and astrocyte activation, inadequate lysosomal functionality, and impaired control of oxidative stress (Kam et al. 2020). Exposure to pesticides including paraquat and rotenone has been linked to an increased likelihood of developing PD (Chen et al. 2018). These substances cause oxidative stress and microglial activation, contributing to neuroinflammation. Neuroinflammation in PD is closely associated with α-synuclein aggregation, mitochondrial dysfunction, and oxidative stress (Castillo-Rangel et al. 2023).

5.1 *Neuroinflammation and PD Progression*

Neuroinflammation involves numerous cellular (particularly neurons, macroglia, and microglia) and molecular mechanisms in the central nervous system (CNS). Its etiology can be broadly classified into infectious and non-infectious causes. Infectious agents that may induce neuroinflammation include bacteria (e.g., *Listeria monocytogenes* and *Neisseria meningitidis*), viruses (e.g., dengue, West Nile virus, measles, and Rubella virus), fungi (Cryptococcus), and parasites (*Plasmodium*,

Schistosomiasis, and *Toxoplasma gondii*) (Alloo et al. 2022; Chekrouni et al. 2022; Nurmukanova et al. 2024). Non-infectious diseases include neurotrauma, autoimmune diseases, and neurodegenerative disorders (Kölliker-Frers et al. 2021). Furthermore, metabolic problems and toxic and chemical exposure (lead, mercury, and pesticides) (Nabi and Tabassum 2022) may lead to a neuroinflammatory response.

Neuroinflammation has been linked to the stimulation of microglia, adaptive immune responses, and astrocytes in the brain parenchyma, as well as the existence of inflammatory mediators, such as cytokines, chemokines, nitric oxide synthase (NOS), and reactive oxygen species (ROS) (Kölliker-Frers et al. 2021). Microglial activation and intricate neuroinflammatory cascades are linked to various variables such as genetic makeup, triggering insult (pathogen/tissue trauma), and age (Gao et al. 2023). Nuclear factor-kappaB (NF-κB) activation is one of the most significant pathways because it causes the production of cyclooxygenase (COX)-2, an essential component for enhancing neuroinflammation following the release of pro-inflammatory cytokines, chemokines, and inducible NOS (iNOS) (Selvakumar et al. 2020). The severity of neuroinflammation influences the occurrence of chronic/remitting neurological illnesses such as PD.

Neuroinflammation in Parkinsonism includes glial cell stimulation, such as microglia, astrocytes, and oligodendrocytes (Castillo-Rangel et al. 2023). Microglia, the immune cells (macrophages) that inhabit the CNS (comprising approximately 10–15% of all brain cells) (Xavier et al. 2014), are crucial for controlling neuroinflammatory reactions. In Parkinsonism, microglia are activated in response to different stimuli, including alpha-synuclein aggregation, neuronal damage, and environmental contaminants. Activated microglia secrete pro-inflammatory cytokines, such as TNF-α, IL-1β, and IL-6, as well as reactive oxygen species (ROS), which contribute to the destruction and death of neurons (Azevedo et al. 2013). Astrocytes, the most prevalent type of glial cell, secrete cytokines, chemokines, and other inflammatory mediators. These substances can affect microglial cell activity and neuronal survival. In addition, astrocytes play a role in preserving the blood–brain barrier (BBB) and controlling synaptic function (Manu et al. 2023), both of which are disturbed in Parkinsonism. With significant breakdown of the BBB, immune cells can penetrate the CNS and bring inflammatory substances into it (Brandl and Reindl 2023). While the specific impact of oligodendrocyte dysfunction and death on neuroinflammation in Parkinsonism has not been well studied, it can worsen the degeneration of neurons by hindering their connection and function.

5.2 *Alpha-Synuclein Pathways*

Alpha-synuclein is a fundamental protein in PD etiology. PD is characterized by misfolded and aggregated alpha-synuclein known as Lewy bodies. These aggregates can cause inflammation by interacting with the microglia and astrocytes (Mhyre et al. 2012). Alpha-synuclein stimulates microglia via toll-like receptors

(TLRs), specifically TLR2 and TLR4, causing the release of pro-inflammatory cytokines and ROS, ultimately leading to a loop of neuroinflammation and neurodegeneration (Deyell et al. 2023). Furthermore, alpha-synuclein clumps can alter mitochondrial activity, cause oxidative stress, and amplify inflammatory reactions (Choi et al. 2022). Mitochondrial malfunction and oxidative stress interact with the inflammatory pathways, resulting in a vicious cycle that exacerbates brain damage. Furthermore, damaged mitochondria emit damage-associated molecular pattern (DAMPs) molecules such as mitochondrial DNA (mtDNA), which are identified by pattern recognition receptors (PRRs) on microglia, triggering them and initiating an inflammatory reaction (Wilkins et al. 2017). NLRP3 inflammasome, a multiprotein complex that regulates innate immune responses, has been linked to alpha-synuclein-induced inflammation. When the NLRP3 inflammasome is activated, it releases IL-1β, which contributes to the inflammatory environment in parkinsonism (Li et al. 2018; Lyra et al. 2023).

5.3 *Inflammasomes*

Inflammasomes are multiprotein complexes that play essential roles in the innate immune system. These include intracellular pattern recognition receptors (PRRs) and cellular stress-induced damage-associated molecular patterns (DAMPs). Inflammasomes are primarily composed of three types of proteins: sensors, adaptors, and effector molecules (Dai et al. 2023). The best-studied inflammasome complex, NLRP3, is composed of a sensor molecule and an adaptor molecule that attracts pro-caspase-1, which when in proximity, autoproteolytically cleaves into caspase-1. Caspase-1 activates and cleaves pro-IL-1β and pro-IL-18 precursors, which result in the production of active cytokines. Inflammasome activation is indicated by the higher mRNA and protein levels of IL-1β and IL-18 (Fan et al. 2020). However, these levels are inconclusive, as other inflammatory pathways may also cause elevation. Nevertheless, in vitro and in vivo studies have linked inflammasome activation to inflammation in CNS illnesses, such as PD (Ying et al. 2022).

5.4 *Role of Microglial Activation and Peripheral Immune System Involvement in PD*

Microglia have a ramified form when inactive, with extensive motile processes that constantly inspect the microenvironment. Microglia operate in two functional states: pro-inflammatory (M1) and anti-inflammatory (M2) (Badanjak et al. 2021). The M1 phenotype causes neurotoxicity and apoptosis by releasing pro-inflammatory cytokines, such as TNF-α and IL-1β, as well as ROS and NO, and its chronic activation in PD cases leads to progressive dopaminergic neuron loss. Meanwhile, the M2 phenotype secretes anti-inflammatory cytokines (e.g., IL-10 and TGF-β) and neurotrophic factors (e.g., brain-derived neurotrophic factor [BDNF]), promoting tissue repair and inflammation resolution. In PD, the M1-M2 transition is frequently disrupted, resulting in an imbalance that favors chronic inflammation (Guo et al. 2022; Isik et al. 2023). When pathogenic triggers, such as misfolded proteins (alpha-synuclein in PD) or neuronal damage, are detected, microglia undergo morphological conversion into an amoeboid structure (Tu et al. 2021), which triggers a cascade of immunological responses.

Microglial activation is the earliest sign of neuroinflammation. When activated, microglia round up, move, proliferate, phagocytize, provide antigens to T cells, emit oxidants, and stimulate proteins and genes such as pro-inflammatory cytokines, iNOS, ROS, COX-1, COX-2, and neurotoxic substances that trigger neuronal dysfunction and death, contributing to neurodegeneration in cases of chronic activation (Isik et al. 2023). Several routes of microglial activation have been identified. Toll-like receptors (TLRs), specifically TLR2 and TLR4, recognize misfolded α-synuclein. Binding of alpha-synuclein to TLRs activates signaling cascades involving MyD88, NF-κB, and MAPKs, generating pro-inflammatory cytokines and chemokines (Sanchez-Guajardo et al. 2013). Furthermore, receptor for advanced glycation end products (RAGE) interacts with misfolded proteins and AGEs. Binding of alpha-synuclein to RAGE on microglia activates NF-κB, resulting in the release of inflammatory mediators (Long et al. 2022). Furthermore, activated microglia produce excessive amounts of ROS and NO, which induce the oxidative destruction of neuronal membranes, mitochondria, and DNA, resulting in neurodegeneration (Simpson and Oliver 2020; Huang et al. 2022). Its response can also be characterized by reactive gliosis, which is indicated by increased glial fibrillary acidic protein (GFAP) expression, increased chemokines (e.g., MCP-1/CCL2 and CXCL10), pro-inflammatory cytokine production, and glutamate dysregulation (resulting in excitotoxicity and neuronal injury) (Yang and Wang 2015) (Fig. 3).

Immune cell activation in the periphery may result in leukocyte infiltration and neuroinflammation, a process similar to that observed at peripheral sites. However, this effect does not occur unless the BBB is damaged (Muldoon et al. 2013). Monocytes are attracted to the CNS in response to chemokine release, where they develop into macrophages and participate in a localized inflammatory setting, exacerbating neuroinflammation (Zhang et al. 2023). Moreover, T cell infiltration can trigger further harm to the neurological system. For example, CD4+ T cells can cross the BBB and develop into pro-inflammatory Th1 and Th17 subsets, producing interferon-γ (IFN-γ), rousing microglia, and encouraging M1 polarization (Sica and Mantovani 2012). Th17 cells release IL-17, contributing to neuroinflammation by increasing systemic vascular resistance and sympathetic impulses and aggravating neuronal injury. IL-17A stimulates immunological responses and tissue inflammation by acting as an inflammatory regulator, amplifying a variety of inflammatory mediators, such as cytokines, chemokines, adhesion proteins, and matrix metalloproteinases (Ying et al. 2022). CD8+ T lymphocytes can directly trigger neuronal death by releasing perforin and granzyme (Meuth et al. 2009; Mracsko et al. 2014;

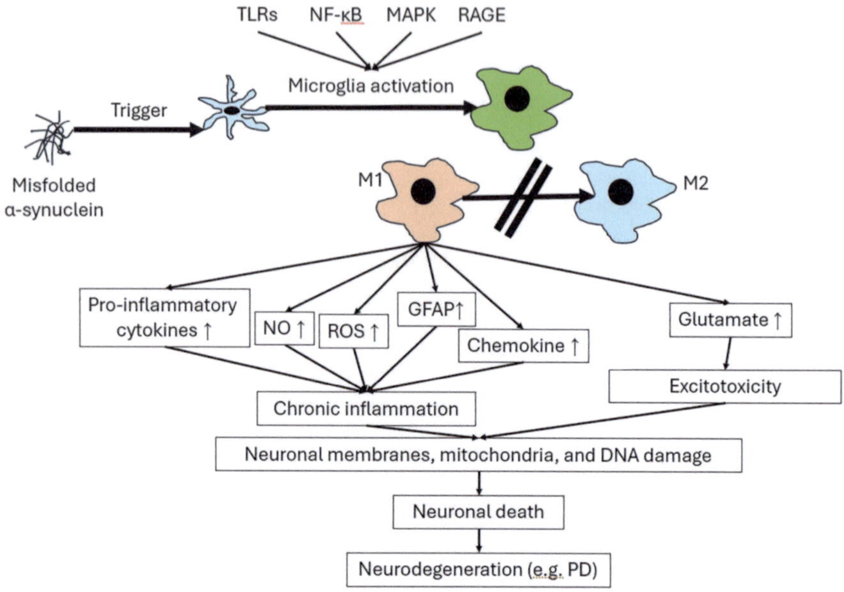

Fig. 3 Microglial activation in PD disease. Abbreviation: DNA = deoxyribonucleic acid, GFAP = glial fibrillary acidic protein, MAPK = mitogen-activated protein kinase, NF-κB = nuclear factor kappa B, NO = nitric oxide, PD = Parkinson's disease, RAGE = receptor for advanced glycation end products, ROS = reactive oxygen species, TLRs = toll-like receptors

Golstein and Griffiths 2018). CD8+ T-cell counts are elevated in the substantia nigra of patients (Brochard et al. 2008; Galiano-Landeira et al. 2020).

Numerous pro-inflammatory cytokines, including TNF-α, IL-1, and IL-6, are involved in neuroinflammation. In CNS neurodegeneration, microglia generate two forms of TNF-α: soluble and membrane-bound. Two receptors are involved in TNF-α signaling: TNFR1 (tumor necrosis factor receptor-1), which is broadly distributed in cells, and TNFR2, which is present in glial cells and several neuronal subtypes at lower levels (Raffaele et al. 2020). Cell survival can be achieved by activating NF-κB or promoting proapoptotic proteins through TNF-α binding to TNFR1 (Guadagno et al. 2013). After binding, TNFR's intracellular death domain engages in an interaction with TRADD, enlisting more signaling complexes (Park et al. 2014). Complex I, one of these complexes, is made up of TNF receptor-associated factor (TRADD), TRAF2/5, RIP (receptor-interacting protein), c-IAP 1/2 (cellular inhibitor of apoptosis protein), and Ubc13 (Naudé et al. 2011), which are essential for inflammation and cell viability. When TNFR1 and TRADD interact with the Fas-associated death domain protein (FADD), another mechanism linked to complex II formation takes place (Pobezinskaya and Liu 2012). When this complex is formed, procaspases 8 and 10's homologous domains become attached to FADD's N-terminal death effector domain (DED), which subsequently activates them (Yao et al. 2007). By opposing the kinase- and NF-κB-mediated activities of complex I, complex II triggers apoptosis (Pobezinskaya and Liu 2012).

In neuroinflammation activity, IL-1 and IL-6 are the two main cytokines. Chemokines, inflammation-associated molecules, and other pro-inflammatory cytokines may be triggered by the IL-1 family of cytokines, which includes IL-1α, IL-1β, and IL-1 receptor antagonists (IL-1RA) (Dinarello 2018). Many ligands and receptor components of the IL-1 family, including as IL-1α, IL-1β, IL-1ra, IL-1R1, IL-1RII, and AcP, are found in the brain (Rothwell 2000). In healthy people, however, they circulate at modest amounts. The main source of IL-1 in response to infections or wounds is microglia. Pro-IL-1β is broken down by the microglia-produced caspase-1, which then converts it into the physiologically active IL-1β. In reaction to damage, microglia release ROS, RNS, cytokines, chemokines, proteases, and prostaglandins. IL-1 causes both adaptive and inappropriate responses from these cells (Li et al. 2018). Studies have shown that IL-1β plays an integral role in both chronic neurodegenerative disorders and acute neuroinflammatory reactions (Labus et al. 2014). Meanwhile, IL-6 is routinely generated in the brain during physiological settings at low concentrations; however, IL-6 release is elevated in neurological diseases, such as PD (Rothaug et al. 2016). Microglia, astrocytes, and neurons secrete IL-6. However, its receptor is only present in microglial cells. IL-6-mediated neurodegeneration depends on trans-signaling in the CNS, whereas IL-6 classical signaling plays a regenerative role. IL-6 activates gp-130-mediated signaling in microglia (Kummer et al. 2021). It is pertinent to emphasize that classical signaling does not stimulate microglial cells in pathological conditions, as microglia become more pro-inflammatory when exposed to hyper IL-6 compared to IL-6 (Rothaug et al. 2016) (Table 6.2).

6 Concept of Manipulating Gut Microbiota to Treat PD

Manipulation of the gut microbiota to treat Parkinson's disease (PD) is based on the understanding that gut dysbiosis (an imbalance in gut microbiota) is closely linked to PD pathogenesis. The gut–brain axis, a bidirectional communication pathway, plays a crucial role in this relationship, in which the gut microbiota can influence neuroinflammation and neurodegeneration (Jia et al. 2024; Li et al. 2024; Wang et al. 2024).

6.1 Potential Strategies

Probiotics

- Probiotics, such as *Bifidobacterium longum*, have shown potential neuroprotective effects by reducing neuroinflammation and restoring the gut microbiota balance (Li et al. 2024; Yao et al. 2024).

Table 6.2 Traditional versus microbiota-based therapies

Aspect	Traditional therapies	Microbiota-based therapies	References (with DOI)
Mechanism of action	Traditional therapies, such as **levodopa** and **dopamine agonists**, primarily aim to compensate for dopamine loss in the brain by either increasing dopamine levels or mimicking its effects **Levodopa** converts into dopamine after crossing the blood–brain barrier **Dopamine agonists** stimulate dopamine receptors directly **MAO-B inhibitors** reduce the breakdown of dopamine	**Microbiota-based therapies** target gut–brain communication via the **gut microbiota** to reduce systemic inflammation and potentially alter neurodegeneration Probiotics and **fecal microbiota transplantation (FMT)** modulate microbial populations Aim to influence metabolites like **short-chain fatty acids (SCFAs)** that affect the brain Reduction in **gut dysbiosis** and systemic inflammatory markers is central to these therapies	Pistollato et al. (2018) and Sampson et al. (2016)
Focus of treatment	**Symptom management** by improving motor function. Primarily focuses on addressing the **motor symptoms** of PD like tremor, bradykinesia, and rigidity, without addressing the underlying disease progression Medications aim to reduce motor complications temporarily, but do not halt or reverse the disease process	Aims to **alter disease progression** by modifying the gut–brain axis and reducing neuroinflammation, which is hypothesized to play a role in the development and progression of PD Targets the root cause of PD by rebalancing gut microbiota to prevent chronic inflammation that may contribute to neurodegeneration	Hill-Burns et al. (2017) and Keshavarzian et al. (2015)
Outcomes	Provides **immediate symptom relief** for motor issues, but long-term use often leads to motor complications such as **dyskinesia** (uncontrolled movements) and **wearing-off** (reduced efficacy over time) Does not affect the progression of non-motor symptoms such as **constipation** or **sleep disturbances**	**Early-stage research** suggests microbiota-based therapies may improve both motor and **non-motor symptoms** (e.g., gastrointestinal issues, sleep disorders, cognitive decline) by reducing systemic inflammation and gut-derived neurotoxins The goal is to **slow disease progression**, though definitive clinical evidence is still lacking	Scheperjans et al. (2015)

(continued)

Table 6.2 (continued)

Aspect	Traditional therapies	Microbiota-based therapies	References (with DOI)
Treatment process	Daily oral medications or transdermal patches (dopamine agonists, MAO-B inhibitors), often requiring **dose adjustments** and combinations over time to manage complications **Surgical options** such as **deep brain stimulation (DBS)** are introduced in advanced PD to manage severe motor fluctuations	**Probiotics**, **prebiotics**, and **FMT** are the primary interventions used in microbiota-based therapy Dietary modifications to promote beneficial bacteria (e.g., high-fiber diet) can also play a role These approaches are less invasive and could become part of early disease management Clinical trials on long-term effects are still ongoing	Poewe et al. (2017) and Privitera et al. (2021)
Long-term efficacy	Effectiveness **wanes over time** due to the progression of PD and the development of motor complications like dyskinesias, requiring higher doses and more complex medication regimens Motor fluctuations often worsen after several years of treatment	Potential for **long-term benefits** by addressing underlying neuroinflammation and gut dysbiosis. While current studies are in early phases, these therapies may offer a **disease-modifying** approach rather than just symptom management However, **long-term clinical efficacy** in large patient populations remains unproven	Sampson et al. (2016) and Huang et al. (2019)
Side effects and risks	Common side effects include **nausea**, **hallucinations**, and **orthostatic hypotension** (low blood pressure when standing) Long-term use often leads to **motor complications** (e.g., dyskinesia, motor fluctuations) Neuropsychiatric side effects (e.g., impulse control disorders) are common, particularly with dopamine agonists	Potential for fewer **systemic side effects** compared to traditional medications. However, risks associated with **FMT** include infection or the transfer of pathogens Probiotics and prebiotics are generally well tolerated, but more research is needed to establish safety, efficacy, and standardized treatment protocols	Poewe et al. (2017) and Privitera et al. (2021)
Potential advantages	**Well-established** and effective in providing **symptomatic relief**, particularly for motor symptoms Allows for immediate improvement in QoL in early PD Surgical options like **DBS** offer significant benefits in carefully selected advanced-stage patients Decades of clinical experience	May **address the root cause** of PD by targeting gut dysbiosis and chronic neuroinflammation **Potential to modify disease progression** rather than only managing symptoms Could provide **long-term benefits** with fewer motor complications Offers promise for improving both **motor** and **non-motor** symptoms	Pistollato et al. (2018)

- Probiotics can enhance gut health, reduce inflammation, and maintain immune homeostasis, thereby contributing to a neuroprotective environment (Wang et al. 2021; Yao et al. 2024).
- The gut microbiota of people with Parkinson's disease (PD) is significantly dysregulated, according to recent experimental and clinical findings, which may be a factor in the pathophysiology of the disease (Li et al. 2024).
- Dysregulation of the microbiome has been linked to gastrointestinal dysfunction in PD, and may contribute to the pathogenesis of the disease (Wang et al. 2024).
- Gut dysbiosis is associated with behavioral and pathophysiological hallmarks in animal models of PD, indicating its potential as a therapeutic target.

Prebiotics

- Prebiotics can increase beneficial gut bacteria and their metabolites, such as short-chain fatty acids, which have anti-inflammatory properties and improve gut barrier integrity (Menozzi and Schapira 2024; Yao et al. 2024).

Fecal Microbiota Transplant (FMTs)

- FMT involves transferring stool from a healthy donor to a patient with PD to restore the healthy gut microbiota. Preliminary studies have shown that FMT can improve both motor and non-motor symptoms in patients with PD (Vendrik et al. 2023; Jia et al. 2024; Menozzi and Schapira 2024).

6.2 *Ongoing Research*

- Research to explore the safety and efficacy of FMT in patients with PD is ongoing, with studies indicating potential improvements in motor complications and gut microbiota composition (Vendrik et al. 2023).
- Studies have also investigated the role of specific probiotics and prebiotics in modulating the gut–brain axis and their impact on PD symptoms (Wang et al. 2021; Li et al. 2024; Yao et al. 2024).
- The interaction between the gut microbiota and PD medications, such as levodopa, is being studied to optimize treatment strategies (Menozzi and Schapira 2023, 2024) (Table 6.3).

Table 6.3 Condenses the research results on FMT, prebiotics, and probiotics

Strategy	Mechanism	Research findings
Probiotics	Reduce neuroinflammation, restore gut balance	Shown to alleviate PD symptoms in animal models
Prebiotics	Increase beneficial bacteria, improve gut barrier	Positive effects on gut health and inflammation
FMTs	Restore healthy gut microbiota	Preliminary studies show symptom improvement

This emerging field holds promise for the development of new therapeutic strategies to manage PD by targeting gut microbiota. Further research and clinical trials are essential to validate these approaches and understand their long-term effects.

Mechanisms of Probiotics, Prebiotics, and FMTs in PD

- Interventions targeting the gut microbiota in PD, including the use of prebiotics, probiotics, and fecal microbiota transplantation (FMT), have shown efficacy in treating PD. Probiotics and FMT have been suggested to influence the gut–brain axis, potentially modulating neuroinflammation and motor outcomes in PD. Probiotic supplementation has been shown to confer health benefits by modulating the gut microbiota and improving the motor and non-motor symptoms of PD (Zhong et al. 2023).

Ongoing Research and Challenges

- Ongoing clinically focused studies are examining whether certain members of the microbiome and/or whole restoration of the gut microbiome community can provide therapeutic benefits for people living with PD (Li et al. 2024). The causal relationship between gut microbiota and PD remains unclear, and further research is needed to understand the complex interactions and mechanisms involved. The therapeutic and diagnostic applications of gut microbiome-modifying approaches in PD are still being elucidated and further targeted studies are warranted to understand the mechanisms of action of probiotics in PD (Vendrik et al. 2023; Yao et al. 2024).

Gut microbiota has been implicated in the pathophysiology of PD, and interventions targeting the gut microbiota, such as probiotics, prebiotics, and FMT, hold promise as potential therapeutic strategies. Ongoing research is focused on understanding the complex interactions between the gut microbiota and PD as well as addressing the challenges associated with gut microbiota-based therapies for PD.

7 Conclusion

This chapter explores the link between Parkinson's disease (PD) and the gut microbiota, highlighting how gastrointestinal health and microbial balance may influence PD pathogenesis and progression. Traditional PD treatments primarily address symptoms without altering the underlying neurodegenerative processes. Gut dysbiosis, a microbial imbalance, could play a significant role in PD development through mechanisms such as alpha-synuclein propagation, neuroinflammation, and disruptions in the gut–brain axis signaling. Specific microbial imbalances, such as an overabundance of pro-inflammatory species and a decrease in short-chain fatty acid producers, can exacerbate both motor and non-motor symptoms. Microbiota-based therapies, including probiotics and prebiotics, show promise in modulating the gut environment to enhance GI health and potentially slow PD progression. Future research should clarify the exact mechanisms by which gut dysbiosis influences PD pathogenesis; explore genetic, dietary, and environmental factors contributing to microbiota composition; and develop targeted microbial therapies tailored to individual microbiota profiles.

References

Adams-Carr KL, Bestwick JP, Shribman S, Lees A, Schrag A, Noyce AJ (2016) Constipation preceding Parkinson's disease: a systematic review and meta-analysis. J Neurol Neurosurg Psychiatry 87(7):710–716

Ahlskog JE, Muenter MD (2001) Frequency of levodopa-related dyskinesias and motor fluctuations as estimated from the cumulative literature. Mov Disord 16(3):448–458

Aldred J, Freire-Alvarez E, Amelin AV, Antonini A, Bergmans B, Bergquist F, Bouchard M, Budur K, Carroll C, Chaudhuri KR, Criswell SR, Danielsen EH, Gandor F, Jia J, Kimber TE, Mochizuki H, Robieson WZ, Spiegel AM, Standaert DG, Talapala S, Facheris MF, Fung VSC (2023) Continuous subcutaneous Foslevodopa/Foscarbidopa in Parkinson's disease: safety and efficacy results from a 12-month, single-arm, open-label, Phase 3 Study. Neurol Ther 12(6):1937–1958

Alloo J, Leleu I, Grangette C, Pied S (2022) Parasite infections, neuroinflammation, and potential contributions of gut microbiota. Front Immunol 13:1024998

Antonini A, Poewe W, Chaudhuri KR, Jech R, Pickut B, Pirtošek Z, Szasz J, Valldeoriola F, Winkler C, Bergmann L, Yegin A, Onuk K, Barch D, Odin P, Amalia E, Arnold G, Bajenaru O, Bergmans B, Bjornara KA, Blackie J, Bode M, Bourgeois P, Bohlhalter S, Buraga I, Burkhard PR, Busson P, Calopa M, Clausen J, Danielsen EH, Defebvre L, Delvaux V, Dethy S, Dietrichs E, De Fabregues O, Gerhard R, Gusmaroli G, Hahn K, Hauptmann B, Henriksen T, Hernandez-Vara J, Jeanjean A, Kaiserova M, Kassubek J, Kimber T, Konitsiotis S, Krüger R, Kulisevsky J, Leenders J, Lundqvist C, Ory Magne F, Marano P, Milanov I, Modugno N, Misbahuddin A, Nevrly M, Panayiotis Z, Pedersen KF, Pedersen SW, Perju-Dumbrava L, Ponsen MM, Popescu BO, Rijntjes M, Puente V, Redecker C, Schrader C, Sensi M, Simu M, Spanaki C, Storch A, Storstein A, Tomantschger V, Van Der Linden C, Van Laar T, Viallet F, Witjas T, Wolz M, Zibetti M, Van Zandijcke M (2017) Levodopa-carbidopa intestinal gel in advanced Parkinson's: final results of the GLORIA registry. Parkinsonism Relat Disord 45:13–20

Arotcarena M-L, Dovero S, Prigent A, Bourdenx M, Camus S, Porras G, Thiolat M-L, Tasselli M, Aubert P, Kruse N, Mollenhauer B, Trigo Damas I, Estrada C, Garcia-Carrillo N, Vaikath NN, El-Agnaf OMA, Herrero MT, Vila M, Obeso JA, Derkinderen P, Dehay B, Bezard E (2020) Bidirectional gut-to-brain and brain-to-gut propagation of synucleinopathy in non-human primates. Brain 143(5):1462–1475

Azevedo EP, Ledo JH, Barbosa G, Sobrinho M, Diniz L, Fonseca ACC, Gomes F, Romão L, Lima FRS, Palhano FL, Ferreira ST, Foguel D (2013) Activated microglia mediate synapse loss and short-term memory deficits in a mouse model of transthyretin-related oculoleptomeningeal amyloidosis. Cell Death Dis 4(9):e789–e789

Badanjak K, Fixemer S, Smajić S, Skupin A, Grünewald A (2021) The contribution of microglia to neuroinflammation in Parkinson's disease. Int J Mol Sci 22(9):4676

Berg D, Postuma RB, Adler CH, Bloem BR, Chan P, Dubois B, Gasser T, Goetz CG, Halliday G, Joseph L, Lang AE, Liepelt-Scarfone I, Litvan I, Marek K, Obeso J, Oertel W, Olanow CW, Poewe W, Stern M, Deuschl G (2015) MDS research criteria for prodromal Parkinson's disease: MDS Criteria for Prodromal PD. Mov Disord 30(12):1600–1611

Bloem BR, Okun MS, Klein C (2021) Parkinson's disease. Lancet (London, England) 397(10291):2284–2303

Bougea A (2024) Some novel therapies in Parkinson's disease: a promising path forward or not yet? A systematic review of the literature. Biomedicines 12(3):549

Braak H, Rüb U, Gai WP, Del Tredici K (2003) Idiopathic Parkinson's disease: possible routes by which vulnerable neuronal types may be subject to neuroinvasion by an unknown pathogen. J Neural Transm 110(5):517–536
Brandl S, Reindl M (2023) Blood–brain barrier breakdown in neuroinflammation: current in vitro models. Int J Mol Sci 24(16):12699
Breen DP, Halliday GM, Lang AE (2019) Gut–brain axis and the spread of α-synuclein pathology: vagal highway or dead end? Mov Disord 34(3):307–316
Brochard V, Combadière B, Prigent A, Laouar Y, Perrin A, Beray-Berthat V, Bonduelle O, Alvarez-Fischer D, Callebert J, Launay J-M, Duyckaerts C, Flavell RA, Hirsch EC, Hunot S (2008) Infiltration of CD4+ lymphocytes into the brain contributes to neurodegeneration in a mouse model of Parkinson disease. J Clin Invest:JCI36470
Caballero-Flores G, Pickard JM, Núñez G (2023) Microbiota-mediated colonization resistance: mechanisms and regulation. Nat Rev Microbiol 21(6):347–360
Camacho M, Macleod AD, Maple-Grødem J, Evans JR, Breen DP, Cummins G, Wijeyekoon RS, Greenland JC, Alves G, Tysnes OB, Lawson RA, Barker RA, Williams-Gray CH (2021) Early constipation predicts faster dementia onset in Parkinson's disease. NPJ Parkinson's Dis 7(1):45
Carabotti M, Scirocco A, Maselli MA, Severi C (2015) The gut-brain axis: interactions between enteric microbiota, central and enteric nervous systems. Ann Gastroenterol 28(2):203–209
Castillo-Rangel C, Marin G, Hernández-Contreras KA, Vichi-Ramírez MM, Zarate-Calderon C, Torres-Pineda O, Diaz-Chiguer DL, De La Mora González D, Gómez Apo E, Teco-Cortes JA, Santos-Paez FDM, Coello-Torres MDLÁ, Baldoncini M, Reyes Soto G, Aranda-Abreu GE, García LI (2023) Neuroinflammation in Parkinson's disease: from gene to clinic: a systematic review. Int J Mol Sci 24(6):5792
Cattaneo C, Jost WH (2023) Pain in Parkinson's disease: pathophysiology, classification and treatment. J Integr Neurosci 22(5):132
Cersosimo MG, Raina GB, Pecci C, Pellene A, Calandra CR, Gutiérrez C, Micheli FE, Benarroch EE (2013) Gastrointestinal manifestations in Parkinson's disease: prevalence and occurrence before motor symptoms. J Neurol 260(5):1332–1338
Chalazonitis A, Rao M (2018) Enteric nervous system manifestations of neurodegenerative disease. Brain Res 1693(Pt B):207–213
Chalazonitis A, Rao M, Sulzer D (2022) Similarities and differences between nigral and enteric dopaminergic neurons unravel distinctive involvement in Parkinson's disease. NPJ Parkinson's Dis 8(1):50
Chekrouni N, Van Soest TM, Brouwer MC, Willemse EAJ, Teunissen CE, Van De Beek D (2022) CSF neurofilament light chain concentrations predict outcome in bacterial meningitis. Neurol Neuroimmunol Neuroinflamm 9(1):e1123
Chen H, Zhao EJ, Zhang W, Lu Y, Liu R, Huang X, Ciesielski-Jones AJ, Justice MA, Cousins DS, Peddada S (2015) Meta-analyses on prevalence of selected Parkinson's nonmotor symptoms before and after diagnosis. Neurodegeneration 4(1):1
Chen X, Hu Y, Cao Z, Liu Q, Cheng Y (2018) Cerebrospinal fluid inflammatory cytokine aberrations in Alzheimer's disease, Parkinson's disease and amyotrophic lateral sclerosis: a systematic review and meta-analysis. Front Immunol 9:2122
Choi ML, Chappard A, Singh BP, Maclachlan C, Rodrigues M, Fedotova EI, Berezhnov AV, De S, Peddie CJ, Athauda D, Virdi GS, Zhang W, Evans JR, Wernick AI, Zanjani ZS, Angelova PR, Esteras N, Vinokurov AY, Morris K, Jeacock K, Tosatto L, Little D, Gissen P, Clarke DJ, Kunath T, Collinson L, Klenerman D, Abramov AY, Horrocks MH, Gandhi S (2022) Pathological structural conversion of α-synuclein at the mitochondria induces neuronal toxicity. Nat Neurosci 25(9):1134–1148
Dai Y, Zhou J, Shi C (2023) Inflammasome: structure, biological functions, and therapeutic targets. MedComm 4(5):e391
Damiani F, Cornuti S, Tognini P (2023) The gut-brain connection: exploring the influence of the gut microbiota on neuroplasticity and neurodevelopmental disorders. Neuropharmacology 231:109491

De Freitas TB, Leite PHW, Doná F, Pompeu JE, Swarowsky A, Torriani-Pasin C (2020) The effects of dual task gait and balance training in Parkinson's disease: a systematic review. Physiother Theo Pract 36(10):1088–1096

Deyell JS, Sriparna M, Ying M, Mao X (2023) The Interplay between α-Synuclein and Microglia in α-Synucleinopathies. Int J Mol Sci 24(3):2477

Dinarello CA (2018) Overview of the IL −1 family in innate inflammation and acquired immunity. Immunol Rev 281(1):8–27

Edwards LL, Quigley EMM, Pfeiffer RF (1992) Gastrointestinal dysfunction in Parkinson's disease: frequency and pathophysiology. Neurology 42(4):726–726

Fan Z, Pan Y-T, Zhang Z-Y, Yang H, Yu S-Y, Zheng Y, Ma J-H, Wang X-M (2020) Systemic activation of NLRP3 inflammasome and plasma α-synuclein levels are correlated with motor severity and progression in Parkinson's disease. J Neuroinflammation 17(1):11

Fanciulli A, Leys F, Falup-Pecurariu C, Thijs R, Wenning GK (2020) Management of orthostatic hypotension in Parkinson's disease. J Parkinsons Dis 10(s1):S57–S64

Ferreiro AL, Choi J, Ryou J, Newcomer EP, Thompson R, Bollinger RM, Hall-Moore C, Ndao IM, Sax L, Benzinger TLS, Stark SL, Holtzman DM, Fagan AM, Schindler SE, Cruchaga C, Butt OH, Morris JC, Tarr PI, Ances BM, Dantas G (2023) Gut microbiome composition may be an indicator of preclinical Alzheimer's disease. Sci Transl Med 15(700):eabo2984

Forsyth CB, Shannon KM, Kordower JH, Voigt RM, Shaikh M, Jaglin JA, Estes JD, Dodiya HB, Keshavarzian A (2011) Increased intestinal permeability correlates with sigmoid mucosa alpha-Synuclein staining and endotoxin exposure markers in early Parkinson's disease, C. Oreja-Guevara (ed). PLoS One 6(12):e28032

Franzosa EA, Huang K, Meadow JF, Gevers D, Lemon KP, Bohannan BJM, Huttenhower C (2015) Identifying personal microbiomes using metagenomic codes. Proc Natl Acad Sci 112(22):E2930–E2938

Galiano-Landeira J, Torra A, Vila M, Bové J (2020) CD8 T cell nigral infiltration precedes synucleinopathy in early stages of Parkinson's disease. Brain 143(12):3717–3733

Gao C, Jiang J, Tan Y, Chen S (2023) Microglia in neurodegenerative diseases: mechanism and potential therapeutic targets. Signal Trans Target Ther 8(1):359

Giancola F, Torresan F, Repossi R, Bianco F, Latorre R, Ioannou A, Guarino M, Volta U, Clavenzani P, Mazzoni M, Chiocchetti R, Bazzoli F, Travagli RA, Sternini C, De Giorgio R (2017) Downregulation of neuronal vasoactive intestinal polypeptide in Parkinson's disease and chronic constipation. Neurogastroenterol Motil 29(5)

Giugni JC, Okun MS (2014) Treatment of advanced Parkinson's disease. Curr Opin Neurol 27(4):450–460

Golstein P, Griffiths GM (2018) An early history of T cell-mediated cytotoxicity. Nat Rev Immunol 18(8):527–535

Guadagno J, Xu X, Karajgikar M, Brown A, Cregan SP (2013) Microglia-derived TNFα induces apoptosis in neural precursor cells via transcriptional activation of the Bcl-2 family member Puma. Cell Death Dis 4(3):e538–e538

Guo S, Wang H, Yin Y (2022) Microglia polarization from M1 to M2 in neurodegenerative diseases. Front Aging Neurosci 14:815347

Hawkes CH, Del Tredici K, Braak H (2007) Parkinson's disease: a dual-hit hypothesis. Neuropathol Appl Neurobiol 33(6):599–614

Hayes MT (2019) Parkinson's disease and Parkinsonism. Am J Med 132(7):802–807

Heijtz RD, Wang S, Anuar F, Qian Y, Björkholm B, Samuelsson A, Hibberd ML, Forssberg H, Pettersson S (2011) Normal gut microbiota modulates brain development and behavior. Proc Natl Acad Sci 108(7):3047–3052

Hill-Burns EM, Debelius JW, Morton JT, Wissemann WT, Lewis MR, Wallen ZD, Peddada SD, Factor SA, Molho E, Zabetian CP, Knight R, Payami H (2017) Parkinson's disease and Parkinson's disease medications have distinct signatures of the gut microbiome. Mov Disord 32(5):739–749

Hou Y, Shan C, Zhuang S, Zhuang Q, Ghosh A, Zhu K, Kong X, Wang S, Gong Y, Yang Y, Tao B, Sun L, Zhao H-Y, Guo X, Wang W, Ning G, Gu Y, Li S, Liu J (2021) Gut microbiota-derived propionate mediates the neuroprotective effect of osteocalcin in a mouse model of Parkinson's disease. Microbiome 9(1):34

Huang X, Sterling NW, Du G, Sun D, Stetter C, Kong L, Zhu Y, Neighbors J, Lewis MM, Chen H, Hohl RJ, Mailman RB (2019) Brain cholesterol metabolism and Parkinson's disease. Mov Disord 34(3):386–395

Huang Q, Pereira AC, Velthuis H, Wong NML, Ellis CL, Ponteduro FM, Dimitrov M, Kowalewski L, Lythgoe DJ, Rotaru D, Edden RAE, Leonard A, Ivin G, Ahmad J, Pretzsch CM, Daly E, Murphy DGM, McAlonan GM (2022) $GABA_B$ receptor modulation of visual sensory processing in adults with and without autism spectrum disorder. Sci Transl Med 14(626):eabg7859

Huang B, Chau SWH, Liu Y, Chan JWY, Wang J, Ma SL, Zhang J, Chan PKS, Yeoh YK, Chen Z, Zhou L, Wong SH, Mok VCT, To KF, Lai HM, Ng S, Trenkwalder C, Chan FKL, Wing YK (2023) Gut microbiome dysbiosis across early Parkinson's disease, REM sleep behavior disorder and their first-degree relatives. Nat Commun 14(1):2501

Isik S, Yeman Kiyak B, Akbayir R, Seyhali R, Arpaci T (2023) Microglia mediated neuroinflammation in Parkinson's disease. Cells 12(7):1012

Jang J-H, Yeom M-J, Ahn S, Oh J-Y, Ji S, Kim T-H, Park H-J (2020) Acupuncture inhibits neuroinflammation and gut microbial dysbiosis in a mouse model of Parkinson's disease. Brain Behav Immun 89:641–655

Jankovic J (2008) Current approaches to the treatment of Parkinson's disease. Neuropsychiatr Dis Treat 743

Jankovic J, Poewe W (2012) Therapies in Parkinson's disease. Curr Opin Neurol 25(4):433–447

Ji T, Huang H, Liu J, Peng T, Zhou X, Tan Q, Yuan J, Hua H, Ding S, Liu H (2021) Leveraging sequence-based faecal microbial community survey data to identify alterations in gut microbiota among patients with Parkinson's disease. Eur J Neurosci 53(2):687–696

Jia X, Wang Q, Liu M, Ding J-Y (2024) The interplay between gut microbiota and the brain-gut axis in Parkinson's disease treatment. Front Neurol 15

Johansson H, Folkerts A-K, Hammarström I, Kalbe E, Leavy B (2023) Effects of motor-cognitive training on dual-task performance in people with Parkinson's disease: a systematic review and meta-analysis. J Neurol 270(6):2890–2907

Kalia LV, Lang AE (2015) Parkinson's disease. Lancet 386(9996):896–912

Kam T-I, Hinkle JT, Dawson TM, Dawson VL (2020) Microglia and astrocyte dysfunction in Parkinson's disease. Neurobiol Dis 144:105028

Keshavarzian A, Green SJ, Engen PA, Voigt RM, Naqib A, Forsyth CB, Mutlu E, Shannon KM (2015) Colonic bacterial composition in Parkinson's disease. Mov Disord 30(10):1351–1360

Kim J-S, Sung H-Y (2015) Gastrointestinal autonomic dysfunction in patients with Parkinson's disease. J Mov Disord 8(2):76–82

Kim S, Kwon S-H, Kam T-I, Panicker N, Karuppagounder SS, Lee S, Lee JH, Kim WR, Kook M, Foss CA, Shen C, Lee H, Kulkarni S, Pasricha PJ, Lee G, Pomper MG, Dawson VL, Dawson TM, Ko HS (2019) Transneuronal propagation of pathologic α-Synuclein from the gut to the brain models Parkinson's disease. Neuron 103(4):627–641.e7

Kishimoto Y, Zhu W, Hosoda W, Sen JM, Mattson MP (2019) Chronic mild gut inflammation accelerates brain neuropathology and motor dysfunction in α-Synuclein mutant mice. NeuroMolecular Med 21(3):239–249

Klingelhoefer L, Reichmann H (2015) Pathogenesis of Parkinson disease—the gut–brain axis and environmental factors. Nat Rev Neurol 11(11):625–636

Kölliker-Frers R, Udovin L, Otero-Losada M, Kobiec T, Herrera MI, Palacios J, Razzitte G, Capani F (2021) Neuroinflammation: an integrating overview of reactive-neuroimmune cell interactions in health and disease, M. Giovarelli (ed.). Mediat Inflamm 2021:1–20

Kummer KK, Zeidler M, Kalpachidou T, Kress M (2021) Role of IL-6 in the regulation of neuronal development, survival and function. Cytokine 144:155582

Labus J, Häckel S, Lucka L, Danker K (2014) Interleukin-1β induces an inflammatory response and the breakdown of the endothelial cell layer in an improved human THBMEC-based in vitro blood–brain barrier model. J Neurosci Methods 228:35–45

Lei C, Sunzi K, Dai F, Liu X, Wang Y, Zhang B, He L, Ju M (2019) Effects of virtual reality rehabilitation training on gait and balance in patients with Parkinson's disease: a systematic review. PLoS One 14(11):e0224819

Li Y, Zhang L, Liu Z (2018) Multi-objective de novo drug design with conditional graph generative model. J Cheminformatics 10(1):33

Li T, Zhang C, Wang Q, Wang S, Yu L, Zhai Q, Tian F (2024) Exploring the therapeutic potential of Bifidobacterium longum subsp. longum CCFM1029 in Parkinson's disease: Insights from behavioral, neurophysiological, gut microbiota, and microbial metabolites analysis. eFood 5(5)

Lin C-H, Chen C-C, Chiang H-L, Liou J-M, Chang C-M, Lu T-P, Chuang EY, Tai Y-C, Cheng C, Lin H-Y, Wu M-S (2019) Altered gut microbiota and inflammatory cytokine responses in patients with Parkinson's disease. J Neuroinflammation 16(1):129

Liu B, Fang F, Pedersen NL, Tillander A, Ludvigsson JF, Ekbom A, Svenningsson P, Chen H, Wirdefeldt K (2017) Vagotomy and Parkinson disease: a Swedish register–based matched-cohort study. Neurology 88(21):1996–2002

Long H, Zhang S, Zeng S, Tong Y, Liu J, Liu C, Li D (2022) Interaction of RAGE with α-synuclein fibrils mediates inflammatory response of microglia. Cell Rep 40(12):111401

Lynch SV, Pedersen O (2016) The human intestinal microbiome in health and disease, E. G. Phimister (ed). N Engl J Med 375(24):2369–2379

Lyra P, Machado V, Rota S, Chaudhuri KR, Botelho J, Mendes JJ (2023) Revisiting alpha-Synuclein pathways to inflammation. Int J Mol Sci 24(8):7137

Ma Q, Xing C, Long W, Wang HY, Liu Q, Wang R-F (2019) Impact of microbiota on central nervous system and neurological diseases: the gut-brain axis. J Neuroinflammation 16(1):53

Machado Sotomayor MJ, Arufe-Giráldez V, Ruíz-Rico G, Navarro-Patón R (2021) Music therapy and Parkinson's disease: a systematic review from 2015–2020. Int J Environ Res Public Health 18(21):11618

Manu DR, Slevin M, Barcutean L, Forro T, Boghitoiu T, Balasa R (2023) Astrocyte involvement in blood–brain barrier function: a critical update highlighting novel, complex, neurovascular interactions. Int J Mol Sci 24(24):17146

Mart'yanov SV, Botchkova EA, Plakunov VK, Gannesen AV (2021) The impact of norepinephrine on mono-species and dual-species staphylococcal biofilms. Microorganisms 9(4):820

Menozzi E, Schapira AHV (2024) The gut microbiota in Parkinson disease: interactions with drugs and potential for therapeutic applications. CNS Drugs 38(5):315–331

Meuth SG, Herrmann AM, Simon OJ, Siffrin V, Melzer N, Bittner S, Meuth P, Langer HF, Hallermann S, Boldakowa N, Herz J, Munsch T, Landgraf P, Aktas O, Heckmann M, Lessmann V, Budde T, Kieseier BC, Zipp F, Wiendl H (2009) Cytotoxic CD8 + T cell–neuron interactions: perforin-dependent electrical silencing precedes but is not causally linked to neuronal cell death. J Neurosci 29(49):15397–15409

Mhyre TR, Boyd JT, Hamill RW, Maguire-Zeiss KA (2012) Parkinson's disease. In: Harris JR (ed) Protein aggregation and fibrillogenesis in cerebral and systemic amyloid disease, subcellular biochemistry, vol 65. Springer, Dordrecht, pp 389–455

Minguez-Castellanos A, Chamorro CE, Escamilla-Sevilla F, Ortega-Moreno A, Rebollo AC, Gomez-Rio M, Concha A, Munoz DG (2007) Do α-synuclein aggregates in autonomic plexuses predate Lewy body disorders?: a cohort study. Neurology 68(23):2012–2018

Monaghan AS, Gordon E, Graham L, Hughes E, Peterson DS, Morris R (2023) Cognition and freezing of gait in Parkinson's disease: a systematic review and meta-analysis. Neurosci Biobehav Rev 147:105068

Montanari M, Imbriani P, Bonsi P, Martella G, Peppe A (2023) Beyond the microbiota: understanding the role of the enteric nervous system in Parkinson's disease from mice to human. Biomedicines 11(6):1560

Mracsko E, Liesz A, Stojanovic A, Lou WP-K, Osswald M, Zhou W, Karcher S, Winkler F, Martin-Villalba A, Cerwenka A, Veltkamp R (2014) Antigen dependently activated cluster of differentiation 8-positive T cells cause Perforin-mediated neurotoxicity in experimental stroke. J Neurosci 34(50):16784–16795
Muldoon LL, Alvarez JI, Begley DJ, Boado RJ, Del Zoppo GJ, Doolittle ND, Engelhardt B, Hallenbeck JM, Lonser RR, Ohlfest JR, Prat A, Scarpa M, Smeyne RJ, Drewes LR, Neuwelt EA (2013) Immunologic privilege in the central nervous system and the blood–brain barrier. J Cereb Blood Flow Metab 33(1):13–21
Nabi M, Tabassum N (2022) Role of environmental toxicants on neurodegenerative disorders. Front Toxicol 4:837579
Natale G, Ryskalin L, Morucci G, Lazzeri G, Frati A, Fornai F (2021) The baseline structure of the enteric nervous system and its role in Parkinson's disease. Life (Basel, Switzerland) 11(8):732
Naudé PJW, Den Boer JA, Luiten PGM, Eisel ULM (2011) Tumor necrosis factor receptor cross-talk. FEBS J 278(6):888–898
Nishiwaki H, Ito M, Ishida T, Hamaguchi T, Maeda T, Kashihara K, Tsuboi Y, Ueyama J, Shimamura T, Mori H, Kurokawa K, Katsuno M, Hirayama M, Ohno K (2020) meta-analysis of gut Dysbiosis in Parkinson's disease. Mov Disord 35(9):1626–1635
Nurmukanova V, Matsvay A, Gordukova M, Shipulin G (2024) Square the circle: diversity of viral pathogens causing neuro-infectious diseases. Viruses 16(5):787
Olanow CW, Schapira AHV, Rascol O (2000) Continuous dopamine-receptor stimulation in early Parkinson's disease. Trends Neurosci 23:S117–S126
Olanow CW, Watts RL, Koller WC (2001) An algorithm (decision tree) for the management of Parkinson's disease (2001): treatment guidelines. Neurology 56(suppl_5)
Oyanagi K, Wakabayashi K, Ohama E, Takeda S, Horikawa Y, Morita T, Ikuta F (1990) Lewy bodies in the lower sacral parasympathetic neurons of a patient with Parkinson's disease. Acta Neuropathol 80(5):558–559
Palacios N, Wilkinson J, Bjornevik K, Schwarzschild MA, McIver L, Ascherio A, Huttenhower C (2023) Metagenomics of the gut microbiome in Parkinson's disease: prodromal changes. Ann Neurol 94(3):486–501
Park Y-H, Jeong MS, Jang SB (2014) Death domain complex of the TNFR-1, TRADD, and RIP1 proteins for death-inducing signaling. Biochem Biophys Res Commun 443(4):1155–1161
Pistollato F, Iglesias RC, Ruiz R, Aparicio S, Crespo J, Lopez LD, Manna PP, Giampieri F, Battino M (2018) Nutritional patterns associated with the maintenance of neurocognitive functions and the risk of dementia and Alzheimer's disease: a focus on human studies. Pharmacol Res 131:32–43
Pobezinskaya YL, Liu Z (2012) The role of TRADD in death receptor signaling. Cell Cycle 11(5):871–876
Poewe W, Mahlknecht P (2009) The clinical progression of Parkinson's disease. Parkinsonism Relat Disord 15:S28–S32
Poewe W, Seppi K, Tanner CM, Halliday GM, Brundin P, Volkmann J, Schrag A-E, Lang AE (2017) Parkinson disease. Nat Rev Dis Prim 3(1):17013
Prajjwal P, Inban P, Natarajan B, Gadam S, Marsool MDM, Tariq H, Paras P, Vora N, Al-Aish ST, Marsool ADM, Amir Hussin O (2023) Remyelination in multiple sclerosis, along with its immunology and association with gut dysbiosis, lifestyle, and environmental factors. Ann Med Surg 85(9):4417–4424
Privitera G, Pugliese D, Rapaccini GL, Gasbarrini A, Armuzzi A, Guidi L (2021) Predictors and early markers of response to biological therapies in inflammatory bowel diseases. J Clin Med 10(4):853
Raffaele S, Lombardi M, Verderio C, Fumagalli M (2020) TNF production and release from microglia via extracellular vesicles: impact on brain functions. Cells 9(10):2145
Rascol O, Brooks DJ, Korczyn AD, De Deyn PP, Clarke CE, Lang AE (2000) A five-year study of the incidence of dyskinesia in patients with early Parkinson's disease who were treated with ropinirole or levodopa. N Engl J Med 342(20):1484–1491

Raza C, Anjum R, Shakeel NUA (2019) Parkinson's disease: mechanisms, translational models and management strategies. Life Sci 226:77–90

Reich SG, Savitt JM (2019) Parkinson's disease. Med Clin N Am 103(2):337–350

Rhee SH, Pothoulakis C, Mayer EA (2009) Principles and clinical implications of the brain-gut-enteric microbiota axis. Nat Rev Gastroenterol Hepatol 6(5):306–314

Riederer P, Laux G (2011) MAO-inhibitors in Parkinson's disease. Experimental Neurobiology 20(1):1–17

Romano S, Savva GM, Bedarf JR, Charles IG, Hildebrand F, Narbad A (2021) Meta-analysis of the Parkinson's disease gut microbiome suggests alterations linked to intestinal inflammation. NPJ Parkinson's Dis 7(1):27

Rothaug M, Becker-Pauly C, Rose-John S (2016) The role of interleukin-6 signaling in nervous tissue. Biochimica et Biophysica Acta (BBA) Mol Cell Res 1863(6):1218–1227

Rothwell N (2000) Interleukin 1 in the brain: biology, pathology and therapeutic target. Trends Neurosci 23(12):618–625

Rukavina K, Leta V, Sportelli C, Buhidma Y, Duty S, Malcangio M, Ray Chaudhuri K (2019) Pain in Parkinson's disease: new concepts in pathogenesis and treatment. Curr Opin Neurol 32(4):579–588

Sampson TR, Debelius JW, Thron T, Janssen S, Shastri GG, Ilhan ZE, Challis C, Schretter CE, Rocha S, Gradinaru V, Chesselet M-F, Keshavarzian A, Shannon KM, Krajmalnik-Brown R, Wittung-Stafshede P, Knight R, Mazmanian SK (2016) Gut microbiota regulate motor deficits and neuroinflammation in a model of Parkinson's disease. Cell 167(6):1469–1480.e12

Sanchez-Guajardo V, Barnum CJ, Tansey MG, Romero-Ramos M (2013) Neuroimmunological processes in Parkinson's disease and their relation to α-Synuclein: microglia as the referee between neuronal processes and peripheral immunity. ASN Neuro 5(2):AN20120066

Scheperjans F, Aho V, Pereira PAB, Koskinen K, Paulin L, Pekkonen E, Haapaniemi E, Kaakkola S, Eerola-Rautio J, Pohja M, Kinnunen E, Murros K, Auvinen P (2015) Gut microbiota are related to Parkinson's disease and clinical phenotype. Mov Disord 30(3):350–358

Selvakumar GP, Ahmed ME, Iyer SS, Thangavel R, Kempuraj D, Raikwar SP, Bazley K, Wu K, Khan A, Kukulka K, Bussinger B, Zaheer S, Burton C, James D, Zaheer A (2020) Absence of glia maturation factor protects from axonal injury and motor behavioral impairments after traumatic brain injury. Exp Neurobiol 29(3):230–248

Sica A, Mantovani A (2012) Macrophage plasticity and polarization: in vivo veritas. J Clin Invest 122(3):787–795

Simpson DSA, Oliver PL (2020) ROS generation in microglia: understanding oxidative stress and inflammation in neurodegenerative disease. Antioxidants 9(8):743

Sivanandy P, Leey TC, Xiang TC, Ling TC, Wey Han SA, Semilan SLA, Hong PK (2021) Systematic review on Parkinson's disease medications, emphasizing on three recently approved drugs to control Parkinson's symptoms. Int J Environ Res Public Health 19(1):364

Stocchi F (2006) The levodopa wearing-off phenomenon in Parkinson's disease: pharmacokinetic considerations. Expert Opin Pharmacother 7(10):1399–1407

Stocchi F, Antonini A, Barone P, Tinazzi M, Zappia M, Onofrj M, Ruggieri S, Morgante L, Bonuccelli U, Lopiano L, Pramstaller P, Albanese A, Attar M, Posocco V, Colombo D, Abbruzzese G (2014) Early DEtection of wEaring off in Parkinson disease: The DEEP study. Parkinsonism Relat Disord 20(2):204–211

Stocchi F, Vacca L, Radicati FG (2015) How to optimize the treatment of early stage Parkinson's disease. Transl Neurodegener 4(1):4

Sulzer D (2007) Multiple hit hypotheses for dopamine neuron loss in Parkinson's disease. Trends Neurosci 30(5):244–250

Svensson E, Horváth-Puhó E, Thomsen RW, Djurhuus JC, Pedersen L, Borghammer P, Sørensen HT (2015) Vagotomy and subsequent risk of Parkinson's disease. Ann Neurol 78(4):522–529

Tansey MG, Wallings RL, Houser MC, Herrick MK, Keating CE, Joers V (2022) Inflammation and immune dysfunction in Parkinson disease. Nat Rev Immunol 22(11):657–673

Tomlinson CL, Patel S, Meek C, Herd CP, Clarke CE, Stowe R, Shah L, Sackley C, Deane KHO, Wheatley K, Ives N (2012) Physiotherapy intervention in Parkinson's disease: systematic review and meta-analysis. BMJ (Clinical Research ed) 345:e5004

Truijen S, Abdullahi A, Bijsterbosch D, van Zoest E, Conijn M, Wang Y, Struyf N, Saeys W (2022) Effect of home-based virtual reality training and telerehabilitation on balance in individuals with Parkinson disease, multiple sclerosis, and stroke: a systematic review and meta-analysis. Neurol Sci 43(5):2995–3006

Tu H, Yuan B, Hou X, Zhang X, Pei C, Ma Y, Yang Y, Fan Y, Qin Z, Liu C, Hu L (2021) α-synuclein suppresses microglial autophagy and promotes neurodegeneration in a mouse model of Parkinson's disease. Aging Cell 20(12):e13522

Ulusoy A, Phillips RJ, Helwig M, Klinkenberg M, Powley TL, Di Monte DA (2017) Brain-to-stomach transfer of α-synuclein via vagal preganglionic projections. Acta Neuropathol 133(3):381–393

Vendrik KEW, Chernova VO, Kuijper EJ, Terveer EM, Van Hilten JJ, Contarino MF (2023) Safety and feasibility of faecal microbiota transplantation for patients with Parkinson's disease: a protocol for a self-controlled interventional donor-FMT pilot study. BMJ Open 13(10)

Vos T, Allen C, Arora M, Barber RM, Bhutta ZA, Brown A, Carter A, Casey DC, Charlson FJ, Chen AZ, Coggeshall M, Cornaby L, Dandona L, Dicker DJ, Dilegge T, Erskine HE, Ferrari AJ, Fitzmaurice C, Fleming T, Forouzanfar MH, Fullman N, Gething PW, Goldberg EM, Graetz N, Haagsma JA, Hay SI, Johnson CO, Kassebaum NJ, Kawashima T, Kemmer L, Khalil IA, Kinfu Y, Kyu HH, Leung J, Liang X, Lim SS, Lopez AD, Lozano R, Marczak L, Mensah GA, Mokdad AH, Naghavi M, Nguyen G, Nsoesie E, Olsen H, Pigott DM, Pinho C, Rankin Z, Reinig N, Salomon JA, Sandar L, Smith A, Stanaway J, Steiner C, Teeple S, Thomas BA, Troeger C, Wagner JA, Wang H, Wanga V, Whiteford HA, Zoeckler L, Abajobir AA, Abate KH, Abbafati C, Abbas KM, Abd-Allah F, Abraham B, Abubakar I, Abu-Raddad LJ, Abu-Rmeileh NME, Ackerman IN, Adebiyi AO, Ademi Z, Adou AK, Afanvi KA, Agardh EE, Agarwal A, Kiadaliri AA, Ahmadieh H, Ajala ON, Akinyemi RO, Akseer N, Al-Aly Z, Alam K, Alam NKM, Aldhahri SF, Alegretti MA, Alemu ZA, Alexander LT, Alhabib S, Ali R, Alkerwi A, Alla F, Allebeck P, Al-Raddadi R, Alsharif U, Altirkawi KA, Alvis-Guzman N, Amare AT, Amberbir A, Amini H, Ammar W, Amrock SM, Andersen HH, Anderson GM, Anderson BO, Antonio CAT, Aregay AF, Ärnlöv J, Artaman A, Asayesh H, Assadi R, Atique S, Avokpaho EFGA, Awasthi A, Quintanilla BPA, Azzopardi P, Bacha U, Badawi A, Balakrishnan K, Banerjee A, Barac A, Barker-Collo SL, Bärnighausen T, Barregard L, Barrero LH, Basu A, Bazargan-Hejazi S, Beghi E, Bell B, Bell ML, Bennett DA, Bensenor IM, Benzian H, Berhane A, Bernabé E, Betsu BD, Beyene AS, Bhala N, Bhatt S, Biadgilign S, Bienhoff K, Bikbov B, Biryukov S, Bisanzio D, Bjertness E, Blore J, Borschmann R, Boufous S, Brainin M, Brazinova A, Breitborde NJK, Brown J, Buchbinder R, Buckle GC, Butt ZA, Calabria B, Campos-Nonato IR, Campuzano JC, Carabin H, Cárdenas R, Carpenter DO, Carrero JJ, Castañeda-Orjuela CA, Rivas JC, Catalá-López F, Chang J-C, Chiang PP-C, Chibueze CE, Chisumpa VH, Choi J-YJ, Chowdhury R, Christensen H, Christopher DJ, Ciobanu LG, Cirillo M, Coates MM, Colquhoun SM, Cooper C, Cortinovis M, Crump JA, Damtew SA, Dandona R, Daoud F, Dargan PI, Das Neves J, Davey G, Davis AC, Leo DD, Degenhardt L, Gobbo LCD, Dellavalle RP, Deribe K, Deribew A, Derrett S, Jarlais DCD, Dharmaratne SD, Dhillon PK, Diaz-Torné C, Ding EL, Driscoll TR, Duan L, Dubey M, Duncan BB, Ebrahimi H, Ellenbogen RG, Elyazar I, Endres M, Endries AY, Ermakov SP, Eshrati B, Estep K, Farid TA, Farinha CSES, Faro A, Farvid MS, Farzadfar F, Feigin VL, Felson DT, Fereshtehnejad S-M, Fernandes JG, Fernandes JC, Fischer F, Fitchett JRA, Foreman K, Fowkes FGR, Fox J, Franklin RC, Friedman J, Frostad J, Fürst T, Futran ND, Gabbe B, Ganguly P, Gankpé FG, Gebre T, Gebrehiwot TT, Gebremedhin AT, Geleijnse JM, Gessner BD, Gibney KB, Ginawi IAM, Giref AZ, Giroud M, Gishu MD, Giussani G, Glaser E, Godwin WW, Gomez-Dantes H, Gona P, Goodridge A, Gopalani SV, Gotay CC, Goto A, Gouda HN, Grainger R, Greaves F, Guillemin F, Guo Y, Gupta R, Gupta R, Gupta V, Gutiérrez RA, Haile D, Hailu AD, Hailu GB, Halasa YA, Hamadeh RR, Hamidi S, Hammami M, Hancock J, Handal AJ, Hankey GJ, Hao Y, Harb HL, Harikrishnan S, Haro JM,

Havmoeller R, Hay RJ, Heredia-Pi IB, Heydarpour P, Hoek HW, Horino M, Horita N, Hosgood HD, Hoy DG, Htet AS, Huang H, Huang JJ, Huynh C, Iannarone M, Iburg KM, Innos K, Inoue M, Iyer VJ, Jacobsen KH, Jahanmehr N, Jakovljevic MB, Javanbakht M, Jayaraman SP, Jayatilleke AU, Jee SH, Jeemon P, Jensen PN, Jiang Y, Jibat T, Jimenez-Corona A, Jin Y, Jonas JB, Kabir Z, Kalkonde Y, Kamal R, Kan H, Karch A, Karema CK, Karimkhani C, Kasaeian A, Kaul A, Kawakami N, Keiyoro PN, Kemp AH, Keren A, Kesavachandran CN, Khader YS, Khan AR, Khan EA, Khang Y-H, Khera S, Khoja TAM, Khubchandani J, Kieling C, Kim P, Kim C, Kim D, Kim YJ, Kissoon N, Knibbs LD, Knudsen AK, Kokubo Y, Kolte D, Kopec JA, Kosen S, Kotsakis GA, Koul PA, Koyanagi A, Kravchenko M, Defo BK, Bicer BK, Kudom AA, Kuipers EJ, Kumar GA, Kutz M, Kwan GF, Lal A, Lalloo R, Lallukka T, Lam H, Lam JO, Langan SM, Larsson A, Lavados PM, Leasher JL, Leigh J, Leung R, Levi M, Li Y, Li Y, Liang J, Liu S, Liu Y, Lloyd BK, Lo WD, Logroscino G, Looker KJ, Lotufo PA, Lunevicius R, Lyons RA, Mackay MT, Magdy M, Razek AE, Mahdavi M, Majdan M, Majeed A, Malekzadeh R, Marcenes W, Margolis DJ, Martinez-Raga J, Masiye F, Massano J, McGarvey ST, McGrath JJ, McKee M, McMahon BJ, Meaney PA, Mehari A, Mejia-Rodriguez F, Mekonnen AB, Melaku YA, Memiah P, Memish ZA, Mendoza W, Meretoja A, Meretoja TJ, Mhimbira FA, Millear A, Miller TR, Mills EJ, Mirarefin M, Mitchell PB, Mock CN, Mohammadi A, Mohammed S, Monasta L, Hernandez JCM, Montico M, Mooney MD, Moradi-Lakeh M, Morawska L, Mueller UO, Mullany E, Mumford JE, Murdoch ME, Nachega JB, Nagel G, Naheed A, Naldi L, Nangia V, Newton JN, Ng M, Ngalesoni FN, Nguyen QL, Nisar MI, Pete PMN, Nolla JM, Norheim OF, Norman RE, Norrving B, Nunes BP, Ogbo FA, Oh I-H, Ohkubo T, Olivares PR, Olusanya BO, Olusanya JO, Ortiz A, Osman M, Ota E, Pa M, Park E-K, Parsaeian M, De Azeredo Passos VM, Caicedo AJP, Patten SB, Patton GC, Pereira DM, Perez-Padilla R, Perico N, Pesudovs K, Petzold M, Phillips MR, Piel FB, Pillay JD, Pishgar F, Plass D, Platts-Mills JA, Polinder S, Pond CD, Popova S, Poulton RG, Pourmalek F, Prabhakaran D, Prasad NM, Qorbani M, Rabiee RHS, Radfar A, Rafay A, Rahimi K, Rahimi-Movaghar V, Rahman M, Rahman MHU, Rahman SU, Rai RK, Rajsic S, Ram U, Rao P, Refaat AH, Reitsma MB, Remuzzi G, Resnikoff S, Reynolds A, Ribeiro AL, Blancas MJR, Roba HS, Rojas-Rueda D, Ronfani L, Roshandel G, Roth GA, Rothenbacher D, Roy A, Sagar R, Sahathevan R, Sanabria JR, Sanchez-Niño MD, Santos IS, Santos JV, Sarmiento-Suarez R, Sartorius B, Satpathy M, Savic M, Sawhney M, Schaub MP, Schmidt MI, Schneider IJC, Schöttker B, Schwebel DC, Scott JG, Seedat S, Sepanlou SG, Servan-Mori EE, Shackelford KA, Shaheen A, Shaikh MA, Sharma R, Sharma U, Shen J, Shepard DS, Sheth KN, Shibuya K, Shin M-J, Shiri R, Shiue I, Shrime MG, Sigfusdottir ID, Silva DAS, Silveira DGA, Singh A, Singh JA, Singh OP, Singh PK, Sivonda A, Skirbekk V, Skogen JC, Sligar A, Sliwa K, Soljak M, Søreide K, Sorensen RJD, Soriano JB, Sposato LA, Sreeramareddy CT, Stathopoulou V, Steel N, Stein DJ, Steiner TJ, Steinke S, Stovner L, Stroumpoulis K, Sunguya BF, Sur P, Swaminathan S, Sykes BL, Szoeke CEI, Tabarés-Seisdedos R, Takala JS, Tandon N, Tanne D, Tavakkoli M, Taye B, Taylor HR, Ao BJT, Tedla BA, Terkawi AS, Thomson AJ, Thorne-Lyman AL, Thrift AG, Thurston GD, Tobe-Gai R, Tonelli M, Topor-Madry R, Topouzis F, Tran BX, Truelsen T, Dimbuene ZT, Tsilimbaris M, Tura AK, Tuzcu EM, Tyrovolas S, Ukwaja KN, Undurraga EA, Uneke CJ, Uthman OA, Van Gool CH, Varakin YY, Vasankari T, Venketasubramanian N, Verma RK, Violante FS, Vladimirov SK, Vlassov VV, Vollset SE, Wagner GR, Waller SG, Wang L, Watkins DA, Weichenthal S, Weiderpass E, Weintraub RG, Werdecker A, Westerman R, White RA, Williams HC, Wiysonge CS, Wolfe CDA, Won S, Woodbrook R, Wubshet M, Xavier D, Xu G, Yadav AK, Yan LL, Yano Y, Yaseri M, Ye P, Yebyo HG, Yip P, Yonemoto N, Yoon S-J, Younis MZ, Yu C, Zaidi Z, Zaki MES, Zeeb H, Zhou M, Zodpey S, Zuhlke LJ, Murray CJL (2016) Global, regional, and national incidence, prevalence, and years lived with disability for 310 diseases and injuries, 1990–2015: a systematic analysis for the Global Burden of Disease Study 2015. Lancet 388(10053):1545–1602

Wakabayashi K, Takahashi H, Takeda S, Ohama E, Ikuta F (1988) Parkinson's disease: the presence of Lewy bodies in Auerbach's and Meissner's plexuses. Acta Neuropathol 76(3):217–221

Wallen ZD, Demirkan A, Twa G, Cohen G, Dean MN, Standaert DG, Sampson TR, Payami H (2022) Metagenomics of Parkinson's disease implicates the gut microbiome in multiple disease mechanisms. Nat Commun 13(1):6958

Wang Q, Luo Y, Ray Chaudhuri K, Reynolds R, Tan E-K, Pettersson S (2021) The role of gut dysbiosis in Parkinson's disease: mechanistic insights and therapeutic options. Brain 144(9):2571–2593

Wang Y, Xie R, Jia S, Li Y, Qiao Y (2024) Research progress in the relationship between gut microbiota and Parkinson's disease. Acta Microbiol Sin 64(10):3610–3619

Wilkins HM, Weidling IW, Ji Y, Swerdlow RH (2017) Mitochondria-derived damage-associated molecular patterns in neurodegeneration. Front Immunol 8:508

Wirdefeldt K, Odin P, Nyholm D (2016) Levodopa-carbidopa intestinal gel in patients with Parkinson's disease: a systematic review. CNS Drugs 30(5):381–404

Wlodarska M, Luo C, Kolde R, Hennezel E, Annand JW, Heim CE, Krastel P, Schmitt EK, Omar AS, Creasey EA, Garner AL, Mohammadi S, O'Connell DJ, Abubucker S, Arthur TD, Franzosa EA, Huttenhower C, Murphy LO, Haiser HJ, Vlamakis H, Porter JA, Xavier RJ (2017) Indoleacrylic acid produced by commensal Peptostreptococcus species suppresses inflammation. Cell Host Microbe 22(1):25–37.e6

Wolters EC, Braak H (2006) Parkinson's disease: premotor clinico-pathological correlations. In: Riederer P, Reichmann H, Youdim MBH, Gerlach M (eds) Parkinson's disease and related disorders. Springer, Vienna, pp 309–319

Xavier AL, Menezes JRL, Goldman SA, Nedergaard M (2014) Fine-tuning the central nervous system: microglial modelling of cells and synapses. Philos Trans R Soc B Biol Sci 369(1654):20130593

Xu X, Fu Z, Le W (2019) Exercise and Parkinson's disease. Int Rev Neurobiol 147:45–74

Yang Z, Wang KKW (2015) Glial fibrillary acidic protein: from intermediate filament assembly and gliosis to neurobiomarker. Trends Neurosci 38(6):364–374

Yang D, Zhao D, Ali Shah SZ, Wu W, Lai M, Zhang X, Li J, Guan Z, Zhao H, Li W, Gao H, Zhou X, Yang L (2019) The role of the gut microbiota in the pathogenesis of Parkinson's disease. Front Neurol 10:1155

Yao Z, Duan S, Hou D, Heese K, Wu M (2007) Death effector domain DEDa, a self-cleaved product of caspase-8/Mch5, translocates to the nucleus by binding to ERK1/2 and upregulates procaspase-8 expression via a p53-dependent mechanism. EMBO J 26(4):1068–1080

Yao L, Yang Y, Yang X, Rezaei MJ (2024) The interaction between nutraceuticals and gut microbiota: a novel therapeutic approach to prevent and treatment Parkinson's disease. Mol Neurobiol 61(11):9078–9109

Ying Z-M, Lv Q-K, Yao X-Y, Dong A-Q, Yang Y-P, Cao Y-L, Wang F, Gong A-P, Liu C-F (2022) BAG3 promotes autophagy and suppresses NLRP3 inflammasome activation in Parkinson's disease. Ann Transl Med 10(22):1218–1218

Yuan H, Zhang Z-W, Liang L-W, Shen Q, Wang X-D, Ren S-M, Ma H-J, Jiao S-J, Liu P (2010) Treatment strategies for Parkinson's disease. Neurosci Bull 26(1):66–76

Zesiewicz TA, Bezchlibnyk Y, Dohse N, Ghanekar SD (2020) Management of early Parkinson disease. Clin Geriatr Med 36(1):35–41

Zhang X, Tang B, Guo J (2023) Parkinson's disease and gut microbiota: from clinical to mechanistic and therapeutic studies. Neurodegeneration 12(1):59

Zhong Z, Ye M, Yan F (2023) A review of studies on gut microbiota and levodopa metabolism. Front Neurol 14

Biomarker Role in Drug Delivery: Microbiota Implications and Nanotechnology for Personalized Therapeutics

Laura-Dorina Dinu, Emanuel Vamanu, and Ionela Avram

Abstract Recent research about the human microbiome has transformed our understanding of its role in human health while it is progressively recognized as a key factor in drug effectiveness and safety. Therefore, this chapter explores the role of microbiota biomarkers in biomarker-driven drug therapy and how they will shape the future of precision medicine and drug design. Understanding individuals as holobiont and knowledge about the complex multidirectional interaction between non-antibiotic drugs, gut microbiota, and host are critical to developing microbial biomarkers that enhance the drug's outcomes and safety. Microbiota biomarker-driven drug therapy can help predict individual responses to specific drugs, personalize medicine dosage, guide microbiota-targeted interventions, and monitor the drug efficacy, in case of the personalized treatments. Progress in nanotechnology, high-throughput next-generation sequencing, machine learning, artificial intelligence, and building comprehensive microbiome databases are expected to enhance the predictability and applicability of this approach. While there are challenges to overcome, the ongoing advancements in technology provide a hopeful outlook for the future of microbiota biomarker-driven precision medicine that will make therapies more precise, effective, and patient-specific.

Keywords Pattern · Dysbiosis · Biomarker · SCFAs · TMA

L.-D. Dinu · E. Vamanu (✉)
Department of Industrial Biotechnology, Faculty of Biotechnology, University of Agronomic Sciences and Veterinary Medicine of Bucharest, Bucharest, Romania
e-mail: emanuel.vamanu@bth.usamv.ro

I. Avram
Faculty of Biology, University of Bucharest, Bucharest, Romania

S. K. Singh et al. (eds.), *Microbiota Profiling for Precision Medicine*, Interdisciplinary Biotechnological Advances,
https://doi.org/10.1007/978-981-95-2104-3_7

1 Introduction

Microbiota is a complex community of microorganisms with a characteristic pattern, located inside and outside of the individual's body and plays a crucial role in health and disease progression. The human microbiota pattern contains many bacteria, Archaea, viruses, fungi, and parasites that have a dynamic fingerprint, and most of these microbes are present in the gastrointestinal tract. The human gut microbiome (collection of the microbiota genomes) varies across populations due to dietary habits, genetic patterns, environmental influences, and lifestyle habits. The gut microbiota (GM) includes both beneficial and symbiotic microbes that are able to promote pathology when specific conditions happen in the host, known as pathobionts. The microbial activity produces many bioactive molecules with low molecular weight that affect the overall health of the host. These molecules include short-chain fatty acids (SCFAs), mainly, acetate, propionate, and butyrate, the primary metabolites synthesized during the anaerobic fermentation of the intake fibers; SCFAs play a crucial role in the physiology of various host systems. Also, branched-chain amino acids, bile acid (BAs) derivatives, trimethylamine (TMA), tryptophan, and indole derivatives are important metabolites produced by intestinal microbiota. Microorganisms from the human gut have a common evolution and they have developed a series of mutual relationships with human cells based on a complex environment regarding metabolism, communication, and intercellular regulation. Therefore, GM should be viewed as an endocrine organ of the human body since it has been demonstrated to regulate the release of some hormones or interfere with their metabolism (Clemente-Suárez et al. 2024). It is well-established that the composition and the balance of the GM play a crucial role in the overall health of the host, while different strategies will support the rebalance to a healthy microbiota, as summarized in Fig. 1. Accumulated data that present the role of this unbalance, known as dysbiosis, are connected to a wide range of metabolical and physiological problems, including obesity, diabetes, inflammatory bowel disease (IBD), cardiovascular diseases, and mental health disorders, such as depression and anxiety (Brown and Taylor 2023). The changes in microbiota fingerprint from a homeostatic state to an unbalanced state can be established using microbiota biomarkers.

A clinical marker is a defined characteristic that is measured as an indicator of normal biological processes, pathogenic processes, or responses to an exposure or intervention, including therapeutic interventions. Molecular, histologic, radiographic, or physiologic characteristics are types of biomarkers as described by the BEST (Biomarkers, EndpointS and other Tools) resource (https://www.ncbi.nlm.nih.gov/pubmed/27010052). Thus, biomarkers are quantifiable substances whose presence or absence can be used to predict or indicate disease. The traditionally acknowledged biomarkers are derived from biological materials (e.g., blood, urine, soft tissue). In parallel, there is strong support for the development of digital and microbiota biomarkers that will likely play an increasingly important role in shaping the future of healthcare and drug development, streamlining personalized therapeutics.

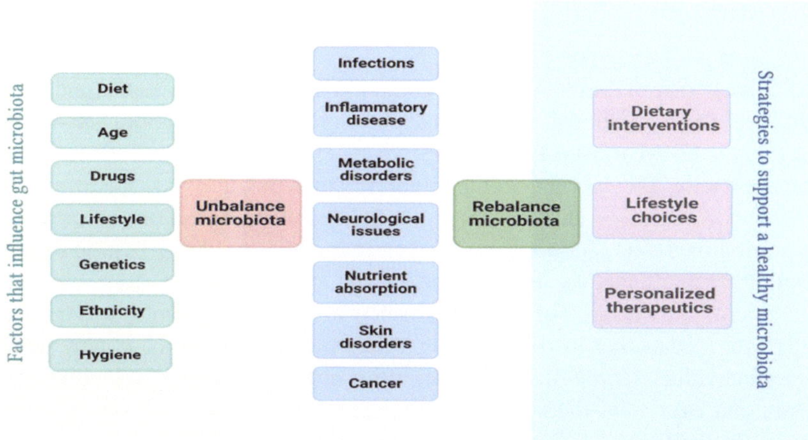

Fig. 1 Relationship between gut microbiota and dietary, genetic, environmental, and lifestyle factors

The microbiota biomarkers are defined as non-invasive and measurable indicators of the microbiome's state and can be genetic (e.g., specific DNA sequences), metabolic (signaling metabolites involved in communication with the host), or proteomic (proteins or enzymes). Although GM is the most extensively studied due to its dense population and diverse functions, the microbial biomarkers can be found in various body sites (human tissues and biofluids), including skin, oral cavity, respiratory, and urinary tract. Furthermore, microbiome signature can be exploited as a predictive, diagnostic, or prognosis biomarker in a variety of diseases, such as complex inflammatory, metabolic, or neurodegenerative diseases, cancer, or drug therapy (Hajjo et al. 2022). Microbiota biomarkers are typically identified through advanced sequencing technologies, metagenomics, and metabolomics, which allow for the comprehensive analysis of microbial DNA, RNA, proteins, and metabolites.

In this chapter, we focus on the role of the microbiota biomarkers to predict and assess the drug's efficacy and safety, tailoring therapies based on individual's needs in precision medicine. To explore the role of microbiome (viriome, bacteriome, and mycobiome) biomarkers, the holobiont concept and proof of the interaction between non-antibiotic orally administered drugs, gut microbiota, and host are essential. Also, this knowledge can be used to implement more effective strategies for developing reliable and robust microbiota-based biomarkers for drug design. Moreover, the implications of nanotechnology tools and microbiota-based biomarkers in drug delivery to successfully implement personalized therapies based on highly effective drugs with fewer adverse effects will be investigated.

2 Microbiota Biomarker-Driven Drug Therapy Concept

Biomarker-driven drug delivery is a paradigm shift in drug therapy, fundamentally different from traditional methods, which often rely on a "one-drug-fits-all" strategy. In the last years, this fundamental concept in pharmacology has been promptly replaced with "the right drug for the right patient at the right dose and time" as treatments become more customized and patient-centered based on the principles of precision medicine (Qahwaji et al. 2024). Individual variability in drug response is caused by many factors, mainly genetics and environmental factors, lifestyle, and gut microbiota impact. Therefore, the term pharmacomicrobiomics was proposed in 2010 for studies that investigate drug–microbiome interaction and how intra- and inter-individual microbiome variations affect drug absorption, distribution, metabolism, and excretion, thereby influencing the drug efficacy and safety (Rizkallah et al. 2010). Since 1937, when researchers first discovered the GM activating of antimicrobial drug prontosil many studies focused on gut microbiota and drug interaction (Dikeocha et al. 2022). However, many common orally administered drugs are still formulated and prescribed without adequate consideration of the impact of drug–gut microbiota interaction. Thus, it has been shown that up to 10% of patients have specific strains of *Eggerthella lenta* in GM, responsible for digoxin inactivation, a drug used to treat arrhythmia (Haiser et al. 2013). Also, the antimicrobial activity of non-antibiotic medicine on GM and its consequences on host health was well documented, proving the non-negligible role of intestinal microbiota in drug efficacy and toxicity (Pryor et al. 2020; Zhao et al. 2023). In the case of drugs that are non-orally administered but released by biliary excretion in the duodenum, they still can go through GM metabolism (Feng et al. 2020). The meaningful observation from pharmacomicrobiomics studies leads to the idea of using gut microbiota as a biomarker in medicine therapy.

The microbiota biomarker-driven drug therapy offers valuable information about the microbiome pattern to guide and improve drug treatment outcomes and safety. The holobiont concept is the main concept when considering microbiota-based therapy. That is a holistic ecological and evolutionary approach that views an organism and its associated microbial communities as a single, integrated unit named holobiont (Simon et al. 2019). At the same time, hologenome refers to the collective genome of the host and the "secondary genome," which incorporates all of its symbiotic microorganisms, such as bacteria, archaea, viruses, fungi, and sometimes even other multicellular organisms. The human gut holobiont includes the microbiome (mainly viriome, bacteriome, and mycobiome) which plays a critical role in the modulation of host health, while a close connection between the human microbiome and health or disease is largely supported by scientific evidence. Moreover, the multiple problems observed in germ-free animals proved that not only the presence but also the proper activity of these symbiotic microbes is required for the correct functioning of the entire human body. Additionally, selective pressure (e.g., drugs) acting on holobiont components has likely impacted both the host and microbiome, shaping the microbial community, as it was proved by studies about complex interaction drug–microbiota–host interaction. Such results highlight the importance of understanding individuals (human, animal, or otherwise) as holobiont and have further implications for diagnosis and therapy.

Recently, the interaction between the drug, gut microbiota, and the host has received considerable attention while knowledge in this field is progressing, as the experimental procedures and analysis techniques (such as multi-omic approaches) do. In the last years, numerous quality reviews outlined the advances in pharmacomicrobiomics and revealed up-to-date data (Lam et al. 2019; Pryor et al. 2020; Dikeocha et al. 2022; Zhao et al. 2023). A graphical summary of the key effects produced by the implications of microbes in the drug–gut microbiota–host complex interaction is provided in Fig. 2. Overall, it has been proved that oral drug administration affects GM both directly and indirectly via the host. The metagenomic analyses revealed extensive changes in the gut microbiome's diversity, composition, and function, as well as antimicrobial resistance (AMRs) occurrence after antibiotic therapy have been reported (Hajjo et al. 2022). Non-antibiotic drug effects on intestinal microbiota extended to a wide range of drug classes and chemistry, as demonstrated by a recent study which in vitro tested over 1000 host-targeted medicines and revealed that 24% inhibited the growth of representative gut bacterial strains (Maier et al. 2018). Importantly, medications found to have a significant effect on microbial growth included those used for the treatment of mental health disorders, type 2 diabetes, and cancer. Non-antibiotic medicine exhibits antimicrobial properties directly or indirectly, as they alter the gastrointestinal environment. Additionally, changes (e.g., inhibition, elimination) in the gut microbial community and the gastrointestinal habitat can lead to the colonization of bacteria with pathogenic potential (Feng et al. 2020). The drug-induced dysbiosis has been linked to the alteration of gut mobility, including gut barrier function that induces systematic inflammation. Lately, the unbalanced microbiome could disrupt the bidirectional communication between the microbes and the host cells that involves microbial metabolites (e.g., SCFAs, secondary BAs, TMA) or miRNAs, which are small, single-stranded,

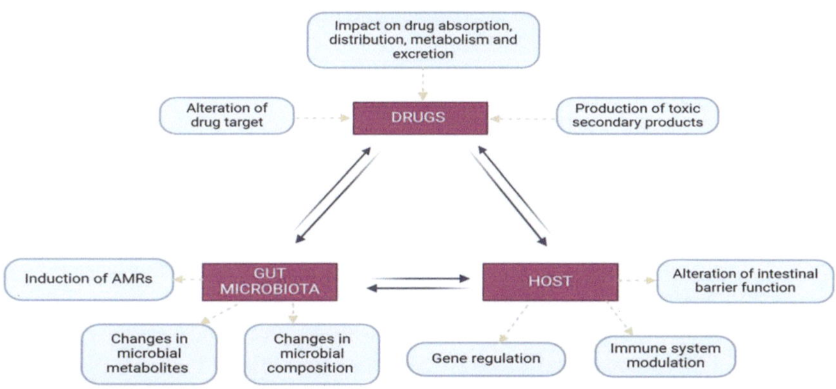

Fig. 2 Key effects of the microorganisms impact in the interaction non-antibiotic drug—gut microbiota—host. (Figure created with BioRender.com)

non-coding RNA molecules involved in the regulation of gene expression. Thus, SCFAs have proved to modulate the host immune system, while other studies have shown that several miRNAs from host cells can modify the microbiome, inducing dysbiosis and vice versa, as these can target the same intracellular signaling pathways (Nucera et al. 2024). The complex cross-talk between host miRNA and GM could lead to pathogenesis in inflammatory bowel disease (IBD) (Casado-Bedmar and Viennois 2022). Similarly, *Fusobacterium nucleatum* provoked decreasing in the expression of miR-18a and miR-4802 that aggravate chemoresistance, while the gut habitats which have been claimed to drive colorectal cancer *Fusobacterium* and *Escherichia coli* were promoted by host microRNAs (miR-515-5p and miR-1226-5p) (Li et al. 2020). It is important to mention that on some occasions drugs positively modify gut dysbiosis with subsequent consequences on host health. Thus, it is considered that the pleiotropic effects of metformin, a known anti-diabetic and cardioprotector drug, are related to changes in the gut microbial composition and metabolites (e.g., SCFAs, BAs) (Luo et al. 2019; Sun et al. 2018).

On the other hand, drug–microbiota interaction is bidirectional and countless reports proved that host and gut microbes can influence drug responses and efficacy by impacting the drug absorption, distribution, metabolism, excretion, and bioavailability, leading to different adverse effects on the host. Additionally, competition between GM metabolites and drugs might affect the efficacy of the medicine and exaggerate their toxicity. The microbiota impact on drugs demonstrated strong associations with microbial features, such as specific enzymes and lipopolysaccharides production that altered drug metabolism in the host (Mohr et al. 2022). Still, studies have proved that GM might enhance drug activation for a low number of drugs (Zhao et al. 2023).

Therefore, a deep understanding of the drug–gut microbiota–host interaction, especially the involved mechanisms, is important to design and validate microbial biomarkers that can predict drug efficiency and toxicity and subsequently improve the drug treatment outcome. These biological biomarkers will help identify patients most likely to respond to the treatment, optimize therapeutic intervention, or guide the development of personalized approaches that increase therapy effectiveness and decrease drug toxicity or adverse effects.

3 The Role of Microbiota Biomarkers in Drug Therapy

Given the increasing recognition of gut microbiota's role in drug metabolism in the last few years, some comprehensive reviews highlighted the importance of gut microbial biomarkers that correlate with drug responsiveness and proposed different strategies to maximize drug outcomes without unwanted secondary effects (Feng et al. 2020; Kamath and Joyce 2025). These microbiome biomarkers will help achieve "gut neutrality," a strategy that represents an opportunity to revitalize therapy, and ultimately improve healthcare and treatment outcomes for patients (Kamath and Joyce 2025). To increase the ability of a drug (initially aimed at targeting host cells) to be fully effective is crucial to identify the microbiota fingerprint from drug-responsive patients and then compare it with the one from non-responders. This is caused by individual variability in drug response, where the host's genetic background, lifestyle, and environmental factors are partially responsible for variability, depending on the disease model (Zhao et al. 2023). For non-responder patients, microbiota biomarker-driven drug therapy will help guide additional treatments to optimize the clinical outcomes of a given therapy.

In light of the shift towards a more patient-centered approach in drug therapy and based on the drug–gut microbiota bidirectional interaction, biomarkers could be divided into two categories. Screening of the microbiota-mediated toxicity with biomarkers that predict drug modification in the early stage of the preclinical investigations will help mitigate the adverse drug-gut microbiota interventions. The microbiota analysis during the treatment will provide information about the microbial modulation by treatment response. These biomarkers that predict and inform about changes in the microbiota signature will help optimize the therapeutic intervention (e.g., drug dosage) and will have a substantial impact on reducing adverse drug reactions and, subsequently on healthcare costs.

3.1 Gut Microbiota Biomarkers to Predict Drug Modification

In the host, the primary site of drug metabolism is the liver. However, drugs that are administered orally have to first-pass interaction with GM which has strong metabolic capability, then to be absorbed and reach the liver. The complex enzymatic machinery of these microbes proved to be involved in drug inactivation, activation, or changes in their bioavailability (Dikeocha et al. 2022). This alteration can either enhance or reduce the concentration of the drug that reaches the systemic circulation, thereby influencing its effectiveness (Dikeocha et al. 2022; Zhao et al. 2023). Alterations in drug pharmacokinetics and pharmacodynamics caused by microbial biotransformation have been proved for more than 180 drugs (McCoubrey et al. 2022). Prontosil, an antibacterial of the sulfonamide group, sulfasalazine used for treating rheumatoid arthritis and ulcerative colitis in the clinic, lovastatin that lowers cholesterol, and non-steroidal anti-inflammatory drug sulindac are well-known drugs activated by gut microbiota. In vitro prontosil exhibited no antibacterial activity, while in vivo GM activation changed the drug that become a potent antibacterial (Wainwright and Kristiansen 2010). Studies have shown that the probiotic strains (e.g., *Lactobacillus acidophilus* L10, *Bifidobacterium lactis* B94, and *Streptococcus salivarius* K12) that possessed azoreductase activity increased the ability of sulfasalazine to be enhanced by the GM (Lee et al. 2012). This medicine is considered an efficient and inexpensive therapy for ulcerative colitis, but some patients have reported secondary adverse effects when using it, such as nausea, skin, rash, and anorexia (Crouwel et al. 2020). Most likely these effects are associated with variations in an individual's microbiome profile regarding the drug metabolism, thus in this case microbial biomarkers could be used to predict the drug toxicity. Lovastatin is a prodrug metabolized by the gut microbiome to its bioactive metabolite, while the bioavailability of this lipid-lowering drug was suggested to be influenced by the GM (Yoo et al. 2014; Kim et al. 2011).

Most common drugs used I different therapies are broken down by microbiota. Therefore, specific bacterial strains, enzymes, and genes involved in this process were proposed as predictive microbial biomarkers of drug inactivation. For many years is known that digoxin is converted into cardio-inactive product dihydrodigoxin by *Eggerthella lenta*, bacteria initially suggested to be used as a predictive biomarker for digoxin inactivation (Saha et al. 1983; Haiser et al. 2013). Later research proved that the strain cannot accurately predict the drug metabolism in the gut, while detection of the *cgr2* gene associated with drug reduction was proposed in predicting the digoxin response (Koppel et al. 2018). *E. lenta* and *Enterococcus faecalis* have demonstrated to be involved in drug converting and producing adverse reactions after levodopa administration, a medicine used for the treatment of Parkinson's disease (Maini Rekdal et al. 2019; van Kessel et al. 2019). In this situation, a bacterial interspecies pathway alters the drug, while enzymes involved were suggested to predict the drug metabolism (Maini Rekdal et al. 2019). Similarly, irinotecan, an anticancer drug that was modified by a high level of enzyme β-glucuronidase (GUSs) produced by some bacterial strains, such as *Escherichia coli*, *Staphylococcus* spp., and *Clostridium* spp. (Stringer et al. 2008). In the case of irinotecan, specific potential pathogenic strains are responsible for its deglucuronidation into toxic metabolites, leading to severe diarrhea in some cancer patients (Yauw et al. 2018). To decrease their abundance different strategies were investigated, including the use of antibiotics (Kehrer et al. 2001), selective inhibition of GUSs (Cheng et al. 2019), and probiotic administration (e.g., *Enterococcus faecium*, *Lactobacillus*, and *Bifidobacterium*) (Takasuna et al. 1996; Gallotti et al. 2021). Moreover, Chamseddine et al. (2019) proposed to use the GUSs enzymatic activity as a predictive biomarker of irinotecan-induced diarrhea severity that will help select patients most likely to respond to the chemotherapy (Chamseddine et al. 2019). Similarly, microbial inactivation of regorafenib, a multikinase inhibitor with excellent antitumor effects, has proved to be induced by GUSs synthesized by microbes from the intestinal tract and this was the major cause of the drug gastrointestinal toxicity (Ervin et al. 2019). Screening a collection of gut microbial isolates for their ability to inactivate the widely used drug for carcinoma doxorubicin, researchers noted that under anaerobic conditions the drug was inactivated by a strain of Gram-negative bacterium *Raoultella planticola*, and degraded by *Klebsiella pneumoniae* and *E. coli* BW25113 (Yan et al. 2017). It has proved that heterotrophic microbial consortia are capable of degrading the antidepressant drug fluoxetine, although the medicine showed antimicrobial effects on gut microflora (Velázquez and Nacheva 2017). Another in vitro example of drug inactivation was demonstrated for immunosuppressant tacrolimus and involved the gut bacteria *Faecalibacterium prausnitzii* (Guo et al. 2019). This could be correlated with results obtained in a clinical study where patients with kidney transplants required higher doses as they have an increased relative abundance of *F. prausnitzii* than stable dose controls (Lee et al. 2015). Moreover, a recent pilot cross-sectional study proved that GM is an important contributor to tacrolimus inter- and intra-pharmacokinetic variability, not explained by the patient's genetic background or by the pathophysiological status (Degraeve et al. 2024). Based on the multiple linear regression model, the research revealed that specific bacterial taxa (e.g., less abundant Gram-negative bacillus *Bilophila*) and some bacterial units (amplicon sequence variant ASV 1508, ASV 1982, and ASV 664) could partially be linked to the interindividual variability in drug dose-response, proving the importance of detection at the species and strain level. Also, a decrease in α-diversity of GM after kidney transplant was noted compared with healthy controls and pre-transplant analysis. Another research reported 37 taxa significantly enriched, including a known SCFAs producer *Subdoligranulum* genera and members of the Lachnospiraceae family, whereas *Bacteroides* was less abundant among patients with heart transplant that require high doses of tacrolimus (Jennings et al. 2020). Thus, specific strains could be used as biomarkers for patient stratification based on individual variability in tacrolimus response, and for an increase of the dosage to achieve the proper therapeutic level.

Detection of microbial metabolites produced in the gastrointestinal tract that compete with drugs and impact their efficacy or promote their toxicity could be used as predictive biomarkers. This competition has been revealed in the case of acetaminophen and p-cresol produced by some gut bacteria and exaggerates its toxicity, as well as between simvastatin and secondary BAs that compete for the host drug transporters SLCO1B1 (Kaddurah-Daouk et al. 2011; Laine et al. 2009).

Although there is a long and rich history of research about GM and drug interaction, limited information is about the drug's impact on the microbiome from other body habitats. A clear example was provided by a study that compared the vaginal microbiota of patients treated with anti-retroviral drug tenofovir. Enhanced efficacy of the tenofovir (61%) was reported for vaginal microbiota with an abundance of *Lactobacillus* spp. compared to microbiota dominated by *Gardnerella vaginalis*, which seems to inactivate the drug (Klatt et al. 2017).

Another experiment that raised questions indicated that bacterial contamination of cancer cell lines contributes to resistance to the anti-cancer drug gemcitabine, used to treat pancreatic ductal adenocarcinoma (Geller et al. 2017). Moreover, intratumoral injections of the enzyme cytidine deaminase, known to be produced by gut bacterial isolates from Gammaproteobacteria, impaired the treatment of disease in the mouse model, while co-treatment with antibiotic abrogated this resistance (Geller et al. 2017). These results highlighted the importance of microbial biomarkers for drug modification in cancer therapy.

Some supportive evidence to be used in the development of microbiota biomarkers for non-antibiotic drug modification was provided in Table 1. Research efforts are underway to identify gut microbiome biomarkers involved in the drug alteration or in the drug metabolite competition. In this case, the most feasible and scalable biomarkers seem to be the metabolic or enzymatic activity and microbial genes encoding these enzymes/metabolites (Feng et al. 2020). Additionally, microbial detection at the strain level using whole-genome shotgun sequencing or more comprehensive techniques should be considered. That has been proved relevant for some drugs (e.g., irinotecan and tacrolimus) inactivated by specific gut strains and

Table 1 Example of microbiota-based biomarkers involved in non-antibiotic drug modification

Drugs	Microbial species/metabolites/enzymes involved	Model/conditions/interaction	References
Digoxin	Actinobacteria *Eggerthella lenta*	Mouse model Predictive microbial biomarker of digoxin inactivation	Haiser et al. (2013)
	Detection of *cgr2* gene associated with digoxin reduction	In vitro study Compared the genomes of 25 *E. lenta* strains and close relatives	Koppel et al. (2018)
Levodopa	Enzyme tyrosine decarboxylases	Rat model and human fecal sample analysis Bacterial enzymes efficiently convert levodopa to dopamine	van Kessel et al. (2019)
	Enzymes (pyridoxal phosphate-dependent tyrosine decarboxylase, molybdenum-dependent dehydroxylase) Produced by *Enterococcus faecalis*, *Eggerthella lenta*	Drug conversion to dopamine by *E. faecalis,* followed by transformation of dopamine to m-tyramine by *E. lenta* Enzymes predict drug inactivation	Maini Rekdal et al. (2019)
Irinotecan	Enzyme β-glucuronidase Strains *E. coli*, *Staphylococcus* spp., and *Clostridium* spp.	Rat model Bacterial strains that express high levels of enzyme were involved in drug inactivation	Stringer et al. (2008)
	Enzymes	Human trial Enzymatic activity as a predictive biomarker of irinotecan-induced diarrhea severity to select responsive patients to chemotherapy	Chamseddine et al. (2019)
	Enzyme β-glucuronidase	Enzyme inhibitor	Cheng et al. (2019)
Doxorubicin	*Raoultella planticola* *Klebsiella pneumoniae and, E. coli BW25113*	In vitro study Strains inactivated or degraded the drug	Yan et al. (2018)

Tacrolimus	*Faecalibacterium prausnitzii* and other gut Clostridiales	In vitro study Strains are capable of metabolizing the drug to an inactive metabolite	Guo et al. (2019)
	Faecalibacterium prausnitzii	Clinic study with 5 patients in dose escalation group post-kidney transplantation compared to control group Fecal bacterial abundance was positively correlated with tacrolimus dosing	Lee et al. (2015)
	Bilophila spp. and ASV 1508, ASV 1982 (*Veillonella*/unclassified *Sporomusaceae*), ASV 664 (unclassified *Oscillospiraceae*)	Clinic study 93 patients post-kidney transplant (CYP3A5 gene non-expressers) Specific strains and ASV associated with the drug pharmacokinetic	Degraeve et al. (2024)
	Subdoligranulum genera *Lachnospiraceae* family increased *Bacteroides* less abundant	Clinic study 24 patients post-heart transplant	Jennings et al. (2020)
Gemcitabine	Enzyme cytidine deaminase produced by Gammaproteobacteria	Mouse model Bacteria might contribute to drug resistance of these tumors	Geller et al. (2017)
Acetaminophen	Metabolite p-cresol	Competition between drug and microbial metabolites	Laine et al. (2009)
Simvastatin	Metabolite secondary bile acids		Kaddurah-Daouk et al. (2011)
Regorafenib	Enzyme β-glucuronidase	Enzyme inhibitor	Ervin et al. (2019)
Diclofenac	Enzyme β-glucuronidase		Yauw et al. (2018)
Antiviral tenofovir	*Gardnerella vaginalis*	Clinic study Bacteria from vaginal microbiota involve in the drug inactivation	Klatt et al. (2017)

lead to interindividual differences in drug pharmacokinetic (bioavailability, effectiveness, and toxicity) and dosage modification. However, it has to be mentioned that most available research findings are based on the small cohort analysis, while the development of clinical microbial biomarkers that correlate with drug responsiveness requires large-scale validation studies, regulatory approval, and standardization methodology.

3.2 Gut Microbiota Biomarkers for Drug-Induced Modulation

Accumulating evidence has suggested the importance of GM in predicting the drug effectiveness in case of oral drug administration. Most non-antibiotic drugs have powerful antimicrobial effects that induce changes in the microflora diversity, composition, and functioning with subsequent consequences on the host. In parallel, there is strong support for the crucial role of GM in enhancing the drug's outcomes in some cases, while understanding this role will help develop microbiota-based biomarkers. Causality between drug effects and microbiota modulation is difficult to draw; yet in some cases, specific microbiome fingerprints could be linked to different treatment consequences. These findings highlight the potential for GM to serve as a predictive biomarker for drugs efficacy and toxicity, as well as an informative biomarker during medicine treatment.

Several common metrics were investigated to predict the drug treatment outcome, especially the microbial community diversity, and define relative diversities expressed as the ratio for two major bacterial phyla, *Firmicutes* and *Bacteroidetes* (F/B ratio), as well as the microbial richness. Increased F/B ratio was linked to dysbiosis associated with diabetes, obesity, or cardiovascular diseases, while decreased F/B ratio in inflammatory bowel disease. The hypoglycemic drug vildagliptin decreased the F/B ratio and improved glucose homeostasis in diabetic rats (Zhang et al. 2017a, b). Conversely, another drug that lowers blood glucose, sitagliptin reduced *Bacteroidetes* and increased *Firmicutes* and Tenericutes phyla in obese and diabetic rats (Yan et al. 2016). The anti-diabetic drug liraglutide increased the *Bacteroidetes*-to-*Firmicutes* ratio in a trial with both overweight and diabetes-overweight patients (Zhao et al. 2018), while in another study with normally-weighed mice, the drug had higher F/B ratio (Wang et al. 2016). Surprisingly, saxagliptin effects on GM have been compared to liraglutide but no substantial changes in the composition and diversity of GM were noted (Wang et al. 2016). Differences between reported results could be attributed to the difference in the target populations and experimental conditions, while this is a clear indication of the complexity of the mechanisms involved. Some works have proved that increased microbiota richness improved the kinetic activity of the drug and its effectiveness (Zhou and Xu 2020). Thus, a recent clinical study found that patients with more complex gut bacterial signatures and high richness of beneficial bacteria *Akkermansia* spp. and *Bifidobacterium* spp. had better drug response outcome and lesser toxicity profile (Shuaib et al. 2024). Also, a recent review highlighted the potential of increased α-diversity and abundance of *Faecalibacterium* and *Roseburia* species to be predictors of antidepressant treatment response (Borgiani et al. 2025).

Another direction was to study the importance of microbial metabolic outputs as biomarkers to predict drug efficiency. Thus, enalapril used to treat cardiovascular diseases (e.g., arterial hypertension, heart failure) has been shown to reduce blood pressure by lowering the level of TMA, a gut bacterial metabolite that produces hypertension (Velasquez et al. 2016; Konop et al. 2018). Absorbed TMA is transformed by the liver in TMA-oxide and is responsible for an increased risk of cardiovascular disease. Microbial metabolites, such as SCFAs that help immune response, were associated with better therapeutic outcomes (Bachem et al. 2019). In diabetic rats, vildagliptin treatment increased butyrate-producing bacteria, including *Bacteroides* and *Erysipelotrichaceae* (Zhang et al. 2017a, b). Another study showed that SCFAs-producing *Lactobacillus* spp. was increased by vildagliptin, as well as by the drugs widely used to treat acid-induced diseases, for example, gastroesophageal reflux (Olivares et al. 2018). More anti-diabetic drugs have proved to increase SCFAs, such as sitagliptin (Yan et al. 2016), and liraglutide that selects higher populations of *Bifidobacterium* spp., *Bacteroides* spp., and *Lachnospiraceae* spp. (Wang et al. 2016; Zhao et al. 2018; Zhang et al. 2018). On the other hand, research indicated that lower fecal butyrate concentrations and decreased relative abundance of SCFAs- and butyrate-producing taxa may be predictive biomarkers of non-response to advance therapies in the case of IBD (Wang et al. 2018; Lee et al. 2021). SCFA levels varied depending on the drug treatment in the case of some psychotropics drugs, while those changes could be linked to bacterial SCFA producers (e.g., Bacteroidetes or *Clostridium* spp.), but the connection with the drug's effect is unclear (Pryor et al. 2020).

The bidirectional regulation of some drugs that increase the SCFAs-producing bacteria and decrease potentially harmful species has been investigated. The proliferation of pathobionts leads to severe delays in drug response as they proved to aggravate inflammatory manifestations, trigger autoimmune diseases, and weaken barrier function (Maier et al. 2018). Thus, the anti-diabetic miglitol increased SCFA-producing and decreased harmful bacteria (Kishida et al. 2017). A drug used to treat hyperlipidemia, acarbose enhanced the abundance of beneficial and SCFAs-producing strains *Prevotella* spp., *Lactobacillus* spp., *Faecalibacterium* spp., *Bifidobacterium* spp., *Dialister*, and inhibit Enterobacteriaceae, especially *E. coli*, Bacteroidaceae, and lecithinase-positive *Clostridium* (Brunkhorst et al. 1999; Su et al. 2015). It is known that treatment with probiotic lactobacilli decreases glycemia but the possible role of taxon *Dialister* in this glucose metabolism has to be demonstrated.

The presence and/or the absence of specific beneficial bacterial strains were used as microbial biomarkers. It could target the presence of microbial species that promote an anti-inflammatory environment and gut integrity or enhanced immune intestinal system and increase the efficacy of drugs, proven for anti-cancer medicine. Most research has demonstrated that specific gut bacteria, including lactic acid bacteria, *Akkermansia muciniphila,* and *Bacteroides*, contribute to gut integrity by modulating mucin production and influencing the vagus nerve function (Conn et al. 2024). Statins dramatically higher the amount of certain antiinflammation-related species (e.g., *Bacteroides*, *Butyricimonas*, and *Mucispirillum*), which help to ameliorate hyperglycemia (Kim et al. 2019). Research with another anti-diabetic drug, liraglutide, proved that colonization of germ-free mice with GM from liraglutide-treated diabetic mice regulated the intestinal immune system and improved insulin secretion (Charpentier et al. 2021). In a clinical trial with 40 patients with type 2 diabetes treated with liraglutide, results suggest that baseline of hemoglobin, blood urea nitrogen, and changed microbial interaction network could be associated with the beneficial effect of the drug (Shang et al. 2021). In another study, captopril used to treat hypertension induced improvements in gut pathology and permeability and decreased systolic blood pressure, and the results correlated with the increase in *Allobaculum*, *Parabacteroides*, and *Mucispirillum* (Yang et al. 2019).

The microbiota-mediated mechanisms that involve changes in the GM might explain the multiple positive effects of metformin, a widespread anti-diabetic drug, known to reduce obesity and decrease the incidence of cardiovascular diseases (Sun et al. 2018; Luo et al. 2019). The gut bacteria *Akkermansia muciniphila* is a key factor in the glucose-lowering effect of the drug (Minamii et al. 2018). In another study, the anti-hyperglycemic effect of metformin was associated with its ability to reduce another gut habitant *Bacteroides fragilis* (Sun et al. 2018). Moreover, changes in gut microbial composition and metabolites (e.g., SCFAs, BAs) were associated with the metformin pleiotropic effect. Conversely, some drugs have both positive and negative effects on beneficial microflora. Thus, methotrexate used to treat autoimmune diseases lowers the abundance of enterobacterial, especially *E. faecium*, and higher beneficial *Lachnospiraceae* spp. However, the decrease in other beneficial strains, especially commensal *Bacteroides fragilis* that are essential to healthy gastrointestinal function (e.g., mucosal immunity and host nutrition), accentuates the gastrointestinal secondary effects (Zhou et al. 2018).

A major concern in drug therapy should be the variation in the individual's GM profile that could lead to alteration in the drug efficacy and toxicity. In a clinical study about R-warfarin, a vitamin K antagonist was used to treat thromboembolic complications; almost 35% of patients had a delay in responding to the drug linked to the abundance of specific genus *Bacteroides*, *Klebsiella*, *Escherichia*, and *Shigella*, while *Enterococcus* increased reaction to this drug (Xue et al. 2017; Wang et al. 2020). Thus, identifying individual microbiota biomarkers that predict adverse reactions can help personalize drug therapy to minimize these risks.

Selection of the appropriate microbiome-based interventions before chemo- and immunotherapy proved to be an effective approach to improve the efficacy of immunotherapy (Yan et al. 2024). In cancer patients, a greater alfa-diversity, intact commensal microbiota, or the presence of a particular bacteria, such as beneficial *Bifidobacterium* spp., was reported to improve anti-tumor responses (Vétizou et al. 2015; Gopalakrishnan et al. 2018). Moreover, recolonization with specific commensal bacteria from *Bacteroides* (*B. fragilis, B. thetaiotaomicron*) and *Burkholderia cepacia* enhanced the anti-tumor response (Vétizou et al. 2015). The beneficial gut populations of *A. muciniphila* and *Lachnospiraceae* spp. increased in patients treated with abiraterone acetate for prostate cancer, while higher α-diversity and *A. muciniphila* were related to better efficacy. (Daisley et al. 2020). Immunogenic gut bacteria *Bacteroides fragilis* and *Erysipelotrichaceae* increase the anti-cancer immune response and play a significant role in the late immunomodulatory effects of oxaliplatin (Iida et al. 2013). *Akkermansia muciniphila*, *Bacteroides fragilis*, and *Faecalibacterium prausnitzii* were linked to enhanced immune activation and better responses to immune checkpoint blockade therapies in cancer (Yan et al. 2024). Also, *Bifidobacterium* spp. improved the anti-tumor efficacy of immune checkpoint inhibitors (Sivan et al. 2015). On the other hand, in a mouse model, gemcitabine that is used to treat pancreatic cancer has been demonstrated to decrease both *Firmicutes* and *Bacteroidetes* and increase Proteobacteria; all these effects were associated with the inflammatory process that contributes to detrimental side effects and reduced drug outcomes (Panebianco et al. 2018).

Recently findings revealed that microbiota from other body parts could be used as biomarkers and be incorporated into therapeutic management to improve the drug response. Thus, some drugs decreased the abundance of GM but increased oral microbiota affecting 133 metabolic pathways (Cammarota et al. 2020). The antiretroviral treatment of HIV patients increased the relative abundance of anaerobic bacteria oral commensals, such as *Prevotella* spp. and serum SCFAs, especially butyrate, which was associated with increased susceptibility to tuberculosis (Segal et al. 2017).

Given the inconsistencies in the literature, it is difficult to make a strong correlation between specific metrics (e.g., diversity, F/B ratio, specific strains) and the response to all drugs. However, several conclusions could be drawn about the microbiota biomarkers that will enhance the drug effectiveness, as is presented in Table 2. Increased diversity and richness, the abundance of commensal and beneficial bacteria (e.g., SCFAs-producing species, anti-inflammatory and gut integrity promoting species or microbes that enhance the immune system), and decreasing the pathobionts are the most important microbial biomarkers to assess the drug efficacy and toxicity. A consistent number of investigations have proved that no single microbe has been found to be responsible for causing dysbiosis. Therefore, an index of several gut microbial species should be considered for a more reliable microbial biomarker.

Recently works suggested that specific microbial networks rather than microbial diversity alone may be a more important factor to focus on in further research about predictors of drug response. In the case of inflammatory bowel disease, Lee et al. (2021) demonstrated that a microbial community with higher abundances of the Bacillota phyla (including *Faecalibacterium*, *Eubacterium*, *Ruminococcus,* and *Roseburia*) was associated with higher rates of clinical remission (Lee et al. 2021). Similarly, Yilmaz et al. (2019) proved that a network containing SCFAs-producing taxa from *Lachnospiraceae* and *Ruminococcaceae* families was associated frequently with poorer response to IBD drug therapy (Yilmaz et al. 2019). Also, the microbial interactions within the network and their functional capacity seem to be very important (Meade et al. 2023).

Table 2 Example of the gut microbiota-based biomarkers linked to the positive drug outcomes

Diseases therapy	Gut microbiota biomarkers for drug-induced modulation	Role
Cancer	Immunotherapy—Presence of certain bacterial species like *Akkermansia muciniphila, Bacteroides fragilis,* Bifidobacterium spp., and *Faecalibacterium prausnitzii*	These bacteria are linked to enhanced immune activation and better responses to immune checkpoint blockade therapies in cancer.
	Chemotherapy—Specific gut microbial patterns with high diversity	Patients with these patterns often respond better to chemotherapies or radiotherapies due to enhanced immune modulation and gut health
Metabolic disorders	Level of SCFA producers and F/B ratio	Restoring the microbial pattern with higher SCFA producers can improve outcomes in obesity, type 2 diabetes, and other metabolic syndrome.
Cardiovascular diseases	Level of TMA producers and F/B ratio	Restoring the microbial pattern with lower TMA producers can decrease blood pressure.
Mental health and neurological disorders	Abundance of *Lactobacillus* and *Bifidobacterium,* both associated with the gut–brain axis	Pre- and probiotics targeting these microbial biomarkers can improve outcomes in anxiety, depression, and neurodegenerative disorders.
Liver diseases	Presence of the pathogenic bacteria like Enterobacteriaceae and SCFA producers	Altering these biomarkers through diet, probiotics, or antibiotics can improve conditions like non-alcoholic fatty liver disease.
Inflammatory and autoimmune diseases	Abundance of *Faecalibacterium prausnitzii* and *Akkermansia muciniphila*	These biomarkers help guide interventions in conditions like Crohn's disease, ulcerative colitis, and rheumatoid arthritis

The role of multi-kingdom (e.g., mycobiome, virome) intestinal microbiota is poorly reported and should be considered in further studies. Mycobiome represents just 0.1% of the gut microflora but is intrinsically interconnected with bacterial communities by various metabolic pathways (Liu et al. 2022). Both bacteriome and mycobiome are influenced by drugs and studies on different types of tumors (melanoma, gastrointestinal cancer, non-small cell lung cancer and, renal carcinoma) indicated that fungal–bacterial interaction, as well as fungal composition are good predictors of immunotherapy (Huang et al. 2023). However, in another work that characterized multi-kingdom microbiota in patients with hepatocellular carcinoma treated with an immune checkpoint inhibitor was observed that fungi are more stable markers comparable to bacteria (Zhu et al. 2024). In this case, the microbial differences between durable and non-durable clinical benefit groups were compared and bacteria (2 bacterial species and 1 metabolite) proved to be a better predictive and prognostic marker for immunotherapy benefits. However, the effect of fungi on bacteria that can influence the therapy outcomes has not been considered in that study. In the last years, another way to improve immunotherapy responsiveness was to modulate the viriome (e.g. by fecal microbial transplant from immunotherapy responders to non-responders, bacteriophages) that helped the response to immunotherapy (Broecker and Moelling 2021). Also, the potential of the mycobiome to predict the response to drugs has been explored. Thus, a clinical trial with 68 patients with inflammatory bowel disease evaluated mycobiome predictors for infliximab response after 12 weeks of drug therapy (Ventin-Holmberg et al. 2021). Although baseline fungal diversity was not shown to predict the therapeutic response, an increase in *Candida* spp. was observed in non-responders, while the relationship between selected genera in the bacteriome and mycobiome correlated with response. In another smaller study (18 patients), disease severity was correlated with certain fungal genera, but significant differences in abundance were only demonstrated after infliximab therapy (Wang et al. 2018).

The microbiota biomarker-driven approach offers several advantages over conventional drug delivery methods, including increased efficacy as orally administered drugs are more likely to achieve the desired therapeutic effect in the presence of "favorable/neutral" gut microbiota. Another important advantage is related to reduced side effects compared to non-targeted drug therapies which often see drug alteration and toxicity. Biomarkers can also streamline the drug development process by providing early indicators of drug efficacy and safety. This can lead to faster clinical trials and regulatory approvals.

Despite its promising potential, microbiota biomarker-driven drug therapy is not without challenges. Therefore, several important challenges need to be addressed, such as a limited understanding of complex mechanisms involved in gut microbiota–drug interactions, the vast amount of data generated that requires advanced bioinformatics tools and expertise for integration and interpretation, and the lack of standardized methodology. One of the primary obstacles is the complexity of identifying reliable and reproducible microbiome biomarkers (e.g., specific strains, metabolites, or genes). Moreover, a single biomarker may not be sufficient to accurately predict a patient's response to a treatment and a network of different types of microbial biomarkers will be required. Another challenge lies in the variability of microbiota biomarkers among different populations, necessitating comprehensive validation across diverse demographic groups. Moreover, the development of these biomarkers requires significant investment in research and technology, including comprehensive microbiome databases. Although biomarker-driven therapies may have higher upfront costs due to the need for molecular testing, they can ultimately be more cost-effective by reducing the incidence of ineffective treatments and hospitalizations due to adverse effects. Additionally, the regulatory landscape for biomarker-driven therapies is still evolving, and there is a need for standardized guidelines to ensure the reliability and reproducibility of biomarker tests (Amos et al. 2020; Andreoletti et al. 2024).

The future of biomarker-driven drug therapy is promising, with ongoing research aimed at discovering new biomarkers and improving drug design. Machine learning algorithms, for example, can analyze large datasets to identify novel biomarkers and predict patient responses more accurately than traditional methods (Kraneveld et al. 2024). Moreover, the integration of multi-omics data (genomics, proteomics, metabolomics, etc.) is likely to provide a more comprehensive understanding of disease mechanisms and patient variability, enabling the development of even more personalized therapies.

4 Implication of Microbiota Biomarker-Driven Therapy and Nanotechnology in Precision Medicine

Precision medicine, also known as personalized medicine, is an innovative approach that focuses on new treatments and medical strategies for prevention, diagnosis, and treatment that consider individual variability in genes, environmental factors, and lifestyle. Traditional medicine has always used a standardized method that does not adequately consider the intricate differences among people, while precision medicine's purpose is to provide the most effective therapeutic options by minimizing the adverse effects and addressing each patient's needs. This comprehensive approach enhances the effectiveness of treatment and helps advance a more individualized, patient-focused, and reliable healthcare system.

In the context of precision medicine, microbiome biomarkers can provide valuable information beyond what is obtainable from human genomic data alone. For instance, while genetic predisposition is important, the microbiome's dynamic nature means it can respond to lifestyle, diet, environment, and treatment in ways that human genetics cannot. This responsiveness makes microbiome biomarkers particularly useful for developing personalized treatment plans that consider an individual's unique microbial composition and its interaction with the host. Therefore, in case of a drug proven to interact with gut microflora is important to consider a network of biomarkers that integrate data about an individual's variability of the human genome and the gut microbiome for more personalized drug treatment (Feng et al. 2020).

Microbiome biomarkers, measurable indicators of the microbiome's state, hold great promise for advancing precision medicine and drug therapy by providing new tools to predict drug effects or to guide personalized treatment (Fig. 3). Microbiota biomarker-driven drug delivery will provide valuable insight about drug–gut microbiota–host interaction, ensuring the efficacy and safety of the drug treatment. By integrating microbiota biomarkers into drug therapy, healthcare providers can identify patient subgroups (responders) that are most likely to benefit from a given therapy, thereby optimizing clinical outcomes. This patient stratification will allow changing and/or optimizing the drug dosage, and protocols or establishing adjacent interventions to enhance the medicine efficacy for non-responders patients. Also, these biological biomarkers could replace or reduce the need for using other invasive techniques (e.g., biomicroscopy and endoscopy). Moreover, microbiota biomarkers that correlate with drug responsiveness are essential for improving drug design. Thus, in 2024 Đanić et al. using pharmacoinformatic tools screened the physicochemical and pharmacokinetic properties of ten gliclazide metabolites produced by the GM to identify compounds with improved pharmacological profiles

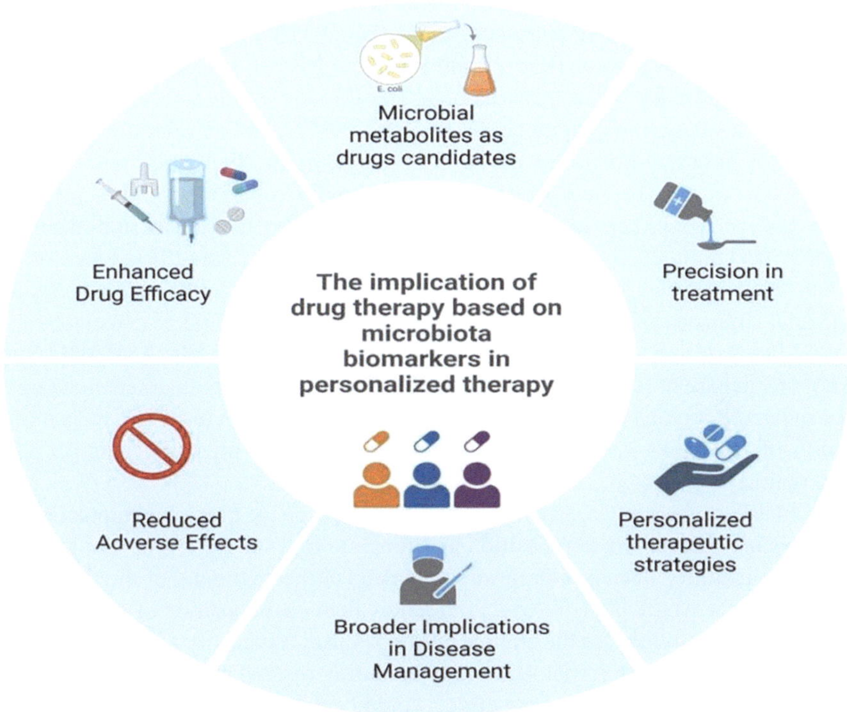

Fig. 3 Implications of microbiota biomarker-driven therapy in precision medicine. (Figure created with BioRender.com)

compared to the parent drug (Đanić et al. 2024). The drug gliclazide is a second-generation sulfonylurea derivative, largely used as a second-line treatment for type 2 diabetes mellitus. The implication of gut microbiota-mediated metabolism of gliclazide has been proved and explained by the inter-individual differences in the drug bioavailability, leading to variations in therapeutic responses among patients. The in silico analyses revealed that some hydroxylated metabolites bind more stable with the target receptor compared to gliclazide, which paves the way for further research about their pharmacological properties (Đanić et al. 2024).

Microbiota biomarkers that predict drug modification are important in precision medicine to improve drug efficacy, as there is compelling evidence that the intestinal microbiome has the metabolic power to inactivate, or even toxify a drug to a level of clinically relevant significance. In this case, biomarkers could be specific enzymes, such as β-glucuronidase—GUSs known to be involved in a lot of drug inactivations, changing their bioavailability, efficacy, and toxicity. This is an inducible enzyme produced by different anaerobic bacteria (e.g., *E. coli*, *Bacteroides*, and *Clostridium*), as well as the host in the liver (Dashnyam et al. 2018). Moreover, recent work found an *E. coli* specific GUSs inhibitor (TCH-3562) that exhibited efficacious inhibitory potency of the bacterial enzyme, without effect on human

GUS (Cheng et al. 2019). In a mouse model, TCH-3562 used as an adjuvant treatment showed protective effects on irinotecan-induced diarrhea, while the drug maintained its anti-tumor efficacy. Similarly, the microbial glucuronidase inhibitor Inh1 proved to reduce diclofenac-induced anastomotic leakage, leak severity, and anastomotic strength in a rat model (Yauw et al. 2018). Another inhibitor prevented anti-tumor drug regorafenib reactivation in the gastrointestinal tract, both in vitro and ex vivo studies (Ervin et al. 2019). The level of microbial metabolites can serve as predictive biomarkers in the case of drugs that compete with these molecules, such as acetaminophen or simvastatin (Kaddurah-Daouk et al. 2011; Laine et al. 2009). However, biomarkers based on the detection and quantification of specific bacterial strains could be used only in particular cases and should be detected at the species level, for instance, *F. prausnitzii*, *Bilophila* spp., ASV 1508, ASV 1982, and ASV 664 in case of tacrolimus. Also, the detection of genetic biomarkers that target key molecules involved in metabolic pathways of inactivation (e.g., qPCR quantification of *cgr* genes associated with digoxin reduction) might be a more reliable option.

Biomarkers about microbiome modulation can guide the precise manipulation of the intestinal microbiota composition and function to a status of optimal biodiversity and signature. Detection of these biomarkers at the beginning of the therapy is important as it was demonstrated that the individual's composition of gut microbiome profoundly influences the clinical drug responses, especially in cancer therapy. Moreover, changes in microbiota fingerprint during disease progression associated with the drug's secondary adverse effects could be detected. Thus, these biological biomarkers can be used to predict and tailor drug protocols and dosages to individual patients, ensuring that optimal drug levels are achieved without causing toxicity. This is particularly important for medicine formulations with narrow therapeutic windows, where precise dosing is critical.

However, finding biomarkers for microbiota therapy has been challenging. These drawbacks have enhanced the interest in using new technologies, machine learning, and deep learning models, powerful toolkits capable of forecasting drug–microbiota interactions at the individual patient level (McCoubrey et al. 2022). This could enhance personalized medicine approaches by tailoring treatments based on a patient's unique microbiome profile. Recently, have proposed a graph-based neural network model (with 93% accuracy) as a rapid and efficient method for screening drugs susceptible to GM depletion. This encourages the improvement in patient-specific dosage responses and formulations.

Researchers aim to establish a domain of precision oncology whereby microbiota-targeted medicines strengthen the efficacy of chemo- and immunotherapy, mitigate adverse effects, and optimize patient outcomes by utilizing microbiome biomarkers in cancer treatment frameworks. Accumulated data suggested that there is a close relationship between GM from cancer patients and clinical response to immunotherapy, while high responders had lower concentrations of drugs in plasma, proving that microbiota increases the drug's bioavailability. However, it is challenging to designate generalized microbiome-based biomarkers. Therefore, Gunjur et al. (2024) first performed deep shotgun metagenomic sequencing of gut microbiome from a cohort of patients with different rare cancers treated with immune checkpoint blockade (ICB) drugs and then used a machine learning model to predict ICB response and perform a meta-analysis of gut genomic with data from other comparable studies (Gunjur et al. 2024). They discovered specific strain-level fingerprints linked to ICB response that may be generalizable across cancer types and geographic locations and are important in developing gut microbial biomarkers. In another study, using transcriptome and microbiome data from patients with colorectal cancer Jin et al. (2024) developed a predictive model to accurately forecast the drug outcomes of patients after chemotherapy and immunotherapy (Jin et al. 2024). This model correlated data about the microbiome (obtained from openly accessible repositories) and the tumor immune microenvironment. Thus, the tumors were divided into two subcategories, C1 with a better prognosis compared to those from the C2 subtype that had higher gut microbial richness. Moreover, genera such as *Robiginitomaculum* and *Myxococcus* were prevalent in the subtype C1, while *Sutterella* and *Zymomonas* in the subtype were linked to poor outcomes. The authors have suggested that the difference in microbial abundance can be used as microbial biomarkers for tumor progression and patient survival. Moreover, the model predicted the response to chemotherapy and immunotherapy and concluded that patients with the C2 subtype respond better to drugs like RO-3306, Tozasertib, Doramapimod, and NU7441 used in immunotherapy.

Conventional routes to target the gut microbiota community have shown great potential for precision medicine but they often have drawbacks. These approaches included fecal microbial transplant, the use of prebiotics, probiotics, synbiotics, and postbiotics, as well as dietary and lifestyle changes (Al-Fakhrany and Elekhnawy 2024; Avram et al. 2023; Dinu et al. 2023; Vamanu et al. 2023). These measures lack specificity as they are not based on the individual's microbiota profile and can lead to changes in untargeted microbial populations. Nanotechnology is reshaping drug and probiotic delivery by enhancing precision, efficacy, and targeting in various ways that can effectively overcome the shortcomings of traditional microbiota-based interventions (Li et al. 2023, 2024; Ma et al. 2023). Thus, nanotechnology enables targeted drug delivery for harmful pathogens based on pH-responsive and stimuli-sensitive nanoparticles that release drugs in specific gut regions based on pH, enzymes, or microbiota composition. Han et al. (2022) built copolymer nanoparticles that can achieve antibiotic-targeted delivery for pathogen infection (Han et al. 2022). This nanomaterial promoted targeted therapy of ciprofloxacin in response to a microbial product (hydrogen sulfide) released by *Salmonella* spp. This not only reduces pathogen bacteria colonization in the intestine tract but also significantly maintains gut microbial homeostasis. Additionally, nanomaterials and nanocarriers have been used to modulate disease-associated dysbiosis and reshape the immune microenvironment (Li et al. 2024). To treat colitis, nanocarriers loaded with 6-gingerol were developed that continuously release polyphenolic substance near the colonic lesion and modulate *Firmicutes*/*Bacteroidetes* ratio (Tian et al. 2022). A nanosynthetic particle containing hyaluronic acid and bilirubin was used to ameliorate the symptoms of IBD as it was able to significantly increase populations of *Akkermansia muciniphila* and *Clostridium* XIVα (Lee et al. 2020).

Surface-deacetylated chitin nanofibers stimulate the beneficial bacteria *Blautia* spp., while reducing the abundance of harmful bacteria (*Morganella* spp. and *Prevotella* spp.) in spontaneously hypertensive rats (Goto et al. 2020). Nanomaterials and nanocarriers help regulate commensal microbiota, increasing beneficial bacteria, or decreasing pathogenic flora before or during the immunotherapy and chemotherapy (Icriverzi et al. 2024). A hybrid system with bacteriophage and dextran nanoparticles was used to carry irinotecan, first-line chemotherapy for colorectal cancer, and proved enhanced efficacy by modulating the proliferation of harmful bacteria and beneficial bacteria in the mouse intestine to reduce the side effects of chemotherapy (Zheng et al. 2019).

Numerous works have demonstrated the close relationship between GM and cancer prevention, development, or treatment, as well as the importance of probiotics to modulate cancer-associated dysbiosis to improve treatment outcomes (Ansari et al. 2020). The nanomaterials and nanocarriers (e.g., liposomes, polymeric nanoparticles) can deliver specific probiotics to the gut microbiome, protecting beneficial

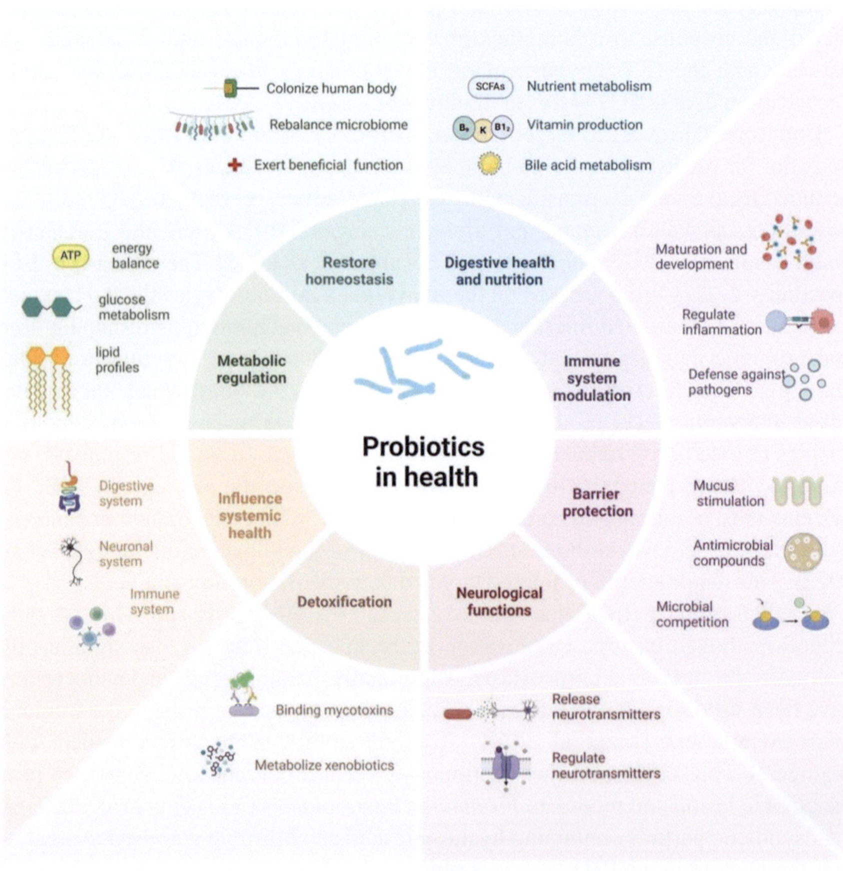

Fig. 4 The effect of probiotics on host health. (Figure created with BioRender.com)

bacteria while targeting microbes with pathogenic potential. Using these approaches to deliver probiotics can enhance their activity and viability during the transit from the noxious stomach environment to the intestine and sustain them when reaching the target site, with subsequent health effects on the host (Fig. 4). Also, nanomaterials could be utilized as delivery systems for prebiotics to enhance their solubility and bioavailability (Ma et al. 2023). A nanocomposite that encapsulates the typical probiotic *Escherichia coli* Nissle 1917 was reported to significantly improve the treatment of IBD (Chen et al. 2023). This self-assembled nitroreductase labile peptidic hydrogel protected probiotics and enabled their release under controlled conditions because of the upregulated nitroreductase in the intestine, which can trigger the decomposition of the hydrogel. The *Bacillus*-loaded nanomaterial and integrated multispecies probiotics significantly improved the therapeutic effect on colitis compared to the administration of free bacterial spores or probiotic strains (Alkushi et al. 2022a, b). Recent studies have proved that nanoparticles can influence signaling pathways by delivering molecules like inhibitors or nucleic acids in a targeted manner (Basu et al. 2024). Thus, nanoparticles could incorporate inhibitors released when triggered by microbiota biomarkers, such as bacterial enzymes (e.g., β-glucuronidase) involved in drug inactivation.

The emerging field of nanotechnology has considerable potential for further development as have been proved that nanoproducts often exhibit superior effects compared to their individual components. The combination of knowledge from microbiota biomarker-drug delivery with these innovative delivery systems will greatly advance the development of gut microbiota-related therapies. Furthermore, collaborating data about biomarkers from an individual's microbiota profile with nanotechnology tools will enable more precise, effective, and personalized medicine. However, some concerns regarding the safety of these nanotechnology tools, especially over longer periods, have not been very clearly taken into consideration. For instance, exposure to graphene has proved to significantly disrupt human GM and intestinal homeostasis (Bantun et al. 2022). On the other hand, nanobiosensors that detect microbiota-derived biomarkers for drug delivery (e.g., SCFAs or other metabolites, microbial DNA, or enzymes) with high sensitivity will allow early detection of drug inactivation, as well as monitoring the drug efficacy and toxicity. Therefore, strategies based on nanotechnology deserve greater attention and more in-depth research.

5 Concluding Remarks

Microbiota biomarkers have emerged as important tools in biomarker-driven drug therapy, providing valuable insights into how individual variations in microbiota composition can influence drug efficacy, metabolism, and safety. The utilization of microbiota biomarkers in drug therapy involves understanding the complex gut microbiota's role in drug interactions, while its potential as a biomarker leads to personalized treatment strategies, ultimately enhancing drug outcomes. Microbiota biomarkers to manage drug therapy are essential in predicting individual response to drug, as well as in guiding and monitoring the therapy effectiveness and safety. The combination of microbiota biomarker-driven drug delivery with the emerging field of nanotechnology could lead to innovative delivery systems that further enhance targeting capabilities and therapeutic efficacy. Moreover, advances in microbiome databases and technologies such as high-throughput next-generation sequencing, CRISPR-based gene editing, machine learning, and artificial intelligence are expected to play a crucial role in enhancing the predictability and applicability of this approach. While the potential of microbiota biomarkers that correlate with drug responsiveness is immense, translating these biomarkers into clinical practice and developing personalized intervention strategies require more research and validation trials in large cohorts, as well as standardization protocols. Overall, microbiome biomarker-driven drug therapy will undoubtedly play an increasingly important role in shaping the future of medicine and drug development, ultimately leading to improved patient outcomes and a deeper understanding of human health.

References

Al-Fakhrany OM, Elekhnawy E (2024) Next-generation probiotics: the upcoming biotherapeutics. Mol Biol Rep 51:505

Alkushi AG, Abdelfattah-Hassan A, Eldoumani H, Elazab ST, Mohamed SAM, Metwally AS, El-Shetry ES, Saleh AA, ElSawy NA, Ibrahim D (2022a) Probiotics-loaded nanoparticles attenuated colon inflammation, oxidative stress, and apoptosis in colitis. Sci Rep 12:5116

Alkushi AG, Elazab ST, Abdelfattah-Hassan A, Mahfouz H, Salem GA, Sheraiba NI, Mohamed EAA, Attia MS, El-Shetry ES, Saleh AA et al (2022b) Multi-strain-probiotic-loaded nanoparticles reduced colon inflammation and orchestrated the expressions of tight junction, NLRP3 Inflammasome and Caspase-1 genes in DSS-induced colitis model. Pharmaceutics 14:1183

Amos A, Anwar S, Fritzsche M, Mate R, Bleazard T, Rijpkema S (2020) Developing standards for the microbiome field. Microbiome 8:98

Andreoletti M, Haller L, Vayena E, Blasimme A (2024) Mapping the ethical landscape of digital biomarkers: a scoping review. PLOS Digit Health 3(5):e0000519

Ansari F, Pourjafar H, Tabrizi A, Homayouni A (2020) The effects of probiotics and prebiotics on mental disorders: a review on depression, anxiety, Alzheimer, and autism spectrum disorders. Curr Pharm Biotechnol 21(15):1455–1462

Avram I, Pelinescu D, Gatea F, Ionescu R, Barcan A, Rosca R, Zanfirescu A, Vamanu E (2023) Boletus edulis extract—a new modulator of dysbiotic microbiota. Life 13(7):1481

Bachem A, Makhlouf C, Binger KJ, de Souza DP, Tull D, Hochheiser K, Whitney PG, Fernandez-Ruiz D, Dähling S, Kastenmüller W, Jönsson J, Gressier E, Lew AM, Perdomo C, Kupz A, Figgett W, Mackay F, Oleshansky M, Russ BE, Parish IA, Kallies A, McConville MJ, Turner SJ, Gebhardt T, Bedoui S (2019) Microbiota-derived short-chain fatty acids promote the memory potential of antigen-activated CD8+ T cells. Immunity 51(2):285–297.e5. https://doi.org/10.1016/j.immuni.2019.06.002

Bantun F, Singh R, Alkhanani MF, Almalki AH, Alshammary F, Khan S, Haque S, Srivastava M (2022) Gut microbiome interactions with graphene based nanomaterials: challenges and opportunities. Sci Total Environ 830:154789

Basu S, Biswas P, Anto M, Singh N, Mukherjee K (2024) Nanomaterial-enabled drug transport systems: a comprehensive exploration of current developments and future avenues in therapeutic delivery. 3 Biotech 14:289

BEST (Biomarkers, EndpointS, and other Tools) Resource (2016) BEST (Biomarkers, EndpointS, and other Tools) Resource. Retrieved from https://www.ncbi.nlm.nih.gov/pubmed/27010052. Accessed 2 Feb 2025

Borgiani G, Possidente C, Fabbri C, Oliva V, Bloemendaal M, Arias Vasquez A, Dinan TG, Vieta E, Menchetti M, De Ronchi D, Serretti A, Fanelli G (2025) The bidirectional interaction between antidepressants and the gut microbiota: are there implications for treatment response? Int Clin Psychopharmacol 40(1):3–26. https://doi.org/10.1097/YIC.0000000000000533

Broecker F, Moelling K (2021) The roles of the virome in cancer. Microorganisms 9(12):2538

Brown R, Taylor C (2023) Current understanding of dysbiosis in disease in human and animal models. Inflamm Bowel Dis 22(5):1137–1150

Brunkhorst C, Andersen C, Schneider E (1999) Acarbose, a pseudooligosaccharide, is transported but not metabolized by the maltose-maltodextrin system of Escherichia coli. J Bacteriol 181:2612–2619

Cammarota G, Ianiro G, Ahern A, Carbone C, Temko A, Claesson MJ, Gasbarrini A, Tortora G (2020) Gut microbiome, big data and machine learning to promote precision medicine for cancer. Nat Rev Gastroenterol Hepatol 17(10):635–648. https://doi.org/10.1038/s41575-020-0327-3

Casado-Bedmar M, Viennois E (2022) MicroRNA and gut microbiota: tiny but mighty—novel insights into their cross-talk in inflammatory bowel disease pathogenesis and therapeutics. J Crohn's Colitis 16(6):992–1005

Chamseddine AN et al (2019) Intestinal bacterial beta-glucuronidase as a possible predictive biomarker of irinotecan-induced diarrhea severity. Pharmacol Ther 199:1–15

Charpentier J, Briand F, Lelouvier B, Servant F, Azalbert V, Puel A, Christensen JE, Waget A, Branchereau M, Garret C, Lluch J, Heymes C, Brousseau E, Burcelin R, Guzylack L, Sulpice T, Grasset E (2021) Liraglutide targets the gut microbiota and the intestinal immune system to regulate insulin secretion. Acta Diabetol 58(7):881–897. https://doi.org/10.1007/s00592-020-01657-8

Chen J, Zhang P, Wu C, Yao Q, Cha R, Gao Y (2023) Reductase-labile peptidic supramolecular hydrogels aided in oral delivery of probiotics. ACS Appl Mater Interfaces 15:31214–31223

Cheng KW, Tseng CH, Tzeng CC et al (2019) Pharmacological inhibition of bacterial β-glucuronidase prevents irinotecan-induced diarrhea without impairing its antitumor efficacy in vivo. Pharmacol Res 139:41–49

Clemente-Suárez VJ, Redondo-Flórez L, Rubio-Zarapuz A, Martín-Rodríguez A, Tornero-Aguilera JF (2024) Microbiota implications in endocrine-related diseases: from development to novel therapeutic approaches. Biomedicine 12(1):221

Conn KA, Borsom EM, Cope EK (2024) Implications of microbe-derived ɣ-aminobutyric acid (GABA) in gut and brain barrier integrity and GABAergic signaling in Alzheimer's disease. Gut Microbes 16(1):2371950. https://doi.org/10.1080/19490976.2024.2371950

Crouwel F, Buiter HJC, de Boer NK (2020) Gut microbiota-driven drug metabolism in inflammatory bowel disease. J Crohns Colitis 15:307–315

Daisley BA, Chanyi RM, Abdur-Rashid K, Al KF, Gibbons S, Chmiel JA, Wilcox H, Reid G, Anderson A, Dewar M, Nair SM, Chin J, Burton JP (2020) Abiraterone acetate preferentially enriches for the gut commensal Akkermansia muciniphila in castrate-resistant prostate cancer patients. Nat Commun 11(1):4822. https://doi.org/10.1038/s41467-020-18649-5

Đanić M, Pavlović N, Zaklan D, Stanimirov B, Lazarević S, Al-Salami H, Mikov M (2024) Computational studies for pre-evaluation of pharmacological profile of gut microbiota-produced gliclazide metabolites. Front Pharmacol 15:1492284

Dashnyam P, Mudududdla R, Hsieh TJ et al (2018) β-Glucuronidases of opportunistic bacteria are the major contributors to xenobiotic-induced toxicity in the gut. Sci Rep 8:16372

Degraeve AL, Bindels LB, Haufroid V, Moudio S, Boland L, Delongie KA, Dewulf JP, Eddour DC, Mourad M, Elens L (2024) Tacrolimus pharmacokinetics is associated with gut microbiota diversity in kidney transplant patients: results from a pilot cross-sectional study. Clin Pharmacol Ther 115(1):104–115

Dikeocha IJ, Al-Kabsi AM, Miftahussurur M, Alshawsh MA (2022) Pharmacomicrobiomics: influence of gut microbiota on drug and xenobiotic metabolism. FASEB J 36(6):e22350. https://doi.org/10.1096/fj.202101986R

Dinu L-D, Gatea F, Roaming Israel F, Lakicevic M, Dedović N, Vamanu E (2023) The modulation effect of a fermented bee pollen postbiotic on cardiovascular microbiota and therapeutic perspectives. Biomedicine 11(10):2712. https://doi.org/10.3390/biomedicines11102712

Ervin SM, Hanley RP, Lim L, Walton WG, Pearce KH, Bhatt AP, James LI, Redinbo MR (2019) Targeting regorafenib-induced toxicity through inhibition of gut microbial β-glucuronidases. ACS Chem Biol 14(12):2737–2744. https://doi.org/10.1021/acschembio.9b00663

Feng W, Liu J, Ao H, Yue S, Peng C (2020) Targeting gut microbiota for precision medicine: focusing on the efficacy and toxicity of drugs. Theranostics 10(24):11278–11301

Gallotti B et al (2021) Effects of dietary fiber intake in chemotherapy-induced mucositis in murine model. Br J Nutr 126:853–864

Geller LT, Barzily-Rokni M, Danino T, Jonas OH, Shental N, Nejman D, Gavert N, Zwang Y, Cooper ZA, Shee K, Thaiss CA, Reuben A, Livny J, Avraham R, Frederick DT, Ligorio M, Chatman K, Johnston SE, Mosher CM, Brandis A, Fuks G, Gurbatri C, Gopalakrishnan V, Kim M, Hurd MW, Katz M, Fleming J, Maitra A, Smith DA, Skalak M, Bu J, Michaud M, Trauger SA, Barshack I, Golan T, Sandbank J, Flaherty KT, Mandinova A, Garrett WS, Thayer SP, Ferrone CR, Huttenhower C, Bhatia SN, Gevers D, Wargo JA, Golub TR, Straussman R (2017) Potential role of intratumor bacteria in mediating tumor resistance to the chemotherapeutic drug gemcitabine. Science 357(6356):1156–1160. https://doi.org/10.1126/science.aah5043

Gopalakrishnan V, Spencer CN, Nezi L, Reuben A, Andrews MC, Karpinets TV, Prieto PA, Vicente D, Hoffman K, Wei SC, Cogdill AP, Zhao L, Hudgens CW, Hutchinson DS, Manzo T, Petaccia de Macedo M, Cotechini T, Kumar T, Chen WS, Reddy SM, Szczepaniak Sloane R, Galloway-Pena J, Jiang H, Chen PL, Shpall EJ, Rezvani K, Alousi AM, Chemaly RF, Shelburne S, Vence LM, Okhuysen PC, Jensen VB, Swennes AG, McAllister F, Marcelo Riquelme Sanchez E, Zhang Y, Le Chatelier E, Zitvogel L, Pons N, Austin-Breneman JL, Haydu LE, Burton EM, Gardner JM, Sirmans E, Hu J, Lazar AJ, Tsujikawa T, Diab A, Tawbi H, Glitza IC, Hwu WJ, Patel SP, Woodman SE, Amaria RN, Davies MA, Gershenwald JE, Hwu P, Lee JE, Zhang J, Coussens LM, Cooper ZA, Futreal PA, Daniel CR, Ajami NJ, Petrosino JF, Tetzlaff MT, Sharma P, Allison JP, Jenq RR, Wargo JA (2018) Gut microbiome modulates response to anti-PD-1 immunotherapy in melanoma patients. Science 359(6371):97–103. https://doi.org/10.1126/science.aan4236

Goto M, Iohara D, Michihara A, Ifuku S, Azuma K, Kadowaki D, Maruyama T, Otagiri M, Hirayama F, Anraku M (2020) Effects of surface-deacetylated chitin nanofibers on non-alcoholic steatohepatitis model rats and their gut microbiota. Int J Biol Macromol 164:659–666

Gunjur A, Shao Y, Rozday T et al (2024) A gut microbial signature for combination immune checkpoint blockade across cancer types. Nat Med 30:797–809. https://doi.org/10.1038/s41591-024-02823-z

Guo Y, Crnkovic CM, Won K-J, Yang X, Lee JR, Orjala J, Lee H, Jeong H (2019) Commensal gut bacteria convert the immunosuppressant tacrolimus to less potent metabolites. Drug Metab Dispos 47:194–202

Haiser HJ, Gootenberg DB, Chatman K, Sirasani G, Balskus EP, Turnbaugh PJ (2013) Predicting and manipulating cardiac drug inactivation by the human gut bacterium *Eggerthella lenta*. Science 341(6143):295–298

Hajjo R, Sabbah DA, Al Bawab AQ (2022) Unlocking the potential of the human microbiome for identifying disease diagnostic biomarkers. Diagnostics (Basel) 12(7):1742

Han L, Liu XW, Zang T, Ren H, Liang DS, Bai SC, Li C, Liao XP, Liu YH, Zhang C, Sun J (2022) H2S responsive PEGylated poly (lipoic acid) with ciprofloxacin for targeted therapy of Salmonella. J Control Release 351:896–906. https://doi.org/10.1016/j.jconrel.2022.09.060

Huang X, Hu M, Sun T et al (2023) Multi-kingdom gut microbiota analyses define bacterial-fungal interplay and microbial markers of pan-cancer immunotherapy across cohorts. Cell Host Microbe 31:1930–1943

Icriverzi M, Florian PE, Bonciu A, Dumitrescu LN, Moldovan A, Pelinescu D, Ionescu R, Avram I, Munteanu CVA, Sima LE, Dinca V, Rusen L, Roseanu A (2024) Hybrid bio-nanoporous peptide loaded-polymer platforms with anticancer and antibacterial activities. Nanoscale Adv 6(1):947

Iida N, Dzutsev A, Stewart CA, Smith L, Bouladoux N, Weingarten RA, Molina DA, Salcedo R, Back T, Cramer S, Dai RM, Kiu H, Cardone M, Naik S, Patri AK, Wang E, Marincola FM, Frank KM, Belkaid Y, Trinchieri G, Goldszmid RS (2013) Commensal bacteria control cancer response to therapy by modulating the tumor microenvironment. Science 342(6161):967–970. https://doi.org/10.1126/science.1240527

Jennings DL, Bohn B, Zuver A, Onat D, Gaine M, Royzman E, Hupf J, Brunjes D, Latif F, Restaino S, Garan AR, Topkara VK, Takayama H, Takeda K, Naka Y, Farr M, Nandakumar R, Uhlemann AC, Colombo PC, Demmer RT, Yuzefpolskaya M (2020) Gut microbial diversity, inflammation, and oxidative stress are associated with tacrolimus dosing requirements early after heart transplantation. PLoS One 15(5):e0233646

Jin S, Zhong W, Li B, Wang K, Lai D (2024) Multidimensional analysis of the impact of Gemmatimonas, Rhodothermus, and Sutterella on drug and treatment response in colorectal cancer. Front Cell Infect Microbiol 14:1457461

Kaddurah-Daouk R, Baillie RA, Zhu H, Zeng ZB, Wiest MM, Nguyen UT et al (2011) Enteric microbiome metabolites correlate with response to simvastatin treatment. PLoS One 6:e25482

Kamath S, Joyce P (2025) A critical need for 'gut neutrality': mitigating adverse drug-microbiome interactions. Expert Opin Drug Metab Toxicol 21(1):1–4

Kehrer DF et al (2001) Modulation of irinotecan-induced diarrhea by cotreatment with neomycin in cancer patients. Clin Cancer Res 7:1136–1141

Kim J, Lee H, An J, Song Y, Lee CK, Kim K, Kong H (2019) Alterations in Gut Microbiota by Statin Therapy and Possible Intermediate Effects on Hyperglycemia and Hyperlipidemia. Front Microbiol 10:1947. https://doi.org/10.3389/fmicb.2019.01947

Kim K, Yoon I, Chun I, Lee N, Kim T, Gwak HS (2011) Effects of bile salts on the lovastatin pharmacokinetics following oral administration to rats. Drug Deliv 18(1):79–83

Kishida Y, Okubo H, Ohno H, Oki K, Yoneda M (2017) Effect of miglitol on the suppression of nonalcoholic steatohepatitis development and improvement of the gut environment in a rodent model. J Gastroenterol 52:1180–1191

Klatt NR, Cheu R, Birse K, Zevin AS, Perner M, Noël-Romas L, Grobler A, Westmacott G, Xie IY, Butler J, Mansoor L, McKinnon LR, Passmore JS, Abdool Karim Q, Abdool Karim SS, Burgener AD (2017) Vaginal bacteria modify HIV tenofovir microbicide efficacy in African women. Science 356(6341):938–945. https://doi.org/10.1126/science.aai9383

Konop M, Radkowski M, Grochowska M, Perlejewski K, Samborowska E, Ufnal M (2018) Enalapril decreases rat plasma concentration of TMAO, a gut bacteria-derived cardiovascular marker. Biomarkers 23(4):380–385

Koppel N, Bisanz JE, Pandelia ME, Turnbaugh PJ, Balskus EP (2018) Discovery and characterization of a prevalent human gut bacterial enzyme sufficient for the inactivation of a family of plant toxins. elife 7:e33953

Kraneveld AD, Tonda A, Oberski DL, Garssen J, Lopez-Rincon A (2024) Methodology for biomarker discovery with reproducibility in microbiome research using machine learning. BMC Bioinform 25:26

Laine JE, Auriola S, Pasanen M, Juvonen RO (2009) Acetaminophen bioactivation by human cytochrome P450 enzymes and animal microsomes. Xenobiotica 39:11–21

Lam KN, Alexander M, Turnbaugh PJ (2019) Precision medicine goes microscopic: engineering the microbiome to improve drug outcomes. Cell Host Microbe 26(1):22–34

Lee HJ, Zhang H, Orlovich DA, Fawcett JP (2012) The influence of probiotic treatment on sulfasalazine metabolism in rat. Xenobiotica 42:791–797

Lee JR, Muthukumar T, Dadhania D, Taur Y, Jenq RR, Toussaint NC, Ling L, Pamer E, Suthanthiran M (2015) Gut microbiota and tacrolimus dosing in kidney transplantation. PLoS One 10:e0122399

Lee Y, Sugihara K, Gillilland MG, Jon S, Kamada N, Moon JJ (2020) Hyaluronic acid–bilirubin nanomedicine for targeted modulation of dysregulated intestinal barrier, microbiome and immune responses in colitis. Nat Mater 19:118–126

Lee JWJ, Plichta D, Hogstrom L, Borren NZ, Lau H, Gregory SM, Tan W, Khalili H, Clish C, Vlamakis H et al (2021) Multi-omics reveal microbial determinants impacting responses to biologic therapies in inflammatory bowel disease. Cell Host Microbe 29(8):1294–1304.e4

Li M, Chen WD, Wang YD (2020) The roles of the gut microbiota–miRNA interaction in the host pathophysiology. Mol Med 26:101

Li X, Wei H, Qi J, Ma K, Luo Y, Weng L (2023) Interactions of nanomaterials with gut microbiota and their applications in cancer therapy. Sensors (Basel) 23(9):4428

Li W, Zhan M, Wen Y, Chen Y, Zhang Z, Wang S, Tian D, Tian S (2024) Recent progress of oral functional nanomaterials for intestinal microbiota regulation. Pharmaceutics 16(7):921

Liu NN, Jiao N, Tan JC et al (2022) Multi-kingdom microbiota analyses identify bacterial–fungal interactions and biomarkers of colorectal cancer across cohorts. Nat Microbiol 7:238–250

Luo F et al (2019) Metformin in patients with and without diabetes: a paradigm shift in cardiovascular disease management. Cardiovasc Diabetol 18:54

Ma Y, Zhang J, Yu N, Shi J, Zhang Y, Chen Z, Jia G (2023) Effect of nanomaterials on gut microbiota. Toxics 11(4):384

Maier L et al (2018) Extensive impact of non-antibiotic drugs on human gut bacteria. Nature 555:623–628

Maini Rekdal V, Bess EN, Bisanz JE, Turnbaugh PJ, Balskus EP (2019) Discovery and inhibition of an interspecies gut bacterial pathway for levodopa metabolism. Science 364:eaau6323

McCoubrey LE, Gaisford S, Orlu M, Basit AW (2022) Predicting drug-microbiome interactions with machine learning. Biotechnol Adv 54:107797

Meade S, Liu Chen Kiow J, Massaro C, Kaur G, Squirell E, Bressler B, Lunken G (2023) Gut microbiome-associated predictors as biomarkers of response to advanced therapies in inflammatory bowel disease: a systematic review. Gut Microbes 15(2):2287073

Minamii T, Nogami M, Ogawa W (2018) Mechanisms of metformin action: in and out of the gut. J Diabetes Investig 9:701–703

Mohr AE, Crawford M, Jasbi P, Fessler S, Sweazea KL (2022) Lipopolysaccharide and the gut microbiota: considering structural variation. FEBS Lett 596(7):849–875. https://doi.org/10.1002/1873-3468.14328

Nucera F, Ruggeri P, Spagnolo CC, Santarpia M, Ieni A, Monaco F, Tuccari G, Pioggia G, Gangemi S (2024) MiRNAs and microbiota in non-small cell lung cancer (NSCLC): implications in pathogenesis and potential role in predicting response to ICI treatment. Int J Mol Sci 25(12):6685

Olivares M, Neyrinck AM, Pötgens SA, Beaumont M, Salazar N, Cani PD (2018) The DPP-4 inhibitor vildagliptin impacts the gut microbiota prevents disruption of intestinal homeostasis induced by a Western diet in mice. Diabetologia 61:1838–1848. https://doi.org/10.1007/s00125-018-4647-6

Panebianco C, Adamberg K, Jaagura M, Copetti M, Fontana A, Adamberg S, Kolk K, Vilu R, Andriulli A, Pazienza V (2018) Influence of gemcitabine chemotherapy on the microbiota of pancreatic cancer xenografted mice. Cancer Chemother Pharmacol 81(4):773–782. https://doi.org/10.1007/s00280-018-3549-0

Pryor R, Martinez-Martinez D, Quintaneiro L, Cabreiro F (2020) The role of the microbiome in drug response. Annu Rev Pharmacol Toxicol 60:417–435. https://doi.org/10.1146/annurev-pharmtox-010919-023612

Qahwaji R, Ashankyty I, Sannan NS, Hazzazi MS, Basabrain AA, Mobashir M (2024) Pharmacogenomics: a genetic approach to drug development and therapy. Pharmaceuticals 17:940

Rehman MU, Hussain I, Tayara H, Chong KT (2024) A graph neural network approach for predicting drug susceptibility in the human microbiome. Comput Biol Med 179:108729

Rizkallah MR, Saad R, Aziz RK (2010) The human microbiome project, personalized medicine, and the birth of pharmacomicrobiomics. Curr Pharmacogenomics Pers Med 8:182–193

Saha JR, Butler VP Jr, Neu HC, Lindenbaum J (1983) Digoxin-inactivating bacteria: identification in human gut flora. Science 220:325–327
Segal LN, Clemente JC, Li Y, Ruan C, Cao J, Danckers M, Morris A, Tapyrik S, Wu BG, Diaz P, Calligaro G, Dawson R, van Zyl-Smit RN, Dheda K, Rom WN, Weiden MD (2017) Anaerobic bacterial fermentation products increase tuberculosis risk in antiretroviral-drug-treated HIV patients. Cell Host Microbe 21(4):530–537.e4. https://doi.org/10.1016/j.chom.2017.03.003
Shang J, Liu F, Zhang B, Dong K, Lu M, Jiang R, Xu Y, Diao L, Zhao J, Tang H (2021) Liraglutide-induced structural modulation of the gut microbiota in patients with type 2 diabetes mellitus. PeerJ 9:e11128. https://doi.org/10.7717/peerj.11128
Shuaib F, Ali A, Khan I, Ahmad A, Khan Z, Amanullah. (2024) Influence of gut microbiota on drug metabolism and therapeutic efficacy. South East Eur J Pub Health XXV(S1):2191–2195. https://doi.org/10.70135/seejph.vi.2352
Simon JC, Marchesi JR, Mougel C, Selosse MA (2019) Host-microbiota interactions: from holobiont theory to analysis. Microbiome 7(1):5. https://doi.org/10.1186/s40168-019-0619-4
Sivan A, Corrales L, Hubert N, Williams JB, Aquino-Michaels K, Earley ZM, Benyamin FW, Lei YM, Jabri B, Alegre ML, Chang EB, Gajewski TF (2015) Commensal bifidobacterium promotes antitumor immunity and facilitates anti-PD-L1 efficacy. Science 350(6264):1084–1089. https://doi.org/10.1126/science.aac4255
Stringer AM et al (2008) Faecal microflora and beta-glucuronidase expression are altered in an irinotecan-induced diarrhea model in rats. Cancer Biol Ther 7:1919–1925
Su B et al (2015) Acarbose treatment affects the serum levels of inflammatory cytokines and the gut content of bifidobacteria in Chinese patients with type 2 diabetes mellitus. J Diabetes 7:729–739
Sun L et al (2018) Gut microbiota and intestinal FXR mediate the clinical benefits of metformin. Nat Med 24:1919–1929. https://doi.org/10.1038/s41591-018-0222-4
Takasuna K et al (1996) Involvement of beta-glucuronidase in intestinal microflora in the intestinal toxicity of the antitumor camptothecin derivative irinotecan hydrochloride (CPT-11) in rats. Cancer Res 56:3752–3757
Tian W, Wang H, Zhu Y, Wang Q, Song M, Cao Y, Xiao J (2022) Intervention effects of delivery vehicles on the therapeutic efficacy of 6-Gingerol on colitis. J Control Release 349:51–66
Vamanu E, Gatea F, Avram I, Radu GL, Singh SK (2023) Dysbiotic gut microbiota modulation by aronia fruits extract administration. Life 13(1):32. https://doi.org/10.3390/life13010032
van Kessel SP, Frye AK, El-Gendy AO, Castejon M, Keshavarzian A, van Dijk G, El Aidy S (2019) Gut bacterial tyrosine decarboxylases restrict levels of levodopa in the treatment of Parkinson's disease. Nat Commun 10(1):310
Velasquez MT, Ramezani A, Manal A, Raj DS (2016) Trimethylamine N-oxide: the good, the bad and the unknown. Toxins 8:326. https://doi.org/10.3390/toxins8110326
Velázquez YF, Nacheva PM (2017) Biodegradability of fluoxetine, mefenamic acid, and metoprolol using different microbial consortiums. Environ Sci Pollut Res 24(7):6779–6793
Ventin-Holmberg R, Eberl A, Saqib S et al (2021) Bacterial and fungal profiles as markers of infliximab drug response in inflammatory bowel disease. J Crohns Colitis 15:1019–1031
Vétizou M, Pitt JM, Daillère R, Lepage P, Waldschmitt N, Flament C, Rusakiewicz S, Routy B, Roberti MP, Duong CP, Poirier-Colame V, Roux A, Becharef S, Formenti S, Golden E, Cording S, Eberl G, Schlitzer A, Ginhoux F, Mani S, Yamazaki T, Jacquelot N, Enot DP, Bérard M, Nigou J, Opolon P, Eggermont A, Woerther PL, Chachaty E, Chaput N, Robert C, Mateus C, Kroemer G, Raoult D, Boneca IG, Carbonnel F, Chamaillard M, Zitvogel L (2015) Anticancer immunotherapy by CTLA-4 blockade relies on the gut microbiota. Science 350(6264):1079–1084. https://doi.org/10.1126/science.aad1329
Wainwright M, Kristiansen JE (2010) On the 75th anniversary of Prontosil. Dyes Pigments 88:231–234
Wang L, Li P, Tang Z, Yan X, Feng B (2016) Structural modulation of the gut microbiota and the relationship with body weight: compared evaluation of liraglutide and saxagliptin treatment. Sci Rep 6:33251

Wang Y, Gao X, Ghozlane A, Hu H, Li X, Xiao Y, Li D, Yu G, Zhang T (2018) Characteristics of faecal microbiota in paediatric Crohn's disease and their dynamic changes during infliximab therapy. J Crohn's Colitis 12:337–346

Wang L, Liu L, Liu X, Xiang M, Zhou L, Huang C, Shen Z, Miao L (2020) The gut microbes, enterococcus and Escherichia-Shigella, affect the responses of heart valve replacement patients to the anticoagulant warfarin. Pharmacol Res 159:104979. https://doi.org/10.1016/j.phrs.2020.104979

Xue L, Holford N, Ding XL, Shen ZY, Huang CR, Zhang H, Zhang JJ, Guo ZN, Xie C, Zhou L, Chen ZY, Liu LS, Miao LY (2017) Theory-based pharmacokinetics and pharmacodynamics of S- and R-warfarin and effects on international normalized ratio: influence of body size, composition, and genotype in cardiac surgery patients. Br J Clin Pharmacol 83(4):823–835

Yan X, Feng B, Li P, Tang Z, Wang L (2016) Microflora disturbance during progression of glucose intolerance and effect of sitagliptin: an animal study. J Diabetes Res 2016:2093171

Yan C, Luo L, Guo CY, Goto S, Urata Y, Shao JH, Li TS (2017) Doxorubicin-induced mitophagy contributes to drug resistance in cancer stem cells from HCT8 human colorectal cancer cells. Cancer Lett 388:34–42. https://doi.org/10.1016/j.canlet.2016.11.018

Yan A, Culp E, Perry J, Lau JT, MacNeil LT, Surette MG, Wright GD (2018) Transformation of the anticancer drug doxorubicin in the human gut microbiome. ACS Infect Dis 4(1):68–76. https://doi.org/10.1021/acsinfecdis.7b00166

Yan J, Yang L, Ren Q, Zhu C, Du H, Wang Z, Qi Y, Xian X, Chen D (2024) Gut microbiota as a biomarker and modulator of anti-tumor immunotherapy outcomes. Front Immunol 15:1471273. https://doi.org/10.3389/fimmu.2024.1471273

Yang T, Aquino V, Lobaton GO, Li H, Colon-Perez L, Goel R, Qi Y, Zubcevic J, Febo M, Richards EM, Pepine CJ, Raizada MK (2019) Sustained captopril-induced reduction in blood pressure is associated with alterations in gut-brain axis in the spontaneously hypertensive rat. J Am Heart Assoc 8(4):e010721

Yauw STK, Arron M, Lomme RMLM, van den Broek P, Greupink R, Bhatt AP, Redinbo MR, van Goor H (2018) Microbial glucuronidase inhibition reduces severity of diclofenac-induced anastomotic leak in rats. Surg Infect 19(4):417–423. https://doi.org/10.1089/sur.2017.245

Yilmaz B, Juillerat P, Øyås O, Ramon C, Bravo FD, Franc Y, Fournier N, Michetti P, Mueller C, Geuking M et al (2019) Microbial network disturbances in relapsing refractory Crohn's disease. Nat Med 25(2):323–336

Yoo DH, Kim IS, Van Le TK, Jung IH, Yoo HH, Kim DH (2014) Gut microbiota-mediated drug interactions between lovastatin and antibiotics. Drug Metab Dispos 42(9):1508–1513. https://doi.org/10.1124/dmd.114.058354

Zhang X, Fang Z, Zhang C, Xia H, Jie Z, Han X, Chen Y, Ji L (2017a) Effects of acarbose on the gut microbiota of prediabetic patients: a randomized, double-blind. Controlled Crossover Trial Diabetes Ther 8(2):293–307. https://doi.org/10.1007/s13300-017-0226-y

Zhang Q, Xiao X, Li M, Yu M, Ping F, Zheng J, Wang T, Wang X (2017b) Vildagliptin increases butyrate-producing bacteria in the gut of diabetic rats. PLoS One 12(10):e0184735. https://doi.org/10.1371/journal.pone.0184735

Zhang Q, Xiao X, Zheng J, Li M, Yu M, Ping F, Wang T, Wang X (2018) Featured article: structure moderation of gut microbiota in liraglutide-treated diabetic male rats. Exp Biol Med (Maywood) 243(1):34–44. https://doi.org/10.1177/1535370217743765

Zhao L et al (2018) A glucagon-like peptide-1 receptor agonist lowers weight by modulating the structure of gut microbiota. Front Endocrinol (Lausanne) 9:233

Zhao Q, Chen Y, Huang W, Zhou H, Zhang W (2023) Drug-microbiota interactions: an emerging priority for precision medicine. Signal Transduct Target Ther 8(1):386. https://doi.org/10.1038/s41392-023-01619-w

Zheng DW, Dong X, Pan P, Chen KW, Fan JX, Cheng SX, Zhang XZ (2019) PhageGuided modulation of the gut microbiota of mouse models of colorectal cancer augments their responses to chemotherapy. Nat Biomed Eng 3:717–728

Zhou Q, Xu H (2020) Impact of gut microbiota on drug metabolism and absorption. Pharm J 10(3):120–126

Zhou B et al (2018) Induction and amelioration of methotrexate-induced gastrointestinal toxicity are related to immune response and gut microbiota. EBioMedicine 33:122–133

Zhu C, Zhang C, Wang S, Xun Z, Zhang D, Lan Z, Zhang L, Chao J, Liang Y, Pu Z, Ning C, Sang X, Yang X, Wang H, Jiang X, Zhao H (2024) Characterizations of multi-kingdom gut microbiota in immune checkpoint inhibitor-treated hepatocellular carcinoma. J Immunother Cancer 12(6):e008686. https://doi.org/10.1136/jitc-2023-008686

Advancing Precision Therapeutics Integrating Microbiota Biomarkers and Nanotechnology for Targeted Drug Delivery: Insights from Case Studies

Pankaj Kalia, Pooja Thakur, and Rajesh Kumar

Abstract Targeted drug delivery with advanced precision therapeutics emerges at a fast rate through the integrating microbiota biomarkers and nanotechnology innovations. The research examines how combination diagnostics based on microbiome analysis work together with nanotechnology systems to develop powerful drugs which deliver more effective results while being safer. Drug metabolism significantly relies on the gut microbiota because it directly affects both pharmacological drug absorption patterns and therapeutic response. Touching microbiota-derived biomarkers helps researchers build individualized medical approaches which improves drug effectiveness and restricts adverse reactions. Scientists draw conclusions in this review about how microbiota-guided nanocarriers demonstrate effective results for treating oncology, infectious diseases, and metabolic disorders from studied case examples. Microbiota-guided nanoparticles deliver probiotics to treat diseases in the gastrointestinal tract and function as cancer therapy carriers and act as lipid-based systems that control immune system responses. The identified cases strongly demonstrate how a microbiome–nanotechnology combination can handle differences in therapeutic outcomes between patients. The adoption of microbiome–nanotechnology faces three main obstacles: regulatory barriers, complex formulation development, and the necessity of standardized microbiome analysis methods. Future research needs to optimize predictive models of microbiota and develop safe nanomaterials together with animal and human research translations to reach brand new levels of specificity that will establish next-generation personalized medicine.

Keywords Microbiota · Biomarkers · Nanotechnology · Targeted drug delivery · Case studies

---

P. Kalia · P. Thakur · R. Kumar (✉)
Department of Pharmacology, Maharaja Agrasen School of Pharmacy, Maharaja Agrasen University, Solan, Himachal Pradesh, India

S. K. Singh et al. (eds.), *Microbiota Profiling for Precision Medicine*,
Interdisciplinary Biotechnological Advances,
https://doi.org/10.1007/978-981-95-2104-3_8

1 Introduction

Human health and disease management heavily depend on the gut microbiota, which represents a complex and diverse microbial community established inside the gastrointestinal (GI) tract. What we know about this microbial ecosystem has improved significantly through advances in metagenomics combined with metabolomics and proteomics research. This work demonstrates the ecosystem's important roles for digestion and immune function and metabolism of the whole system (Gebrayel et al. 2022). The gut microbiota studies of recent times have established links between the microbial community and obesity, diabetes, inflammatory bowel disease (IBD), neurodegenerative disorders, and cancer. The analysis of gut microbiota biomarkers provides investigators with meaningful data about host–microbe relationships since they include specific microbial organisms combined with short-chain fatty acids (SCFAs) and microbial genetic information (Lamas et al. 2020). The assessment of IBD disease activity, along with mucosal healing through replacement of endoscopy, now uses non-invasive fecal biomarkers such as calprotectin and lactoferrin. Research conducted in 2022 revealed that particular bacterial strains, including *Akkermansia muciniphila*, improve disease response to immune checkpoint inhibitors in cancer immunotherapy. A study from 2023 identified major dysbiosis along with reduced *Prevotella* levels in Parkinson's disease patients, showing a connection between the gut–brain pathway and PD disease formation. The implementation of microbiome analysis as a biomarker became possible in 2021 through large-scale studies, which showed environmental elements, along with eating habits and everyday practices that create specific microbial communities related to disease susceptibilities (Lamas et al. 2020; Bairamian et al. 2022). Metaproteomics analyzes the functional changes of microbiota through its ability to show increases in SCFA-producing enzymes following dietary interventions, thus demonstrating its metabolic variability. A growing number of research investigations in 2024 established that young individuals who experience biodiverse environments develop better gut microbial diversity combined with improved immune system regulation, because of their natural ecosystem contact. Mounting evidence supports using the gut microbiota as a therapeutic target through experiments with probiotics alongside fecal microbiota transplantation (FMT), as they demonstrate potential for dysbiosis management. Medical studies demonstrate that FMT has successfully produced remission in patients with *Clostridioides difficile* infections, while medical experts investigate its potential to treat ulcerative colitis and metabolic syndromes. Research discoveries show that microbial metabolites, including indole derivatives and bile acids, act as signaling molecules which manipulate host metabolism and immunity establish the potential use of the microbiota in personalized medicine approaches (Daniel et al. 2021). Moving such research discoveries into clinical practice demands uniform processes for biological material gathering and sequencing and data analysis methods to guarantee research repeatability. The current research faces difficulties when differentiating between microbial changes that have causative effects from those that show only associative relationships.

Technological advancements like artificial intelligence and machine learning enable better analysis of large datasets and localization of stable microbial biomarkers within studies of the microbiome. The advancements track the increasing research in microbiome-based detections and therapeutic methodologies for disease avoidance and treatment (Zhao et al. 2023a). Biomedical research now focuses on the gut microbiota for unparalleled possibilities to study and control human health. The application of nanotechnology techniques represents a revolutionary method to deliver drugs specifically into gut microbiota dysregulated diseases while reducing side effects throughout the system. Homeostasis depends on gut microbiota as a diverse microbial ecosystem, which functions fundamentally for health since its imbalance connects with inflammatory bowel disease (IBD), along with colorectal cancer (CRC), obesity and diabetes, and neurodegenerative disorders (Daniel et al. 2021; Zhao et al. 2023a).

Researchers have dedicated their work in nanotechnology advances through 2025 for developing drug delivery nanocarriers that respond to stimuli to reach specific areas in diseased gut tissues. Nanoparticles designed for pH responses use the different pH conditions across the gastrointestinal (GI) tract as their operating mechanism. The drug release mechanism of poly (lactic-co-glycolic acid) (PLGA)-based nanoparticles commences in the slightly alkaline intestine because the acidic stomach environment does not impact anti-inflammatory agents, which remain protected. Scientific research has focused on ligand-functionalized nanoparticles because these systems specifically seek target cells or microorganisms (Rahman et al. 2022). Selective drug delivery to target areas in the gut and pathogenic bacteria is accomplished through nano-sized platforms, which have been functionally prepared with folic acid, mannose, or antibody-based ligands. The research conducted in 2022 demonstrated that mannose-coated liposomes acted successfully against *Escherichia coli* biofilms, thus demonstrating their potential in treating biofilm-associated diseases. Nanocarriers that react to bacterial enzymes such as azoreductase and β-glucuronidase display selective bacterial pathogen targeting capabilities while maintaining healthy bacterial communities. Research into biofilm-penetrating nanoparticles represents an active approach to treat diseases caused by gut dysbiosis (Martín Giménez et al. 2023). Biofilms protect pathogenic bacteria through protective barriers, so nanocarriers with coatings like zwitterionic and cationic modifications achieve superior drug delivery penetration functions. Researchers used silica nanoparticles with developed biofilm-disrupting functions to improve pharmaceutical drug distribution in IBD patients, according to a 2024 scientific report. The application of nanotechnology beyond microbial objectives enables improved delivery of treatment agents for colorectal cancer, including chemotherapeutic drugs and immunomodulatory medications. Scientists demonstrated that aptamer-functionalized gold nanoparticles demonstrated precise targeting ability for cancer cells with reduced harm to healthy tissue units in 2023 research (Daniel et al. 2021; Rahman et al. 2022; Martín Giménez et al. 2023). Nanoparticle systems that deliver probiotics alongside short-chain fatty acid (SCFA) precursor substances demonstrate effectiveness in maintaining proper gut microbial equilibrium and controlling metabolic processes in diabetes patients alongside those with obesity (Ayan et al. 2023).

Modern research focuses on developing degradable nanomaterials to enhance both safety features and extended biological compatibility because researchers also pursue smart nanorobotic systems that can recognize dysbiotic bacteria and deactivate them. Artificial intelligence (AI), along with microbiome sequencing, drives the development of personalized nanomedicine through its ability to manufacture nanocarriers that adapt to each patient's unique microbial profile. The encapsulation of probiotics by nanotechnology advances at a fast rate because it allows microorganisms to survive the GI tract and achieve targeted gut delivery (Ayan et al. 2023; Mosquera et al. 2024). Nanotechnology will transform gut microbiota disease treatment through significant advancements from 2020 to 2025 that create important frameworks for developing precise medical strategies and personalized therapies. Studies involving gut microbiota have shown its connections to health conditions and disease states. The researchers used fecal microbiota transplantation (FMT) to cure recurrent *Clostridioides difficile* infection and obtained a 90% success rate. The metabolic changes caused by obesity result in decreased *Bacteroidetes* and elevated *Firmicutes* which leads to increased energy harvest according to scientific research. Scientific research on inflammatory bowel disease (IBD) found that patients suffer from dysbiosis, which reduces their beneficial bacteria levels while decreasing *Faecalibacterium prausnitzii* counts (Chen et al. 2023). Novel scientific research showed that eating certain diets modifies the microbiota, which influences metabolic syndrome and glucose intolerance. These clinical investigations highlight how the microbiota presents itself as a targeted therapeutic approach that can treat different diseases ranging from GI disorders to metabolic conditions. Advanced human health receives benefits from the intersection of nanotechnology and gut microbiota through targeted drug delivery and probiotics and diagnostic functions (Lu et al. 2023). Nanotechnology provides a mechanism to encapsulate probiotics, which results in enhanced probiotic survival within the gut. The analysis presents two cases where butyrate delivery through nanoparticles serves to control gut dysbiosis in inflammatory bowel disease patients and gold nanoparticles monitor gut microbiota-derived metabolites in real time (Li et al. 2023).

2 Role of Gut Microbiota in Health and Disease

The gut microbiota functions as a vital network of about 10–100 trillion microorganisms which resides mainly in the large intestine to advance human health while controlling the functionality of metabolic pathways and immune systems and neurological systems (Fig. 1). Metagenomics and metabolomics research during 2025 has elevated scientific understanding regarding these intricate microorganisms and their health and disease functions in ecosystems (Table 1) (Hou et al. 2022). Smooth fermentation of complex polysaccharides into short-chain fatty acids (SCFAs) including butyrate and propionate and acetate functions as the main energy source

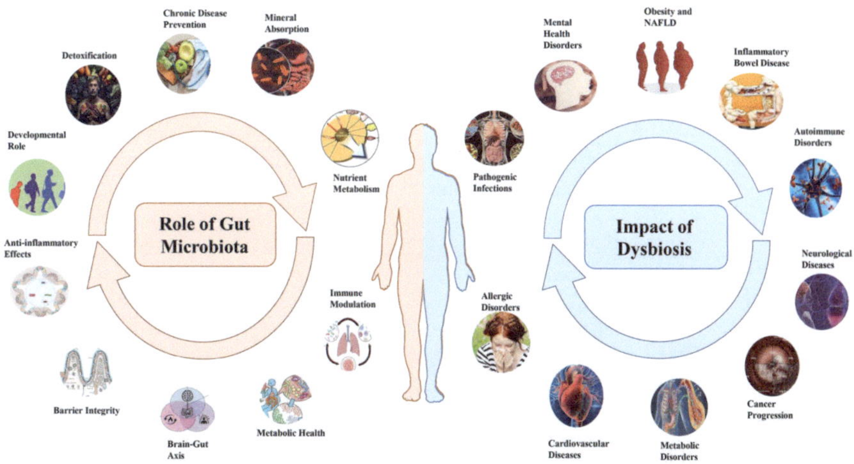

Fig. 1 Role of gut microbiota in health and disease

for colonocytes while reducing inflammation and tightening the barrier between intestinal cells by boosting tight junction protein expression. The compound butyrate stands as a crucial element that helps avoid intestinal wall leaks and reduces symptoms of inflammatory bowel disease (IBD). In microbial deconjugation reactions, the microbiota transforms primary bile acids to secondary acids that improve lipid metabolic processes and control glucose homeostasis through interaction with FXR and GPBAR1. In addition to bile acid transformations, the microbiota creates vitamin K alongside various B vitamins (B12, B6, riboflavin, and folate) through its synthetic functions (Zhao et al. 2023b). The immune system develops through and gets modulated by gut microbiota, which teaches immune cells to identify harmful versus non-harmful antigens to establish immune tolerance. Regulatory T cell (Treg) differentiation occurs through polysaccharide A (PSA) secretion from *Bacteroides fragilis*, which also lowers pro-inflammatory responses. At the same time, microbial metabolites like SCFAs use G-protein-coupled receptors (GPCRs) on immune cells to manage cytokine production and decrease systemic inflammation (Zhao et al. 2023b; Akasu et al. 2022).

The study from 2024 showed that propionate activation of GPCR43 leads to increased numbers of Treg populations. The microbiota creates immune protection by making antimicrobial peptides and bacteriocins to stop pathogenic bacteria from growing and through its enhanced mucosal barrier functionality. The gut–brain axis serves as evidence of microbiota control since it establishes two-way communication paths which extend across the brain and gastrointestinal systems until they reach the central nervous system (CNS) (Akasu et al. 2022). Several bacterial species, including *Lactobacillus* and *Bifidobacterium*, create central nervous system-regulating neurotransmitters like serotonin and dopamine and gamma-aminobutyric acid (GABA) that control mood, together with cognition and stress functions. Microbial actions lead to the transformation of tryptophan into serotonin

Table 1 Role of gut microbiota in health and disease

Health aspect	Role of gut microbiota	Targeted diseases	Impact of dysbiosis
Nutrient metabolism	Synthesizes essential vitamins (e.g., B vitamins, vitamin K) Enhances digestion of complex carbohydrates and production of SCFAs like butyrate, acetate, and propionate	Pathogenic infections	Imbalances can lead to overgrowth of pathogens like *Clostridium difficile*, *helicobacter pylori*, or *salmonella*, increasing infection risk
Mineral absorption	Improves absorption of minerals such as calcium, magnesium, and iron by maintaining gut pH and aiding in solubilization	Inflammatory bowel disease	Dysbiosis is implicated in Crohn's disease and ulcerative colitis, disrupting gut barrier function and triggering chronic inflammation
Immune modulation	Promotes immune tolerance and development of regulatory T-cells Protects against harmful pathogens by stimulating immune responses	Autoimmune disorders	Altered microbial profiles are linked to autoimmune diseases such as rheumatoid arthritis, multiple sclerosis, and type 1 diabetes
Barrier integrity	Reinforces the intestinal barrier by promoting tight junction protein expression and producing SCFAs	Cancer progression	Microbial metabolites like secondary bile acids may promote colon cancer, while protective metabolites like SCFAs are reduced in dysbiosis
Detoxification	Degrades and removes harmful xenobiotics, environmental toxins, and metabolic by-products	Obesity and NAFLD	Dysbiosis alters energy extraction, contributing to obesity, insulin resistance, and non-alcoholic fatty liver disease (NAFLD)
Brain–gut axis	Produces neurotransmitter precursors (e.g., serotonin, dopamine) and regulates stress responses through the HPA axis	Mental health disorders	Linked to anxiety, depression, and autism spectrum disorders due to disrupted brain–gut axis communication
Metabolic health	Regulates blood glucose and lipid metabolism, reducing the risk of metabolic syndrome and cardiovascular diseases	Cardiovascular diseases	Imbalances may lead to increased production of trimethylamine-N-oxide (TMAO), a metabolite associated with atherosclerosis
Anti-inflammatory effects	Produces SCFAs and other bioactive metabolites that suppress systemic inflammation	Allergic disorders	Dysbiosis is associated with increased susceptibility to allergies, such as asthma and atopic dermatitis, due to altered immune responses

(continued)

Table 1 (continued)

Health aspect	Role of gut microbiota	Targeted diseases	Impact of dysbiosis
Developmental role	Contributes to early development of the immune system and gut–brain communication in neonates	Neurological diseases	Dysbiosis contributes to neurodegenerative diseases like Alzheimer's and Parkinson's through systemic inflammation and gut–brain axis disruptions
Chronic disease prevention	Associated with lower risks of diseases like type two diabetes, obesity, and certain neurodegenerative conditions due to anti-inflammatory and immunomodulatory effects	Metabolic disorders	Linked to type two diabetes and metabolic syndrome due to impaired SCFA production and increased endotoxemia

precursors, including 5-hydroxytryptophan (5-HTP), at the levels of approximately 90% of body-produced serotonin. The occurrence of dysbiosis, as an improper microbial community distribution among individuals, correlates with the development of various mental conditions, including anxiety and depression, along with Parkinson's disease progression (Shi et al. 2021). A fundamental clinical study in 2025 demonstrated how *Lactobacillus plantarum* supplements reduced depression while managing cortisol metabolism through improved gut defenses. Newborn exposure to germs during development plays a vital role in brain construction because germ-free mice with altered stress reactions develop neural and behavioral irregularities because of incomplete microbiota influence. The gut microbiota controls insulin functions and body fat regulation through its essential metabolic role. Physiologically active short-chain fatty acids (SCFAs) utilize AMP-activated protein kinase (AMPK) activation in liver and muscle tissues to improve insulin sensitivity, and microbial metabolites work to affect body fat development, together with fat retention processes (Melo-González et al. 2022; George et al. 2022).

The presence of dysbiosis causes metabolic disorders such as obesity and type 2 diabetes and metabolic syndrome because it decreases beneficial species, including *Akkermansia muciniphila* and *Faecalibacterium prausnitzii*. Present-day research (2025) shows that restoration of specific microbial populations with dietary change or probiotic supplementation enhances both metabolic markers of fasting glucose levels and lipid profile outcomes. Harmful metabolic actions occur in the gut microbiota when they process dietary choline and L-carnitine, which becomes trimethylamine (TMA) before liver cells convert it to trimethylamine N-oxide (TMAO) (Shi et al. 2021; George et al. 2022). Elevated TMAO levels are linked to atherosclerosis and cardiovascular disease development. Social scientists examine microbial-based treatment strategies to minimize TMAO production and lessen cardiovascular disease risk levels. The human gut microbiota helps develop chronic diseases through its effect on inflammatory conditions and cancer development. In IBD, dysbiosis exacerbates inflammation, with reductions in anti-inflammatory species like *F. prausnitzii* and overgrowth of pathogenic Enterobacteriaceae. Colorectal cancer development depends on dysbiosis, which leads to carcinogenic hydrogen sulfide production in certain bacteria, but butyrate and its effect on histone deacetylase inhibition promote cancer cell apoptosis (Szablewski 2022). In 2025, researchers found fecal microbiota transplantation succeeded in treating recurrent *Clostridioides difficile* infections at a 75% success rate, thus driving future studies attempting to utilize it for IBD and metabolic disorder treatment. The microbiota serves as a key factor in drug processing and detoxification processes, where it activates sulfasalazine prodrugs and converts polyphenolic nutrients into beneficial compounds that possess anti-inflammatory and antioxidant functions (Shi et al. 2021; Wang et al. 2023).

Microbial metabolism creates toxic metabolites including p-cresol and ammonia which demonstrate the complex nature of bacterial breakdown activity by both helping and harming metabolic processes and hepatic encephalopathy development. Dysbiosis creates widespread medical complications which affect people through NAFLD development, along with asthma and allergy progression and autoimmune disease manifestation. NAFLD patients demonstrate gut microbiota dysbiosis primarily through increased Firmicutes compared to Bacteroidetes, which leads to hepatic lipid accumulation based on the findings reported in 2025 (George et al. 2022; Szablewski 2022; Wang et al. 2023). Inflammatory diseases involving immune dysfunction worsen when dysbiosis. Boosts systemic inflammation through its ability to increase gut permeability, which enables blood entry of microbial endotoxins such as lipopolysaccharides (LPS), resulting in "metabolic endotoxemia". The consistent low-grade inflammatory situation leads to weight problems along with insulin resistance, followed by multiple metabolic health issues. Proper diet choices, together with life habits, play a vital role in preserving healthy gut microbiota. Prebiotic ingredients inulin and fructooligosaccharides (FOS), along with high-fiber diets, feed beneficial bacteria *Bifidobacterium* and *Lactobacillus* to grow effectively (Berding et al. 2021). The field of precision prebiotics is advancing rapidly and will undergo clinical trials beginning in 2025. The clinical implementation of fecal microbiota transplantation (FMT) as routine treatment against recurrent *C. difficile* infections has paved the way to use this method for metabolic and neuropsychiatric disorder management. Preserving a healthy intestinal environment depends on preventing exposure to excessive antibiotics, alcohol, plus processed foods, which both harm microbial diversity while reducing functionality. Human physiology regulation through the gut microbiota stands as the central controlling force that affects all metabolic and immune system functions, alongside the neurological systems. Ongoing research in 2025 about microbiota complexity enables the development of precision-interventional therapies that show significant potential for disease prevention and treatment across various medical conditions, thereby establishing microbiota as fundamental to precision medicine and therapeutic innovation (Wang et al. 2023; Berding et al. 2021; Martínez et al. 2021).

The gut microbiota functions as a vital part of host health because it contains trillions of microorganisms that include bacteria and archaea, fungi as well as viruses, and protozoa. The microbial consortium mainly found in the colon regulates essential procedures, including immune regulation, metabolic control, and neurohumoral activity through the gut–brain axis. The pathologic condition known as dysbiosis refers to microbial imbalances that link with many disease developments. The immune system obtains three essential elements from commensal microbes, which enable proper gut-associated lymphoid tissue (GALT) development while regulating mucosal immunity through cytokine equilibrium maintenance (Boisseau et al. 2022). The pathogenesis of inflammatory bowel disease (IBD), along with rheumatoid arthritis and multiple sclerosis, results from dysbiosis that causes damage to tight junction proteins and promotes microbial translocation and continuous inflammatory responses. During IBD, bacterial imbalances lead to two major changes, including the loss of anti-inflammatory *Faecalibacterium prausnitzii* together with increased numbers of pathogenic bacteria, which worsen mucosal inflammation by releasing inflammatory mediators. The effects of dysbiosis spread beyond gastrointestinal diseases since it strongly shapes metabolic disorders (Berding et al. 2021; Martínez et al. 2021; Boisseau et al. 2022). The microbiome found within the gut exhibits connections to T2DM along with obesity since it manages both insulin sensitivity and energy regulation, as well as fat tissue accumulation. Short-chain fatty acids (SCFAs) developed through microbial fermentation of dietary fibers activate G-protein-coupled receptors (e.g., GPR41 and GPR43) to generate systemic metabolic effects, among which butyrate, propionate, and acetate belong. Dysbiosis leads to decreased SCFA production while increasing intestinal permeability to allow lipopolysaccharides (LPS) transfer to blood circulation, which results in metabolic endotoxemia and insulin resistance. NAFLD produces negative effects due to dysbiotic shifts that create increased gut permeability, allowing inflammatory bacterial metabolites to reach the liver through toll-like receptors (TLR) and result in the progression of steatosis and fibrosis (George et al. 2022; Szablewski 2022; Boisseau et al. 2022).

Microbial metabolites such as SCFAs and tryptophan catabolites established through the gut–brain axis communicate with the CNS and modify both neuroinflammation and blood–brain barrier (BBB) integrity and neurotransmitter creation mechanisms. Studies link dysbiosis to neurodegenerative and neuropsychiatric disorders, which include Parkinson's disease (PD), Alzheimer's disease, and depression, because these conditions develop from changes in serotonergic signaling and stronger neuroinflammatory responses and PD-specific microbial alpha-synuclein aggregation. The oncogenic aspects of dysbiosis have been profoundly investigated in colorectal cancer (CRC) because specific microbial organisms, including *Fusobacterium nucleatum* and *Escherichia coli*, create genotoxic substances coupled with pro-inflammatory metabolites that boost epithelial cell growth while enabling them to evade immune responses (George et al. 2022; Berding et al. 2021; Boisseau et al. 2022). *Akkermansia muciniphila* among other specific taxa significantly improves the effectiveness of immune checkpoint inhibitors according to studies on dysbiotic profile effects on immune checkpoint inhibitor responses. Microbial products, including TMAO, which forms from dietary choline and carnitine metabolism, induce dysbiosis to develop cardiovascular conditions by causing atherogenesis and endothelial dysfunction and promoting thrombosis. The therapeutic approach of probiotics, which uses actively existing friendly microorganisms, demonstrates potential for treating IBD and repeated *Clostridioides difficile* infections. The use of nondigestible fibers called prebiotics enhances both beneficial microbial growth and the production of SCFAs and strengthens intestinal barriers. FMT produces beneficial results by transferring stool from healthy donors to dysbiotic recipients and shows promise to benefit refractory infections, as well as potential use for autoimmune and metabolic diseases. Those diets rich in fermentable fibers and polyphenols lead to beneficial changes in gut bacterial composition alongside lowering inflammation (Cheung et al. 2023).

Emerging precision therapies for diseases include engineered probiotics, phage therapy, and microbial metabolite inhibitors, aiming to target specific disease pathogenesis pathways in a targeted manner. Therefore, the commensal gut microbiota, which contributes to multiple immunological, metabolic, and neurological processes, is a key determinant of health and disease, and dysbiosis is a key driver of pathophysiological states and a potential target for therapeutic interventions (Boisseau et al. 2022; Cheung et al. 2023).

3 Synthetic Nanotechnology for Microbial Engineering

Synthetic nanotechnology enables unparalleled microbial engineering possibilities because it provides programs and augmentation mechanisms that operate at the nanoscale. Modern synthetic nanotechnology research has the capacity to bring advancements in biotechnology, as well as sustainable development and medical sciences through its current challenge-solving approaches (Villaverde 2010). Partnerships between nanoscale science and synthetic biology through synthetic nanotechnology in microbial engineering enable precise system enhancement and reprogramming of microbes for multiple biotechnological and healthcare applications, environmental solutions, and energy production purposes (Okeke et al. 2022; Ma et al. 2024). Natal Technology uses nanomaterial properties of quantum dots together with metallic particles and carbon-based constructs to enhance microbial systems by enabling foreign biochemical processes that improve metabolic efficiencies along with improved electron transport, which boosts microbial fuel cell outputs (Huang et al. 2024; Adir et al. 2022). Nanozymes developed as artificial enzyme mimics function as effective tools for microbial metabolic engineering to generate new biochemical synthesis possibilities through reaction enhancement. Gene-editing tools like CRISPR-Cas9 receive precise delivery using nanoscale systems made from polymeric nanoparticles and liposomes thus they can target specific microbial pathways while minimizing collateral activity (Dhanker et al. 2021; Sultana et al. 2021). DNA origami nanostructures enable scaffold-based precision through their ability to direct synthetic gene circuit assembly, which programs complex genetic operations in engineered microbes. Biotechnology utilizes nanotechnology through quantum dots and plasmonic nanoparticles to boost the signal output of biosensors, thus creating highly sensitive systems that detect environmental pollutants and biochemical markers. Real-time monitoring capabilities advanced due to synthetic reporter systems working in combination with these recent developments in the field (Silva et al. 2023; Singh et al. 2023a). Engineered microbes through microbial engineering applications enable sustainable production of quantum dots alongside metal nanoparticles and nanofibers while allowing exceptional dimensions control (Nasr et al. 2020; Munir et al. 2020).

Probiotics modified with nanoscale systems have demonstrated promise for overcoming the challenges of gastrointestinal degradation, targetable and sustained delivery of therapeutic agents, and modulation of the host microbiome, in drug delivery and other applications (Soni et al. 2023; Hasanzadeh et al. 2020). On top of that, engineered microbes with nanomaterials, for example, silver nanoparticles, function as strong antimicrobial agents to fight against multidrug-resistant pathogens. For environmental applications, microbial nanotechnology hybrids have been demonstrated to be successful for bioremediation, where functionalized microbes with magnetic nanoparticles can separate the unwanted compound away from the area and subsequently recuperate and reuse it (Hasanzadeh et al. 2020; Kumar et al. 2022a). Ever since, these advancements have been employed for the production of bioenergy in the form of hydrogen, through interfacing photosynthetic microorganisms, such as cyanobacteria, with light-harvesting nanomaterials, to enhance light absorption and conversion efficiency. As in synthetic nanotechnology, microbial consortia are also engineered through synthetic nanotechnology to mediate intercellular communication and metabolic resource exchange for multi-step processes for carbon capture or biomanufacturing. However, despite its immense potential, realizing its potential is hampered by challenges, including nanomaterial toxicity, which can be a product of metabolic perturbation due to interactions with microbial systems and scalability issues where lab-scale innovations need robust and cost-effective methods to translate to industrial scale (Malik et al. 2023; Mamo et al. 2010). Hurdles are also due to ethical and regulatory concerns, especially regarding the application of nanotechnology modified microbes in natural or clinical environments and the need to observe extravagant safety protocols to mend unpleasant ecological or health by-effects. Challenges associated with nanomaterials (NM) are poised to be overcome through the use of AI-driven design tools that predict and optimize NM microbe interactions and the development of biodegradable NM that decrease environmental and biological risk (Mamo et al. 2010; Harsh Pandey et al. 2018).

Advanced targeting mechanisms, such as receptor-mediated nanocarriers and programmable, nanoscale devices, will also further integrate into the precision and efficacy of microbial systems for applications from targeted drug delivery to complex biosynthetic pathways. Additionally, synthetic microbial consortia possessing nanoscale tools will be created that can mutually perform cognate or synergistic pollutant degradation, or synergistic biosynthesis of high-value compounds for innovations in sustainability and bioproduction (Sakamoto et al. 2010). Synthetic nanotechnology in this field will continue to transform the field of microbial engineering and redefine the boundaries and create unprecedented opportunities for global challenges in energy, medicine, environmental stewardship and will push the forefront of scientific advancement (Blanco-Fernandez et al. 2021).

4 Microbiota Biomarkers

Blood biomarkers that show signs of microbial dysbiosis enable a deep understanding of both disease formation and systematic physiological deterioration (Table 2). Lipopolysaccharide (LPS) stands as the main endotoxin marker originating from the outer membrane structures of Gram-negative bacteria. The link between endotoxemia and elevated LPS blood levels results in simultaneous gut dysbiosis at the same time as increased intestinal permeability alongside systemic low-grade inflammation that leads to metabolic syndrome and type 2 diabetes and cardiovascular diseases (Leng et al. 2021). LPS-binding protein (LBP) functions as a biomarker that helps the immune system recognize LPS and detects microbial translocation into systemic circulation through the gut wall. Hence, it serves as a sensitive marker for systemic inflammatory states originating from the gut. The regulatory protein Zonulin works as an essential marker of impaired intestinal barrier integrity because it controls tight junctions in the intestinal epithelium. Zonulin levels rise frequently in celiac disease patients along with those in irritable bowel syndrome (IBS) and inflammatory bowel disease (IBD) cases because it measures increased epithelial permeability that enables microbial translocation (Zhang et al. 2022). Systemic inflammation markers including C-reactive protein (CRP) tend to increase when gut dysbiosis causes barriers of the intestinal epithelium to break down, although CRP shows non-specific results because it helps measure inflammation levels connecting to gut dysbiosis and systemic diseases. The intestinal metabolic compounds known as short-chain fatty acids (SCFAs) function as essential markers that show how healthy the gut is. Butyrate and propionate together with acetate emerge from dietary fiber fermentation executed by commensal bacteria. The bloodstream concentration of SCFAs directly correlates with dysbiosis severity and metabolic disorders in inflammatory conditions because these metabolites support gut barrier function and immune regulation and metabolism control (Ke et al. 2024; Singh et al. 2024a).

While butyrate is one of these, it does so by improving colonic epithelium health and inhibiting proinflammatory processes, and its depletion is a hallmark of intestinal inflammation and leaky gut syndrome. Conversely, metabolites such as trimethylamine-N-oxide (TMAO), a product of microbial metabolism of dietary choline and carnitine, represent pathogenic biomarkers. TMAO levels are higher in individuals with cardiovascular disease pathology as it promotes atherosclerotic plaque formation and endothelial dysfunction (Zhang et al. 2022; Ke et al. 2024). Gut-derived uremic toxins such as indoxyl sulfate and p-cresyl sulfate, which are metabolites of tryptophan and tyrosine, respectively, also accumulate in systemic circulation during dysbiosis and are closely linked to chronic kidney disease (CKD) and oxidative stress. In the case of the gut–brain axis, biomarkers, mainly serotonin (5-HT), synthesized by enterochromaffin cells under the control of gut microbes, reflect disruptions. Both gut microbiota and neural pathways are crucial to the interplay between gut microbiota and gut motility disorders, IBS, and neuropsychiatric conditions, as altered serotonin levels in the bloodstream are associated (Petakh

Table 2 Biomarkers associated with the gut microbiota

Sl. no.	Biomarker	Associated condition	Level (↑/↓)	Clinical significance
1.	Lipopolysaccharide (LPS)	Gut dysbiosis, leaky gut syndrome	↑	Elevated LPS indicates increased gut permeability and systemic endotoxemia
2.	LPS-binding protein (LBP)	Gut dysbiosis, metabolic syndrome	↑	Reflects LPS translocation into circulation; a marker of microbial translocation
3.	Short-chain fatty acids (SCFAs)	Obesity, type 2 diabetes	↓	Reduced levels indicate disrupted fermentation by gut microbes and poor gut health
4.	Zonulin	Increased intestinal permeability	↑	A marker of leaky gut; elevated levels are linked to celiac disease and inflammatory conditions
5.	Trimethylamine-N-oxide (TMAO)	Cardiovascular disease, dysbiosis	↑	Elevated TMAO is a metabolite from gut microbes linked to atherosclerosis
6.	Indoxyl sulfate	Chronic kidney disease, dysbiosis	↑	A gut-derived toxin; elevated in dysbiosis and kidney dysfunction
7.	p-Cresyl sulfate	Chronic kidney disease, gut dysbiosis	↑	Gut microbiota-derived metabolite associated with systemic inflammation and toxicity
8.	C-reactive protein (CRP)	Inflammatory bowel disease (IBD), dysbiosis	↑	Non-specific marker of inflammation; elevated in systemic effects of gut inflammation
9.	Serotonin (5-HT)	Gut–brain axis dysregulation, IBS	↑/↓	Altered levels linked to gut dysfunction and mood disorders; 90% of serotonin is gut-derived
10.	Bile acids (primary/secondary)	Non-alcoholic fatty liver disease (NAFLD), gut dysbiosis	↑/↓	Altered levels reflect gut microbiota dysregulation affecting bile acid metabolism
11.	Interleukin-6 (IL-6)	Gut inflammation, dysbiosis	↑	Pro-inflammatory cytokines elevated in gut-associated systemic inflammation
12.	Tumor necrosis factor-α (TNF-α)	IBD, gut dysbiosis	↑	Elevated in gut-related inflammatory diseases; contributes to systemic inflammation
13.	Butyrate	Gut dysbiosis, inflammation	↓	Reduced levels indicate decreased beneficial microbial activity in the gut

(continued)

Table 2 (continued)

Sl. no.	Biomarker	Associated condition	Level (↑/↓)	Clinical significance
14.	Histamine	Food intolerances, dysbiosis	↑	Elevated in conditions where gut microbiota produce excess histamine
15.	Folate (vitamin B9)	Dysbiosis, malabsorption	↓	Reduced levels linked to dysbiosis impairing microbial synthesis of B vitamins
16.	Lactate	Gut fermentation disorders	↑	Elevated in dysbiosis affecting lactate-utilizing microbial pathways

et al. 2024; Kumar et al. 2022b). Other key indexes are a bioactive amine derived from microbial histidine decarboxylation, called histamine, which is elevated in cases of microbial overgrowth or food intolerances. Systemic inflammation, gastrointestinal discomfort, allergic responses, and more are associated with elevated levels of histamine spurred by the gut microbiota's action on immune regulation. Another metabolic set induced by microbes is bile acids, both primary and secondary. Disruption of bile acid metabolism by dysbiosis produces bile acid profiles that are associated with metabolic disorders, nonalcoholic fatty liver disease (NAFLD), and inflammation of the gut. Thus, microbial dysregulation in host lipid metabolism is reflected by elevated primary bile acids and reduced secondary bile acids (Zhang et al. 2022; Guo et al. 2022).

Systemic inflammation markers are mainly characterized by two pro-inflammatory cytokines: interleukin-6 (IL-6) and tumor necrosis factor-alpha (TNF-α). Blood tests indicate immune activation because of microbial translocation, which strongly links to IBD, metabolic syndrome, and autoimmune diseases (Kumar et al. 2021a). The cytokines increase inflammatory processes, which simultaneously lead to widespread tissue injuries while worsening the state of disease. The microbial functionality can be evaluated through folate (vitamin B9) measurements, which the gut bacteria partially generate (Petakh et al. 2024; Kang et al. 2022). The condition of dysbiosis results in low bloodstream folate that causes deficiencies affecting protein synthesis and cell multiplication processes and systemic metabolic processes (Kumar et al. 2021b). The presence of high lactate amounts in blood indicates problems with fermentation pathway regulation by microbial communities. During dysbiosis, lactate accumulates within the body because lactate-consuming bacteria fail to properly utilize this intermediate, thus leading to acidosis along with multiple metabolic problems (Guo et al. 2022; Kang et al. 2022).

The biomarkers demonstrate how gut microbiota dysregulation produces extensive effects on systemic systems, which both affect gut-specific issues and contribute to multiple diseases through metabolic pathways and immune systems and neuroendocrine mechanisms (Martín Giménez et al. 2023; Landberg and Hanhineva 2020). The biomarkers function as vital diagnostic tools for disease evaluation by connecting microbial communities to systemic health, so medical practitioners can both diagnose and manage microbial imbalance-related diseases while uncovering treatment approaches for microbial restoration (Rob et al. 2022; Dai et al. 2024; Pandey et al. 2019).

5 Nanoparticle-Based Drug Delivery Systems

Nanoparticles have established themselves as innovative therapeutic systems, which tackle diseases caused by disturbances in gut microbiota, but they address dysbiosis of microbial populations found in the gastrointestinal (GI) tract. The composition of gut microbiota functions as a critical regulator of physiological balance, while it controls metabolic activities and immune response in addition to communicating through the gut–brain axis (Xue et al. 2023). Medical evidence suggests that pathologic gut microbial imbalances referred to as dysbiosis contribute to the development of Crohn's disease and ulcerative colitis, together with irritable bowel syndrome (IBS), colorectal cancer, diabetes and obesity and Parkinson's disease, and Alzheimer's disease. Standard therapies for these diseases present therapeutic problems because of their systemic adverse effects alongside poor substance availability and non-targeted properties, which result in subpar therapeutic effects. The nanometer-scale dimensions alongside controllable physical properties of nanoparticles resolve fundamental healthcare delivery issues by providing precise delivery of therapeutic substances both locally and at targeted locations while achieving sustained action (Yang and Merlin 2020; Dhiman et al. 2016a; Vijaya Durga 2015). The nanoparticulate carriers protect various bioactives, including drugs and biologics, including RNA therapeutic compounds, and probiotics and prebiotics, through their encapsulation mechanism while ensuring precise delivery to specific gut sections and microbial communities despite gastric conditions. Researchers use liposomes and solid lipid nanoparticles (SLNs) as lipid-based nanoparticles which successfully encapsulate drugs with different water solubility while simultaneously improving medication stability and treatment accessibility through the body (Wang et al. 2021a; Perumal et al. 2021). The medical use of liposomes has proved valuable for delivering antibiotics and probiotics throughout the colon, and this strategic method protects the native microbial communities. Mucosa-adhering polymer-based nanoparticles, which integrate chitosan along with poly (lactic-co-glycolic acid) (PLGA), demonstrate effective controlled drug release functions, thus enhancing the treatment of IBD using anti-inflammatory agents like curcumin and corticosteroids (Perumal et al. 2021; Yang et al. 2020).

The surface area and pore functionality of mesoporous silica nanoparticles enable antibiotic or bioactive encapsulation alongside antimicrobial disruption of pathogenic bacterial membranes achieved with metallic nanoparticles, including gold and silver. Nanoparticles enabled with stimuli-responsive capabilities allow medical professionals to deliver drugs based on particular GI environment conditions, including pH changes and enzymatic activity, which demonstrates progress in precision medicine. Nanoparticles have two main applications for microbiota control through their ability to protect probiotic bacteria from stomach acid and protect prebiotics from digestion while delivering both to the colon (Gao et al. 2024; Garg et al. 2022). Microbiota restoration strategies together with improved outcomes are possible when using these techniques for IBS and metabolic syndrome treatment. Research indicates that nanoparticle systems possess the ability to control gut-brain communication by using delivery systems of short-chain fatty acids (SCFAs) or serotonin precursors for neurological and psychiatric condition management related to digestive tract health. Various obstacles stand in the way of nanoparticle-based therapies finding common usage in clinical settings (Yang and Merlin 2019; Singh et al. 2024b). Safety evaluations through preclinical and clinical studies become essential to address issues about possible toxicity levels, along with the potential accumulation of nanoparticles in bodies. The manufacturing scale-up process for functionalized nanoparticles faces major installation barriers because entire production methods tend to remain complex and demanding (Abed et al. 2022). Multiple factors in the way individuals' gut microbiota differ from one another make it difficult to create standardized therapeutic plans, which underscores the importance of moving toward personalized medication approaches. New safety solutions emerge from the combination of bioinspired and biodegradable materials in nanoparticle design and the implementation of artificial intelligence (AI) and machine learning technology for developing personalized nanoparticle systems based on individual microbiome profiles (Karavolos and Holban 2016; Singh et al. 2023b). Nanorobots and smart nanocarriers, which function with sensory abilities and targeted delivery systems and dynamic population control of gut microorganisms, shall transform the healthcare field in future applications. The combination of technological advancements creates better treatment options for gut microbiota diseases and leads to patient-specific, microbiome-targeted interventions for the future (Liu et al. 2024a).

6 Active Targeting Mechanisms

Nanotechnology serves as a transformative technology to precisely control gut microbiota populations because it combines active targeting methods that engage with particular microbial populations and their metabolic processes. The targeting function of nanocarriers works through surface attachment of ligands (antibodies, peptides, or small molecules), which bind exclusively to bacterial receptors and unique surface molecules (Guo and Wu 2017; Dhiman et al. 2015a). Nanoparticles with mannose functionalities recognize and target *Escherichia coli* strains by binding to their mannose-binding lectins (Dhiman et al. 2016b). Drug delivery systems based on responsive nanocarriers achieve controlled medication delivery through environmental triggers that exist in the gut, such as pH, temperature, and enzymatic activity levels (Upreti et al. 2024). Nanoparticles coated with polydopamine possess a built-in mechanism which releases antibiotics or prebiotics based on limitations of pH within the gastrointestinal tract for targeted therapeutic impact (Wang et al.

2022; Kumar et al. 2024a). Nanocarriers made from polymers provide enhanced protection to beneficial microbial species through their survival mechanisms and enable customized interactions with targeted gut areas, including the ileum and colon segments. Scientists have developed the targeted application of quorum-sensing inhibitors (QSIs) through nanotechnology to selectively disrupt bacterial communication networks, which restricts pathogenic biofilm development without harming beneficial commensal bacteria (Lin et al. 2021; Covasa et al. 2019). The combination of gold nanoparticles and furanone derivatives functions as an effective pathogenic potential reducer by blocking quorum-sensing pathways in *Pseudomonas aeruginosa* bacteria. Nanotechnology-based delivery systems offer a promising solution to distribute prebiotics or metabolites which activate beneficial bacteria, including *Bifidobacteria* and *Lactobacilli.* Selectivity in microbial carrier uptake can be improved through the addition of receptor-specific ligands to the functionalized carriers. Emerging strategies in nanotechnology offer even greater specificity and therapeutic potential (Zhu et al. 2023; Hamid et al. 2024). Nanocarriers that employ DNA and RNA as building blocks present CRISPR-Cas systems, allowing genome editing of both harmful genes and beneficial microbial functions. Nanoparticles containing Cas13a have been developed to target specific antibiotic-resistance genes in pathogenic bacteria, thus offering an innovative antibiotic-resistance solution. Biodegradable nanomaterials consisting of PLA and PCL offer a secure method for releasing medicinal agents within the gut while responding to enzymatic degradation found in this section of the body (Jia et al. 2008; Kumar et al. 2024b).

Both precise targeting and efficient delivery of bioactive cargos are provided by hybrid nanocarriers that consist of inorganic nanoparticles such as silica or gold with organic coatings. These innovations offer promise for a variety of health outcomes related to gut microbiota. Use of nanotechnology to modulate gut microbiota finds application across combating antimicrobial resistance, microbial imbalance in gut dysbiosis, and systemic diseases, depending on the microbiota (Kumar et al. 2024c). Delivering antibiotics or antimicrobial peptides targeted to trigger microbes reduces "drug side effects," such as the development of resistance by unwanted microbes (Zam 2018; James et al. 2017). Active targeting is used to suppress pathogenic strains, but promote growth of beneficial bacteria, including butyrate-producing bacteria, which are known to be anti-inflammatory and anti-carcinogenic. For the management of chronic conditions involving a high incidence of microbial imbalances like inflammatory bowel disease (IBD) and metabolic diseases like obesity and diabetes, this strategy is especially applicable. In addition, modulation of gut microbiota has shown promise in improving systemic health by mechanisms of improving SCFAs, microbial-derived metabolites that are known to have protective effects on the host physiology (Jia et al. 2008; Ma et al. 2017). Prebiotic-loaded nanocarriers promote the abundance of butyrate-producing bacteria that play an anti-carcinogenic role as well as improve the therapeutic outcomes in cancer therapy. However, there remains a set of challenges that have to be overcome in order to harness the remarkable potential of these approaches (Kumar et al. 2022c, 2024d). Although precise targeting is desirable, such efforts are obstructed by the complexity and dynamic nature of gut microbiota and by individual variability in microbial composition. Essential for long-term applications is developing nanocarriers that can travel such complexity while being biocompatible and safe. There are regulatory hurdles too, that is, novel nanocarriers have to go through stringent tests to pass medical and safety standards. To meet these challenges, it will require interdisciplinary collaboration and advanced technologies like artificial intelligence (AI), machine learning (ML), and others (Jia et al. 2008; Woo et al. 2022; Zhu et al. 2022).

Such tools can further optimize the design of nanocarriers and predict microbiota–nanocarriers interactions more accurately, leading to more effective and personalized therapeutic interventions. In the future, it would be possible to integrate nanotechnology along with omics technologies such as metagenomics and metabolomics to harness our understanding of gut microbiota and therapeutics based on it. However, by mapping microbial genomes and metabolic pathways at the genome scale, researchers can now design nanocarriers that precisely target, or disrupt, specific microbial functions or metabolites. On top of that, intelligent, multifunctional nanocarriers that can concurrently target, treat, and monitor will improve the field. For example, nanocarriers could be made to deliver therapeutic agents only to respond to disease-associated microbial signals while offering real-time feedback on the efficacy of treatment (Feng et al. 2020; Kumar et al. 2021c). Ongoing research into biodegradable nanomaterials and hybrid delivery systems, combined with these advancements, are on the verge of revolutionizing the field of gut microbiota modulation with safe, efficacious, and extremely targeted therapeutic options. In the end, the integration of nanotechnology into modern scientific knowledge promises the development of effective approaches to solve the problem of a variety of gut microbiota-related health problems and to enhance patient outcomes (Kumar et al. 2021c; Xia et al. 2021).

7 Responsive Nanocarriers for Targeted Drug Delivery

The development of responsive nanocarriers has brought a significant breakthrough to drug delivery for gut dysbiosis by utilizing diverse environmental factors and biological triggers for site-specific therapeutic control. The modern delivery systems respond to the internal contextual changes of pH and enzymatic modifications and oxidative pressure along with external stimuli including light, ultrasound, and magnetic fields. The pH-sensitive nanocarriers encompass Eudragit-coated nanoparticles as well as alginate/chitosan hydrogel systems and zein nanoparticles, which release their therapeutic agents within gut tissue acidic environments (Ganta et al. 2008). The release of drugs through enzyme-sensitive nanocarriers occurs precisely by using β-glucuronidase-sensitive micelles together with gelatin nanoparticles and dextran-based nanoparticles, which depend on overexpressed enzymes present in dysbiotic conditions. The redox-sensitive nanocarrier group includes disulfide-linked polymeric micelles and ROS-sensitive mesoporous silica nanoparticles

(MSNs) and polydopamine-coated nanoparticles because they use enhanced ROS concentration in inflamed regions for precise targeting (Wang et al. 2020; Kumar et al. 2021d). The combination of external stimulus-responsive nanocarriers using gold nanoparticles (AuNPs) activated by light, while magnetic iron oxide nanoparticles (IONPs) follow magnetic guidance, and ultrasound-responsive lipid nanobubbles bring additional functionality yet face obstacles regarding the tissue penetration of light illumination in the gastrointestinal tract (Kumari et al. 2016; Majumder and Minko 2021). Nanocarrier effectiveness depends heavily on how their surfaces are functionally modified and what materials make up their composition. The controlled release capabilities combined with engineered pH and enzyme sensitivity of poly (lactic-co-glycolic acid) (PLGA) nanoparticles and chitosan-based nanoparticles, and polysaccharide-based nanoparticles are available in polymeric systems. Lipid-based carriers composed of liposomes and solid lipid nanoparticles (SLNs) and nanostructured lipid carriers (NLCs) support drug delivery processes in a sustainable fashion through the integration of functional groups that reach distinct gut territories. Earth-based nanocarriers comprising cerium oxide-based nanozymes and silver nanoparticles (AgNPs), as well as zinc oxide nanoparticles, shield against pathogenic bacteria and various strains of oxidative stress, which characterize gut dysbiosis (Sawant et al. 2006).

The surface modification of these nanoparticles by folic acid-long chains and peptide-long chains and antibody-coated nanoparticles creates receptor pathways for better inflamed gut epithelial cell access (Kumar et al. 2022c). The nanocarriers help resolve dysbiosis problems by treating inflammation while managing the microbiome and delivering drugs to specific areas. Doctors use dendrimers and PLGA nanoparticles containing vancomycin together with cyclodextrin-based nanoparticles to deliver antibiotics by targeting harmful bacterial strains and reducing drug side effects and limiting the development of antibiotic resistance (Sawant et al. 2006; Mura et al. 2013). The delivery of probiotics and prebiotics through nanoparticles coated in alginate, as well as synbiotic nanocarriers and pectin-based nanoparticles, protects beneficial microbes while permitting their regeneration of gastrointestinal microbiota balance. Anti-inflammatory treatments employ curcumin-loaded dextran nanoparticles together with budesonide-carrying pH-sensitive liposomes and anti-inflammatory drugs encapsulated by cyclodextrin in order to decrease oxidative stress and locally deliver corticosteroids without causing systemic adverse effects. Emerging gene therapy methods use lipid nanoparticles (LNPs), chitosan-modified MSNs, together with polymer-based siRNA nanoparticles to provide molecular solutions for dysbiosis by delivering mRNA or other nucleic acids that modify genes in microorganisms or epithelial cells (Martín Giménez et al. 2021; Kumar et al. 2017). The clinical application of responsive nanocarriers remains limited due to their inadequate ability to overcome both mucus obstruction and enzyme degradative conditions. PEGylated liposomes and mucoadhesive nanoparticles together with pH-sensitive hydrogels enable better penetration of mucus barriers and detailed nanocarrier optimization through artificial intelligence design systems, which use individual microbiome profiles for personalized therapy. The next stage of development will incorporate microbiota-sensing elements that use short-chain fatty acids to activate nanocarriers for targeting particular therapeutic outcomes (Zhou et al. 2017; Dhiman et al. 2015b). Eco-friendly and secure drug delivery solutions are provided by biodegradable materials that include lignin-based nanoparticles, which solve long-term accumulation issues. Medical science has developed magnetically controlled nanorobots, which provide the most advanced form of precise delivery systems for accessing difficult-to-reach gut regions. The continuous research in material science and artificial intelligence and microbiome study has led to responsive nanocarriers such as PLGA nanoparticles, liposomes, nanozymes, dendrimers, SLNs, NLCs, and similar types to revolutionize gut dysbiosis treatments through better performance and fewer adverse reactions while maintaining ecological sustainability in digestive health (Fleige et al. 2012; Mi 2020).

8 Integration of Microbiota and Nanotechnology

Gut microbiota integration with nanotechnology has created a disruptive scientific frontier that both medicine and health services, and biotechnology techniques now depend upon. Modern nanotechnology produces novel methods for controlling microbiome analysis while utilizing this biological system both for treatment and diagnostics. Nanocarriers, including chitosan and PLGA or liposomes, function to encapsulate probiotics for targeted drug delivery using a protective mechanism that stops gastric environment degradation while releasing the probiotics in controlled-release patterns (Zhang et al. 2021; Dhiman et al. 2015c). Prebiotic nanoparticles enhance beneficial microbial growth through specific activation, but antimicrobial carriers use antibiotics or peptides to kill pathogenic bacteria without disturbing the gastrointestinal microbiome balance. Diagnostic techniques have improved through the development of a nano biosensor comprising gold nanoparticles and quantum dots for sensitive detection of the vital gut health indicators short-chain fatty acids (SCFAs) and lipopolysaccharides (LPS) (Giménez and Manucha 2023; Fuentes-Chust et al. 2021). The use of magnetic nanoparticles allows fast metagenomic analysis and disease detection when scientists can extract genomic material and biomarkers from sources. Nanotechnology uses its capabilities to actively reshape the composition and activity of gut microbiota with the goal of achieving balanced microbiota in diseases linked to dysbiosis. Engineered nanoparticles show application in detrimental microbe inhibition, along with targeted microbial metabolism modification, while functionalized nanomaterials with ligands or antibodies treat particular microbial communities to boost gut health. Nanorobots and related emerging technologies present choices for exact medical treatments through gastrointestinal bacterium extraction while managing pH. The management of inflammatory bowel disease undergoes investigation using nanotechnology-based formulations which deliver anti-inflammatory agents or probiotics through specifically targeted gut areas (Liu et al. 2024b; Rosales et al. 2022).

Gut microbiota modulation through nanotechnology treatment improves immunotherapy performance in cancer therapy while minimizing chemotherapy-caused side effects. Scientists investigate nanocarriers with microbiota-modulation properties to manage conditions regarding both metabolism and obesity and metabolic disorders. Namely, several distinct applications show the extensive impact of nanotechnology combined with gut microbiota (Jiang et al. 2017). Nanoparticles work as delivery methods for engineered bacteria colonization and as encapsulation platforms to boost metabolic potential through microbial enzymes and for developing nanogel structures to serve as microbial gut growth platforms. Expert scientists have explored multiple nanotech applications which use small-scale surface coatings to make postbiotic treatments more effective, as well as nanocarrier platforms to transport gut-derived immune-regulating substances and hybrid nano-components that protect both host cells and modify microbial populations (Wei et al. 2024). Nanoparticle technologies combine functional peptides with targeted delivery mechanisms to specifically address gut bacteria receptors, and researchers deploy nanoemulsion systems to transport lipophilic compounds that affect microbial ecological diversity. Nanotechnology-based systems contribute to bacteriophage delivery, helping doctors target pathogenic bacteria in the gut while bio-responsive nanoparticles release therapeutic components when microbial enzymes activate them (Vijayaram et al. 2024). Personalized and precision medicine represent a promising area of intersection between research on gut microbiota and nanotechnology applications. Current research focuses on developing personalized nanoparticle forms that match individual gut microbiome profiles in order to maximize treatment success. Research engages with nanotechnology delivery systems through individualized nutrition to optimize microbiota-regulating substances in addition to strengthening their stability and accessibility (Sofia and Cordeiro 2022).

Various obstacles continue to exist even with current advancements. Scientists need to invest more research into understanding how nanomaterials affect the gut microbiota as well as host physiology and their overall safety characteristics. Determining standardized nanoparticle formulations that scale up easily, alongside creating reproducible procedures, represents a multifaceted problem, while existing regulatory standards for nanomedicine-biomedical approaches targeting the microbiota require comprehensive preclinical testing and clinical study assessments before approval (Ye et al. 2021; Bondarenko et al. 2021). Research teams work to develop environmentally friendly nanomaterials for safety purposes, alongside developing CRISPR-Cas systems to edit the microbiome through nanotechnology platforms, while establishing AI-based analysis platforms to enhance nanotechnology applications in microbiome management. Interdisciplinary cooperation between fields will transform both gut disease management approaches and modify comprehensive knowledge of microbiome operations and nanotechnology applications, thus creating a new era of biological systems advancement (Fei et al. 2022; Qi et al. 2025; Bouwmeester et al. 2018).

9 Bacterial Nanotechnology in Cancer Therapy

Bacterial nanotechnology creates a novel juncture of microbiology and nanoscience, and oncology develops advanced therapeutic strategies for specific cancer treatment (Table 3). Researchers employ bacterial tumor-targeting qualities by using bacterial strains including *Escherichia coli*, *Salmonella typhimurium*, *Clostridium novyi*, *Bifidobacterium breve*, *Listeria monocytogenes*, *Pseudomonas aeruginosa*, *Streptococcus pyogenes*, *Mycobacterium bovis* (BCG strain), *Yersinia enterocolitica*, and *Lactobacillus casei* (Gharatape et al. 2016; Dunn et al. 2011). Bacteria work in both natural and genetically modified forms for therapy enhancement. One essential feature of this technology allows it to utilize the hypoxic tumor microenvironment, which remains resistant to standard therapeutic methods. *Clostridium novyi* belongs to the obligate anaerobic group of bacteria that survives well without oxygen in the tumor cores where it targets only cancer cells without harming surrounding healthy tissue. Obligate anaerobes function perfectly as drug transporters because they can carry therapeutic nanoparticles along with plasmids containing anticancer agents directly to tumor locations. Scientists are currently exploring nanomaterial-functionalized bacteria as a new pathway within the field of study. The combination of gold nanoparticles and superparamagnetic iron oxide nanoparticles (SPIONs) with *Salmonella typhimurium* achieves improved cancer tumor targeting together with photothermal abilities that generate localized heat energy for cancer cell destruction (Kunjalwar et al. 2024; Zhang et al. 2023). By acquiring tumor-targeted plasmids *Listeria monocytogenes* has been engineered to bring about immune checkpoint blockade which improves tumor-targeted immune response. Hybrid systems unite bacterial tumor-homing nature with nanomaterial therapeutic capabilities to provide multi-therapeutic approaches. The combination of photodynamic therapy with light-activated nanoparticles to produce reactive oxygen species and magnetically targeted drug delivery platforms that enhance precision represents two examples of such techniques. Tumors can be influenced by bacterial treatments which help control immune activity at the cellular level. The biological agent *Mycobacterium bovis* (BCG) functions in bladder cancer management by stimulating dendritic cells together with T-cells to produce significant immune reactivity (Lu et al. 2024; Chen et al. 2020).

The fields of synthetic biology have enabled scientists to modify bacteria such as *Escherichia coli* and *Pseudomonas aeruginosa* so they generate immune-modulatory molecules like cytokines together with tumor necrosis factor-alpha (TNF-α) specifically at tumor sites. Research indicates that the probiotics *Bifidobacterium breve* and *Lactobacillus casei* work with the immune systems of the gut microbiota to regulate their function and thus indirectly affect tumor development. Due to their capability to specifically infiltrate oxygen-deprived tumor regions as well as their natural replication ability, bacteria demonstrate advantages over standard chemotherapeutic medicines. The field of employing bacteria for cancer therapy faces multiple obstacles (Erdal et al. 2012). The main obstacles to bacterial cancer therapy differ between sepsis and uncontrolled bacterial multiplication and immune

Table 3 Bacterial nanotechnology in cancer therapy

Category	Description	Bacterial image
Targeting the tumor microenvironment	Exploits tumor hypoxia and unique metabolic features for selective bacterial colonization.	*Clostridium novyi*, *salmonella typhimurium*, *Escherichia coli (E. coli)*, and *Bifidobacterium breve*.
Drug delivery systems	Bacteria engineered to produce or deliver therapeutic agents directly at tumor sites.	Genetically modified *E. coli* secreting cytotoxic proteins, *salmonella typhimurium* engineered to produce prodrug-activating enzymes like cytosine deaminase.
Outer membrane vesicles (OMVs)	Nano-sized vesicles secreted by bacteria, functionalized to carry drugs, RNA, or immunostimulatory molecules to tumors.	OMVs from *Escherichia coli* or *Pseudomonas aeruginosa* loaded with chemotherapeutics or siRNA.
Bacterial biosensors	Engineered bacteria detect tumor-specific markers or metabolites and respond by producing therapeutic molecules or signaling outputs.	*E. coli* or *salmonella* engineered to sense hypoxia and release therapeutic proteins.
Nanoparticle-bacteria hybrids	Bacteria combined with nanoparticles for imaging or therapy.	Gold nanoparticle-functionalized *E. coli* or *salmonella* for fluorescence imaging, magnetotactic *Magnetospirillum gryphiswaldense* for targeted delivery.
Magnetotactic bacteria	Bacteria naturally produce magnetic nanoparticles, enabling external control for targeted therapy.	*Magnetospirillum magneticum*, *Magnetospirillum gryphiswaldense*.
Immunotherapy enhancements	Bacterial components act as immune adjuvants or deliver immune checkpoint inhibitors, stimulating immune responses against cancer cells.	*Listeria monocytogenes* delivers tumor-associated antigens, bacterial lipopolysaccharides (LPS), or peptidoglycans for immune stimulation.
Photothermal/ photodynamic therapy	Bacteria conjugated with light-absorbing nanoparticles deliver these to tumors, generating heat or ROS upon light activation to destroy cancer cells.	*E. coli* conjugated with gold nanoshells or carbon dots for photothermal therapy.
Gene therapy	Bacterial systems enable the delivery of genetic editing tools like CRISPR/Cas9 or siRNA to modify cancer-associated genes.	*E. coli* engineered to deliver CRISPR/Cas9 systems targeting oncogenes.
Combination therapies	Synergistic approaches integrating bacterial therapy with existing modalities like chemotherapy, radiotherapy, or immunotherapy.	*Salmonella typhimurium* secreting enzymes to activate chemotherapeutic prodrugs in combination with checkpoint inhibitors or radiotherapy.

(continued)

Table 3 (continued)

Category	Description	Bacterial image
Bacteria-derived nanomaterials	Biomaterials extracted from bacteria, such as lipopolysaccharides or proteins, serve as biocompatible nanocarriers for drug delivery or immunomodulation.	Lipopolysaccharides from *Pseudomonas aeruginosa*, bacterial proteins like flagellin from *Salmonella typhimurium*.

system clearance. Advancements in synthetic biology, specifically through CRISPR-Cas9-mediated precision engineering, have led to the development of safer pathogen-free bacterial strains. Disease specialists have developed *Clostridium novyi-NT* as a non-harmful strain that enables tumor-specific targeting without affecting overall body health. The safety, along with proper functional capability of these therapies in clinical practice, depends on comprehensive regulatory approval and preclinical validation procedures (Liang et al. 2022; Allemailem 2021). Modern discoveries are leading the medical sector toward practical clinical applications. The hybridization of bacterial cells with SPIONs leads to improved tumor tracking efficiency when an external magnetic field is applied. Researchers are developing genetically modified Salmonella typhimurium strains which can produce prodrug-converting enzymes, including cytosine deaminase, for local chemotherapy applications. An antitumor systemic immune response emerges through the delivery of tumor-specific antigens by *Listeria monocytogenes* with immune checkpoint inhibitors. Experimental findings indicate bacterial-based treatments work effectively, while multiple clinical trials involving bacterial treatments are currently in early phases (Ashu et al. 2019). Due to their integration in oncology, bacterial nanotechnological systems deliver three major benefits, which include high tumor targeting performance coupled with limited systemic toxicity while maintaining continuous drug delivery (Azzawi et al. 2016).

Bacterial nanotechnology presents itself as a promising new therapeutic approach because of its advantageous features, which specifically benefit therapy-resistant tumor patients. The successful implementation of bacterial nanotechnology faces three main obstacles because it requires uniform bacterial function and a reduction in unwanted side effects, and inadequate immune response detection. Improvements in bioengineering, along with nanotechnology and immune regulation methods, will be needed to handle these challenges (Shabib Akhtar et al. 2024; Song et al. 2019). The future of cancer therapy appears transformative due to the tremendous potential of bacterial nanotechnology despite existing technical challenges. Bacterial nanotechnology shows great promise for precision oncology as researchers develop it into an essential aspect of modern medical care for difficult cancers (Basu et al. 2022).

10 Nanotechnology in Colorectal Cancer Treatment

Motorized technology has developed into an evolutionary therapeutic method for colorectal cancer (CRC) treatments by delivering precise medicine through improved drug action rates and toxic substance reduction with site-specific delivery capabilities (Fig. 2). Nanoparticles like dendrimers, polymeric nanoparticles, and liposomes have changed how therapy drugs are delivered through their role as drug delivery systems (Namasivayam et al. 2021). The FDA-approved treatment Onivyde (liposomal irinotecan) utilizes lipid vesicles to encase irinotecan so it releases drugs in a controlled manner and enhances the pharmacokinetic profile while decreasing harmful effects on non-target cells (Table 4). The delivery of oncogene inhibitors and tumor suppressor function restorers through nanoparticles carrying small interfering RNA (siRNA) or microRNA (miRNA) shows potential for gene therapy applications. The delivery of KRAS mutation-specific siRNA molecules through gold nanoparticles enables excellent cellular uptake while safeguarding genetic content from nucleases and specifically prevents tumor expansion (Fortina et al. 2007; Freitas et al. 2023). PTT and PDT illustrate prominent examples of cutting-edge strategies that result from nanotechnology applications. Minimal invasiveness can be achieved with localized hyperthermia using photothermal conversion under near-infrared (NIR) light of gold nanorods and carbon-based nanoparticles. PDT sets the basis for using porphyrin-functionalized systems, which deliver reactive oxygen species (ROS) after light activation, leading to apoptosis in drug-resistant

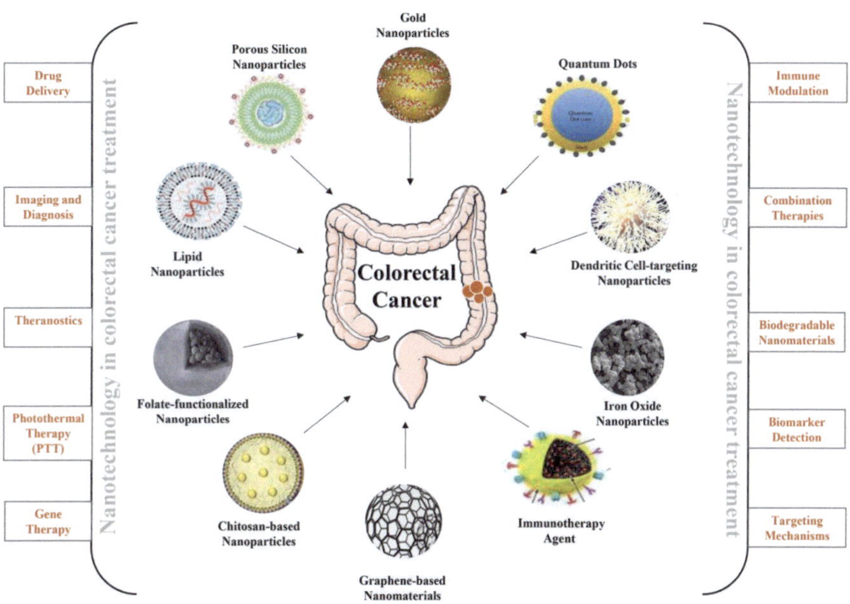

Fig. 2 Nanotechnology in colorectal cancer treatment

Table 4 Nanotechnology in colorectal cancer treatment

Techniques	Description	Applications
Drug delivery	Utilization of nanoparticles for targeted drug delivery to colorectal tumors	Liposomes, polymeric nanoparticles, porous silicon nanoparticles (pSiNPs)
Imaging and diagnosis	Nanoparticles for improved imaging and early detection of colorectal cancer	Quantum dots, gold nanoparticles, magnetic nanoparticles
Theranostics	Combined diagnostic and therapeutic functionalities in a single nanoparticle platform	Iron oxide nanoparticles for MRI-guided hyperthermia therapy
Photothermal therapy (PTT)	Use of nanoparticles to convert light energy into heat, destroying cancer cells	Gold nanoshells, graphene-based nanomaterials
Gene therapy	Delivery of genetic material to colorectal cancer cells using nanocarriers	Lipid nanoparticles for siRNA delivery, CRISPR/Cas9-loaded nanoparticles
Immune modulation	Nanotechnology-based platforms to enhance anti-cancer immune response	Nano vaccines, dendritic cell-targeting nanoparticles
Combination therapies	Integration of multiple treatment modalities using nanoparticle systems	pSiNPs co-loaded with chemotherapy and immunotherapy agents
Biomarker detection	Use of nanoparticles to detect specific biomarkers for colorectal cancer	Gold nanoparticles for carcinoembryonic antigen (CEA) detection
Targeting mechanisms	Surface functionalization of nanoparticles with ligands or antibodies to target colorectal tumors	Folate-functionalized pSiNPs, RGD peptide-conjugated nanoparticles
Biodegradable nanomaterials	Development of environmentally friendly and safe nanocarriers for colorectal cancer treatment	Poly (lactic-co-glycolic acid) (PLGA) nanoparticles, chitosan-based nanoparticles

cancer cells through oxidative stress (Choukaife et al. 2022; Brar et al. 2021). Immunotherapy benefits from nanotechnology through an enhanced ability to change the tumor environment, along with improved activation of anti-cancer immune responses. The delivery of mRNA vaccines encapsulated in lipid nanoparticles, which contain tumor-associated antigens including carcinoembryonic antigen (CEA), activates dendritic cells to boost T-cell activity against tumors (Kasi et al. 2023). Nanoparticles that combine immune checkpoint inhibitors lead to enhanced immune response that addresses the immune evasion techniques frequently observed in CRC cases. Multifunctional nanoparticles now enable combination therapy because they deliver therapeutic agents together as 5-fluorouracil (5-FU) and oxaliplatin under a single platform to combat chemoresistance (Cisterna et al. 2016).

PLGA (poly (lactic-co-glycolic acid)) nanoparticles represent this method through synergistic therapeutic mechanisms that reduce the adverse effects traditionally associated with single medications. These nanoscale platforms help deliver diagnostic information jointly with therapeutic treatment functions for continuous assessment of therapeutic outcomes. Laboratory detection and imaging tools for CRC have evolved through nanotechnology by implementing nano biosensors and imaging agents for early-stage recognition and visual presentation (Alrushaid et al. 2023; Santosh et al. 2021). Magnetic nanoparticles improve magnetic resonance imaging (MRI) contrast, which enables quantum dots to create tumor maps for colonoscopy through their highly specific fluorescent properties. Targeted delivery systems reach their peak efficiency through the usage of ligand-functionalized nanoparticles like folic acid-functionalized silica nanoparticles that make use of folate receptors, which are overexpressed on CRC cells. The process provides efficient cell uptake together with precise binding as well as improved treatment effectiveness, which decreases systemic toxicities (Viswanath et al. 2016). The introduction of hafnium oxide nanoparticles (NBTXR3) serves as radiosensitizers by enhancing cancer cell DNA damage while protecting healthy tissue, so radiation doses for treatment can be decreased. Such innovations leverage the specific mechanisms by which nanoparticles utilize the enhanced permeability and retention (EPR) effect to accumulate in tumor blood vessels and surface-active targeting ligands that target biomarkers like epithelial growth factor receptors (EGFR) or integrins (Deng et al. 2022).

The designed nanoscale structure improves system stability while providing an extended blood circulation time and controlled drug delivery, which enhances pharmacological properties. Several noteworthy nanotechnology examples demonstrate the benefits and versatility of this approach through the combination chemotherapy approach using PLGA nanoparticles and liposomal irinotecan, while PTT receives help from gold nanorods and PDT from porphyrin nanoparticles with hafnium oxide acting as a radiosensitizing agent (Tiwari et al. 2020; Arshad et al. 2020). The treatment outcomes and side-effect reduction capabilities of these advancements lead to CRC patients getting an improved quality of life. Nanotechnology's future development will enable refined treatment strategies and precision oncology improvements, which will boost CRC patient survival outcomes and diagnosis results (Gogoi et al. 2022; Ren et al. 2023).

11 Major Obstacles in Microbiome-Nanotechnology

The combination of microbiome complexity with the precision of nanotechnology through microbiome-nanotechnology represents a developing interdisciplinary field which shows great promise for medical use and agricultural and environmental applications. Achieving the entire potential of this field encounters numerous hurdles throughout scientific, technical, ethical, and regulatory domains. The main difficulty originates from the intricate way in which microorganisms respond to nanoparticles (Tiwari and Kumar 2025). Nanoparticles interact with diverse microbial communities that make up microbiomes when influenced by several environmental factors, such as pH as well as ionic strength, and organic material concentration. Multiple factors, such as ecosystem heterogeneity and environmental variations, create significant barriers for assessing how nanoparticles impact microbial systems and their properties. Microbial gene expression, together with metabolic functions and inter-species relationships, experience modification from nanoparticles, though scientists still lack knowledge about lasting effects (Addissouky et al. 2024). Safety and biocompatibility problems worsen these complications because antimicrobial nanoparticles such as silver and zinc oxide create broad antifoulant activities, which damage helpful microbes while breaking essential microorganism functions and upsetting ecological equilibrium. These negative effects disrupt the process of nutrient cycling alongside pathogen resistance and immune modulation mechanisms, while nanoparticles that accumulate in the environment result in risks to microbial populations of soil and aquatic environments. Careful evaluation must be done to study nanoparticle compatibility with living systems and their impact on environmental sustainability because such secondary effects exist (Vikesland 2017; Mohammed et al. 2024). Microbiome-specific application design of nanoparticles introduces extra technical challenges. The process of specifying microbial strain targeting demands identifying special biomarkers because microbial species display significant variation between themselves (Mohanta et al. 2022).

The stability and operational capability of nanoparticles within intricate biological settings matter due to how aggregation and degradation will degrade their effectiveness. Large-scale production of nanoparticles poses technical and economic challenges for maintaining uniformity in nanoparticle properties because standardization processes are difficult to implement on a large scale. Significant hurdles arise from ethical regulations along with other obstacles. The present risk evaluation structures for microbiome-nanotechnology show inadequacy, which leads to uncertainty when determining the safety profile and effectiveness of contemporary interventional methods (Qadri et al. 2020; Sosnowski et al. 2020). Regulatory agencies experience difficulties since they lack standardized procedures to evaluate nanoparticle-based microbiome modifications, which creates complications in product approvals and slows down innovation. The ethical consequences become most significant for human health applications that use genetic engineering and synthetic biology approaches for microbiome manipulation since they raise uncertainties about modifying the natural microbial communities linked to human bodies (Nawaz et al. 2024). The need exists for strong regulatory systems along with ethical watch programs to protect safety while technological advancement continues. The current technical limitations prevent scientific advancement since metagenomics with metabolomics and transcriptomics fail to properly identify nanoparticle–microbiome interactions. Advanced real-time monitoring methods must develop quickly because they help scientists fully understand how nanoparticles influence microbial communities in dynamic ways (Stater et al. 2021).

Computational models that forecast nanoparticle interactions remain in their developmental stage, which restricts the ability to create foreseeable and efficient nanoparticle-based solutions. The interaction patterns between host biological systems and microbiomes, together with nanoparticles, represent an advanced level of complexity (Fig. 3). Multiple host-specific factors that control immune responses and microbiome make-up have major effects on the way nanoparticles behave within human bodies (Takallu et al. 2024). The intended pathogenic microbe targeting function of nanoparticles might activate host immune responses, which subsequently change microbiome operations and distribution patterns. The biodegradability, along with the biocompatibility of nanoparticles, must be ensured for both reducing harmful residue accumulation and minimizing adverse effects. Safe nanomaterials based on biodegradable substances that leave harmless end-products during degradation offer effective solutions to safety concerns (Sumida et al. 2021; Chen et al. 2025). Bioinspired nanomaterials served with ligand-functionalized nanoparticles and machine learning models used for predicting nanoparticle–microbiome interactions provide potential solutions to the existing obstacles. The field needs unified research between microbiology teams and specialists from nanotechnology, as well as toxicology and computational biology disciplines, to resolve present challenges and speed up advancements (Sadr et al. 2023). Research into microbiome complexity combined with nanoparticle safety guarantee and functional performance, along with the elimination of technical and legislative and ethical barriers, enables microbiome-nanotechnology to create a transformative influence on personalized medicine and sustainable agricultural systems and environmental treatment methods, which reshapes our microbial ecosystem management in society (Domingues et al. 2023).

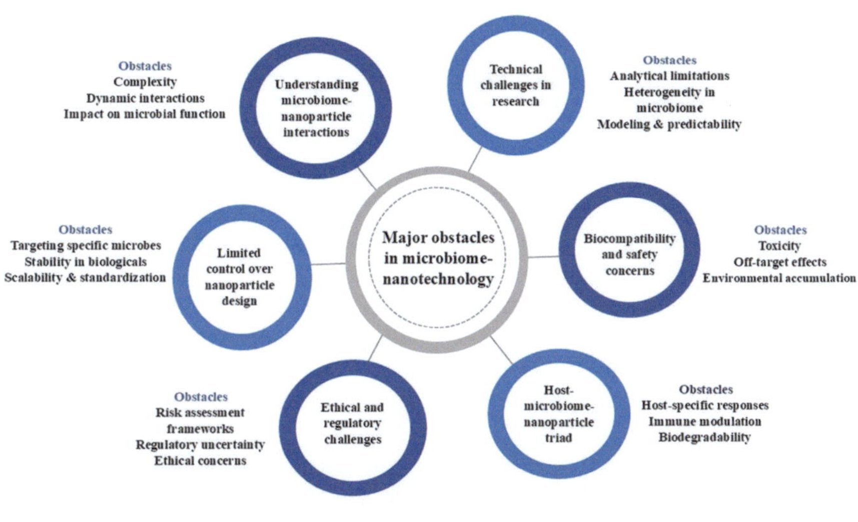

Fig. 3 Major obstacles in microbiome–nanotechnology

12 Clinical Trials and Case Studies

The gut microbiome research regarding brain function has intensified rapidly because scientists now understand its potential part in neurological conditions like Alzheimer's disease, as well as autism, multiple sclerosis, Parkinson's, and stroke. Although clinical research reveals potential relations between disease pathophysiology and microbiota changes, it becomes difficult to interpret data because of food consumption and physical activities. The research field is currently at an early stage because it needs long-term analysis and controlled experiments to validate associations with therapy possibilities (Cryan et al. 2020).

"Omic" technology advancements during the last two decades have improved our knowledge of gut microbiota while showing its major impact on human health. The microbiome functions as an organ system that keeps close contact with the host body to support its wellness, yet plays a role in some medical conditions. Researchers have conducted a review that examines how gut disorders affect metabolic syndrome and liver disease and inflammatory bowel disease, and colorectal cancer, as well as proposes strategies including antibiotics and prebiotics, probiotics, polyphenols, and fecal microbiota transplantation (Marchesi et al. 2016).

A research study establishes that type 1 diabetes results in major modifications of gut microbiota structure. Children diagnosed with diabetes displayed a lower Firmicutes to Bacteroidetes ratio, together with reduced *Bifidobacterium* and *Lactobacillus* levels, while showing increased numbers of potentially dangerous bacteria such as *Clostridium*. The microbial shifts that occurred during this study were associated with glycemic measurements, which could indicate their contribution to disease development. The reduced availability of gut-integrity supporting bacteria shows that microbiome changes might relate to diabetes development. Gut microbiota modulation can become a potential treatment option for type 1 diabetes because of the research discoveries made between diabetes and microbial changes (Murri et al. 2013).

Another research evidence demonstrates that the SIFR® (Systemic Intestinal Fermentation Research) technology demonstrates predictive potential for monitoring gut microbiota changes. The ex vivo prebiotic simulation model of the SIFR® technology demonstrated fast clinical forecasting of microbial changes and metabolic alterations before actual patient results. The technological process displayed unique microbial reactions towards inulin, resistant dextrin, and 2′-fucosyllactose, thus showing it improves existing laboratory outcomes. This method works as a fundamental link between preclinical and clinical research because it repeatedly demonstrates results which correspond directly to live human gut microorganisms. Medical treatments aimed at the microbiome will gain effectiveness through this approach, while clinical trials achieve better success rates (Van den Abbeele et al. 2023).

The systematic review detected major changes in gut microbiome structure between MDD patients and healthy participants, alongside modifications in particular bacterial organism levels. The researchers found no changes to alpha diversity, yet they detected alterations in microbial composition patterns. The use of probiotic and synbiotic treatments demonstrated a small and positive effect on decreasing depressive symptoms within a study period of 4–9 weeks, but prebiotic treatments showed no notable results. The research demonstrates that the gut microbiome has both disease modeling and therapeutic potential in treating MDD, but additional studies must evaluate microbiome therapies relative to antidepressants and assess durability for clinical outcomes (Alli et al. 2022).

The researchers analyzed bacterial modifications inside obese pregnant patients together with their link to gestational diabetes mellitus (GDM). Regulation of the gut microbiota among women who did not develop GDM produced marked transformations during pregnancy, but the women who developed GDM displayed minimal changes. The research did not detect particular bacterial strains that caused GDM development, yet *Ruminococcus obeum* existed in increased numbers during late pregnancy among patients with GDM who had the supplement combination. The study findings indicate that gestational diabetes mellitus successfully restricts both the range of possible gut microbial adaptations and the dietary adjustments they can make (Mokkala et al. 2021).

The phase Ib/II study demonstrates that the combination of regorafenib with toripalimab for colorectal cancer treatment achieves an ORR of 15.2% and a disease control rate of 36.4% at the recommended phase II dosage. The PFS rate for this treatment combination is 2.1 months, while overall survival reaches 15.5 months. Survey results show that liver metastases patients have a reduced response rate in comparison to other patients, and every patient with lung metastases experiences a favorable outcome. People with high levels of *Fusobacterium* in their initial gut microbiota patterns experience poorer treatment results, and their disease progresses within 2.1 months (Wang et al. 2021b).

Research examines the relation between gastrointestinal bacteria and necrotizing enterocolitis development in very low birthweight infants. The research team observed major bacterial variations between NEC cases and controls through the evaluation of 3586 stool samples obtained from 166 infants who participated across various study cohorts beginning at age 1 month. The relationships found here were most noticeable among infants who were born prematurely before week 27. Research shows that the composition of gut microbiota acts as a risk factor for NEC, which enables the identification of potential preventive targets for very young preterm babies (Martín Giménez et al. 2023).

The research demonstrates that children with early-onset multiple sclerosis (MS) exhibit major differences in their gut microbiomes compared to regular controls. A difference in bacterial beta diversity did not relate to MS, but therapy exposure affected these results. In Multiple Sclerosis (MS), studies show a distinct gut microbial shift with increased Desulfovibrionaceae, Bilophila, and Desulfovibrio genera, along with elevated Christensenellaceae, while Lachnospiraceae and Ruminococcaceae are reduced. The results of functional analysis demonstrated that the glutathione metabolism pathways were enriched, which is linked to neurodegeneration and inflammation. Results indicate that pediatric MS pathogenesis may be influenced by gut microbiota because they seem to create settings that promote inflammation (George et al. 2022).

The expanding study of gut microbiota in human health requires improved intervention methods to create favorable microbiome changes. The combination of innovative technology with measurements and microbiota–host interaction understanding has improved despite the challenges posed by varying research methods and parameters across study populations. A review presents systemized instructions about studying and analyzing gut microbiota interventions with focus on designing experimental plans and competence evaluations as well as statistical requirements and methodological assessment approaches (Swann et al. 2020).

Research results reveal alterations in the gut microbiome structure as observed between CKD5 patients under dialysis or without dialysis and normal subjects. The patient population showed substantial microbial variations because *Neisseria*, *Lachnoclostridium*, and *Bifidobacterium* increased, while *Faecalibacterium* decreased in abundance. The amounts of uremic toxins indoxyl sulfate (IS) and p-cresyl sulfate (PCS) increased as the microbial changes occurred. The results indicate gut microbiota may play a part in CKD advancement while establishing an urgent need to investigate probiotic therapy options for managing CKD-related problems (Li et al. 2019).

The authors demonstrate that gut microbiome variation shows substantial temporal changes because a person's bacterial composition differs more across time than between different individuals. Research analysis of the microbiome throughout 6 weeks in 20 women revealed major changes in microbial genera, which caused abundance variations reaching up to 100-fold. Stool moisture functions as the main factor that leads to microbial abundance changes with diet acting as a secondary source, according to their findings. Research findings indicate that both multiple sample collection across time and wide community-based microbiome descriptions might improve microbiome research reliability through better diagnostic and therapeutic outcomes (Vandeputte et al. 2021).

13 Conclusion

Nanotechnology and microbiome biomarkers work together to improve therapeutic delivery and drug-associated adverse effects, advancing precision medicine. Research teams that are able to use insights acquired from microbiota can develop personalized medications that improve therapeutic outcomes and drug processing. Through their roles as targeted transportation systems and immunoregulatory components, the use of microbiota-guided nanocarriers offers promising applications in the treatment of infectious diseases, cancer, and metabolic conditions. Probiotics are delivered to the gastrointestinal tract with the use of nanoparticles, improving therapy outcomes. Although microbiome research has advanced, widespread adoption is hampered by the lack of appropriate regulatory requirements, advanced formulation strategies, and reliable microbiome measuring methodologies. Research needs to enhance predictive microbiota technology while creating safer nanomaterials and establishing better connections between research tests and clinical studies. The elimination of these barriers through better integration of microbiome technology with nanotechnology will create novel specifications that enable individualized treatments from personalized medicine approaches.

Conflict of Interest The authors declare no conflict of interest.

References

Abed OA, Attlassy Y, Xu J, Han K, Moon JJ (2022) Emerging nanotechnologies and microbiome engineering for the treatment of inflammatory bowel disease. Mol Pharm 19(12):4393–4410. https://doi.org/10.1021/acs.molpharmaceut.2c00222

Addissouky TA, El Sayed IET, Ali MMA, Alubiady MHS, Wang Y (2024) Harnessing innovation for the future of breast cancer management. Clin Res Oncol 1(1):10–17. https://doi.org/10.46439/oncology.1.004

Adir O et al (2022) Synthetic cells with self-activating optogenetic proteins communicate with natural cells. Nat Commun 13(1). https://doi.org/10.1038/s41467-022-29871-8

Allemailem KS (2021) Innovative approaches of engineering tumor-targeting bacteria with different therapeutic payloads to fight cancer: a smart strategy of disease management. Int J Nanomedicine 16:8159–8184. https://doi.org/10.2147/IJN.S338272

Alli SR, Gorbovskaya I, Liu JCW, Kolla NJ, Brown L, Müller DJ (2022) The gut microbiome in depression and potential benefit of prebiotics, probiotics and Synbiotics: a systematic review of clinical trials and observational studies. Int J Mol Sci 23(9). https://doi.org/10.3390/ijms23094494

Alexander J, Wilson I, Teare J et al (2017) Gut microbiota modulation of chemotherapy efficacy and toxicity. Nat Rev Gastroenterol Hepatol 14:356–365. https://doi.org/10.1038/nrgastro.2017.20

Alrushaid N, Khan FA, Al-Suhaimi EA, Elaissari A (2023) Nanotechnology in cancer diagnosis and treatment. Pharmaceutics 15(3). https://doi.org/10.3390/pharmaceutics15031025

Arshad U et al (2020) Critical considerations for targeting colorectal liver metastases with nanotechnology. Wiley Interdiscip Rev Nanomed Nanobiotechnol 12(2):1–23. https://doi.org/10.1002/wnan.1588

Ashu EE, Xu J, Yuan ZC (2019) Bacteria in cancer therapeutics: a framework for effective therapeutic bacterial screening and identification. J Cancer 10(8):1781–1793. https://doi.org/10.7150/jca.31699

Ayan E, DeMirci H, Serdar MA, Palermo F, Baykal AT (2023) Bridging the gap between gut microbiota and Alzheimer's disease: a metaproteomic approach for biomarker discovery in transgenic mice. Int J Mol Sci 24(16). https://doi.org/10.3390/ijms241612819

Azzawi M, Seifalian A, Ahmed W (2016) Nanotechnology for the diagnosis and treatment of diseases. Nanomedicine 11(16):2025–2027. https://doi.org/10.2217/nnm-2016-8000

Bairamian D et al (2022) Microbiota in neuroinflammation and synaptic dysfunction: a focus on Alzheimer's disease. Mol Neurodegener 17(1):1–23. https://doi.org/10.1186/s13024-022-00522-2

Basu A, Singh R, Gupta S (2022) Bacterial infections in cancer: a bilateral relationship. Wiley Interdiscip Rev Nanomed Nanobiotechnol 14(3):1–29. https://doi.org/10.1002/wnan.1771

Berding K et al (2021) Diet and the microbiota-gut-brain Axis: sowing the seeds of good mental health. Adv Nutr 12(4):1239–1285. https://doi.org/10.1093/advances/nmaa181

Blanco-Fernandez B, Castaño O, Mateos-Timoneda MÁ, Engel E, Pérez-Amodio S (2021) Nanotechnology approaches in chronic wound healing. Adv Wound Care 10(5):234–256. https://doi.org/10.1089/wound.2019.1094

Boisseau N, Barnich N, Koechlin-Ramonatxo C (2022) The nutrition-microbiota-physical activity triad: an inspiring new concept for health and sports performance. Nutrients 14(5):1–25. https://doi.org/10.3390/nu14050924

Bondarenko O et al (2021) Nanotoxicology and nanomedicine: the Yin and Yang of nano-bio interactions for the new decade. Nano Today 39. https://doi.org/10.1016/j.nantod.2021.101184

Bouwmeester H, van der Zande M, Jepson MA (2018) Effects of food-borne nanomaterials on gastrointestinal tissues and microbiota. Wiley Interdiscip Rev Nanomed Nanobiotechnol 10(1):1–12. https://doi.org/10.1002/wnan.1481

Brar B et al (2021) Nanotechnology in colorectal cancer for precision diagnosis and therapy. Front Nanotechnol 3(September):1–21. https://doi.org/10.3389/fnano.2021.699266

Chen Y, Du M, Yu J, Rao L, Chen X, Chen Z (2020) Nanobiohybrids: a synergistic integration of bacteria and nanomaterials in cancer therapy. BIO Integr 1(1):25–36. https://doi.org/10.15212/bioi-2020-0008

Chen BY et al (2023) Roles of oral microbiota and oral-gut microbial transmission in hypertension. J Adv Res 43:147–161. https://doi.org/10.1016/j.jare.2022.03.007

Chen A et al (2025) Navigating a challenging path: precision disease treatment with tailored oral nano-armor-probiotics. J Nanobiotechnol 23(1). https://doi.org/10.1186/s12951-025-03141-3

Cheung MK et al (2023) Extent of oral–gut transmission of bacterial and fungal microbiota in healthy Chinese Adults. Microbiol Spectrum 11(1):1–11. https://doi.org/10.1128/spectrum.02814-22

Choukaife H, Seyam S, Alallam B, Doolaanea AA, Alfatama M (2022) Current advances in chitosan nanoparticles based oral drug delivery for colorectal cancer treatment. Int J Nanomed 17:3933–3966. https://doi.org/10.2147/IJN.S375229

Cisterna BA, Kamaly N, Il Choi W, Tavakkoli A, Farokhzad OC, Vilos C (2016) Targeted nanoparticles for colorectal cancer. Nanomedicine 11(18):2443–2456. https://doi.org/10.2217/nnm-2016-0194

Covasa M, Stephens RW, Toderean R, Cobuz C (2019) Intestinal sensing by gut microbiota: targeting gut peptides. Front Endocrinol 10(Feb). https://doi.org/10.3389/fendo.2019.00082

Cryan JF, O'Riordan KJ, Sandhu K, Peterson V, Dinan TG (2020) The gut microbiome in neurological disorders. Lancet Neurol 19(2):179–194. https://doi.org/10.1016/S1474-4422(19)30356-4

Dai JH, Tan XR, Qiao H, Liu N (2024) Emerging clinical relevance of microbiome in cancer: promising biomarkers and therapeutic targets. Protein Cell 15(4):239–260. https://doi.org/10.1093/procel/pwad052

Daniel N, Lécuyer E, Chassaing B (2021) Host/microbiota interactions in health and diseases—time for mucosal microbiology! Mucosal Immunol 14(5):1006–1016. https://doi.org/10.1038/s41385-021-00383-w

de Melo NCO, Cuevas-Sierra A, Fernández-Cruz E, de la VO, Martínez JA (2023) Fecal microbiota composition as a metagenomic biomarker of dietary intake. Int J Mol Sci 24(5). https://doi.org/10.3390/ijms24054918

Deng S et al (2022) Application of nanotechnology in the early diagnosis and comprehensive treatment of gastrointestinal cancer. J Nanobiotechnol 20(1):1–29. https://doi.org/10.1186/s12951-022-01613-4

Dhanker R, Hussain T, Tyagi P, Singh KJ, Kamble SS (2021) The emerging trend of bioengineering approaches for microbial nanomaterial synthesis and its applications. Front Microbiol 12(March). https://doi.org/10.3389/fmicb.2021.638003

Dhiman R, Durga V, Rio AJ (2015a) Bioanalytical method for lornoxicam determination in human plasma by using piroxicam as internal standard by LC-MS/MS. J Chem Pharm Res 7(8):54–60

Dhiman R, Durga V, Jayasankar K, Raj S (2015b) Solid phase extraction: potential carbocisteine bioanalysis using carbocisteine 13C3 stable isotope: LC-MS/MS technique with pharmacokinetics on healthy human subjects. Der Pharm Lett 7(8):126–134

Dhiman R et al (2015c) Novel LC-MS/MS bioanalytical method validation for gabapentin in human plasma in accordance with current regulatory guidances. Int J Pharm Chem Sci 4(4):462–468. [Online]. Available: http://www.ijpcsonline.com/files/08-10-15/07-1035.pdf

Dhiman R, Kumar D, Kumar B, Pandey BL (2016a) Quantitative determination of atorvastatin, ortho-hydroxy atorvastatin, para-hydroxy atorvastatin in human plasma using rosuvastatin as internal standard by LC-MS/MS. Int J Pharm Chem Sci 4(4):487–500

Dhiman R, Kumar B, Kumar D, Kumar A (2016b) Liquid-liquid extraction method for ziprasidone (ZRS) bioanalysis by using ZRS-D 8 (stable isotope) as internal standard. J Chem Pharm Res 8(3):722–727

Domingues C, Cabral C, Jarak I, Veiga F, Dourado M, Figueiras A (2023) The debate between the human microbiota and immune system in treating aerodigestive and digestive tract cancers: a review. Vaccine 11(3):1–30. https://doi.org/10.3390/vaccines11030492

Dunn AM, Hofmann OS, Waters B, Witchel E (2011) Cloaking malware with the trusted platform module. In: Proceedings of the 20th USENIX security symposium, pp 395–410

Erdal E et al (2012) Preparation and characterization of magnetically responsive bacterial polyester based nanospheres for cancer therapy. J Biomed Nanotechnol 8(5):800–808. https://doi.org/10.1166/jbn.2012.1431

Fei Y, Ma Y, Zhang H, Li H, Feng G, Fang J (2022) Nanotechnology for research and treatment of the intestine. J Nanobiotechnol 20(1):1–20. https://doi.org/10.1186/s12951-022-01517-3

Feng W, Liu J, Ao H, Yue S, Peng C (2020) Targeting gut microbiota for precision medicine: focusing on the efficacy and toxicity of drugs. Theranostics 10(24):11278–11301. https://doi.org/10.7150/thno.47289

Fleige E, Quadir MA, Haag R (2012) Stimuli-responsive polymeric nanocarriers for the controlled transport of active compounds: concepts and applications. Adv Drug Deliv Rev 64(9):866–884. https://doi.org/10.1016/j.addr.2012.01.020

Fortina P et al (2007) Applications of nanoparticles to diagnostics and therapeutics in colorectal cancer. Trends Biotechnol 25(4):145–152. https://doi.org/10.1016/j.tibtech.2007.02.005

Freitas SC et al (2023) New Frontiers in colorectal cancer treatment combining nanotechnology with photo- and radiotherapy. Cancers (Basel) 15(2). https://doi.org/10.3390/cancers15020383

Fuentes-Chust C et al (2021) The microbiome meets nanotechnology: opportunities and challenges in developing new diagnostic devices. Adv Mater 33(18). https://doi.org/10.1002/adma.202006104

Ganta S, Devalapally H, Shahiwala A, Amiji M (2008) A review of stimuli-responsive nanocarriers for drug and gene delivery. J Control Release 126(3):187–204. https://doi.org/10.1016/j.jconrel.2007.12.017

Gao J, Li J, Luo Z, Wang H, Ma Z (2024) Nanoparticle-based drug delivery systems for inflammatory bowel disease treatment. Drug Des Dev Ther 18:2921–2949. https://doi.org/10.2147/DDDT.S461977

Garg R et al (2022) COVID-19: a challenge towards the sustainability of health in platform era. NeuroQuantology 20(11):2503–2519. https://doi.org/10.47750/pnr.2022.13

Gebrayel P et al (2022) Microbiota medicine: towards clinical revolution. J Transl Med 20(1):1–20. https://doi.org/10.1186/s12967-022-03296-9

George S et al (2022) Bacterial gut microbiota and infections during early childhood. Front Microbiol 12(January):1–23. https://doi.org/10.3389/fmicb.2021.793050

Gharatape A, Davaran S, Salehi R, Hamishehkar H (2016) Engineered gold nanoparticles for photothermal cancer therapy and bacteria killing. RSC Adv 6(112):111482–111516. https://doi.org/10.1039/c6ra18760a

Giménez VMM, Manucha W (2023) Innovative approaches for intestinal dysbiosis: integrating nanotechnology into pharmacological interventions. Curr Pharm Des 29(21):1629–1631. https://doi.org/10.2174/1381612829666230803111610

Gogoi P, Kaur G, Singh NK (2022) Nanotechnology for colorectal cancer detection and treatment. World J Gastroenterol 28(46):6497–6511. https://doi.org/10.3748/wjg.v28.i46.6497

Guo P, Wu C (2017) Gut microbiota brings a novel way to illuminate mechanisms of natural products in vivo. Chin Herb Med 9(4):301–306. https://doi.org/10.1016/s1674-6384(17)60109-6

Guo X et al (2022) Gut microbiota is a potential biomarker in inflammatory bowel disease. Front Nutr 8(January):1–11. https://doi.org/10.3389/fnut.2021.818902

Hamid A et al (2024) Artificial intelligence (AI) based colonoscopy. In: Nanobiotechnology and artificial intelligence in gastrointestinal diseases. IOP Publishing, Bristol. https://doi.org/10.1088/978-0-7503-6134-7CH12

Harsh Pandey YBT, Srivastava S, Kumar R (2018) Preclinical acute and repeated dose toxicity of Pueraria tuberosa (PTWE) on Charles foster rats. Int J Pharm Sci Res 9(11):4572–4581. https://doi.org/10.13040/IJPSR.0975-8232.9(11).4572-81

Hasanzadeh A, Alamdaran M, Ahmadi S (2020) Since January 2020 Elsevier has created a COVID-19 resource centre with free information in English and Mandarin on the novel coronavirus COVID-19. The COVID-19 resource centre is hosted on Elsevier Connect, the company's public news and information, January 2020

Hou K et al (2022) Microbiota in health and diseases. Signal Transduct Target Ther 7(1). https://doi.org/10.1038/s41392-022-00974-4

Huang Y et al (2024) Nanotechnology's frontier in combatting infectious and inflammatory diseases: prevention and treatment. Signal Transduct Target Ther 9(1). https://doi.org/10.1038/s41392-024-01745-z

Ito Z, Koido S, Kato K, Odamaki T, Horiuchi S, Akasu T, Saruta M, Hata T, Kumagai Y, Fujioka S, Misawa T, Xiao J-z, Sato N, Ohkusa T (2022) Dysbiosis of the Fecal and Biliary Microbiota in Biliary Tract Cancer. Cancers, 14(21):5379. https://doi.org/10.3390/cancers14215379

Jia W, Li H, Zhao L, Nicholson JK (2008) Gut microbiota: a potential new territory for drug targeting. Nat Rev Drug Discov 7(February):123–129

Jiang HJ, Underwood TC, Bell JG, Ranjan S, Sasselov D, Whitesides GM (2017) Mimicking lighting-induced electrochemistry on the early earth. Proc Natl Acad Sci 120:2017. https://doi.org/10.1073/pnas

Kang GU et al (2022) Exploration of potential gut microbiota-derived biomarkers to predict the success of fecal microbiota transplantation in ulcerative colitis: a prospective cohort in Korea. Gut Liver 16(5):775–785. https://doi.org/10.5009/gnl210369

Karavolos M, Holban A (2016) Nanosized drug delivery systems in gastrointestinal targeting: interactions with microbiota. Pharmaceuticals 9(4):1–15. https://doi.org/10.3390/ph9040062

Kasi PB, Mallela VR, Ambrozkiewicz F, Trailin A, Liška V, Hemminki K (2023) Theranostics nanomedicine applications for colorectal cancer and metastasis: recent advances. Int J Mol Sci 24(9). https://doi.org/10.3390/ijms24097922

Ke Y, Yue J, He J, Liu G (2024) Integrating machine learning algorithms and single-cell analysis to identify gut microbiota-related macrophage biomarkers in atherosclerotic plaques. Front Cell Infect Microbiol 14(April):1–12. https://doi.org/10.3389/fcimb.2024.1395716

Kumar R, Kumar B, Kumar A (2017) Pharmacokinetic profile of antitubercular drugs and their clinical applications in therapeutic drug monitoring. Asian J Pharm 3(5):145–154. [Online]. Available: http://ajpp.in/uploaded/p71.pdf

Kumar R, Kumar B, Kumar A, Kumar A, Singh M (2021a) GC-MS analysis of Phytocomponents in the methanol extract of Premna latifolia Roxb. Pharm Res 14(1):19–23. https://doi.org/10.5530/pres.14.1.4

Kumar A et al (2021b) Antiurolithiasis, antioxidant, anti-inflammatory, analgesic, and diuretic activity of Ethanolic extract of seeds of Caesalpinia bonducella. Int J Pharm Investig 11(3):306–311. https://doi.org/10.5530/ijpi.2021.3.54

Kumar R et al (2021c) HPTLC and GC-MS analyses of biologically active extracts and fractions from *Premna latifolia* Roxb leaves reveal anti-inflammatory and analgesic effects. Syst Rev Pharm 12(10):811–818

Kumar R, Kumar B, Kumar A, Kumar A, et al. (2021d) HPTLC and GC-MS Analyses of Biologically Active Extracts and Fractions from Premna Latifolia Roxb Leaves. J Neurol Disord 9:454

Kumar A et al (2022a) Acute and subacute toxicity study of ethanolic extract of *Calotropis procera* (Aiton) Dryand flower in Swiss albino mice. Phytomed Plus 2(2):100224. https://doi.org/10.1016/j.phyplu.2022.100224

Kumar A et al (2022b) Toxicity (acute and subacute) assessment and in-vivo antiurolithiatic activity of ethanolic extract of *Caesalpinia bonducella* seed in albino Wistar rat. J Appl Pharm Sci 12(2):187–197. https://doi.org/10.7324/JAPS.2022.120220

Kumar A et al (2022c) Assessment of pain-ameliorative potential of ethanolic extract of *Calotropis procera* and its active components characterization by HPTLC and GC-MS analysis. Syst Rev Pharm 13(09):581–586. https://doi.org/10.31858/0975-8453.13.6.659-664

Kumar R, Dutt S, Tripathi AD, Singh AK, Chaturvedi VK, Singh SK (2024a) In: Gautam V, Kumar R, Das Manandhar K, Kamble SC (eds) Navigating safety and toxicity challenges in nanomedicine: strategies, assessment, and mitigation BT – nanomedicine: innovations, applications, and breakthroughs in the quest for health and medicine's future. Springer Nature, Cham, pp 15–37. https://doi.org/10.1007/978-3-031-72467-1_2

Kumar R, Dutt S, Goyal A, Kumar A, Kumar B (2024b) Nanotechnology in gastrointestinal endoscopy. In: Nanobiotechnology and artificial intelligence in gastrointestinal diseases. IOP Publishing, Bristol. https://doi.org/10.1088/978-0-7503-6134-7CH5

Kumar R, Singh Y, Piplani M, Bhateja P, Garg R, Kumar B (2024c) The chemistry and pharmacological study of *Copaifera* spp. oils. Chem Inside Spices Herbs Res Dev 4:79–116. https://doi.org/10.2174/9789815196832124040006

Kumar A, Singh P, Kumar R, Dutt S (2024d) Novel drug delivery systems for inflammatory bowel disease. In: Nanobiotechnology and artificial intelligence in gastrointestinal diseases. IOP Publishing, Bristol. https://doi.org/10.1088/978-0-7503-6134-7CH4

Kumari P, Ghosh B, Biswas S (2016) Nanocarriers for cancer-targeted drug delivery. J Drug Target 24(3):179–191. https://doi.org/10.3109/1061186X.2015.1051049

Kunjalwar R, Keerti A, Chaudhari A, Sahoo K, Meshram S (2024) Microbial therapeutics in oncology: a comprehensive review of bacterial role in cancer treatment. Cureus 16(10). https://doi.org/10.7759/cureus.70920

Lamas B, Martins Breyner N, Houdeau E (2020) Impacts of foodborne inorganic nanoparticles on the gut microbiota-immune axis: potential consequences for host health. Part Fibre Toxicol 17(1):1–22. https://doi.org/10.1186/s12989-020-00349-z

Landberg R, Hanhineva K (2020) Biomarkers of a healthy nordic diet—from dietary exposure biomarkers to microbiota signatures in the metabolome. Nutrients 12(1). https://doi.org/10.3390/nu12010027

Leng Q, Holden VK, Deepak J, Todd NW, Jiang F (2021) Microbiota biomarkers for lung cancer. Diagnostics 11(3):1–15. https://doi.org/10.3390/diagnostics11030407

Li Y et al (2019) Dysbiosis of the gut microbiome is associated with CKD5 and correlated with clinical indices of the disease: a case-controlled study. J Transl Med 17(1):1–13. https://doi.org/10.1186/s12967-019-1969-1

Li X, Wei H, Qi J, Ma K, Luo Y, Weng L (2023) Interactions of nanomaterials with gut microbiota and their applications in cancer therapy. Sensors 23(9). https://doi.org/10.3390/s23094428

Liang S, Wang C, Shao Y, Wang Y, Xing D, Geng Z (2022) Recent advances in bacteria-mediated cancer therapy. Front Bioeng Biotechnol 10(October):1–20. https://doi.org/10.3389/fbioe.2022.1026248

Lin TL et al (2021) Role of gut microbiota in identification of novel TCM-derived active metabolites. Protein Cell 12(5):394–410. https://doi.org/10.1007/s13238-020-00784-w

Liu T et al (2024a) A nanoparticle-based sonodynamic therapy reduces helicobacter pylori infection in mouse without disrupting gut microbiota. Nat Commun 15(1). https://doi.org/10.1038/s41467-024-45156-8

Liu J, Yuan S, Bremmer A, Hu Q (2024b) Convergence of nanotechnology and bacteriotherapy for biomedical applications. Adv Sci 11(16):1–18. https://doi.org/10.1002/advs.202309295

Lu G et al (2023) Global trends in research of pain–gut-microbiota relationship and how nutrition can modulate this link. Nutrients 15(17). https://doi.org/10.3390/nu15173704

Lu Y et al (2024) Bacteria-based nanoprobes for cancer therapy. Int J Nanomed 19:759–785. https://doi.org/10.2147/IJN.S438164

Ma J, Zhou Q, Li H (2017) Gut microbiota and nonalcoholic fatty liver disease: insights on mechanisms and therapy. Nutrients 9(10). https://doi.org/10.3390/nu9101124

Ma X et al (2024) Nanotechnology in healthcare, and its safety and environmental risks. J Nanobiotechnol 22(1). https://doi.org/10.1186/s12951-024-02901-x

Majumder J, Minko T (2021) Multifunctional and stimuli-responsive nanocarriers for targeted therapeutic delivery. Expert Opin Drug Deliv 18(2):205–227. https://doi.org/10.1080/17425247.2021.1828339

Malik S, Muhammad K, Waheed Y (2023) Nanotechnology: a revolution in modern industry. Molecules 28(2):661. https://doi.org/10.3390/MOLECULES28020661

Mamo T et al (2010) Emerging nanotechnology approaches for HIV/AIDS treatment and prevention. Nanomedicine 5(2):269–285. https://doi.org/10.2217/nnm.10.1

Marchesi JR et al (2016) The gut microbiota and host health: a new clinical frontier. Gut 65(2):330–339. https://doi.org/10.1136/gutjnl-2015-309990

Martín Giménez VM, Arya G, Zucchi IA, Galante MJ, Manucha W (2021) Photo-responsive polymeric nanocarriers for target-specific and controlled drug delivery. Soft Matter 17(38):8577–8584. https://doi.org/10.1039/d1sm00999k

Martín Giménez VM, Modrego J, Gómez-Garre D, Manucha W, de las Heras N (2023) Gut microbiota dysbiosis in COVID-19: modulation and approaches for prevention and therapy. Int J Mol Sci 24(15):1–16. https://doi.org/10.3390/ijms241512249

Martínez JE, Vargas A, Pérez-Sánchez T, Encío IJ, Cabello-Olmo M, Barajas M (2021) Human microbiota network: unveiling potential crosstalk between the different microbiota ecosystems and their role in health and disease. Nutrients 13(9). https://doi.org/10.3390/nu13092905

Melo-González F et al (2022) Distal consequences of mucosal infections in intestinal and lung inflammation. Front Immunol 13(April):1–14. https://doi.org/10.3389/fimmu.2022.877533

Mi P (2020) Stimuli-responsive nanocarriers for drug delivery, tumor imaging, therapy and theranostics. Theranostics 10(10):4557–4588. https://doi.org/10.7150/thno.38069

Mohammed S, Raheem A, Jassim AJ, Jasem AJ (2024) Advancements in nanotechnology-based drug delivery systems for Oral Healthcare: from traditional practices to personalized medicine: a review. Al-Bayan J Med Health Sci May:24–37

Mohanta YK et al (2022) Nanotechnology in combating biofilm: a smart and promising therapeutic strategy. Front Microbiol 13(March):1–30. https://doi.org/10.3389/fmicb.2022.1028086

Mokkala K et al (2021) Metagenomics analysis of gut microbiota in response to diet intervention and gestational diabetes in overweight and obese women: a randomised, double-blind, placebo-controlled clinical trial. Gut 70(2):309–318. https://doi.org/10.1136/gutjnl-2020-321643

Mosquera FEC, Guevara-Montoya MC, Serna-Ramirez V, Liscano Y (2024) Neuroinflammation and schizophrenia: new therapeutic strategies through psychobiotics, nanotechnology, and artificial intelligence (AI). J Pers Med 14(4). https://doi.org/10.3390/jpm14040391

Munir MU, Ahmed A, Usman M, Salman S (2020) Recent advances in nanotechnology-aided materials in combating microbial resistance and functioning as antibiotics substitutes. Int J Nanomedicine 15:7329–7358. https://doi.org/10.2147/IJN.S265934

Mura S, Nicolas J, Couvreur P (2013) Stimuli-responsive nanocarriers for drug delivery. Nat Mater 12(11):991–1003. https://doi.org/10.1038/nmat3776

Murri M et al (2013) Gut microbiota in children with type 1 diabetes differs from that in healthy children: a case-control study. BMC Med 11(1):1–12. https://doi.org/10.1186/1741-7015-11-46

Namasivayam SKR, Venkatachalam G, Bharani RSA, Kumar JA, Sivasubramanian S (2021) Molecular intervention of colon cancer and inflammation manifestation by tannin capped biocompatible controlled sized gold nanoparticles from *Terminalia bellirica*: a green strategy for pharmacological drug formulation based on nanotechnology principles. 3 Biotech 11(9):1–20. https://doi.org/10.1007/s13205-021-02944-z

Nasr SM et al (2020) Biodegradable nanopolymers in cardiac tissue engineering: from concept towards nanomedicine. Int J Nanomed 15:4205–4224. https://doi.org/10.2147/IJN.S245936

Nawaz T, Gu L, Fahad S, Saud S, Bleakley B, Zhou R (2024) Exploring sustainable agriculture with nitrogen-fixing cyanobacteria and nanotechnology. Molecules 29(11). https://doi.org/10.3390/molecules29112534

Okeke ES et al (2022) Antibiotic resistance in aquaculture and aquatic organisms: a review of current nanotechnology applications for sustainable management, vol 29, no. 46. Springer, Berlin Heidelberg. https://doi.org/10.1007/s11356-022-22319-y

Pandey H, Srivastava S, Kumar A, Kumar R, Bhusan Tripathi Y (2019) Pharmacokinetics study of puerarin absorption in blood after consumption of *Pueraria tuberosa* water extract (PTWE) by rats. Explor Anim Med Res 9(2):188–196

Perumal K et al (2021) Nanoparticles and gut microbiota in colorectal cancer. Front Nanotechnol 3(July):1–9. https://doi.org/10.3389/fnano.2021.681760

Petakh P, Oksenych V, Kamyshna I, Boisak I, Lyubomirskaya K, Kamyshnyi O (2024) Exploring the interplay between posttraumatic stress disorder, gut microbiota, and inflammatory biomarkers: a comprehensive meta-analysis. Front Immunol 15(February):1–11. https://doi.org/10.3389/fimmu.2024.1349883

Qadri M, Short S, Gast K, Hernandez J, Wong ACN (2020) Microbiome innovation in agriculture: development of microbial based tools for insect Pest management. Front Sustain Food Syst 4(October):1–20. https://doi.org/10.3389/fsufs.2020.547751

Qi Y, Wang Y, Wang X, Zheng H, Lu Y (2025) Tuning gut microbiota by advanced nanotechnology. Mater Futur 4(1). https://doi.org/10.1088/2752-5724/ada62b

Rahman MM et al (2022) The gut microbiota (microbiome) in cardiovascular disease and its therapeutic regulation. Front Cell Infect Microbiol 12(June):1–22. https://doi.org/10.3389/fcimb.2022.903570

Ren SN et al (2023) Application of nanotechnology in reversing therapeutic resistance and controlling metastasis of colorectal cancer. World J Gastroenterol 29(13):1911–1941. https://doi.org/10.3748/WJG.V29.I13.1911

Rob F et al (2022) Association between ustekinumab therapy and changes in specific anti-microbial response, serum biomarkers, and microbiota composition in patients with IBD: a pilot study. PLoS One 17(12):1–24. https://doi.org/10.1371/journal.pone.0277576

Rosales TKO, Hassimotto NMA, Lajolo FM, Fabi JP (2022) Nanotechnology as a tool to mitigate the effects of intestinal microbiota on metabolization of anthocyanins. Antioxidants 11(3):1–22. https://doi.org/10.3390/antiox11030506

Sadr S et al (2023) Challenges and prospective of enhancing hydatid cyst chemotherapy by nanotechnology and the future of nanobiosensors for diagnosis. Trop Med Infect Dis 8(11). https://doi.org/10.3390/tropicalmed8110494

Sakamoto JH, van de Ven AL, Godin B, Blanco E, Serda RE, Grattoni A et al (2010) Enabling individualized therapy through nanotechnology. Pharmacol Res 62(2):57–89. https://doi.org/10.1016/j.phrs.2009.12.011

Santosh F et al (2021) Understanding diverse signaling pathways involved in colorectal carcinoma and potentials of nanotechnology in their diagnosis and therapeutics. Pharmacology 2:1226–1243

Sawant RM et al (2006) 'SMART' drug delivery systems: double-targeted pH-responsive pharmaceutical nanocarriers. Bioconjug Chem 17(4):943–949. https://doi.org/10.1021/bc060080h

Shabib Akhtar M et al (2024) Nanoengineered chitosan functionalized titanium dioxide biohybrids for bacterial infections and cancer therapy. Sci Rep 14(1):1–16. https://doi.org/10.1038/s41598-024-52847-1

Shi XR et al (2021) Microbiota in gut, oral cavity, and mitral valves are associated with rheumatic heart disease. Front Cell Infect Microbiol 11(March):1–12. https://doi.org/10.3389/fcimb.2021.643092

Silva DF et al (2023) Biomedical approach of nanotechnology and biological risks: a mini-review. Int J Mol Sci 24(23). https://doi.org/10.3390/ijms242316719

Singh A, Prasad LB, Shiv K, Kumar R, Garai S (2023a) Synthesis, characterization, and in vitro antibacterial and cytotoxic study of Co(II), Ni(II), Cu(II), and Zn(II) com-

plexes of N-(4-methoxybenzyl) N-(phenylethyl) dithiocarbamate ligand. J Mol Struct 1288(May):135835. https://doi.org/10.1016/j.molstruc.2023.135835
Singh M et al (2023b) Pharmacokinetic screening to estimate the drug likeliness characteristics of selected herbal anticancer drugs. Res J Pharm Technol 16(7):3422–3426. https://doi.org/10.52711/0974-360X.2023.00566
Singh A et al (2024a) Synthesis, crystal structure and screening for anticonvulsant and antianxiety activities of three new Ni(II), Cu(II), and Zn(II) dithiocarbamate complexes. J Mol Struct 1298(P2):137052. https://doi.org/10.1016/j.molstruc.2023.137052
Singh A et al (2024b) Dithiocarbamate-based novel anti-histaminic agents: synthesis, characterization, crystal structure and thermal study. Dalt Trans. https://doi.org/10.1039/d4dt01777c
Sofia J, Cordeiro S (2022) Impact of microbiome in cancer nano-immunotherapy. Mestrado Integrado em Ciências Farmacêuticas
Song W, Anselmo AC, Huang L (2019) Nanotechnology intervention of the microbiome for cancer therapy. Nat Nanotechnol 14(12):1093–1103. https://doi.org/10.1038/s41565-019-0589-5
Soni A et al (2023) Nano-biotechnology in tumour and cancerous disease: a perspective review. J Cell Mol Med 27(6):737–762. https://doi.org/10.1111/jcmm.17677
Sosnowski K, Akarapipad P, Yoon J (2020) The future of microbiome analysis: biosensor methods for big data collection and clinical diagnostics. Med Devices Sensors 3(5):1–19. https://doi.org/10.1002/mds3.10085
Stater EP, Sonay AY, Hart C, Grimm J (2021) The ancillary effects of nanoparticles and their implications for nanomedicine. Nat Nanotechnol 16(11):1180–1194. https://doi.org/10.1038/s41565-021-01017-9
Sultana A, Zare M, Luo H, Ramakrishna S (2021) Surface engineering strategies to enhance the in situ performance of medical devices including atomic scale engineering. Int J Mol Sci 22(21). https://doi.org/10.3390/ijms222111788
Sumida K, Lau WL, Kovesdy CP, Kalantar-Zadeh K, Kalantar-Zadeh K (2021) Microbiome modulation as a novel therapeutic approach in chronic kidney disease. Curr Opin Nephrol Hypertens 30(1):75–84. https://doi.org/10.1097/MNH.0000000000000661
Swann JR et al (2020) Considerations for the design and conduct of human gut microbiota intervention studies relating to foods. Eur J Nutr 59(8):3347–3368. https://doi.org/10.1007/s00394-020-02232-1
Szablewski L (2022) Human gut microbiota in health and Alzheimer's disease. Adv Alzheimer's Dis 9:217–228. https://doi.org/10.3233/AIAD220015
Takallu S et al (2024) Nanotechnology improves the detection of bacteria: recent advances and future perspectives. Heliyon 10(11):e32020. https://doi.org/10.1016/j.heliyon.2024.e32020
Tiwari A, Kumar C (2025) Unravelling the herbal formulation of floating microspheres for Gut microbiome modulation: Curren1Abhinay challenges and future prospects. Curr Pharm Res 1:144–162
Tiwari A, Saraf S, Jain A, Panda PK, Verma A, Jain SK (2020) Basics to advances in nanotherapy of colorectal cancer. Drug Deliv Transl Res 10(2):319–338. https://doi.org/10.1007/s13346-019-00680-9
Upreti P, Devhare LD, Abdulmageed LH, Kumar YG, Kumar R, Dharmamoorthy G (2024) In: Gangwar M, Nath G (eds) Combatting antibiotic resistance: leveraging fecal microbial transplantation for gut health BT - emerging paradigms for antibiotic-resistant infections: beyond the pill. Springer Nature, Singapore, pp 211–232. https://doi.org/10.1007/978-981-97-5272-0_10
Van den Abbeele P et al (2023) Bridging preclinical and clinical gut microbiota research using the ex vivo SIFR® technology. Front Microbiol 14(April):1–11. https://doi.org/10.3389/fmicb.2023.1131662
Vandeputte D et al (2021) Temporal variability in quantitative human gut microbiome profiles and implications for clinical research. Nat Commun 12(1). https://doi.org/10.1038/s41467-021-27098-7

Vijaya Durga RD (2015) Determination of S-amlodipine in human plasma using tizanidine as internal standard by Lc/Ms/Ms method. J Bioanal Biomed 07(03):91–96. https://doi.org/10.4172/1948-593x.1000130
Vijayaram S, Razafindralambo H, Sun YZ, Piccione G, Multisanti CR, Faggio C (2024) Synergistic interaction of nanoparticles and probiotic delivery: a review. J Fish Dis 47(5):1–15. https://doi.org/10.1111/jfd.13916
Vikesland P (2017) Environmental engineering science in the 21st century. Environ Eng Sci 34(1):1–2. https://doi.org/10.1089/ees.2016.0548
Villaverde A (2010) Nanotechnology, bionanotechnology and microbial cell factories. Microb Cell Factories 9:2–5. https://doi.org/10.1186/1475-2859-9-53
Viswanath B, Kim S, Lee K (2016) Recent insights into nanotechnology development for detection and treatment of colorectal cancer. Int J Nanomed 11:2491–2504. https://doi.org/10.2147/IJN.S108715
Wang Q et al (2020) Enhanced and prolonged antitumor effect of salinomycin-loaded gelatinase-responsive nanoparticles via targeted drug delivery and inhibition of cervical cancer stem cells. Int J Nanomed 15:1283–1295. https://doi.org/10.2147/IJN.S234679
Wang H et al (2021a) Update on nanoparticle-based drug delivery system for anti-inflammatory treatment. Front Bioeng Biotechnol 9(February):1–9. https://doi.org/10.3389/fbioe.2021.630352
Wang F et al (2021b) Regorafenib plus toripalimab in patients with metastatic colorectal cancer: a phase Ib/II clinical trial and gut microbiome analysis. Cell Rep Med 2(9):100383. https://doi.org/10.1016/j.xcrm.2021.100383
Wang J et al (2022) The gut microbiota as a target to control hyperuricemia pathogenesis: potential mechanisms and therapeutic strategies. Crit Rev Food Sci Nutr 62(14):3979–3989. https://doi.org/10.1080/10408398.2021.1874287
Wang Y et al (2023) Meat consumption and gut microbiota: a scoping review of literature and systematic review of randomized controlled trials in adults. Adv Nutr 14(2):215–237. https://doi.org/10.1016/j.advnut.2022.10.005
Wei Y et al (2024) The challenge and opportunity of gut microbiota-targeted nanomedicine for colorectal cancer therapy. iMeta 3(4):1–7. https://doi.org/10.1002/imt2.213
Woo CW, Tso P, Yiu JHC (2022) Commensal gut microbiota-based strategies for oral delivery of therapeutic proteins. Trends Pharmacol Sci 43(12):1004–1013. https://doi.org/10.1016/j.tips.2022.08.002
Xia F, Wen L-P, Ge B-C, Li Y-X, Li F-P, Zhou B-J (2021) Gut microbiota as a target for prevention and treatment of type 2 diabetes: mechanisms and dietary natural products. World J Diabetes 12(8):1146–1163. https://doi.org/10.4239/wjd.v12.i8.1146
Xue J, Blesso C, Luo Y (2023) A comprehensive review of nanoparticles for oral delivery in food: biological fate, evaluation models, and gut microbiota influences. Annu Rev Food Sci Technol 14:1–33. https://doi.org/10.1146/annurev-food-060721-025159
Yang C, Merlin D (2019) Nanoparticle-mediated drug delivery systems for the treatment of IBD: current perspectives. Int J Nanomed 14:8875–8889. https://doi.org/10.2147/IJN.S210315
Yang C, Merlin D (2020) Can naturally occurring nanoparticle-based targeted drug delivery effectively treat inflammatory bowel disease? Expert Opin Drug Deliv 17(1):1–4. https://doi.org/10.1080/17425247.2020.1698543
Yang M et al (2020) Nanoparticle-based therapeutics of inflammatory bowel diseases: a narrative review of the current state and prospects. J Bio-X Res 3(4):157–173. https://doi.org/10.1097/JBR.0000000000000078
Ye Z, Liang L, Lu H, Shen Y, Zhou W, Li Y (2021) Nanotechnology-employed bacteria-based delivery strategy for enhanced anticancer therapy. Int J Nanomed 16:8069–8086. https://doi.org/10.2147/IJN.S329855
Zam W (2018, 2018) Gut microbiota as a prospective therapeutic target for curcumin: a review of mutual influence. J Nutr Metab (Figure 1):1367984. https://doi.org/10.1155/2018/1367984

Zhang P, Guo Z, Ullah S, Melagraki G, Afantitis A, Lynch I (2021) Nanotechnology and artificial intelligence to enable sustainable and precision agriculture. Nat Plants 7(7):864–876. https://doi.org/10.1038/s41477-021-00946-6

Zhang C et al (2022) Combined non-invasive prediction and new biomarkers of Oral and fecal microbiota in patients with gastric and colorectal cancer. Front Cell Infect Microbiol 12(May):1–17. https://doi.org/10.3389/fcimb.2022.830684

Zhang Q, Ma X, Zhao X (2023) Hybrid bacteria system with nanomaterials for tumor therapy. Adv NanoBiomed Res 3(7). https://doi.org/10.1002/anbr.202300003

Zhao LY et al (2023a) Role of the gut microbiota in anticancer therapy: from molecular mechanisms to clinical applications. Signal Transduct Target Ther 8(1). https://doi.org/10.1038/s41392-023-01406-7

Zhao Q, Chen Y, Huang W, Zhou H, Zhang W (2023b) Drug-microbiota interactions: an emerging priority for precision medicine. Signal Transduct Target Ther 8(1). https://doi.org/10.1038/s41392-023-01619-w

Zhou M et al (2017) The application of stimuli-responsive Nanocarriers for targeted drug delivery. Curr Top Med Chem 17(20):2319–2334. https://doi.org/10.2174/1568026617666170224121008

Zhu H et al (2022) Network pharmacology exploration reveals gut microbiota modulation as a common therapeutic mechanism for anti-fatigue effect treated with maca compounds prescription. Nutrients 14(8):1533. https://doi.org/10.3390/nu14081533

Zhu LR, Li SS, Zheng WQ, Ni WJ, Cai M, Liu HP (2023) Targeted modulation of gut microbiota by traditional Chinese medicine and natural products for liver disease therapy. Front Immunol 14(February):1–14. https://doi.org/10.3389/fimmu.2023.1086078

Strategies for Biomarker Identification and Validation in Precision Medicine: Microbial Predictors, Data Integration, Machine Learning, and Clinical Trial Insights

Amrita Thakur

Abstract Biomarkers play a pivotal role in precision medicine by enabling disease diagnosis, treatment response monitoring, and personalized therapeutic strategies. This chapter explores key strategies for biomarker identification and validation, with a focus on microbial predictors, data integration, machine learning, and clinical trial insights. Microbial biomarkers have gained prominence due to their role in disease pathogenesis and therapeutic modulation, particularly in oncology, cardiovascular diseases, and inflammatory disorders. Advanced sequencing techniques and bioinformatics tools aid in identifying microbial signatures associated with disease states. Integrating multi-omics data—including genomics, proteomics, and metabolomics—enhances biomarker discovery by providing a comprehensive understanding of molecular interactions. Machine learning algorithms, including supervised and unsupervised models, are revolutionizing biomarker identification by analyzing high-dimensional datasets and uncovering predictive patterns. However, challenges such as data heterogeneity, validation inconsistencies, and computational complexities hinder biomarker translation into clinical practice. Rigorous validation through clinical trials remains essential for establishing biomarker reliability, reproducibility, and clinical utility. Biomarker-driven trials leverage adaptive designs and stratified patient selection to enhance precision medicine outcomes. Emerging technologies such as artificial intelligence, spatial transcriptomics, and digital health monitoring further refine biomarker discovery and validation processes. Despite ongoing challenges, the integration of multi-omics data, machine learning, and robust clinical validation frameworks promises to accelerate the implementation of biomarkers in personalized healthcare. This chapter provides a comprehensive overview of these strategies, addressing current limitations and highlighting future directions in biomarker research to optimize clinical decision-making and improve patient outcomes.

A. Thakur (✉)
School of Pharmacy, Vishwakarma University, Pune, Maharashtra, India

S. K. Singh et al. (eds.), *Microbiota Profiling for Precision Medicine*, Interdisciplinary Biotechnological Advances,
https://doi.org/10.1007/978-981-95-2104-3_9

Keywords Biomarkers · Precision medicine · Microbial predictors · Multi-omics integration · Personalized therapy · Bioinformatics

1 Introduction

The field of precision medicine has gained substantial attention in recent years as healthcare moves towards more personalized and targeted approaches to diagnosis, treatment, and prognosis. A fundamental aspect of precision medicine is identifying and validating biomarkers—measurable indicators that reflect physiological, pathological, or therapeutic responses within the body (Strimbu and Tavel 2010). Biomarkers play a crucial role in predicting disease risk, monitoring therapeutic responses, and tailoring interventions to individual patient profiles, thus enhancing clinical outcomes and reducing adverse effects (Henry and Hayes 2012). Biomarkers serve as the cornerstone of precision medicine, offering invaluable insights into disease mechanisms and therapeutic responses. They enable the stratification of patients based on disease subtypes or molecular characteristics, allowing for the selection of optimal therapeutic strategies. This is particularly relevant in oncology, where molecular biomarkers guide the use of targeted therapies and immune checkpoint inhibitors (Cheng et al. 2020). Additionally, biomarkers facilitate early disease detection, enabling prompt intervention and improved prognosis (Snyderman and Dinan 2010). While biomarkers hold tremendous potential, their clinical utility largely depends on robust identification and rigorous validation processes. The identification of biomarkers often involves integrating large-scale omics data—such as genomics, transcriptomics, proteomics, and metabolomics—with clinical data (Wang et al. 2018a, 2018b). However, merely identifying a potential biomarker is insufficient; validation is essential to confirm its reliability, reproducibility, and accuracy across diverse patient populations. Validation often encompasses analytical validation (accuracy and precision), clinical validation (association with clinical outcomes), and clinical utility (impact on decision-making) (Pepe et al. 2001). The journey from biomarker discovery to clinical implementation is fraught with challenges. One of the most significant obstacles is the reproducibility crisis, where biomarkers identified in exploratory studies fail to hold up under rigorous validation (Ioannidis and Khoury 2011a, 2011b). Additionally, the heterogeneity of patient populations, variations in sample collection, and differences in analytical methods further complicate biomarker validation (Simon 2011a, 2011b). Despite these challenges, advances in high-throughput technologies, bioinformatics, and machine learning offer unprecedented opportunities for discovering novel biomarkers with higher predictive accuracy and clinical relevance (Baker 2019a, 2019b). This

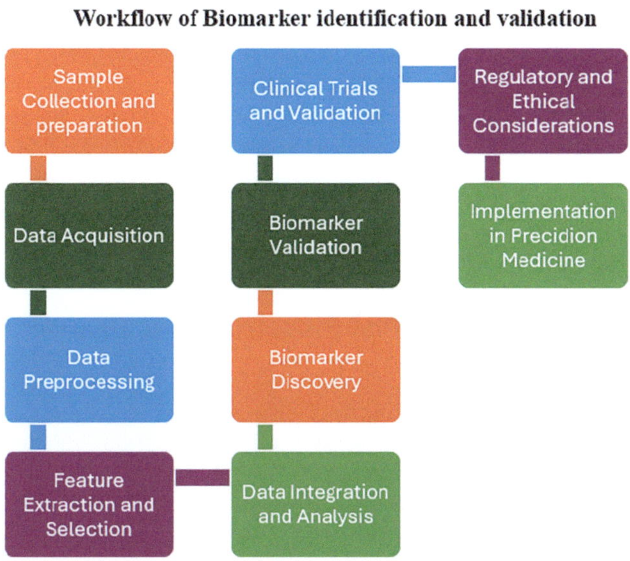

Fig. 1 Flowchart of biomarker identification and validation

chapter aims to comprehensively explore strategies for biomarker identification and validation in the context of precision medicine. It covers the integration of microbial predictors, data integration approaches, machine learning applications, and the role of clinical trials in biomarker validation. By examining both challenges and emerging technologies, this chapter seeks to provide insights into optimizing biomarker discovery processes and translating them into clinically viable solutions. Figure 1 illustrates the comprehensive workflow of biomarker identification and validation, encompassing the stages from discovery to clinical application.

2 Microbial Predictors in Biomarker

The human microbiome, comprising trillions of microorganisms residing on and within the human body, plays a pivotal role in maintaining health and contributing to disease pathogenesis (Lloyd-Price et al. 2016). The microbiome's influence on metabolic functions, immune modulation, and disease susceptibility has made it a critical focus in precision medicine. Increasing evidence suggests that variations in microbial composition can serve as valuable biomarkers for diagnosing, monitoring, and predicting the progression of various diseases (Clemente et al. 2012). Precision medicine leverages microbiome data to develop targeted therapies that cater to individual microbiota profiles. For instance, the gut microbiome is increasingly recognized for its role in influencing responses to cancer immunotherapy. Patients with a higher abundance of specific bacterial genera, such as *Bacteroides* and *Faecalibacterium*, demonstrate improved responses to immune checkpoint inhibitors (Gopalakrishnan et al. 2018). By profiling the microbiome, clinicians can predict therapeutic efficacy and tailor interventions accordingly. Furthermore, microbial biomarkers extend beyond oncology to other chronic conditions, such as inflammatory bowel disease (IBD), diabetes, and cardiovascular diseases. Studies have identified distinct microbial signatures that correlate with disease severity and prognosis, enabling more precise and individualized care (Franzosa et al. 2019).

3 Identification of Microbial Biomarkers for Disease Diagnosis and Therapy

The identification of microbial biomarkers involves high-throughput sequencing techniques such as 16S ribosomal RNA (rRNA) gene sequencing, whole-genome shotgun metagenomics, and meta transcriptomics (Qin et al. 2010). These methods facilitate the comprehensive analysis of microbial communities, enabling the detection of specific taxa or functional genes linked to disease states. One widely adopted approach is the comparative analysis of microbial profiles between healthy and diseased cohorts. For example, significant alterations in the gut microbiome composition have been observed in colorectal cancer patients, with increased levels of *Fusobacterium nucleatum* correlating with tumor development and progression (Kostic et al. 2012). Identifying such microbial biomarkers allows for early detection and monitoring of disease. Advanced bioinformatics tools, including machine learning algorithms, play an essential role in discerning complex microbial patterns. These models integrate multi-omics data, identifying microbial signatures with diagnostic and prognostic potential (Pasolli et al. 2019). Additionally, network-based approaches assess microbial interactions and their collective influence on disease phenotypes, thereby enhancing biomarker discovery efforts (He et al. 2020).

4 Case Studies and Applications of Microbial Biomarkers

1. *Cancer Immunotherapy Response Prediction*: Gopalakrishnan et al. (2018) demonstrated that the gut microbiome composition significantly impacts responses to immune checkpoint inhibitors in melanoma patients. Individuals with higher levels of *Faecalibacterium* exhibited enhanced anti-tumor immunity and prolonged survival. These findings underscore the potential of gut microbial biomarkers in guiding immunotherapy decisions.
2. *Microbial Biomarkers in Cardiovascular Diseases*: Studies have identified an association between *Trimethylamine-N-oxide* (TMAO), a metabolite produced by gut microbes, and cardiovascular risks (Wang et al. 2011). Elevated TMAO levels have been linked to atherosclerosis, and measuring TMAO can help predict cardiovascular events, guiding preventive interventions.

3. *Inflammatory Bowel Disease (IBD) Biomarkers*: Franzosa et al. (2019) reported distinct microbial profiles in IBD patients, with increased levels of *Escherichia coli* and decreased levels of *Faecalibacterium prausnitzii*. These biomarkers can help distinguish between Crohn's disease and ulcerative colitis, facilitating more accurate diagnosis and personalized treatment strategies. Despite the promising potential of microbial biomarkers, several challenges impede their clinical translation. One major hurdle is the inherent variability in microbiome composition across individuals, influenced by factors such as diet, genetics, lifestyle, and geographic location (Ding and Schloss 2014a, 2014b). This variability complicates the standardization of microbial biomarker panels, reducing their reproducibility and clinical utility. Moreover, the dynamic nature of the microbiome poses challenges for longitudinal biomarker studies. Temporal fluctuations in microbial composition may affect the stability of proposed biomarkers, necessitating repeated measurements and longitudinal analyses to ensure reliability (Turnbaugh et al. 2009a, 2009b). Technical challenges also arise from sample collection, preservation, and sequencing methods, which can introduce biases and affect data quality (Costea et al. 2017a, 2017b). Therefore, standardizing methodologies and validation frameworks remains a priority to ensure that microbial biomarkers can be accurately and consistently applied in clinical settings. To elaborate more, Fig. 2 is shown below to describe the role of microbial biomarkers in precision medicine.

5 Data Integration in Biomarker Identification

In the era of precision medicine, the integration of multi-omics data has emerged as a powerful approach for biomarker discovery. Multi-omics refers to the comprehensive analysis of various molecular layers, including genomics, transcriptomics, proteomics, metabolomics, epigenomics, and microbiomics, to gain a holistic understanding of complex biological systems (Hasin et al. 2017). The integration of these diverse datasets provides a more comprehensive view of disease mechanisms and enhances the identification of robust biomarkers, leading to improved diagnostic accuracy and therapeutic precision (Zhou et al. 2019a, 2019b). Single-omics approaches often fall short in capturing the complexity of diseases, as they focus on one aspect of molecular regulation. In contrast, multi-omics integration enables the correlation of genetic variations with gene expression changes, protein abundance, and metabolic alterations, thereby revealing intricate molecular networks associated with disease states (Li et al. 2018). For instance, in cancer research, integrating genomics with transcriptomics and proteomics has revealed biomarkers that predict tumor progression and patient response to targeted therapies (Kim et al. 2016). Furthermore, multi-omics approaches allow for the discovery of disease subtypes that cannot be identified through single-omics data alone. This stratification is crucial for personalized medicine, as it enables the design of more tailored therapeutic

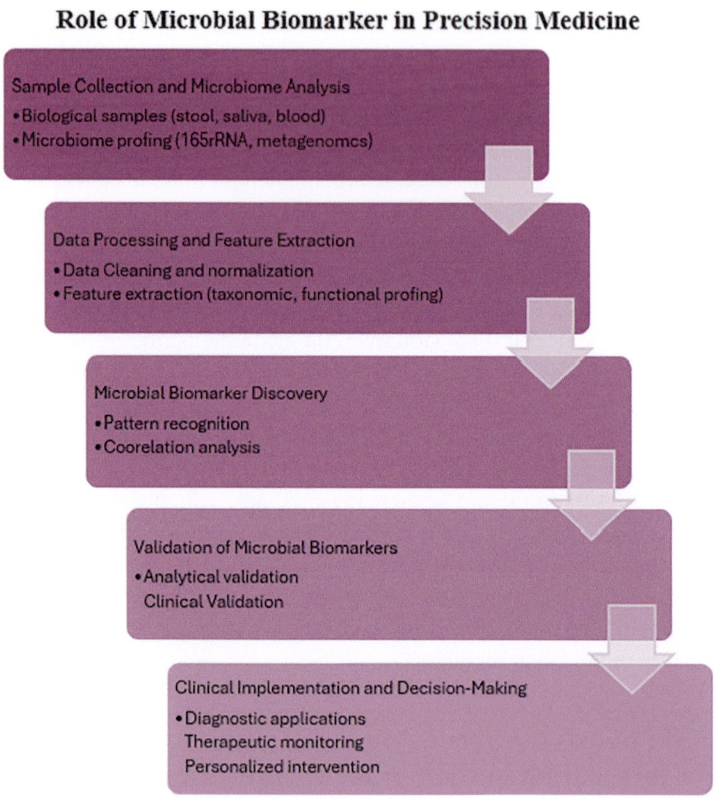

Fig. 2 Role of microbial biomarker in precision medicine

strategies (Huang et al. 2017a, 2017b). Figure 3 describes the workflow of data integration in biomarker discovery.

6 Data Integration Strategies and Computational Approaches

The integration of multi-omics data requires sophisticated computational frameworks capable of handling high-dimensional, heterogeneous datasets. The primary strategies for multi-omics data integration include:

1. *Early Integration (Data-Level Integration)*: This approach combines raw data from different omics platforms before downstream analysis. Techniques like matrix factorization and canonical correlation analysis (CCA) are commonly used to identify co-variation patterns between datasets (Meng et al. 2014a, 2014b).
2. *Intermediate Integration (Feature-Level Integration)*: Here, features such as gene expression levels or protein abundances are integrated after individual anal-

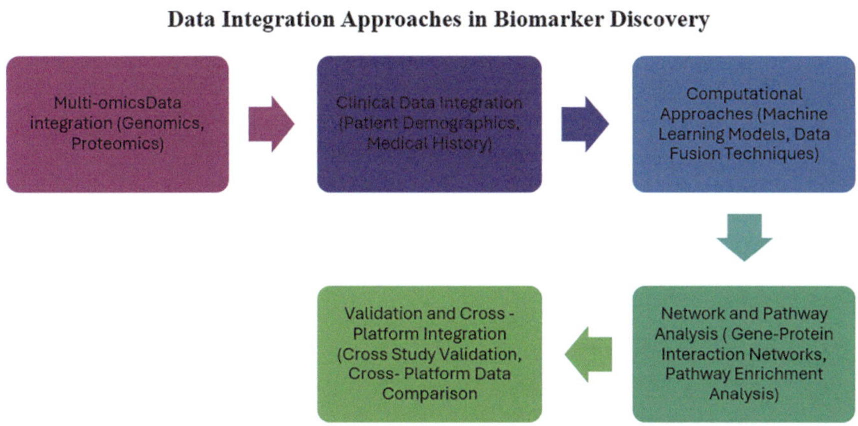

Fig. 3 Data integration approach in biomarker discovery

ysis. Methods like principal component analysis (PCA) and partial least squares (PLS) are often employed to reduce dimensionality and enhance interpretability (van der Kloet et al. 2010).

3. *Late Integration (Decision-Level Integration)*: This method involves integrating results obtained from individual analyses of different omics layers. Ensemble learning techniques and meta-analyses are commonly used to combine predictions from multiple models (Chakraborty et al. 2018a, 2018b).
4. *Network-Based Integration*: Network-based methods construct interaction networks to assess relationships among omics data. For example, gene co-expression networks and protein–protein interaction networks help identify multi-omics biomarkers that exhibit coordinated regulation (Zitnik et al. 2017).
5. *Machine Learning and Deep Learning Approaches*: Recent advances have led to the application of machine learning algorithms, including random forests (RF) and support vector machines, as well as deep learning frameworks like convolutional neural networks (CNNs) and autoencoders for multi-omics integration (Rappoport and Shamir 2018).

7 Case Studies of Successful Data Integration in Precision Medicine

1. *Cancer Subtyping through Multi-Omics Integration*: One notable example of successful data integration is The Cancer Genome Atlas (TCGA) project, which combined genomics, transcriptomics, and proteomics data to classify subtypes of various cancers, including breast and lung cancers. By integrating multiple data types, TCGA was able to identify distinct molecular subtypes with varying prognoses and therapeutic responses (Hoadley et al. 2014).

2. *Biomarker Discovery in Alzheimer's Disease*: In Alzheimer's research, multi-omics integration has led to the identification of novel biomarkers by combining genomic data with proteomic and metabolomic profiles. Xie et al. (2021) used network-based integration to link gene expression data with amyloid-beta and tau protein levels, highlighting potential therapeutic targets.
3. *Integrative Analysis of Cardiovascular Disease Biomarkers*: In cardiovascular research, multi-omics approaches have helped identify biomarkers linked to atherosclerosis by integrating lipidomics with transcriptomics and metabolomics. This comprehensive analysis revealed metabolic pathways associated with cardiovascular risk, allowing for the development of more targeted interventions (Ghosh et al. 2015).

Despite the promising potential of multi-omics data integration, several challenges hinder its practical application:

1. *Data Heterogeneity*: Different omics technologies often generate data with varying scales, formats, and quality, complicating integration and analysis (Misra et al. 2019). Moreover, discrepancies between datasets can arise from batch effects or differences in data preprocessing.
2. *High Dimensionality and Noise*: Multi-omics data typically involve high-dimensional features with significant noise and sparsity, posing challenges for traditional statistical methods. Feature selection and dimensionality reduction techniques are essential but can result in loss of critical information (Meng et al. 2014a, b).
3. *Computational Complexity*: Integrating and analyzing large, multi-omics datasets demand significant computational resources and expertise. Developing scalable and efficient algorithms remains a critical need (Huang et al. 2017a, b).
4. *Interpretability of Integrated Models*: Machine learning models used for multi-omics integration often operate as "black boxes," making it difficult to interpret how different omics layers contribute to the final output. Enhancing model transparency is essential for clinical adoption (Rappoport and Shamir 2018).
5. *Reproducibility and Validation*: A major challenge lies in reproducing results across independent studies and validating biomarkers in diverse patient cohorts. The lack of standardized protocols further exacerbates reproducibility issues (Zhou et al. 2019a, 2019b).

8 Machine Learning in Biomarker Identification and Validation

Machine learning (ML) has emerged as a transformative tool in the field of biomarker discovery and validation, offering advanced computational methods to analyze high-dimensional biological data. Unlike traditional statistical approaches, ML algorithms can identify complex patterns and correlations that are often hidden within vast datasets, making them invaluable for precision medicine (Chicco and

Jurman 2020a, 2020b). The primary objective of ML in biomarker discovery is to generate predictive models that can accurately classify disease states, predict treatment responses, or stratify patients based on their molecular profiles. Various ML techniques, including supervised, unsupervised, and semi-supervised learning, are utilized to analyze multi-omics data and identify reliable biomarkers (Libbrecht and Noble 2015). Supervised learning approaches, such as support vector machines (SVM), random forests (RF), and neural networks (NN), are commonly used for classification and regression tasks. These models learn from labelled data to predict outcomes, such as disease presence or absence, based on a set of features (Xie et al. 2021). On the other hand, unsupervised learning methods like clustering (e.g., *k*-means and hierarchical clustering) help group samples without predefined labels, aiding in the discovery of novel disease subtypes (Meng et al. 2021). Additionally, deep learning algorithms, particularly convolutional neural networks (CNNs) and recurrent neural networks (RNNs), have shown remarkable success in processing large and complex datasets, including genomic sequences and medical images (LeCun et al. 2015). Integrating these techniques with data from proteomics, transcriptomics, and metabolomics enhances biomarker discovery by uncovering intricate relationships between different molecular layers.

9 Predictive Modelling and Feature Selection for Biomarker Identification

Predictive modeling in biomarker identification involves building models that can predict disease outcomes or therapeutic responses based on specific biomolecular features. Selecting the most informative features from high-dimensional datasets is crucial to improving model accuracy and reducing overfitting (Guyon and Elisseeff 2003).

10 Feature Selection Techniques

1. *Filter Methods*: These methods assess the relevance of features independently of the model. Statistical tests (e.g., *t*-tests, analysis of variance [ANOVA]) and correlation-based methods (e.g., Pearson's correlation) are commonly employed. Although computationally efficient, filter methods may overlook feature interactions (Ding and Peng 2005).
2. *Wrapper Methods*: Wrapper methods evaluate feature subsets by training a model and assessing its performance. Techniques such as recursive feature elimination (RFE) and genetic algorithms (GAs) fall under this category. These methods are computationally intensive but offer higher accuracy (Kohavi and John 1997).

3. *Embedded Methods*: Embedded techniques perform feature selection as part of the model training process. For instance, decision trees and LASSO (Least Absolute Shrinkage and Selection Operator) naturally incorporate feature selection during model construction (Tibshirani 1996).
4. *Hybrid Approaches*: Combining multiple methods can enhance feature selection efficiency and accuracy. For example, combining filter and wrapper techniques often balances computational cost and performance (Saeys et al. 2007).

11 Validation Strategies Using Machine Learning Models

Validation is a critical step to ensure that machine learning models generalize well to new data. Various validation techniques are employed to assess the reliability and reproducibility of biomarker-based models:

1. *Cross-Validation*:

Cross-validation, such as *k*-fold or leave-one-out, helps evaluate model performance by partitioning the data into training and testing sets multiple times. This method mitigates the risk of overfitting, especially with limited data (Hastie et al. 2009).

2. *Bootstrapping*:

Bootstrapping generates multiple training sets through resampling, providing robust estimates of model accuracy and variance (Efron and Tibshirani 1993).

3. *External Validation*:

Testing the model on an independent dataset collected from a different population ensures its generalizability and robustness, which is crucial for clinical applications (Schoenfeld et al. 2017).

4. *Performance Metrics*:

Evaluation metrics such as accuracy, sensitivity, specificity, F1-score, and area under the receiver operating characteristic curve (AUC-ROC) provide comprehensive insights into model performance (Saito and Rehmsmeier 2015).

12 Real-World Applications and Case Studies

1. *Breast Cancer Prognostication Using Gene Expression Data*: Deep learning algorithms have been successfully applied to analyze gene expression profiles for predicting breast cancer recurrence. For instance, CNNs have demonstrated high accuracy in stratifying patients based on risk, allowing personalized treatment planning (Yao et al. 2022).
2. *Predicting Treatment Response in Rheumatoid Arthritis*: Machine learning models integrating multi-omics data have been used to identify biomarkers predicting response to tumor necrosis factor-alpha (TNF-α) inhibitors in rheumatoid

arthritis patients. RF-based models showed high precision, outperforming traditional statistical approaches (Bouget et al. 2022).
3. *Alzheimer's Disease Diagnosis through Proteomics*: Using SVMs to analyze cerebrospinal fluid proteomics data, researchers identified biomarkers that distinguish Alzheimer's patients from healthy controls with remarkable accuracy (Hou et al. 2025).

13 Limitations and Future Perspectives of Artificial Intelligence (AI) in Biomarker Validation

Despite its potential, AI-driven biomarker discovery faces several limitations:

1. *Data Quality and Heterogeneity*: Inconsistent and noisy datasets can negatively impact model training and performance. Data preprocessing and quality control remain crucial steps to ensure reliable outcomes (Libbrecht and Noble 2015).
2. *Interpretability Issues*: Deep learning models, while powerful, often function as "black boxes," making it challenging to interpret their predictions. Enhancing model interpretability is essential for clinical translation (Lundberg and Lee 2017).
3. *Lack of Standardization*: Variability in data acquisition, preprocessing, and model evaluation can lead to irreproducible results. Establishing standardized protocols for data integration and model validation is necessary (Hingorani et al. 2019).
4. *Ethical and Legal Concerns*: Privacy issues related to patient data and the ethical implications of algorithm-driven decisions need to be addressed, especially when integrating multi-omics data from diverse sources (Vayena et al. 2018a, 2018b). Integrating AI with domain-specific knowledge is crucial to maximizing machine learning's potential in biomarker discovery. Hybrid models that combine mechanistic and data-driven approaches may improve accuracy while enhancing interpretability. Furthermore, advances in explainable AI (XAI) could offer insights into model decision-making processes, thereby fostering clinical trust and acceptance (Samek et al. 2017a, 2017b).

14 Role of Clinical Trials in Validating Biomarkers

Clinical trials play a pivotal role in validating biomarkers by providing robust evidence of their clinical utility, predictive accuracy, and therapeutic relevance. Biomarker validation ensures that a candidate biomarker can accurately predict clinical outcomes, guide therapeutic decisions, or monitor disease progression. The validation process typically involves assessing the biomarker's sensitivity, specificity, and reproducibility across diverse populations and settings (Dancey et al. 2012a,

2012b). Validated biomarkers are essential in precision medicine, where they are used to identify patients most likely to benefit from targeted therapies, reduce adverse effects, and enhance treatment efficacy. The importance of rigorous validation has increased as biomarkers become integral to oncology, cardiology, and neurodegenerative disease management (Burrell et al. 2021a, 2021b).

15 Design and Implementation of Biomarker-Driven Clinical Trials

Designing biomarker-driven clinical trials requires careful consideration of trial structure, patient selection, and adaptive strategies to ensure robust validation. Some key design strategies include:

1. *Prospective versus Retrospective Studies*: Prospective clinical trials are designed to evaluate biomarkers before patient outcomes are known, reducing biases and enabling real-time assessment. Retrospective studies, on the other hand, analyze existing data to validate biomarkers that have shown potential in preliminary studies. Prospective trials are generally preferred for new biomarkers due to their ability to minimize confounding variables (Woodcock and LaVange 2017a, 2017b).
2. *Randomized Controlled Trials (RCTs) with Biomarker Stratification*: RCTs that incorporate biomarker stratification enable researchers to evaluate the efficacy of targeted therapies in biomarker-positive versus biomarker-negative groups. This stratification improves the trial's power and reduces sample size requirements by focusing on patients most likely to benefit (McShane et al. 2013).
3. *Adaptive Trial Designs*: Adaptive designs allow for modifications during the trial based on interim analyses, such as adjusting the sample size or altering treatment arms. These designs enhance trial efficiency and reduce resource utilization while maintaining statistical validity (Simon et al. 2019).
4. *Enrichment Strategies*: Enrichment strategies focus on selecting trial participants who are more likely to exhibit the biomarker of interest. This approach enhances the likelihood of demonstrating a treatment effect and improves trial outcomes (Freidlin et al. 2012).

16 Regulatory Considerations and Ethical Issues

Biomarker-driven clinical trials must adhere to stringent regulatory requirements and ethical standards to ensure patient safety and data integrity. Key considerations include:

- *Regulatory Guidelines*: Regulatory agencies such as the Food and Drug Administration (FDA) and European Medicines Agency (EMA) have established guidelines for biomarker validation, including assay standardization, analytical validation, and clinical qualification. These guidelines ensure that biomarkers meet predefined criteria before being implemented in clinical practice (FDA 2018).

- *Analytical and Clinical Validation*: Analytical validation confirms that the biomarker accurately measures the intended analyte, while clinical validation demonstrates its relevance to clinical outcomes. Both are essential for regulatory approval and integration into clinical settings (Yusuf et al. 2016a, 2016b).
- *Ethical Challenges*: Ethical issues often arise when mandatory biopsies or genetic testing are required, raising concerns about patient autonomy and safety. Obtaining informed consent and ensuring transparency are crucial in mitigating these challenges (Burrell et al. 2021a, 2021b).

17 Examples of Successful Biomarker Validation Through Clinical Trials

Numerous clinical trials have successfully validated biomarkers, significantly impacting patient care:

- *HER2 in Breast Cancer*: The HER2 biomarker, validated through clinical trials, transformed breast cancer treatment by identifying patients who benefit from trastuzumab therapy, significantly improving survival rates (Slamon et al. 2001).
- *KRAS in Colorectal Cancer*: KRAS mutations were validated as predictive biomarkers for Epidermal Growth Factor Receptor (EGFR)-targeted therapies. Clinical trials demonstrated that patients with wild-type KRAS benefit significantly from anti-EGFR (Epidermal Growth Factor Receptor) monoclonal antibodies, guiding treatment decisions (Amado et al. 2008).
- Programmed Death-Ligand 1(PD-L1) *Expression in Immunotherapy*: The validation of PD-L1 as a biomarker for checkpoint inhibitor therapies, such as pembrolizumab, revolutionized the treatment of various cancers. High PD-L1 (Programmed Death-Ligand 1) expression correlated with better therapeutic responses, as demonstrated in multiple clinical trials (Brahmer et al. 2015).
- *NTRK Fusion in Solid Tumors*: NTRK gene fusions have been validated as biomarkers for response to TRK inhibitors, with clinical trials demonstrating remarkable efficacy in patients harboring these mutations (Drilon et al. 2018). Clinical trials are integral to the successful validation of biomarkers in precision medicine. Rigorous design, ethical considerations, and adherence to regulatory guidelines are essential to achieving clinically meaningful outcomes. Validated biomarkers enable more personalized and effective therapeutic strategies, ultimately advancing patient care and treatment efficacy.

18 Major Challenges in Biomarker Discovery and Validation

Despite significant advancements in biomarker discovery and validation, numerous challenges persist in translating these biomarkers into clinical practice. One of the foremost challenges is the heterogeneity of patient populations, which can result in inconsistent biomarker performance across diverse demographic and clinical settings (Alizadeh et al. 2015). Moreover, variability in sample collection, processing, and storage can lead to discrepancies in biomarker measurements, compromising reproducibility and robustness (Ransohoff 2017a, 2017b). Another significant hurdle is the lack of standardization in data acquisition and processing, particularly in multi-omics studies. Differences in technologies and analytical pipelines can lead to discrepancies in biomarker identification and validation outcomes (Zhang et al. 2020a, 2020b, 2020c). Additionally, high-dimensional data and complex biological networks increase the difficulty of isolating clinically relevant biomarkers from background noise (Budhraja et al. 2023). The ethical and regulatory challenges associated with biomarker-driven research also pose significant obstacles. Obtaining informed consent, particularly for genetic and microbial biomarker studies, raises privacy concerns. Regulatory agencies demand rigorous validation and demonstration of clinical utility, adding to the complexity of biomarker approval and implementation (Burrell et al. 2021a, 2021b). Despite the significant advancements in biomarker discovery and validation, numerous challenges and limitations continue to hinder their clinical translation and practical application. Table 1 summarizes the key challenges, their impacts, proposed solutions, and the emerging technologies aimed at addressing these obstacles.

19 Emerging Technologies and Methodologies

Emerging technologies are reshaping the landscape of biomarker discovery and validation. One promising area is single-cell multi-omics, which enables the simultaneous analysis of gene expression, DNA methylation, and protein profiles at the single-cell level (Stuart and Satija 2019a, 2019b). This approach enhances the resolution of biomarker identification and provides insights into cell-to-cell variability within tumors and tissues. Machine learning and artificial intelligence (AI) are also making significant contributions by enabling predictive modeling and pattern recognition from vast datasets. Techniques such as deep learning and ensemble methods are being employed to identify novel biomarkers from genomic, transcriptomic, and proteomic data (Ang et al. 2021a, 2021b). Furthermore, integrating data from wearable health devices and digital biomarkers is becoming increasingly common in longitudinal studies, enabling real-time health monitoring (Petersen et al. 2022). The advent of spatial transcriptomics and spatial proteomics offers another innovative approach by mapping gene and protein expression directly onto tissue architectures. This integration of spatial data with molecular profiling enhances our understanding of tissue heterogeneity and microenvironmental influences on biomarker expression (Moffitt et al. 2018a, 2018b). In recent years, emerging technologies have revolutionized the field of biomarker discovery, offering innovative solutions to overcome traditional challenges and enhance the accuracy and precision of biomarker identification. Table 2 highlights some of the most promising technologies, including single-cell multi-omics, spatial transcriptomics, AI-driven predictive modelling, federated learning, and organ-on-a-chip models. Each technology is presented with its primary application, advantages, and limitations, demonstrating its potential impact on precision medicine.

20 Future Trends in Precision Medicine and Personalized Healthcare

The future of precision medicine lies in the seamless integration of data from diverse sources, including genomics, epigenomics, proteomics, and metabolomics. Multimodal data fusion, enabled by advanced computational frameworks, will facilitate comprehensive biomarker discovery (Yuan et al. 2022a, 2022b). Additionally, the integration of patient-reported outcomes and digital biomarkers from wearable devices will provide a more holistic view of patient health (Topol 2019). Personalized healthcare will increasingly leverage AI-driven decision support systems that incorporate real-time data, enabling dynamic adjustments to treatment regimens. These systems will be essential for optimizing therapeutic strategies based on continuously evolving patient profiles (Jiang et al. 2020). Furthermore, the use of organ-on-a-chip and patient-derived organoid models will enhance biomarker validation by providing human-like experimental platforms (Zhang et al. 2020a, 2020b, 2020c). As precision medicine becomes more prevalent, efforts to democratize access to advanced diagnostics and therapies will be crucial. Collaboration among academic institutions, healthcare providers, regulatory agencies, and the biotechnology industry will be key to accelerating the adoption of innovative biomarker technologies (Collins and Varmus 2015).

21 Integrating Microbial, Data-Driven, and Machine Learning Approaches

Integrating microbial biomarkers with multi-omics data and machine learning approaches represents a transformative strategy for precision medicine. Microbial signatures often serve as early indicators of disease states, and their integration with genomic and clinical data can enhance predictive accuracy (Vogtmann et al. 2018). Machine learning models can process vast and heterogeneous datasets, identifying

Table 1 Challenges and limitations in biomarker discovery and validation

Challenge	Description	Impact on biomarker validation	Proposed solution	Technology/approach	References
Data heterogeneity	Variations in sample collection, processing, and storage leading to inconsistencies	Reduces reproducibility and reliability	Standardized protocols and harmonization of data acquisition	Multi-omics integration frameworks	Ransohoff (2017a, b) and Zhang et al. (2020a, b, c)
High dimensionality	Excessive number of features compared to sample size	Increases risk of overfitting and reduces model robustness	Feature selection and dimensionality reduction techniques	Principal Component Analysis (PCA), LASSO	Hastie et al. (2009) and Tibshirani (1996)
Lack of standardization	Inconsistent data processing methods across studies	Limits comparability and reproducibility of results	Development of universal guidelines and frameworks	Data normalization and benchmarking	Hingorani et al. (2019) and Zhou et al. (2019a, b)
Computational complexity	High computational demands for processing and analyzing multi-omics data	Limits scalability and accessibility of advanced analytics	High-performance computing and cloud-based solutions	Federated learning, GPU acceleration	Huang et al. (2017a, b) and Meng et al. (2014a, b)
Lack of validation across populations	Biomarkers identified in one population may not be generalizable to others	Reduces clinical applicability and robustness	External validation with diverse cohorts	Multi-center clinical trials	Schoenfeld et al. (2017) and Burrell et al. (2021a, b)
Inadequate interpretability	Complexity of ML and AI models making interpretation difficult	Hinders clinical adoption and acceptance	Development of explainable AI techniques	Shapley values, LIME	Lundberg and Lee (2017) and Samek et al. (2017a, b)
Noise and variability	Background noise masking the signal, affecting biomarker detection	Leads to false positives and unreliable results	Signal enhancement and noise reduction techniques	Signal-to-noise ratio (SNR) optimization	Ding and Schloss (2014a, b) and Costea et al. (2017a, b)
Ethical and privacy concerns	Issues related to patient consent and data protection, especially in genomics	Reduces data availability and public trust	Strict adherence to data privacy regulations (e.g., General Data Protection Regulation (GDPR))	De-identified data collection and secure data storage	Vayena et al. (2018a, b) and Yusuf et al. (2016a, b)

Limited clinical utility	Biomarkers lack clinical relevance despite statistical significance	Fails to impact patient outcomes or decision-making	Emphasis on clinical validation and utility studies	Randomized controlled trials (RCTs)	Woodcock and LaVange (2017a, b) and Dancey et al. (2012a, b)
Data integration challenges	Integrating multi-omics data from different platforms and formats	Introduces biases and inconsistencies	Advanced computational algorithms for data harmonization	Network-based integration, ensemble learning	Zitnik et al. (2017) and Chakraborty et al. (2018a, b)
Overfitting risk	Model learns patterns specific to training data but fails to generalize	Reduces model robustness and reliability	Cross-validation techniques and model regularization	*k*-fold validation, dropout methods	Kohavi and John (1997) and Saeys et al. (2007)
High costs and resource demand	Expensive and time-consuming experimental procedures	Limits the feasibility of large-scale validation studies	Automation and cost-efficient technologies	High-throughput screening	Baker (2019a, b) and Simon (2011a, b)
Variability in biomarker performance	Biomarkers perform inconsistently across patient subgroups	Challenges in clinical translation	Stratified analysis and subgroup validation	Stratification-based modeling	Ioannidis and Khoury (2011a, b) and Wang et al. (2018a, b)
Insufficient data volume	Lack of large datasets with comprehensive biomarker profiles	Limits model training and validation	Data augmentation and synthetic data generation	Data synthesis techniques	LeCun et al. (2015) and Libbrecht and Noble (2015)
Temporal variability	Changes in biomarker levels over time affecting reliability	Reduces predictive accuracy	Longitudinal studies and temporal modeling	Time-series analysis	Turnbaugh et al. (2009a, b) and Petersen et al. (2022)

Table 2 Emerging technologies and their applications in biomarker discovery

Technology	Description	Applications in biomarker discovery	Advantages	Limitations	References
Single-cell multi-omics	Integrates genomics, transcriptomics, and proteomics at single-cell level	Identifying cell-specific biomarkers and cellular heterogeneity	High resolution and detailed molecular profiling	High cost and computational complexity	Stuart and Satija (2019a, b) and Ang et al. (2021a, b)
Spatial transcriptomics	Mapping gene expression data onto tissue architecture	Biomarker localization and tissue heterogeneity analysis	Maintains spatial context, enhancing biomarker relevance	High data volume and limited spatial resolution	Moffitt et al. (2018a, b) and Yuan et al. (2022a, b)
Deep learning and AI models	Data-driven approaches using neural networks and machine learning	Predictive modeling and pattern recognition from complex datasets	Identifies complex interactions and data correlations	Risk of overfitting and lack of interpretability	LeCun et al. (2015) and Chicco and Jurman (2020a, b)
Federated learning	Distributed learning without centralized data storage	Collaborative biomarker discovery while preserving data privacy	Enhances data privacy and allows multi-center studies	Complex implementation and potential data heterogeneity issues	Rieke et al. (2020a, b) and Vayena et al. (2018a, b)
Digital biomarkers	Data collected from wearable and mobile devices	Real-time health monitoring and disease progression prediction	Continuous data collection with real-time insights	Data accuracy and external variability issues	Petersen et al. (2022) and Topol (2019)
Organ-on-a-chip models	Microfluidic devices simulating human organ functions	Validating biomarkers under physiologically relevant conditions	Mimics human physiology better than conventional models	Limited scalability and complexity in organ simulation	Zhang et al. (2020a, b, c) and Huh et al. (2011)
Nanotechnology-based assays	Nanosensors for detecting biomarkers at the molecular level	Ultra-sensitive and rapid biomarker detection	High specificity and sensitivity	Potential toxicity and limited clinical validation	Jain (2005) and Rosi and Mirkin (2005)
CRISPR-based biosensors	Genome editing techniques for precise biomarker detection	Accurate detection of genetic biomarkers	High specificity and rapid response	Off-target effects and challenges in standardization	Zhang et al. (2021) and Wang et al. (2020)

Quantum biosensors	Sensors using quantum effects for ultra-sensitive biomarker detection	Detecting low-abundance biomarkers with high precision	Extremely high sensitivity and precision	Technological complexity and limited real-world application	Degen et al. (2017) and Schirhagl et al. (2014)
Microfluidic devices	Miniaturized devices for analyzing small biological samples	Rapid and high-throughput biomarker screening	Low sample volume and cost-efficient	Risk of clogging and integration issues	Whitesides (2006) and Sia and Kricka (2008)
Mass spectrometry	Analytical technique to identify and quantify biomolecules	Proteomic and metabolomic profiling for biomarker discovery	High accuracy and specificity	High operational cost and skilled personnel required	Aebersold and Mann (2003) and Smith et al. (2006)
Liquid biopsy technologies	Non-invasive methods to detect circulating biomarkers	Early cancer detection and monitoring of treatment response	Minimally invasive with potential for early diagnosis	Limited sensitivity for detecting rare biomarkers	Alix-Panabières and Pantel (2014) and Crowley et al. (2013)
Wearable biosensors	Sensors embedded in wearable devices for continuous health tracking	Detecting metabolic and cardiovascular biomarkers	Real-time monitoring and patient-centric data collection	Calibration challenges and inconsistent data quality	Sun et al. (2020) and Heikenfeld et al. (2018)
Raman spectroscopy	Vibrational spectroscopy technique for molecular characterization	Non-destructive analysis of biochemical changes in tissues	High sensitivity and specificity for molecular profiling	Fluorescence interference and data complexity	Kneipp et al. (1999) and Smith and Dent (2016)
Electrochemical biosensors	Electrochemical techniques for biomarker detection	Real-time monitoring of metabolites and disease markers	High sensitivity and rapid detection	Prone to signal interference and biofouling	Wang (2006) and Katz and Willner (2003)
Next-generation sequencing (NGS)	High-throughput sequencing to analyze genetic variations	Identifying genetic mutations and genomic biomarkers	Comprehensive analysis of genomic data	High cost and large data management challenges	Mardis (2008) and Metzker (2010)
Digital PCR	Precise quantification of nucleic acids	Detecting low-abundance genetic biomarkers with high accuracy	High sensitivity and quantification precision	High cost and complex data interpretation	Hindson et al. (2011) and Baker (2012)

(continued)

Table 2 (continued)

Technology	Description	Applications in biomarker discovery	Advantages	Limitations	References
Biosensing platforms	Integrated systems for real-time biomarker detection	Multiplexed detection of disease-specific biomarkers	High throughput and rapid analysis	Potential cross-reactivity and non-specific binding.	Li et al. (2018) and Turner (2013)
Optical biosensors	Devices that detect changes in optical properties	Real-time detection of biomolecular interactions	High specificity and real-time data acquisition	Prone to optical interference and sensitivity issues	Homola (2003) and Fan et al. (2008)
Immunoassays	Antibody-based techniques for biomarker detection	Quantifying proteins and small molecules in biological samples	High specificity and established protocols	Cross-reactivity and limited multiplexing capability	Wild (2013) and Lequin (2005)

complex relationships between microbial composition, host genetics, and clinical outcomes. For instance, integrating microbiome data with transcriptomic and proteomic profiles has shown potential in identifying biomarkers for gastrointestinal disorders and metabolic diseases (Zhu et al. 2019). Moreover, leveraging federated learning and distributed data processing can enable multi-center collaborations without compromising patient privacy (Rieke et al. 2020a, b). This approach can facilitate large-scale validation of microbial biomarkers and ensure that model predictions generalize across diverse populations. The future of biomarker discovery and validation lies in overcoming existing challenges and embracing emerging technologies. The integration of multi-omics, machine learning, and microbial profiling will enhance precision medicine's potential to deliver personalized and targeted healthcare solutions. Addressing ethical and regulatory challenges while fostering collaborative innovation will be crucial to translating these advancements into clinical practice. Figure 4 will help to understand the approach used for biomarker identification with the help of Machine Learning.

22 Conclusion

Biomarker identification and validation are pivotal to the advancement of precision medicine, enabling the development of targeted therapies and personalized healthcare solutions. This chapter has outlined various strategies and methodologies essential for biomarker discovery, emphasizing the integration of microbial predictors, multi-omics data, machine learning approaches, and clinical trial validation. The role of microbial biomarkers has emerged as a crucial aspect of disease diagnosis and therapeutic decision-making, particularly in areas like oncology, metabolic disorders, and gastrointestinal diseases. Integrating microbial data with host genetic and clinical profiles can significantly enhance predictive accuracy and treatment efficacy. Despite their potential, challenges in standardization, reproducibility, and validation remain, necessitating the development of robust frameworks for microbial biomarker integration. Data integration remains a cornerstone of biomarker discovery, particularly through the fusion of multi-omics data, which offers a comprehensive understanding of disease mechanisms and pathways. However, the complexity and heterogeneity of data require sophisticated computational approaches, including machine learning algorithms that can identify patterns and correlations within vast datasets. Predictive modelling, feature selection, and model validation are key elements in leveraging machine learning for biomarker discovery. Nonetheless, caution must be exercised to address the biases and overfitting that can undermine model reliability. Clinical trials remain the gold standard for biomarker validation, as they provide the necessary evidence of clinical utility and safety. The successful design and implementation of biomarker-driven trials demand meticulous planning and adherence to regulatory standards to ensure that biomarkers translate into real-world clinical benefits. Moreover, the ethical considerations associated with patient data and genetic information continue to pose challenges that

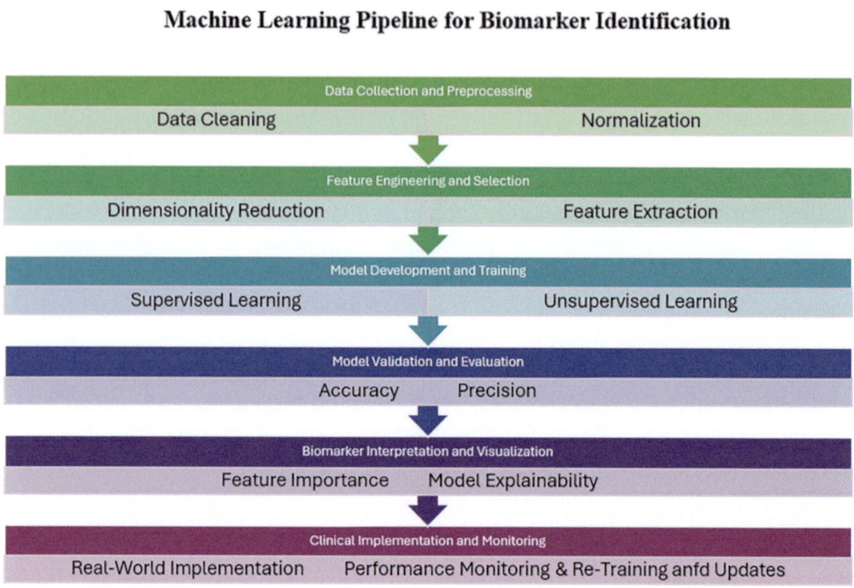

Fig. 4 Machine learning pipeline for biomarker identification

require transparent and patient-centric approaches. Looking ahead, the future of precision medicine will increasingly rely on integrating microbial, data-driven, and machine-learning approaches to achieve robust and clinically relevant biomarker identification. Emerging technologies, such as spatial transcriptomics, single-cell multi-omics, and federated learning, offer exciting prospects for refining biomarker discovery and validation processes. Collaboration between academic institutions, regulatory bodies, healthcare providers, and the biotech industry will be vital in driving these innovations forward. Ultimately, the successful translation of biomarkers into clinical practice will not only improve diagnosis and treatment but also foster a more patient-centered approach to healthcare. Continued investment in cutting-edge technologies and methodological advancements will be essential to unlocking the full potential of precision medicine.

References

Aebersold R, Mann M (2003) Mass spectrometry-based proteomics. Nature 422(6928):198–207

Alix-Panabières C, Pantel K (2014) Liquid biopsy: current status and future perspectives. Mol Oncol 8(4):613–619

Alizadeh AA, Aranda V, Bardelli A, Blanpain C, Bock C, Borowski C (2015) Toward understanding and exploiting tumor heterogeneity. Nat Med 21(8):846–853

Amado RG, Wolf M, Peeters M, Van Cutsem E, Siena S, Freeman DJ (2008) Wild-type KRAS is required for panitumumab efficacy in patients with metastatic colorectal cancer. J Clin Oncol 26(10):1626–1634

Ang Y, Rivas MA, Plotkin JB (2021a) Emerging applications of single-cell technologies. Genome Biol 22(1):55
Ang J, Hamarneh G, Tam R (2021b) Deep learning for multi-omics data integration in precision oncology. Front Genet 12:123–135
Baker, M., 2012. Digital PCR hits its stride. Nature Methods, 9(6), pp.541–544.
Baker M (2019a) Reproducibility crisis: blame it on the antibodies. Nature 575(7781):278–282
Baker M (2019b) High-throughput screening in drug discovery. Nat Biotechnol 37(5):422–428
Bouget, V., et al., 2022. Machine learning predicts response to TNF inhibitors in rheumatoid arthritis: results on the ESPOIR and ABIRISK cohorts. RMD Open, 8(2), e002442.
Brahmer JR, Tykodi SS, Chow LQ, Hwu WJ, Topalian SL, Hodi FS (2015) Safety and activity of anti–PD-L1 antibody in patients with advanced cancer. N Engl J Med 366(26):2455–2465
Budhraja, S., Doborjeh, M., Singh, B., et al., 2023. Filter and Wrapper Stacking Ensemble (FWSE): a robust approach for reliable biomarker discovery in high-dimensional omics data. Briefings in Bioinformatics, 24(6), bbad382.
Burrell RA, McGranahan N, Bartek J, Swanton C (2021a) Standards and challenges for integrating biomarkers in precision oncology. Nat Rev Clin Oncol 18(2):75–76
Burrell RA, McGranahan N, Bartek J, Swanton C (2021b) Strategies for biomarker validation in precision medicine. Nat Rev Drug Discov 20(5):295–310
Chakraborty S, Hsu A, Wen H (2018a) Multimodal data integration and analysis in precision medicine. Front Genet 9:489
Chakraborty S, Sinha A, Ghosh S, Pal R (2018b) Data fusion approaches for multi-omics data integration. J Proteome 188:199–207
Cheng M, Li Y, Yao Z (2020) Biomarker-driven targeted therapy in cancer: advances and challenges. J Clin Oncol 38(9):100–108
Chicco D, Jurman G (2020a) Machine learning in precision medicine. Brief Bioinform 21(6):2067–2078
Chicco D, Jurman G (2020b) Machine learning in bioinformatics. Brief Bioinform 21(3):593–602
Clemente JC, Ursell LK, Parfrey LW, Knight R (2012) The impact of the gut microbiota on human health: an integrative view. Cell 148(6):1258–1270
Collins FS, Varmus H (2015) A new initiative on precision medicine. N Engl J Med 372(9):793–795
Costea PI et al (2017a) Meta-analysis of microbiome studies identifies shared and unique responses to different human diseases. Nat Commun 8(1):1–10
Costea PI, Zeller G, Sunagawa S, Bork P (2017b) Challenges and opportunities in microbiome data analysis. Nat Biotechnol 35(11):1077–1087
Crowley E, Di Nicolantonio F, Loupakis F, Bardelli A (2013) Liquid biopsy: monitoring cancer-genetics in the blood. Nat Rev Clin Oncol 10(8):472–484
Dancey JE, Dobbin KK, Groshen S, Shankar LK (2012a) Biomarker development in the context of cancer therapeutics. Nat Rev Clin Oncol 9(3):147–156
Dancey JE, Dobbin KK, Groshen S, Shankar LK (2012b) Biomarker-driven oncology clinical trials: key challenges and future directions. Nat Rev Clin Oncol 9(4):204–210
Degen CL, Reinhard F, Cappellaro P (2017) Quantum sensing. Rev Mod Phys 89(3):035002
Ding C, Peng H (2005) Minimum redundancy feature selection from microarray gene expression data. J Bioinforma Comput Biol 3(2):185–205
Ding T, Schloss PD (2014a) Dynamics and associations of microbial community composition and function. Nat Rev Microbiol 12(5):322–334
Ding T, Schloss PD (2014b) Dynamics and associations of microbial community types across the human body. Nature 509(7500):357–360
Drilon A, Laetsch TW, Kummar S, DuBois SG, Lassen UN, Demetri GD (2018) Efficacy of larotrectinib in TRK fusion-positive cancers in adults and children. N Engl J Med 378(8):731–739
Efron, B. and Tibshirani, R.J., 1993. An introduction to the bootstrap. New York: Chapman & Hall/CRC.

Fan, X., White, I.M., & Shopova, S.I., 2008. Sensitive optical biosensors for unlabeled targets: A review. Analytica Chimica Acta, 620(1–2), pp.8–26.
Food and Drug Administration (2018) Principles for codevelopment of an in vitro companion diagnostic device with a therapeutic product, FDA guidance documents. Available at: https://www.fda.gov. Accessed 19 Mar 2025
Franzosa EA et al (2019) Gut microbiome structure and metabolic activity in inflammatory bowel disease. Nat Microbiol 4(2):293–305
Freidlin B, McShane LM, Korn EL (2012) Biomarker enrichment strategies: matching trial design to biomarker credentials. Nat Rev Clin Oncol 9(4):215–221
Ghosh S, Vivar JC, Carreras A (2015) Multi-omics data integration for identifying biomarkers in cardiovascular diseases. Bioinformatics 31(19):3152–3160
Gopalakrishnan V et al (2018) Gut microbiome influences efficacy of PD-1–based immunotherapy against epithelial tumors. Science 359(6371):97–103
Guyon, I. and Elisseeff, A., 2003. An introduction to variable and feature selection. Journal of Machine Learning Research, 3, pp.1157–1182.
Hasin Y, Seldin M, Lusis A (2017) Multi-omics approaches to disease. Genome Biol 18(1):83
Hastie T, Tibshirani R, Friedman J (2009) The elements of statistical learning: data mining, inference, and prediction, 2nd edn. Springer, New York
He Y et al (2020) Network analysis reveals distinct dysbiosis patterns in healthy and inflammatory bowel disease patients. Microbiome 8(1):1–11
Heikenfeld J, Jajack A, Rogers J, Gutruf P, Franklin D (2018) Wearable sensors: frontiers in diagnostics. Nat Biotechnol 36(11):927–937
Henry NL, Hayes DF (2012) Cancer biomarkers. Mol Oncol 6(2):140–146
Hindson, C.M., et al., 2011. High-throughput droplet digital PCR system for absolute quantitation of DNA copy number. Analytical Chemistry, 83(22), pp.8604–8610.
Hingorani AD, Kooner JS, Chambers JC, Sofat R (2019) Standardizing the methods for high-throughput biomarker discovery. Nat Biotechnol 37(10):1044–1054
Hoadley KA et al (2014) Multiplatform analysis of 12 cancer types reveals molecular classification within and across tissues of origin. Cell 158(4):929–944
Homola, J., 2003. Present and future of surface plasmon resonance biosensors. Analytical and Bioanalytical Chemistry, 377(3), pp.528–539.
Hou, X., et al., 2025. Machine-learning based strategy identifies a robust protein biomarker panel for Alzheimer's disease in cerebrospinal fluid. Scientific Reports, 15(1), 12345.
Huang S, Chaudhary K, Garmire LX (2017a) More is better: recent progress in multi-omics data integration methods. Front Genet 8:84
Huang S, Chaudhary K, Garmire L (2017b) Computational challenges in multi-omics data integration. Nat Rev Genet 18(5):333–346
Huh D, Hamilton GA, Ingber DE (2011) From 3D cell culture to organs-on-chips. Trends Cell Biol 21(12):745–754
Ioannidis JPA, Khoury MJ (2011a) Improving validation practices in 'omics' research. Science 334(6060):1230–1232
Ioannidis JPA, Khoury MJ (2011b) Improving validation practices in clinical biomarker research. Sci Transl Med 3(125):125
Jain KK (2005) Applications of nanobiotechnology in clinical diagnostics. Clin Chem 51(8):1570–1579
Jiang F, Jiang Y, Zhi H, Dong Y, Li H, Ma S (2020) Artificial intelligence in healthcare: past, present and future. Stroke Vasc Neurol 4(4):230–243
Katz E, Willner I (2003) Biomolecule-functionalized carbon nanotubes: applications in electrochemical biosensors. ChemPhysChem 4(10):1018–1024
Kim JY et al (2016) Integrative analysis of genomic and transcriptomic profiles identifies novel biomarkers in cancer. Cancer Res 76(14):4284–4292
Kneipp K, Kneipp H, Itzkan I, Dasari RR, Feld MS (1999) Surface-enhanced Raman scattering and biophysics. Chem Rev 99(10):2957–2976

Kohavi R, John GH (1997) Wrappers for feature subset selection. Artif Intell 97(1–2):273–324
Kostic AD et al (2012) Fusobacterium nucleatum potentiates intestinal tumorigenesis and modulates the tumor-immune microenvironment. Cell Host Microbe 14(2):207–215
LeCun Y, Bengio Y, Hinton G (2015) Deep learning. Nature 521(7553):436–444
Li X et al (2018) Multi-omics integration for precision oncology. Nat Rev Cancer 18(4):220–235
Libbrecht MW, Noble WS (2015) Machine learning applications in genomics. Nat Rev Genet 16(6):321–332
Lequin, R.M., 2005. Enzyme immunoassay (EIA)/enzyme-linked immunosorbent assay (ELISA). Clinical Chemistry, 51(12), pp.2415–2418.
Lloyd-Price J et al (2016) Multi-omics of the gut microbial ecosystem. Nature 535(7610):65–74
Lundberg SM, Lee S-I (2017) A unified approach to interpreting model predictions. In: Advances in neural information processing systems, vol 30. Curran Associates, Inc
Mardis, E.R., 2008. Next-generation DNA sequencing methods. Annual Review of Genomics and Human Genetics, 9, pp.387–402.
McShane LM, Hayes DF, Saad ED (2013) Criteria for the use of omics-based predictors in clinical trials. Nature 502(7471):317–320
Meng C et al (2014a) Dimension reduction techniques for the integrative analysis of multi-omics data. Brief Bioinform 15(4):660–674
Meng C, Zeleznik O, Guo X, Wang L, Liu Y (2014b) Dimension reduction techniques for multi-omics data integration. Genome Med 6(5):49
Metzker, M.L., 2010. Sequencing technologies — the next generation. Nature Reviews Genetics, 11, pp.31–46.
Misra BB, Langefeld CD, Olivier M (2019) Multi-omics data integration approaches and resources: an assessment of publicly available methods and tools. Front Genet 10:735
Moffitt RA, Marayati R, Flate EL, Volmar KE, Loeza SG, Hoadley KA (2018a) Virtual microdissection identifies distinct tumor- and stroma-specific subtypes of pancreatic ductal adenocarcinoma. Nat Genet 47(10):1168–1178
Moffitt JR, Bambah-Mukku D, Eichhorn SW, Vaughn E, Shekhar K, Zhuang X (2018b) Molecular, spatial, and functional single-cell profiling of the hypothalamic preoptic region. Nature 555(7697):769–774
Pasolli E et al (2019) Machine learning meta-analysis of large metagenomic datasets: tools and biological insights. PLoS Comput Biol 15(1):e1006764
Pepe MS et al (2001) Phases of biomarker development for early detection of cancer. J Natl Cancer Inst 93(14):1054–1061
Petersen C, Round JL, Cheng M (2022) Wearable health technologies and precision medicine: emerging opportunities. J Pers Med 12(1):1–15
Qin J et al (2010) A human gut microbial gene catalogue established by metagenomic sequencing. Nature 464(7285):59–65
Ransohoff DF (2017a) Biomarkers in precision medicine: the challenge of high-dimensional data. Clin Chem 63(6):942–945
Ransohoff DF (2017b) Problems and pitfalls in research synthesis. Am J Public Health 107(1):33–39
Rappoport N, Shamir R (2018) Multi-omics integration—a new perspective for personalized medicine. PLoS Comput Biol 14(2):e1005878
Rieke N, Hancox J, Li W, Milletari F, Roth HR, Albarqouni S (2020a) Federated learning for data privacy preserving in healthcare. Nat Med 26(9):1324–1331
Rieke N, Hancox J, Li W, Milletari F, Roth HR, Albarqouni S, Bakas S, Galtier M, Landman B, Maier-Hein KH, Cardoso MJ (2020b) Federated learning for medical imaging. Nat Mach Intell 2(5):305–315
Rosi NL, Mirkin CA (2005) Nanostructures in biodiagnostics. Chem Rev 105(4):1547–1562
Saeys Y, Inza I, Larrañaga P (2007) A review of feature selection techniques in bioinformatics. Bioinformatics 23(19):2507–2517
Saito T, Rehmsmeier M (2015) The precision-recall plot is more informative than the ROC plot when evaluating binary classifiers on imbalanced datasets. PLoS One 10(3):e0118432

Samek W, Wiegand T, Müller K-R (2017a) Explainable artificial intelligence: understanding, visualizing and interpreting deep learning models, arXiv preprint arXiv:1708.08296
Samek W, Wiegand T, Müller K-R (2017b) Explainable AI: interpreting, explaining and visualizing deep learning models. Digit Signal Process 73:1–15
Schirhagl R, Chang K, Loretz M, Degen CL (2014) Nitrogen-vacancy centers in diamond: nanoscale sensors for physics and biology. Annu Rev Phys Chem 65:83–105
Schoenfeld JD, Ferrone S, Catalano PJ, Ryan DP (2017) Generalizability and robustness of biomarkers in clinical trials. Clin Cancer Res 23(12):2796–2803
Sia SK, Kricka LJ (2008) Microfluidic devices for diagnostic applications. Lab Chip 8(12):1937–1940
Simon R (2011a) Genomic biomarkers in predictive medicine: an update. Cancer Biomark 9(4):173–177
Simon R (2011b) Pitfalls in biomarker discovery and validation. J Natl Cancer Inst 103(1):1–3
Simon R, Stein M, Wainberg Z (2019) Adaptive designs for clinical trials in the era of precision medicine. Annu Rev Med 70:25–41
Slamon, D.J., Leyland-Jones, B., Shak, S., et al., 2001. Use of chemotherapy plus a monoclonal antibody against HER2 for metastatic breast cancer that overexpresses HER2. New England Journal of Medicine, 344(11), pp. 783–792.
Smith E, Dent G (2016) Modern Raman spectroscopy: a practical approach. John Wiley & Sons
Smith RD, Shen Y, Thrall BD (2006) Mass spectrometry for high-throughput proteomics. Chem Rev 106(3):1206–1221
Snyderman R, Dinan MA (2010) Precision medicine: implications for health care reform. JAMA 303(6):663–664
Strimbu K, Tavel JA (2010) What are biomarkers? Curr Opin HIV AIDS 5(6):463–466
Stuart T, Satija R (2019a) Integrative single-cell analysis of transcriptional and epigenetic states in the human brain. Science 364(6440):1054–1056
Stuart T, Satija R (2019b) Integrative single-cell analysis. Nat Rev Genet 20(5):257–272
Sun F, Wei P, Wang L (2020) Wearable biosensors for healthcare monitoring. Adv Healthc Mater 9(12):2000494
Tibshirani R (1996) Regression shrinkage and selection via the lasso. J R Stat Soc Series B Stat Methodol 58(1):267–288
Topol EJ (2019) High-performance medicine: the convergence of human and artificial intelligence. Nat Med 25(1):44–56
Turnbaugh PJ et al (2009a) A core gut microbiome in obese and lean twins. Nature 457(7228):480–484
Turnbaugh PJ, Ley RE, Hamady M, Fraser-Liggett CM, Knight R, Gordon JI (2009b) The human microbiome project: re-thinking the definition of human. Nature 457(7228):480–484
Turner, A.P.F., 2013. Biosensors: Sense and sensibility. Science, 340(6133), pp.1186–1190.
van der Kloet, F.M., Sebastián-León, P., Conesa, A., Smilde, A.K. and Westerhuis, J.A., 2010. Separating common from distinctive variation. BMC Bioinformatics, 17(Suppl 5), p.S195.
Vayena E, Blasimme A, Cohen IG (2018a) Machine learning in precision medicine: addressing ethical challenges. Science 359(6376):421–422
Vayena E, Dzenowagis J, Brownstein JS, Sheikh A (2018b) Ethical challenges of big data in health and medicine. J Law Med Ethics 46(1):119–129
Vogtmann E, Goedert JJ, Ravel J (2018) Microbiome and precision cancer medicine. Genome Med 10(1):1–10
Wang J (2006) Electrochemical biosensors: towards point-of-care cancer diagnostics. Biosens Bioelectron 21(10):1887–1892
Wang Z et al (2011) Gut flora metabolism of phosphatidylcholine promotes cardiovascular disease. Nature 472(7341):57–63
Wang X et al (2018a) Big data integration in precision medicine. J Biomed Inform 83:59–71
Wang Z, Xu M, Li J, Chen W (2018b) Multi-omics data integration in cancer biomarker discovery. Nat Commun 9(1):2435

Wang H, La Russa M, Qi LS (2020) CRISPR/Cas9 in genome editing and beyond. Annu Rev Biochem 89:321–340

Whitesides GM (2006) The origins and the future of microfluidics. Nature 442(7101):368–373

Wild, D., 2013. The Immunoassay Handbook: Theory and Applications of Ligand Binding, ELISA and Related Techniques. 4th ed. Oxford: Elsevier.

Woodcock J, LaVange LM (2017a) Biomarker validation in the context of clinical trials: emerging challenges and opportunities. Clin Trials J 14(3):255–268

Woodcock J, LaVange LM (2017b) Biomarker-driven clinical trials: rationale and design. N Engl J Med 377(7):675–678

Yuan H, Liu H, Ma Y, Zhang P, Zhang Y (2022a) Multi-omics data integration in precision medicine: current progress and future directions. J Transl Med 20(1):89–101

Yuan Q, Li X, Zhang Y (2022b) Spatial transcriptomics for high-resolution tissue biology. Nat Methods 19(1):20–30

Yusuf S, Collins R, Peto R (2016a) Challenges in the development and implementation of biomarkers in cardiovascular disease. J Am Coll Cardiol 68(23):2525–2533

Yusuf S, Joseph P, Rangarajan S (2016b) Data privacy and informed consent in precision medicine. Lancet 388(10057):737–739

Zhang L, Wang Y, Bi J (2020a) Spatial transcriptomics: advances and applications. Nat Methods 17(10):1024–1033

Zhang B, Li W, Wang Y, Chen L, Zhao J, Liu H (2020b) Standardization of data acquisition and processing in multi-omics studies. J Biomed Inform 103:103368

Zhang B, Li W, Liu H (2020c) Organ-on-a-chip: a new paradigm for biomarker validation. J Biomed Eng 42(3):255–268

Zhang F, Abudayyeh OO, Gootenberg JS (2021) CRISPR-based diagnostics. Nat Biotechnol 39(7):775–785

Zhou B et al (2019a) Omics data integration for biomedical research. Nat Biotechnol 37(7):715–726

Zhou W, Xu L, Peng B, Li M (2019b) Data integration strategies in multi-omics analysis. Trends Biotechnol 37(7):698–711

Zhu Q, Gao Y, Guo X, Sun X (2019) Integration of microbiome and multi-omics data for precision medicine in chronic disease. Front Med 6:23–30

Zitnik M, Agrawal M, Leskovec J (2017) Predicting multicellular function through multi-omics networks. Nat Methods 14(12):1150–1158

Xie, L., He, B., Varathan, P., Risacher, S.L., Saykin, A.J. and Shen, L., 2021. Integrative -omics for discovery of network-level disease biomarkers for Alzheimer's disease. Alzheimer's & Dementia, 17(S10), p.e053964.

Meng, L., Avram, D., Tseng, G. and Huo, Z., 2021. Outcome-guided sparse K-means for disease subtype discovery via integrating phenotypic data with high-dimensional transcriptomic data. Bioinformatics, 37(23), pp.4591–4599.

Yao, Y., Lv, Y., Tong, L., Liang, Y., Xi, S., Ji, B., Zhang, G., Li, L., Tian, G., Tang, M., et al., 2022. ICSDA: A multi-modal deep learning model to predict breast cancer recurrence and metastasis risk by integrating pathological, clinical, and gene expression data. Briefings in Bioinformatics, 23(6), pp. bbac448.

Microbiota-Driven Precision Therapies: Applications in Autoimmune Disorders, Optimizing Cancer Therapy and Advancements in Gastrointestinal Disease Treatment

Swati Prakash, Shradha Bisht, and Vishakha Jaiswal

Abstract The human microbiota inhabits diverse microorganisms at different locations inside the body contributing to a complex but stable ecosystem. Among them, the microbiota present in the gut are most studied, because of their involvement in different signalling mechanisms for the maintenance of health. Dysbiosis, an imbalanced gut microbiota, is linked to various diseases, underscoring the importance of the well-balanced microbial system. Their bidirectional communication with different organ systems influences neurological, metabolic and immune responses. The advanced techniques have underlined the potential of microbiota-driven precision therapies to revolutionize treatment across various clinical domains. It is a novel and transformative approach in medicine, utilizing the associations between microbiota and health for the personalized treatment. Based on the individual's genetic and microbiome profiles, microbiota-precision treatment optimizes the therapeutic outcomes by maximizing the efficacy and minimizing the drug toxicity. This chapter will focus on the emergence of the therapeutic utility of gut microbiota in immune-mediated disorders, gastrointestinal complications and oncology. The microbiota and immune system interplay is also discussed for comprehending their role in inflammatory and immune-mediated disorders. However, microbiota-based precision therapy, while promising, is still an emerging field with significant challenges. These include identification of microbial strain, regulatory approvals and other challenges as well. Finding solutions to the above challenges is crucial to optimize microbiota-precision therapies, and formulating a comprehensive roadmap is

S. Prakash (✉) · V. Jaiswal
Amity Institute of Pharmacy, Amity University, Lucknow, Uttar Pradesh, India
e-mail: sprakash@lko.amity.edu; vjaiswal@lko.amity.edu

S. Bisht
Uttaranchal Institute of Pharmaceutical Sciences, Uttaranchal University, Dehradun, Uttarakhand, India
e-mail: shradhabisht@uumail.com

S. K. Singh et al. (eds.), *Microbiota Profiling for Precision Medicine*, Interdisciplinary Biotechnological Advances,
https://doi.org/10.1007/978-981-95-2104-3_10

needed to guide the development of these therapies, ensuring they can fulfil their therapeutic potential in a safe and effective manner.

Keywords Gut microbiota · Dysbiosis · Autoimmune disorders · Cancer · Gastrointestinal health · Precision medicine

Abbreviations

AhR	aryl hydrocarbon receptor
CDI	*Clostridium difficile* infections
CPIs	checkpoint inhibitors
DCs	dendritic cells
FMT	Faecal microbiota transplantation
GBA	gut–brain axis
GI tract	gastrointestinal tract
IAA	indole-3-acetic acid
IBD	inflammatory bowel disease
IBS	irritable bowel syndrome
ILs	Interleukins
LCA	lithocholic acid
LGG	*Lactobacillus rhamnosus* GG
LPS	lipopolysaccharides
MS	multiple sclerosis
NAFLD	non-alcoholic fatty liver diseases
NEC	necrotizing enterocolitis
NGS	next-generation sequencing
NSCLC	non-small cell lung cancer
RA	rheumatoid arthritis
RRMS	Relapsing–Remitting Multiple Sclerosis
SCFA	short chain fatty acids
T1D	type-1 diabetes
TLR-2	toll-like receptor-2
TLRs	toll-like receptors
Treg cells	regulatory T cells

1 Introduction

The human microbiota includes a vast and complex microbial community inhabiting the various regions of the human body such as skin, oral cavity, respiratory tract and chiefly the gut. Among these, densely populated microbiota colonies of the gut are the most diverse and densely populated, with over 1000 different species of bacteria contributing to essential physiological processes (Hou et al. 2022). The composition of the human microbiome varies significantly among individuals and is influenced by genetics, diet, age, environmental factors and medical interventions also (Hou et al. 2022; Hasan and Yang 2019). The dominant microbiota phyla in the gut includes *Firmicutes* and *Bacteroidetes*, along with smaller populations of *Actinobacteria* and *Proteobacteria* (Li and Ma 2020). This dynamic ecosystem regulates health by facilitating digestion, synthesizing vitamins and competing against pathogenic organisms.

The interaction between the microbiota and the host's immune system is critical for immune homeostasis. Gut-associated lymphoid tissues (GALT) are in close communication with gut microbes, modulating immune responses to prevent over-inflammation while defending against pathogens (Mörbe et al. 2021). Additionally, microbiota-produced metabolites, such as short-chain fatty acids (SCFAs), interact with metabolic pathways, influencing energy balance, glucose metabolism and lipid regulation (Portincasa et al. 2022). Beyond metabolic functions, the microbiota significantly impacts host physiology by influencing the gut–brain axis, playing a role in mood regulation, cognitive function and neurological health (Hrncir 2022).

The human microbiota has emerged as a promising therapeutic target owing to its critical functions such as metabolism, integrity of the gut barrier and immune responses. Healthy microbiota composition aids in nutrient absorption, protection against pathogens and production of essential metabolites such as SCFAs, bile acid derivatives, indole derivatives and amino acid derivatives (Fusco et al. 2023; Rauf et al. 2022). Conversely, dysbiosis, an imbalance in microbial composition, has been linked to a variety of conditions, including inflammatory bowel disease (IBD), obesity, diabetes, cancer and autoimmune disorders (Mousa et al. 2022; Gebrayel et al. 2022). For example, in IBD, reduced microbial diversity exacerbates intestinal inflammation, while in metabolic disorders, gut dysbiosis influences insulin resistance and fat storage. Emerging research shows that restoring balance to the microbiota through probiotics, prebiotics and dietary interventions can support immune modulation and disease prevention (Menafra et al. 2024; Valdes et al. 2018) (Fig. 1).

2 Emergence of Microbiota as a Therapeutic Target: Evolution from Traditional Therapies to Microbiota-Driven Interventions

The evolution from traditional therapies to microbiota-driven interventions reflects a significant paradigm shift in medicine. Traditional approaches, such as antibiotics or anti-inflammatory drugs, primarily targeted disease symptoms or pathogens directly. However, these often disrupt the microbiome, leading to long-term health implications. In contrast, microbiota-targeted therapies aim to restore microbial balance and overall health. The mechanism by which microbiome-based interventions work can differ based on the approach used, ranging from reshaping the overall

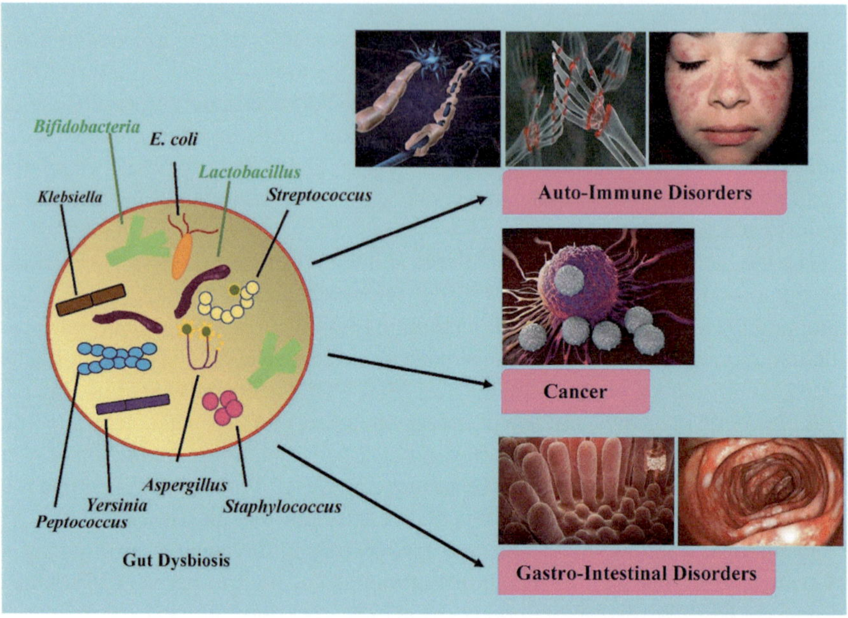

Fig. 1 Dysbiosis of the GI tract induces autoimmune, cancer and gastrointestinal disorders

microbial community or introducing specific species. This variability allows for targeted or broad-spectrum interventions depending on the desired therapeutic outcome. A principal intervention technique in microbiota-driven therapies is the faecal microbiota transplantation (FMT). It involves the transfer of prepared stool from a healthy donor to a diseased recipient, using the lower gastrointestinal (GI) endoscopy (Hitch et al. 2022). However, concerns have been raised about the widespread use of FMT due to the lack of standardization and the complexity of materials involved. Faecal filtrates, designed to remove microbial cells while retaining viruses and small molecules from the donor, offer a potential solution. In fact, faecal filtrate transfer (FFT) has shown promising results in treating *Clostridium difficile infection* (CDI), likely because of the transfer of bacteriophages (Ott et al. 2017). Additionally, comparing the efficacy of faecal sediment and supernatant for *C. difficile* treatment revealed a 27% improvement with the supernatant, highlighting the importance of active components beyond just the bacterial fraction (Varga et al. 2023).

Other strategies, such as precision probiotics, dietary polyphenols and prebiotics, focus on enhancing beneficial microbial populations. These microbiome-based approaches are now being integrated into treatment protocols for autoimmune disorders, metabolic diseases and cancer, highlighting the transition to personalized, microbiota-driven care. Gut microbiome is influenced by different diets, which in turn modulates immune responses. Whole plant foods, such as fruits, vegetables, whole grains, legumes and nuts, are rich sources of dietary fibres (Armet et al.

2022). Fermentable fibres are broken down into SCFAs by the gut microbes that influence the immune system through regulating the inflammatory mediators (Ranjbar et al. 2021a). Among SCFAs, butyrate regulates extrathymic regulatory T cells (Tregs), while propionate stimulates de novo Treg cell production, in mice. Fibre-rich diet maintains a healthy intestinal mucus layer lowering the risk of pathogenic infections by forming a strong barrier (Desai et al. 2016).

Prebiotics are substances that selectively nourish beneficial microorganisms in the gut, leading to health benefits. Though the definition of prebiotics is debated, non-digestible carbohydrates (NDCs) influence the composition and function gut microbiome, whereas probiotics are live microorganisms that enhance gut diversity. These can either include one species or complex mixtures of different species and are taken through food or supplements. A well-known probiotic strain, *Lacticaseibacillus rhamnosus* GG (LGG), is useful for the treatment of C. difficile colitis and improving weight loss in obese mice (Shah et al. 2024). Bacteriophages (bacteria targeting viruses) are being explored for targeted therapies, such as treating Clostridium difficile infections (CDI) (Venhorst et al. 2022). Phage cocktails have been developed for the elimination of selected pathogenic bacteria while thriving the beneficial ones. However, this therapy can have unintended effects on non-target species due to inter-species microbial interactions. Postbiotics are non-living microbiota derived bioactive substances with health benefits. For example, membrane protein, Amuc_1100.2, present in the pasteurized *Akkermansia muciniphila* cells has shown to improve the metabolic profile (He et al. 2023).

3 Microbiota and Autoimmune Disorders

3.1 Autoimmune Diseases

The failure of immune system to distinguish between self and non-self precipitates the autoimmune disorders. Numerous autoimmune conditions are identified and include multiple sclerosis (MS), rheumatoid arthritis (RA), type 1 diabetes (T1D), lupus and many more (Fugger et al. 2020). Disturbed immune responses are observed in the autoimmune conditions, where the T and B cells target the specific organs or tissues. For instance, in T1D, β-cells are destroyed by the T-cells, while in RA, immune cells attack the joints, causing chronic inflammation and joint damage (Yang et al. 2021).

The risk factors for autoimmune diseases include genetics, environment, stress and hormones. Research has shown that dysbiosis is one of the primary causes of autoimmune conditions, where the gut barrier is impaired increasing the intestinal permeability promoting systemic inflammation (Pan et al. 2021). The environmental triggers, such as infections, stress and certain medications, can further exacerbate the autoimmune responses. Therapeutic strategies now focus on modulating immune responses, often using biologics like monoclonal antibodies, alongside emerging microbiota-based therapies that improve dysbiosis and support immune regulation (Alam et al. 2023).

3.2 Gut Microbiota in Autoimmunity

3.2.1 Dysbiosis and Immune Dysregulation

Gut dysbiosis interferes with the delicate balance between immune tolerance and immune activation developing the autoimmune diseases. Intestinal microbiota promotes the production and recruitment of Tregs that prevent overactivation of the immune response maintaining homeostasis. Henceforth, dysbiosis leads to the expansion of pro-inflammatory microbial species, reducing Treg populations and increasing T helper 17 cells, which are implicated in autoimmune responses (Mousa et al. 2022). For instance, IBD and RA are associated with reduced microbial leading to the over activation of immune pathways producing chronic inflammation and tissue damage (Yue et al. 2019; Padoan et al. 2023). The gut barrier also regulates the immune responses, by preventing the harmful microbial products from entering the bloodstream, whereas in autoimmune conditions, dysbiosis increases the gut permeability ('leaky gut'), allowing microbial antigens, lipopolysaccharides (LPS) and other toxins migrate into the circulatory system, triggering systemic inflammation and precipitating autoimmune conditions (Mu et al. 2017).

3.2.2 Microbial Antigens and Molecular Mimicry

One of the primary mechanisms of the induction of autoimmunity via gut dysbiosis is molecular mimicry. This occurs when microbial antigens closely resemble host antigens, and the immune system mistakenly attacks the self tissues.

Hence, the immune system fails to differentiate the pathogen and the host tissues and attacks both. Several autoimmune diseases are thought to arise from this mechanism. For instance, in periodontal disease, the bacterium *Porphyromonas gingivalis* produces antigens that mimic human proteins involved in joint tissues, triggering immune responses that lead to RA (Mahmoudi et al. 2023). In T1D, the microbial antigens mimic the β-cells, further activating the autoreactive T cells and destroying them (Pugliese 2017). Similarly, in MS, certain gut bacteria produce proteins that mimic myelin, a protective sheath surrounding nerve fibres. This can lead to immune-mediated destruction of myelin, causing neurological damage characteristic of MS (Cavallo 2020). The ability of certain microbes to imitate host antigens thus becomes a driving force in the pathogenesis of autoimmune diseases. This growing understanding has led to innovative therapies, such as FMT and precision probiotics, aimed at restoring microbial balance and reducing autoimmune responses by modulating the gut–immune axis.

3.2.3 Immune Regulation Through Microbiota-Based Interventions

The onset of immune development occurs with the inception of the intestinal microbiome during early period of life. Interference in the microbes at this stage is associated with the higher occurrence of immune-mediated disorders. Studies suggest the involvement of maternal factors, like diet and specific microbes (e.g., *Bifidobacterium*), in the microbiota and immune development, in later life (Vandenplas et al. 2020). Bifidobacterium, particularly in human milk, aids in immune programming. Its absence in infants can lead to heightened inflammation, while its presence promotes anti-inflammatory immune responses. Studies show that supplementation with strains like *Bifidobacterium longum* subsp. *infantis* EVC001 enhances immune regulation by modulating T-cell responses, through the interaction of like indole-3-lactic acid (ILA) metabolites, via the aryl hydrocarbon receptor (AhR) (Gavzy et al. 2023). Furthermore, strengthening the epithelial barrier during infancy is crucial in preventing the immune system overstimulation, which may lead to conditions like necrotizing enterocolitis (NEC). Supplementation with Bifidobacterium, especially in preterm infants, has been shown to reduce the incidence of NEC by promoting epithelial regeneration and reducing cell death in the intestines (Sadeghpour Heravi and Hu 2023). These findings underscore the importance of gut microbiota in immunity development and protecting against early-life diseases.

3.2.3.1 Modulation of Innate Immune Function

Probiotics and prebiotics significantly improve the functioning of the epithelial barrier, not only in early life but also in adults. The weakened barrier integrity due to leaky gut allows the microbial products to enter the body and develop diseases. Microbiota-derived SCFAs improve intestinal barrier function. For example, *Bifidobacterium bifidum* produces acetate, which reduces intestinal permeability and susceptibility to colitis. Butyrate, when provided in controlled doses, enhances the expression of proteins like occludin and claudins, which strengthen tight junctions in the gut. However, excessive butyrate can have adverse effects, increasing permeability (Yao et al. 2021).

Probiotic strains like *Lactobacillus rhamnosus GG* (LGG) also play a key role in barrier integrity. LGG-derived proteins prevent intestinal apoptosis, protecting the gut barrier. Moreover, the bacterial protein Amuc_1100 from *Akkermansia muciniphila* enhances TJ proteins and improves the epithelial barrier (Mo et al. 2024). The pasteurized cells of *A. muciniphila* reduce the markers of gut permeability-like plasma LPS levels, highlighting their potential therapeutic benefits (Goldbaum 2022). The mucus layer can be influenced by butyrate due to the increased expression of mucin genes, strengthening the mucus layer, reducing inflammation and enhancing resistance to infections. Specific microbes, including Bifidobacterium and *A. muciniphila*, have been shown to stimulate goblet cell activity, increasing mucus production, supporting a healthier gut lining. Additionally, some phases like phiEG37k can stimulate mucin production, reducing inflammation and maintaining intestinal barrier (El Haddad et al. 2022). Microbiota-based interventions also influence the gut dendritic cells (DCs) and macrophages. Butyrate and conditioned media from probiotics improve the antimicrobial effects of immune cells. The tolerogenic properties in DCs are modulated by specific probiotic strains, protecting against conditions like colitis (Baradaran Ghavami et al. 2021).

3.2.3.2 Modulation of Cellular Immune Function

The adaptive immune system is significantly influenced by microbiome-based interventions, influencing T cells and other immune components. For example, propionate supplementation (1000 mg/day) has been shown to expand Tregs in renal diseases and MS (Gill et al. 2022). It also decreases C-reactive protein (CRP) levels, while increasing interleukin-10 (IL-10) production by Tregs, which inhibits pro-inflammatory T-helper 17 (Th17) cells. Bile acid metabolism by gut bacteria also affects the immune system. Lithocholic acid (LCA) and its derivatives, 3-oxolithocholic acid and isolithocholic acid, inhibit Th17 differentiation without affecting Tregs or Th1 cells (Zhang et al. 2023a). This mechanism, coupled with the ability of LCA to inhibit Th1 cell differentiation, suggests that bile acid metabolism modulates T cell populations. Moreover, certain microbial metabolites, such as IAA produced from tryptophan metabolism, activate AhR pathway, enhancing IL-22 production and promoting gut recovery in colitis models. Other immune cells, including macrophages and DCs, are also modulated by microbiome interventions. For instance, the probiotic consortia GUT-108 reduces CD4+ T cells and inflammatory cytokines in mice, though human trials are needed for validation (Yang et al. 2023). Additionally, gut microbes such as *Lacticaseibacillus casei* produce proteases like lactocepin, which degrades pro-inflammatory cytokines and has shown potential in managing IBD. Cytokine modulation by the microbiome is crucial in immune regulation. Certain species, such as *Bacteroides fragilis*, regulate the immune response to checkpoint inhibitors (CPIs) like CTLA-4 blockade (Dehinsilu et al. 2023). FMT from CPI responders into non-responders has shown success, indicating that specific microbial communities can improve immune responses during cancer therapy.

These findings highlight the capacity of microbiome-based therapies in managing immune-mediated conditions and enhancing the efficacy of immunotherapies, though further research is necessary to refine these approaches for clinical applications (Fig. 2).

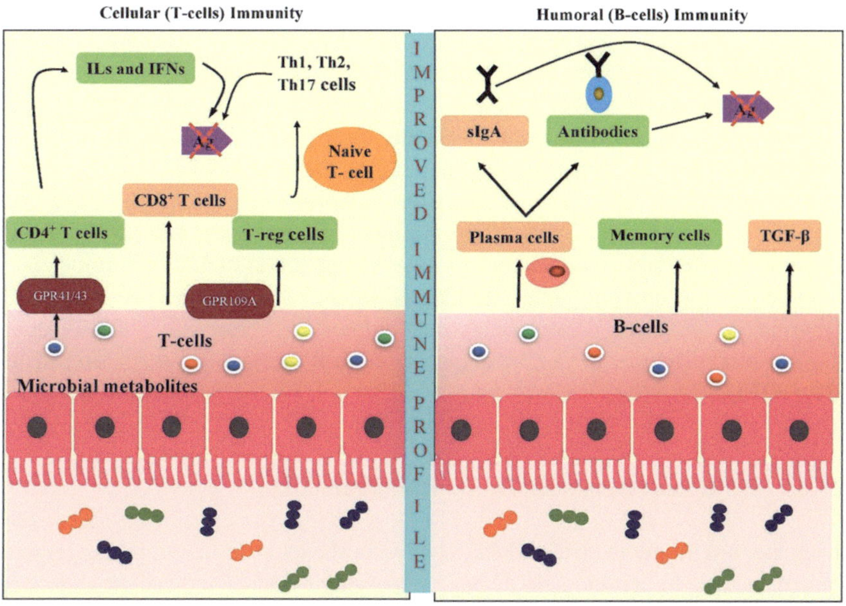

Fig. 2 Modulation of cellular and humoral immunity by microbiota-derived microbial metabolites (SCFAs)

3.3 *Precision Therapies for Autoimmune Diseases*

Precision therapies for autoimmune diseases are revolutionizing the treatment by targeting the underlying biological mechanisms that drive immune dysfunction, often linked to gut dysbiosis. Traditional approaches, such as immune suppressants, typically offer broad immunomodulation but lack specificity, leading to widespread side effects. In contrast, precision therapies influence the immune system in a targeted way, addressing dysregulated immune responses through interventions that restore balance in the gut microbiota. These therapies include probiotics, prebiotics, dietary modifications and advanced techniques like FMT. By enhancing beneficial microbial populations or introducing healthy donor microbiota, these approaches aim to mitigate autoimmune responses while preserving immune tolerance. This shift from generalized treatment to microbiota-driven strategies holds promise for chronic conditions like RA, MS and IBD.

3.3.1 RA: Modulation of Gut Bacteria to Reduce Inflammation

In RA, the immune system chronically attacks the synovial joints inducing inflammation. It is evident from the recent studies that gut dysbiosis induces and exacerbates inflammation in RA. The bacterial species *Prevotella copri* is linked with increased inflammation and more severe RA symptoms (Coradduzza et al. 2023). Modulating the gut microbiota to restore a healthy balance is emerging as a precision therapy for RA (Horta-Baas et al. 2021). Approaches include the use of probiotics that introduce beneficial bacteria, as well as dietary interventions that promote anti-inflammatory gut microbiota. These therapies aim to reducing joint inflammation and slow disease progression by modulating the immune regulatory pathways, such as increasing Tregs and reducing pro-inflammatory cytokines (Horta-Baas et al. 2021; Reyes-Castillo et al. 2021). Results of clinical trials reflect the potential of *Lactobacillus* and *Bifidobacterium* in reducing the systemic inflammation by modulating gut-immune interactions (Kaźmierczak-Siedlecka et al. 2021). Additionally, prebiotic dietary fibres that promote the production of SCFAs are being explored as therapeutic strategies to control immune dysregulation in RA (Kim 2023).

3.3.2 MS: Role of Microbiota in Neuroinflammation

In another autoimmune disease, MS, the myelin sheath is destroyed by the immune system, leading to neurological damage. Emerging evidence indicates a significant role of gut microbes in inducing neuroinflammation in pathogenesis of MS. Gut dysbiosis in MS patients is associated with enhanced intestinal permeability and systemic immune activation, causing CNS inflammation. Studies have identified an altered microbiota composition in MS patients, with a reduction in *Bacteroides* species and increase in pro-inflammatory species like *Akkermansia muciniphila* (Amini et al. 2020; Schepici et al. 2019). Miyake et al. found lower levels of Clostridium (specifically Clostridia clusters XIV and IV) in Relapsing–Remitting Multiple Sclerosis (RR-MS) patients, decreasing SCFAs production. As discussed earlier, SCFAs are crucial for the generation of Tregs and anti-inflammatory cytokine IL-10 (Haghikia et al. 2015). Increased prevalence of Firmicutes genera, like *Blautia* and *Dorea* and *Bacteroidetes* genera like *Pedobacteria* and *Flavobacterium*, was found in MS patients compared to healthy controls (Chen et al. 2016). Conversely, other *Bacteroidetes* genera, including *Parabacteroides*, *Bacteroides* and *Prevotella*, are found in lower abundance in MS patients. Notably, lipodipeptide Lipid 654, produced by Bacteroidetes commensals, are the ligands for toll-like receptor 2 (TLR-2), and their serum levels are observed to be reduced in MS patients. Lipid 654 modulates the immune responses by maintaining TLR-2 and interferon-beta (IFN-β) signalling cascade. In addition to the decline in *Bacteroidetes* genera, reduced *Adlercreutzia* levels are found in RRMS patients. *Adlercreutzia* is involved in the phytoestrogen metabolism modulating anti-inflammatory responses. Bacteria such as *Adlercreutzia* utilize β-glucosidase enzymes to convert phytoestrogens into their monomeric forms (Hu et al. 2024). In individuals with RRMS, a decrease in *Adlercreutzia* leads to a reduced ability to metabolize phytoestrogens, resulting in increased oxidative stress and higher levels of chemokine protein-1 and IL-6 (Baranowska-Wójcik et al. 2024). Thus, increasing *Adlercreutzia* could reduce the inflammatory state in RRMS patients, potentially delaying disease progression.

Therapeutic strategies are now focusing on modulating the gut microbiota to regulate neuroinflammatory responses. Precision therapies using probiotics or dietary interventions restore the normal microbial flora and reduce inflammation in the CNS. They exert neuroprotective and anti-inflammatory effects by increasing the production of SCFAs. Furthermore, clinical studies involving microbiota-based therapies in combination with traditional MS treatments are underway to assess their efficacy in improving patient outcomes (Schepici et al. 2019; Mangalam and Murray 2019).

3.3.3 IBD: Personalized Probiotics and FMT

IBD involves chronic inflammation in the GI tract, where dysbiosis is its characteristic feature. It includes ulcerative colitis and Crohn's disease.

Precision therapies focus on the utilization of personalized probiotics and FMT for reducing inflammation by maintaining a healthy gut microbiome. FMT is now explored for IBD after showing promising results in treating recurrent CDI. By restoring microbial diversity, FMT promotes immune regulation and decreases intestinal inflammation (Baranowska-Wójcik et al. 2024). Personalized probiotics, tailored to the individual's microbiota profile, are being developed to target specific microbial imbalances in IBD patients. These precision therapies improve the immunological responses, enhance intestinal barrier integrity, thus preventing disease flare-ups, offering a more targeted approach compared to traditional anti-inflammatory drugs (Fig. 3).

3.4 Challenges and Opportunities in Precision Therapies for Autoimmune Diseases

1. Identifying Specific Microbial Strains

One of the primary challenges in developing microbiota-driven therapies is identifying specific microbial strains that can modulate immune responses beneficially in autoimmune diseases. Gut microbiota is incredibly diverse, where the same microbial strain can have varying effects depending on the individual's microbiota composition, genetics and environment. For example, while certain *Lactobacillus* and *Bifidobacterium* strains promote Treg cells in conditions like RA, their efficacy can vary significantly across patient populations (Bungau et al. 2021). The task of pinpointing which strains are effective in mitigating autoimmune responses requires large-scale, individualized microbiome analysis and a deeper understanding of how these microbes interact with both local and systemic immune cells.

2. Overcoming the Complex Microbiota–Host Interactions in Autoimmune Therapies

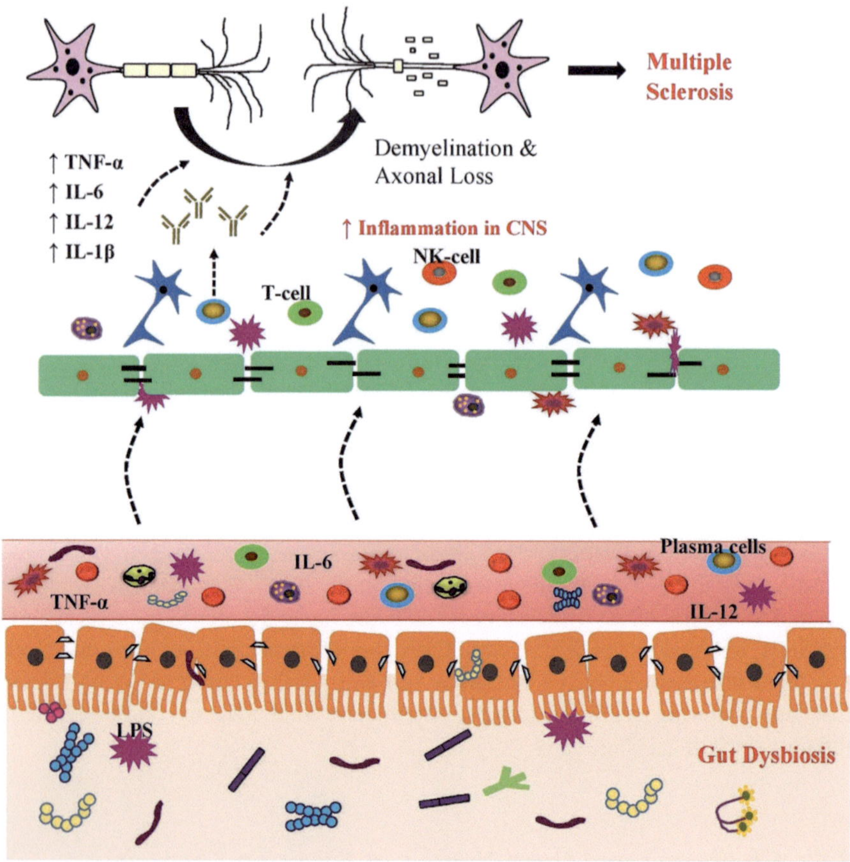

Fig. 3 The epithelial cells of the GI tract are weakened due to gut dysbiosis, infiltrating toxic metabolites and immune cells in the blood. It further crosses the periphery of the blood–brain barrier, where the immune cells consider the myelin sheath as non-self, causing its destruction and precipitating MS

Since the microbiota–host interactions are complex, they pose a significant hurdle in precision autoimmune therapies. It is challenging to forecast the outcomes of microbiome modification since the gut microbiota interacts with the host's immune system, metabolism and CNS. Additionally, the same microbial intervention might have different outcomes depending on the disease stage, patient genetics or environmental factors. For example, while FMT has shown promise in treating conditions like IBD, its efficacy in other autoimmune diseases like MS is still under investigation due to the intricate and poorly understood interactions between gut microbes and neuroinflammation. Despite these complexities, the development of precision diagnostics, such as microbiome sequencing and immune profiling, offers opportunities to refine treatments and personalize therapies, tailoring interventions to individual microbial profiles and immune system needs (Shukla et al. 2024).

4 Microbiota-Driven Approaches in Cancer Therapy

4.1 Brief Overview of Cancer

Unchecked cell proliferation and the ability to invade or spread to other regions of the body are characteristics of the cancers. The environmental factors like smoking, ultraviolet (UV) rays and infections are responsible for the genetic mutations inducing cancer. These mutations disrupt normal cell regulation, allowing cells to proliferate uncontrollably. Based on the grade, location and type of cancer, its treatment typically involves surgery, chemotherapy, radiotherapy and, more recently, immunotherapy.

4.2 Role of Microbiota in Cancer Progression

Intestinal microbiota is significantly involved in cancer development, influencing its promotion and suppression, both. Dysbiosis is associated with inflammation, DNA damage and immune dysfunction, contributing to tumorigenesis (Vimal et al. 2020). *Fusobacterium nucleatum* promotes colorectal cancer progression by activating inflammatory pathways and inhibiting the immune system's ability to attack tumour cells (Ou et al. 2022). Conversely, a healthy microbiota reduces the risk of cancer development through modulating the epithelial barrier integrity, immune responses and anti-inflammatory mechanisms. *Fusobacterium nucleatum*, typically absent in the healthy gut, can damage the colon's epithelial lining and promote cancer (Ranjbar et al. 2021b). Through the *Fusobacterium nucleatum* adhesion (FadA) mechanism, *F. nucleatum* induces the β-catenin signalling pathway by binding to E-cadherin, inducing oncogene expression and altered immune responses. This weakens the epithelial barrier, allowing the bacteria to infiltrate tumours and stimulate cell proliferation. Additionally, *F. nucleatum* induces chronic inflammation and DNA damage, contributing to tumour growth. Similar mechanisms are observed in *Bacteroides fragilis*, which produces a toxin that also activates β-catenin signalling, further promoting cancer development. Both bacterial species induce the release of pro-inflammatory ILs, that is, IL-17 and IL-23, creating an environment conducive to tumour growth (Borroni et al. 2019; Wu et al. 2019).

One species of the gut microbiota is *Enterococcus faecalis*. Its aberrant activation causes colorectal carcinomas and damages DNA. By means of demethylmenaquinone-mediated autoxidation, E. faecalis can generate extracellular superoxide (O*2), hydroxyl radical (OH^-) and H_2O_2. In a mouse model, *E. faecalis* was shown to be capable of causing DNA damage, and it was discovered that these free radicals were the source of the chromosomal instability linked to colorectal cancer (Kumar et al. 2023). The intestinal microbiota in the periphery aids in the development of adaptive immunity in the GI tract. Intestinal homeostasis is controlled by the interaction between intestinal mucosa and bacteria, which is mediated by TLRs. T cells and macrophages trigger a TLR-dependent pathway when the mucosal barrier breaks, which promotes the proliferation of epithelial cells and draws inflammatory cells to the area. Additionally, *Helicobacter pylori* and *Listeria monocytogenes* infections can activate TLRs, which eventually cause carcinogenesis (Table 1).

4.3 Influence on Immune Checkpoint Blockade, Chemotherapy and Radiotherapy Outcomes

The effectiveness of cancer treatments, such as immune checkpoint inhibition, chemotherapy and radiation, is also strongly influenced by the microbiota. By increasing T cell activity, the presence of particular gut bacteria, such as *Bifidobacterium* and *Akkermansia muciniphila*, has been linked to better responses in cancer patients receiving immune checkpoint inhibitors, such as anti-PD-1 (programmed cell death protein-1) or anti-CTLA-4 treatments. (Derosa et al. 2022; Kang et al. 2024). On the other hand, dysbiosis can lead to reduced treatment efficacy or increased side effects. In chemotherapy and radiotherapy, certain microbial species modulate the host's response to treatment, influencing both therapeutic success and the severity of adverse reactions (Li et al. 2023). Understanding how the microbiota affects cancer treatments creates new opportunities to improve treatment effectiveness using microbiome-targeted therapy like FMT or probiotics.

Table 1 List of microorganisms that the International Agency for Research on Cancer (IARC) has classified as Class 1 (carcinogens) (Cancer IAfRo 2012)

Microbes	Cancer sites
Helicobacter pylori	Stomach
Hepatitis B virus (HBV) Hepatitis C virus (HCV) *Opisthorchis viverrini* *Clonorchis sinensis*	Liver
Human papillomavirus (HPV)	Cervix Vagina Vulva Anus Penis Oropharynx
Epstein-Barr virus (EBV)	Nasopharynx Non-Hodgkin lymphoma Hodgkin's lymphoma
Kaposi's sarcoma-associated herpesvirus (KSHV or HHV8)	Kaposi's sarcoma Primary effusion lymphoma
Human T-cell lymphotropic virus type 1 (HTLV-1)	Adult T-cell lymphoma
Schistosoma haematobium	Bladder

4.4 *Enhancing Cancer Therapies Through Microbiota Modulation*

4.4.1 FMT in Immunotherapy

FMT appears to have potential for enhancing the efficacy of immunotherapy in the treatment of cancer. FMT attempts to re-establish a balanced microbiome in cancer patients by transplanting the gut microbiota from healthy donors to cancer patients, which may enhance immune responses. Research has indicated that FMT may improve the disease outcomes in individuals receiving immune checkpoint inhibitors, including anti-PD-1 treatments. For example, FMT enhanced treatment responses and survival rates in individuals with melanoma and non-small cell lung cancer (NSCLC) (Xin et al. 2024). The fundamental process is believed to be immune system regulation, in which advantageous microorganisms boost T cell activation and broaden the range of immune responses against tumours.

4.4.2 Personalized Gut Microbiome Interventions to Improve Patient Response Rates

Personalized interventions targeting the gut microbiome are increasingly recognized for their potential to improve patient response rates to cancer therapies. By analysing an individual's microbiome composition, clinicians can tailor interventions that enhance microbial populations associated with positive therapeutic outcomes. For example, personalized dietary modifications, prebiotic supplementation or specific probiotic strains can be recommended based on a patient's microbiome profile. By improving the gut environment, this strategy seeks to foster the development of advantageous microorganisms that could improve immunological response and treatment effectiveness. Research indicates that such personalized strategies can lead to improved responses in patients receiving chemotherapy or immunotherapy, thereby advancing the concept of precision oncology.

4.4.3 Probiotics and Prebiotics as Adjuncts to Existing Cancer Therapies

Prebiotics and probiotics are being investigated as supplemental therapy to improve the efficacy of traditional cancer treatments. Live beneficial bacteria, or probiotics, can help heal the gut microbiota and reduce the negative consequences of treatment, like chemotherapy-induced mucositis and diarrhoea (Batista et al. 2020). The body's immunological response can be strengthened and the microbiome's general health be supported by prebiotics, which are indigestible fibres that encourage the growth of good gut bacteria. Numerous studies have shown that taking probiotics together with cancer therapies can enhance overall quality of life for patients, reduce the side effects of chemotherapy and increase treatment tolerance and results. As research progresses, the integration of microbiota modulation strategies into cancer treatment protocols holds significant promise for optimizing patient care and improving therapeutic outcomes (Table 2).

4.5 Case Studies and Clinical Trials

Table 3 summarizes the results of different studies illustrating the successful microbiota modulation in cancer treatment, along with key clinical trials and ongoing research in microbiome-cancer therapy integration.

Table 2 Strain specific effect of probiotics on different functions (López-Gómez et al. 2023)

Regulation of dysbiosis	*Lacticaseibacillus casei* *Lacticaseibacillus rhamnosus* *Bifidobacterium bifidum* *Bifidobacterium breve* *Clostridium butyricum*
Modulation of epithelial barrier permeability	*Lacticaseibacillus casei* *Lactobacillus delbrueckii* *Propionibacterium freudenreichii* *Saccharomyces cerevisiae*
Anti-inflammatory effects	*Lacticaseibacillus casei* *Lacticaseibacillus rhamnosus* *Lactobacillus delbrueckii* *Bifidobacterium bifidum* *Bifidobacterium infantis* *Streptococcus thermophilus* *Propionibacterium freudenreichii* *Clostridium butyricum* *Saccharomyces boulardii*
Modulation of host immune response	*Lactobacillus delbrueckii* *Lactobacillus acidophilus* *Bifidobacterium bifidum* *Bifidobacterium infantis* *Streptococcus thermophilus* *Propionibacterium freudenreichii* *Saccharomyces cerevisiae* *Saccharomyces boulardii*
Reduction in oxidative stress	*Saccharomyces cerevisiae* *Saccharomyces boulardii*
Prevention of apoptosis	*Saccharomyces boulardii* *Lacticaseibacillus rhamnosus*

Table 3 Clinical research and case reports illustrating how probiotic intervention affects cancer modulation

Case studies/ clinical trials	Description	Outcomes	References
FMT in melanoma patients	Study involving FMT from healthy donors to melanoma patients undergoing anti-PD-1 therapy	Improved treatment response rates and overall survival compared to controls	Vongsavath et al. (2024) and Baruch et al. (2021)
Probiotics in (NSCLC)	Clinical study investigating the effects of *Lactobacillus* and *Bifidobacterium* probiotics in combination with chemotherapy for NSCLC	Reduced chemotherapy-related side effects and improved immune responses	Wei et al. (2024) and Pizzo et al. (2022)
Personalized microbiome interventions	Personalized dietary modifications and prebiotic supplementation based on individual microbiome profiles in cancer patients	Enhanced treatment efficacy and tolerance in patients receiving immunotherapy	Russo et al. (2020) and Chervin and Gajewski (2020)
IBD and cancer treatment	Effect of gut microbiota modulation on cancer therapy outcomes in IBD	Importance of microbiota balance in improving therapeutic responses and minimizing adverse effects is highlighted	Khan et al. (2019)
Microbiome and radiotherapy	Clinical study investigating how the gut microbiota may improve the effects of radiation therapy in cancer patients	Indications that a healthy microbiome may boost immune responses and reduce toxicity from radiation therapy	Yu et al. (2021)

5 Advancements in Gastrointestinal Disease Treatment Through the Modulation of Gut Microbiota

5.1 Microbiota in Gastrointestinal Health: Implication of Gut–Brain Axis (GBA) in GI Disorders

GBA is a complex network of neurological, hormonal and immunological pathways connecting the brain and GI tract. This relationship is crucial for preserving gastrointestinal health and is becoming more widely acknowledged for its effects on a number of GI tract conditions. According to recent studies, GBA is greatly influenced by gut microbiota, having an impact on mental and physical health. For example, neuroactive substances like SCFAs influence the CNS and change mood and behaviour (Liu et al. 2022a). Dysregulation in this communication can lead various GI tract disorders, such as functional dyspepsia, irritable bowel syndrome (IBS) and IBD. In conditions such as IBS, the GBA may be disrupted, leading to altered gut motility, increased sensitivity and changes in gut permeability (Tang et al. 2021). The emerging therapies such as probiotics and dietary patterns, influence the gut microbiota, restoring their balance and improving both gastrointestinal symptoms and associated psychological factors. Additionally, GBA has opened new avenues for innovative treatment therapies, including the use of microbiota modulation to enhance mental health outcomes in patients with GI tract disorders.

5.2 Dysbiosis in GI Tract Diseases Like IBS, IBD and Celiac Disease

Dysbiosis is deeply associated with functional gastrointestinal diseases, such as IBS, IBD and celiac disease. In IBS, studies have shown a reduced diversity of gut microbiota, representing increased pro-inflammatory bacteria and reduced good gut bacteria such as *Bifidobacterium* and *Lactobacillus* (Napolitano et al. 2023). This contributes to IBS symptoms, including pain, bloating with altered bowel patterns, suggesting that restoring a healthy microbiome may offer therapeutic benefits (Principi et al. 2018). Dysbiosis in IBD has been linked to increased intestinal permeability and flare-ups of inflammation (Ramos and Papadakis 2019). Research indicates that IBD patients often have reduced anti-inflammatory microbes, further complicating the disease process. Two examples of therapy strategies that focus on the microbiota are being investigated as potential treatments to lower inflammation and restore microbial balance are probiotics and FMT. Similar to this, certain microbial strains have been linked to the regulation of immune responses to gluten in celiac disease, underscoring the significance of the microbiome in disease management and possible treatment pathways (Cristofori et al. 2018). Advancements in understanding of intestinal microbiota and its role in GI health are reshaping the landscape of treatment options for GI tract disorders. By targeting dysbiosis through microbiota modulation, new therapies hold promise for improving patient outcomes in conditions like IBS, IBD and celiac disease. Ongoing research into the GBA and microbial interactions continues to uncover novel strategies for enhancing gastrointestinal health.

5.3 Precision Therapies for Gastrointestinal Diseases

5.3.1 Tailored Prebiotics, Probiotics and Synbiotic for Treating GI Conditions

Customized prebiotics, probiotics and synbiotics have been developed as precision therapeutics for GI-related issues because of recent advances in our understanding of the gut flora. Probiotics are live microorganisms that provide health advantages when given in sufficient quantities, whereas prebiotics are dietary fibres that specifically promote the growth of good gut flora. To increase its effectiveness, synbiotics combine probiotics and prebiotics. Studies have shown that tailored formulations can alleviate symptoms in conditions such as IBS, IBD and functional dyspepsia. For instance, it has been demonstrated that some strains of *Lactobacillus* and *Bifidobacterium* lessen bloating and stomach discomfort in IBS patients (Pratt and Campbell 2020). While tailored prebiotic fibres like inulin can support these strains' growth, enhancing their therapeutic effects (Fan et al. 2023). The personalization of these therapies based on individual microbiome profiles and symptomatology is a promising approach. Clinical trials are increasingly focusing on the development of individualized probiotic regimens, which can optimize treatment outcomes and minimize side effects by aligning microbial interventions with the patient's unique gut microbiota composition.

5.4 FMT in Treating CDI and IBD

FMT has demonstrated potential in the management of IBD and has become a potent therapeutic option for recurrent CDI, with success rates exceeding 90% (Goldenberg and Merrick 2021). The restoration of a diverse and healthy microbiota helps to outcompete *C. difficile* for resources, thereby reducing recurrence rates. According to preliminary evidence, FMT may assist patients with Crohn's disease and ulcerative colitis achieve remission in IBD by regulating immune responses and resolving dysbiosis (Zhou et al. 2023a). Nevertheless, more thorough clinical research is required to create standardized procedures and determine the best donor traits and FMT administration techniques.

5.5 Role of Microbial Metabolites in GI Health

Microbial metabolites, especially SCFAs like butyrate, propionate and acetate, aid in controlling inflammation and GI health. They are formed from the fermentation of fibre-rich diet, and have been demonstrated to offer a host of health advantages, including enhancing the function of the gut barrier, controlling immunological responses and lowering inflammation. For example, butyrate serves as colonocytes' main energy source and contains anti-inflammatory qualities that can aid in the treatment of inflammatory diseases like IBD and IBS (Gasaly et al. 2021). Studies highlight the potential of SCFAs in therapeutic applications, emphasizing their ability to influence gene expression and signalling pathways involved in gut health. As such, diets rich in fibre or specific prebiotics that promote SCFA production are being recommended as part of a comprehensive approach for the management of GI disorders (Guarino et al. 2020). The exact mechanisms by which SCFAs work are the subject of ongoing research, with the goal of using these metabolites to create novel treatment approaches for GI disorders.

5.6 *Gut Microbiota in Liver and Pancreatic Disorders*

Microbiota's Influence on Liver Diseases

The pathophysiology of non-alcoholic fatty liver disease (NAFLD) is significantly influenced by gut microbiota. Excessive hepatic fat deposition without alcohol use is the hallmark of NAFLD, which is closely linked to obesity, insulin resistance and metabolic syndrome. Research has shown that by boosting inflammation and changing lipid metabolism, dysbiosis has a role in the onset and progression of NAFLD (Safari and Gérard 2019). For instance, bacterial species including *Firmicutes* and *Bacteroidetes* are increased in individuals with NAFLD, whereas beneficial strains like *Bifidobacterium* and *Lactobacillus* are often reduced (Plaza-Díaz et al. 2020). SCFAs modulate hepatic inflammation and fat storage. Additionally, the gut–liver axis is impacted by the microbiota through processes such as increased intestinal permeability, which permits the translocation of bacteria and their by-products into the bloodstream and exacerbates liver fibrosis and inflammation. Consequently, therapeutic strategies improving dysbiosis are being investigated for their potential to mitigate NAFLD and improve liver health.

Gut–Liver Axis and Therapeutic Strategies Targeting Microbiota for Liver Protection

The gut–liver axis is a critical interface through which the intestinal microbes affect liver function and health. Therapeutic strategies improving dysbiosis aim to restore homeostasis and enhance liver protection. Interventions such as FMT and the administration of specific probiotics have shown promise in preclinical and clinical studies. For instance, FMT has demonstrated efficacy in treating hepatic encephalopathy, by re-establishing a healthy gut microbiota (Wang et al. 2022). Additionally, recent advances in understanding microbial metabolites' roles, particularly SCFAs, have led to the exploration of their therapeutic potential. SCFAs can exert anti-inflammatory effects, enhance intestinal barrier function and improve lipid metabolism, making them key players in protecting the liver from diseases (He et al. 2020). Furthermore, dietary interventions, such as increasing fibre intake to promote SCFA production, are being recommended as complementary strategies for liver disease management. Collectively, these therapeutic approaches highlight the potential for microbiota modulation for protecting hepatic disorders (Watson and Preedy 2019).

6 Future Directions

6.1 Personalized Medicine and the Microbiome (Fig. 4)

6.1.1 Using Microbiome Profiling for Individualized Treatment Plans

One revolutionary method in healthcare is the incorporation of microbiome profiling into personalized therapy, allowing for individualized treatment plans tailored to a person's unique microbial composition. This approach entails examining the gut microbiome using methods like metagenomic sequencing, which sheds light on the variety and functionality of the GI tract's microbial communities. By understanding the specific microbial profiles associated with various health conditions, clinicians can design targeted interventions optimizing treatment efficacy and minimizing adverse effects. For instance, certain gut bacteria may influence the metabolism of drugs, necessitating adjustments in dosages or even the choice of medication based on the patient's microbiome profile (Zhao et al. 2023).

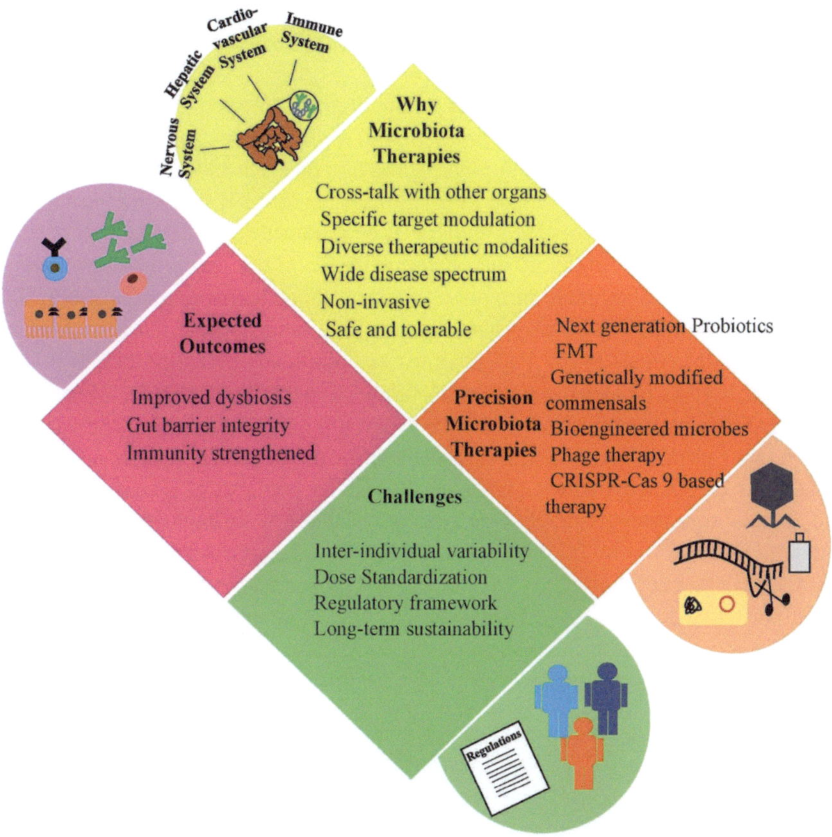

Fig. 4 The need of precision microbiota therapies, their techniques, expected outcomes and further challenges

6.2 Integration of Microbiome Data in Clinical Settings for Precision Medicine

Utilization of microbiome data for clinical use can facilitate a more holistic understanding of disease mechanisms and treatment responses. Integrating this data into routine clinical practice involves creating a framework where microbiome assessments are part of standard diagnostic and therapeutic processes. This could involve using probiotics and prebiotics to restore microbial balance or tailored dietary advice, in patients suffering from conditions such as IBD or IBS. As healthcare continues to move towards precision medicine, it is crucial to establish standardized protocols for microbiome profiling, ensuring that practitioners are equipped to interpret the data effectively and incorporate it into clinical decision-making (Bradburne and Hamosh 2016).

6.3 Ethical and Regulatory Challenges

6.3.1 Ethical Issues Surrounding Microbiome Manipulation

As microbiome-based therapies become more prevalent, ethical concerns regarding their application are growing. One of the primary ethical issues revolves around the potential long-term effects of manipulating the human microbiome. While therapies such as FMT have shown promise in treating recurrent CDI and other conditions, the full spectrum of consequences associated with altering the microbiome remains unknown. Given the complexity and dynamic nature of the microbiota, unintended side effects could arise, including new disease risks or permanent changes to host physiology (Rhodes 2016). Informed consent is another critical issue in microbiome therapy (Ma et al. 2018). Patients must be completely aware of the advantages and possible risks of altering their microbiome, especially in clinical trials where long-term data may be limited. Concerns regarding genetic privacy also exist since microbiome profiling may disclose private information about a person's genetic makeup or health. This raises questions about data ownership, privacy protection and ethical settings of microbiome data in healthcare settings. Another ethical dilemma involves equity in access to microbiome therapies. As microbiome-based treatments advance, disparities may emerge in who can access and benefit from them, particularly if they are expensive or limited to specialized centers. Ensuring equitable access to these innovative interventions is essential to prevent widening existing healthcare inequalities.

6.3.2 Regulatory Frameworks for Microbiota-Based Therapies

There are many obstacles in the constantly changing regulatory environment for microbiota-based treatments. Currently, the U.S. FDA and other global regulatory agencies are working to develop guidelines to govern the use of these therapies. However, one of the key regulatory hurdles is defining the microbiome itself. Unlike traditional pharmaceuticals, microbiota-based therapies often involve live organisms, raising questions about how these should be classified—as drugs, biologics or another category. FMT has sparked regulatory debates about safety and standardization. The variability in microbiota composition between individuals means that it is difficult to create standardized 'doses' of treatment. Regulatory bodies must decide how to ensure the safety of such treatments, especially considering that microbiota populations are affected by dietary patterns, antibiotics and other factors. Another critical challenge is the development of a regulatory framework for the commercialization of microbiome-based products, such as probiotics or genetically engineered bacteria. These products often face less stringent oversight, but as their use expands in precision therapies, more robust regulation may be required for their safety and efficacy. Ensuring ethical standards along with comprehensive regulatory frameworks is essential as microbiome-based therapies transition from experimental to mainstream treatments. The future of these therapies depends on addressing these challenges in a way that balances innovation with patient safety and equitable access.

6.4 The Road Ahead: Next-Generation Therapies in Microbiota Modulation

6.4.1 Synthetic Biology in Microbiota Modulation

Synthetic biology is revolutionizing the field of microbiota modulation by enabling scientists to design and engineer microbial systems that can perform specific, tailored functions within the human body. These engineered organisms are being developed to sense disease states, produce therapeutic molecules and restore microbial balance, potentially transforming treatment strategies for various diseases. For example, engineered bacteria can be programmed to secrete anti-inflammatory compounds in patients with IBD, thereby providing localized treatment without affecting the rest of the body. This precise control over microbial activity potentially minimizes side effects, improves patient outcomes and creates a new class of highly specialized therapies (Bober et al. 2018). Additionally, synthetic biology is being explored to construct microbial consortia that can carry out complex, multi-step processes to combat diseases like cancer or metabolic disorders. By utilizing genetic circuits, scientists can programme microbiota to detect environmental signals and trigger therapeutic responses, offering a dynamic and adaptive approach to treatment (Cubillos-Ruiz et al. 2021). As this field evolves, safety measures to ensure that these engineered organisms function as intended and do not disrupt normal microbiota–host interactions will be critical (Bober et al. 2018).

6.4.2 Genetically Engineered Probiotics and Their Therapeutic Potential

Genetically engineered probiotics represent a significant advancement in microbiota-based therapeutics. Genetically engineered probiotics are an advancement above conventional probiotics, which are good bacteria that support a balanced gut microbiota. By altering their genetic makeup, researchers can enhance their stability, survivability and functional capabilities within the human gut. These probiotics have the potential to generate medicinal compounds like anti-inflammatory drugs or enzymes that break down harmful toxins in the gut, offering potential treatments for numerous diseases like metabolic diseases, infections and autoimmune diseases. For instance, recent studies have shed light on the use of engineered *Lactobacillus* strains that produce immune-modulating cytokines (Llewellyn and Foey 2017), which have shown promise in reducing inflammation in diseases such as RA and IBD. Other applications involve probiotics engineered to secrete antimicrobial peptides or to improve the efficacy of drugs by altering their metabolism within the gut (Mejía-Pitta et al. 2021). As research continues, genetically engineered probiotics may become a key component of personalized medicine, enabling targeted interventions based on an person's microbiota composition (Mazhar et al. 2020).

6.4.3 Future Innovations: Microbiome-Based Diagnostics and Therapeutics

The future of microbiome research holds exciting possibilities for diagnostics and therapeutics. Microbiome-based diagnostics is a non-invasive way to track patient health and determine disease status. By analysing the composition and functional activities of the microbiota, healthcare providers can gain insights into diseases including, neurological problems, diabetes, obesity and cancer. These diagnostic tools offer the potential for earlier detection of diseases and more accurate monitoring of treatment responses. Therapeutically, microbiome-based interventions are moving beyond simple probiotics and prebiotics. Future innovations include the development of microbiome-based drugs that target specific bacterial pathways or metabolites linked to disease. Additionally, FMT is being refined with the aim of developing more precise microbial cocktails tailored to individual patients. As these approaches become more sophisticated, the integration of microbiome data into clinical decision-making will become a cornerstone of precision medicine, enhancing patient care and treatment efficacy (Table 4).

Table 4 List of microbial metabolites and specific biomarkers for precision therapies

Biomarker	Associated microbiota	Diseases	Functions	Therapeutic implications	References
Butyrate	*Faecalibacterium prausnitzii*, *Roseburia*	IBD, colorectal cancer	Anti-inflammatory; supports intestinal barrier integrity	Can inform probiotic or FMT-based therapies targeting gut inflammation	Tan et al. (2014)
Lactate	*Lactobacillus*, *Bifidobacterium*	Metabolic disorders (e.g., obesity, diabetes)	Influences glucose metabolism and insulin sensitivity	Targeted prebiotics or engineered probiotics to enhance metabolic regulation	Aoki et al. (2017)
Bile acids (secondary)	*Clostridium cluster XIVa*	NAFLD, colon cancer	Modulates fat digestion and affects liver metabolism	Personalized dietary interventions or microbiota modulation to regulate bile acids	Wahlström et al. (2016)
LPS	*Proteobacteria* (e.g., *Escherichia coli*)	Cardiovascular disease, metabolic syndrome	Promotes systemic inflammation via toll-like receptor activation	Therapies aimed at reducing endotoxin production or translocation	Croci et al. (2021)
Trimethylamine N-oxide (TMAO)	*Firmicutes*, *Proteobacteria*	Cardiovascular disease	Enhances atherosclerosis by influencing cholesterol metabolism	Microbiota-targeted interventions to reduce TMA production	Thomas and Fernandez (2021) and Jing et al. (2022)
SCFAs	*Bacteroidetes*, *Firmicutes*	IBS, colon cancer	SCFAs regulate inflammation, gut motility and energy metabolism	Use of fibre-based prebiotics to boost SCFA-producing microbiota	Vinelli et al. (2022) and Abdelhalim (2023)
Indole derivatives	*Clostridium*, *Bacteroides*	Autoimmune diseases, colorectal cancer	Regulate gut–immune homeostasis via AhR	Therapeutic targeting of microbiota that produce immunomodulatory indoles	Zhou et al. (2023b)

(continued)

Table 4 (continued)

Biomarker	Associated microbiota	Diseases	Functions	Therapeutic implications	References
Methane	*Methanobrevibacter smithii*	Constipation-predominant IBS	Slows gut transit time through methane production	Methanogenesis inhibitors or selective microbial interventions for IBS management	Stern and Brenner (2018)
Secondary bile acids (e.g., deoxycholic acid)	*Bacteroides*, *Clostridia*	Colon cancer, gallstones	Alters gut lining and increases cancer risk through DNA damage	Interventions to modulate bile acid synthesis or metabolism	Liu et al. (2022b)
Acetate	*Bacteroides*, *Firmicutes*	Type 2 diabetes, cardiovascular diseases	Regulates cholesterol metabolism insulin sensitivity, building muscle and colonic tissue	Prebiotic or synbiotic therapies aimed at boosting acetate production	Roy and Dhaneshwar (2023) and Li et al. (2021)
Phenylacetic acid	*Clostridium*, *Bacteroides*	Autoimmune diseases, IBD	Affects immune response and inflammatory pathways	Therapeutic modulation of microbial pathways producing phenylacetic acid	
Succinate	*Prevotella*, *Bacteroides*	Obesity, type 2 diabetes	Acts as a metabolic intermediary; modulates fat storage and glucose homeostasis	Microbial-targeted therapies regulate succinate levels in metabolic syndromes	Fernández-Veledo and Vendrell (2019)
p-Cresol	*Clostridium difficile*, *Clostridia*	Cardiovascular diseases	Disrupts intestinal barrier function and promotes inflammation	Reduction in p-cresol levels through microbial interventions for cardiovascular diseases	Passmore et al. (2018) and Meijers et al. (2010)
Ammonia	*Bacteroides*, *Clostridia*	Liver diseases (e.g., hepatic encephalopathy)	Contributes to neurotoxicity in liver diseases	Targeted reduction in ammonia-producing microbes for liver protection	Bloom et al. (2021)

7 Navigating the Challenges and Advancing the Solutions in the Microbiota Precision Therapies

Overcoming the barriers in microbiota-driven therapies involves several hurdles spanning scientific, clinical and regulatory domains. With thousands of microbial species in the gut, identification of specific microbial strains is complex due to the diversity and limited understanding of the microbiome. The individualized treatments should address the concerns of the variable host–microbiota interactions that are influenced by genetics and lifestyle. Ethical and regulatory concerns arise with interventions like FMT. With the instability of the microbiota, ensuring long-term efficacy is difficult, increasing recurrence or reducing the effectiveness. Clinical trial design also faces challenges, needing standardized protocols to account for microbiome variations and external factors like diet. To fully achieve the potential of microbiota-based precision medicines, these obstacles must be overcome by further studies, technological developments in microbiome profiling and a multidisciplinary approach. The solutions to the problems are provided below.

7.1 Advanced Microbiome Profiling

Next-generation sequencing (NGS) and metagenomics allow detailed analysis of the microbiome's genetic content, helping to identify previously unknown microbial species and their functions. This approach improves the precision of microbiome therapies by linking specific strains to disease states (Wensel et al. 2022). A comprehensive knowledge of microbiota-host interactions is made possible by multi-omics integration, which combines data from genetics, transcriptomics, proteomics and metabolomics. This facilitates the deciphering of the intricate relationship between host physiology and microbial metabolites, resulting in more focused treatments (Kwoji et al. 2023; Chetty and Blekhman 2024).

7.2 Targeted Microbiota Manipulation

Genetically engineering the specific microbial strains or designing synthetic probiotics can help restore microbiome balance. It is possible to modify these modified strains to generate therapeutic compounds or outcompete harmful microbes, offering a precision-targeted approach to therapy (Bober et al. 2018). Phage therapy involving bacteriophages (viruses that infect bacteria) can be used to selectively target pathogenic bacteria without disturbing beneficial microbes, providing a precise tool for microbiome modulation (Emencheta et al. 2023; Singha et al. 2023).

7.3 Personalized Medicine and Microbiome Data Integration

Personalized formulations of probiotics and prebiotics can be designed based on individual microbiome profiles, dietary patterns and disease conditions. Microbiome analysis using machine learning algorithms enables the customization of these interventions (Papoutsoglou et al. 2023). Overcoming the variability in FMT outcomes, the development of donor screening protocols and microbiome data integration ensures that specific donor microbiota is matched to the patient's condition, improving success rates (Zhang et al. 2023b).

7.4 Ethical and Regulatory Frameworks

Developing clear ethical guidelines and patient consent frameworks is crucial for therapies like FMT or synthetic probiotics. Regulatory bodies like the FDA are beginning to establish guidelines for microbiota-based therapies to ensure safety and efficacy. Regulatory agencies, researchers and healthcare providers must collaborate to ensure patient safety and long-term monitoring while accelerating the approval of microbiota-driven medications.

7.5 Stabilizing Microbiota Modulation

The long-term microbiome monitoring over time after treatment can help understand how to stabilize therapeutic effects. This could involve periodic probiotic or prebiotic supplementation to maintain microbial balance (Si et al. 2022). Administering a defined community of microbes, rather than single-strain probiotics, may improve long-term stability in the microbiome by establishing a balanced microbial ecosystem.

By integrating these methods, we can address the challenges in microbiota-driven therapies and move toward more effective, personalized treatments for diseases involving microbiome dysbiosis.

8 Conclusion

The potential of microbiota-driven therapies is spanning across multiple domains and hence it is emerging as a promising frontier in precision medicine in autoimmune disorders, cancer therapy and gastrointestinal diseases. The increasing depth of human microbiome has opened avenues for innovative therapeutic strategies targeting specific microbial populations, metabolites and host–microbe interactions.

Besides probiotics and prebiotics, advanced therapeutic interventions like FMT, in combination with synthetic biology approaches, offer new possibilities for improving disease outcomes, reducing the toxicity and improving the efficacy of conventional treatments. These therapies hold true potential not only for regulating metabolic disorders but also immune related complications, such as metabolic complications, immune-mediated diseases and cancers. The future of microbiota-driven therapies is promising, with significant advancements anticipated in their application to various diseases. As the field progresses, precision therapies tailored to individual microbiome profiles will likely become mainstream in managing complex conditions, including inflammatory bowel diseases, cancers and metabolic disorders. Ongoing studies into microbiota biomarkers along with next-generation probiotics will further refine these therapies, making them safer and more effective. But there are still issues, especially with relation to moral dilemmas, legal requirements and the requirement for solid clinical evidence to support the treatments' long-term effectiveness and safety. Nonetheless, the future impact of microbiota-based treatments on healthcare is expected to be transformative, revolutionizing the way we understand and treat diseases at the microbial level.

References

Abdelhalim KA (2023) Short-chain fatty acids (SCFAs) from gastrointestinal disorders, metabolism, epigenetics, central nervous system to cancer – A mini-review. Chem Biol Interact 110851

Alam MZ, Maslanka JR, Abt MC (2023) Immunological consequences of microbiome-based therapeutics. Front Immunol 13:1046472

Amini ME, Shomali N, Bakhshi A, Rezaei S, Hemmatzadeh M, Hosseinzadeh R et al (2020) Gut microbiome and multiple sclerosis: new insights and perspective. Int Immunopharmacol 88:107024

Aoki R, Kamikado K, Suda W, Takii H, Mikami Y, Suganuma N et al (2017) A proliferative probiotic Bifidobacterium strain in the gut ameliorates progression of metabolic disorders via microbiota modulation and acetate elevation. Sci Rep 7(1):43522

Armet AM, Deehan EC, O'Sullivan AF, Mota JF, Field CJ, Prado CM et al (2022) Rethinking healthy eating in light of the gut microbiome. Cell Host Microbe 30(6):764–785

Baradaran Ghavami S, Asadzadeh Aghdaei H, Sorrentino D, Shahrokh S, Farmani M, Ashrafian F et al (2021) Probiotic-induced tolerogenic dendritic cells: a novel therapy for inflammatory bowel disease? Int J Mol Sci 22(15):8274

Baranowska-Wójcik E, Winiarska-Mieczan A, Olcha P, Kwiecień M, Jachimowicz-Rogowska K, Nowakowski Ł et al (2024) Polyphenols influence the development of endometrial cancer by modulating the gut microbiota. Nutrients 16(5):681

Baruch EN, Youngster I, Ben-Betzalel G, Ortenberg R, Lahat A, Katz L et al (2021) Fecal microbiota transplant promotes response in immunotherapy-refractory melanoma patients. Science 371(6529):602–609

Batista VL, Da Silva TF, de Jesus LCL, Coelho-Rocha ND, Barroso FAL, Tavares LM et al (2020) Probiotics, prebiotics, synbiotics, and paraprobiotics as a therapeutic alternative for intestinal mucositis. Front Microbiol 11:544490

Bloom PP, Tapper EB, Young VB, Lok AS (2021) Microbiome therapeutics for hepatic encephalopathy. J Hepatol 75(6):1452–1464

Bober JR, Beisel CL, Nair NU (2018) Synthetic biology approaches to engineer probiotics and members of the human microbiota for biomedical applications. Annu Rev Biomed Eng 20(1):277–300

Borroni EM, Qehajaj D, Farina FM, Yiu D, Bresalier RS, Chiriva-Internati M et al (2019) Fusobacterium nucleatum and the immune system in colorectal cancer. Curr Colorectal Cancer Rep 15:149–156

Bradburne C, Hamosh A (2016) Integrating the microbiome into precision medicine. Taylor & Francis, pp 475–477

Bungau SG, Behl T, Singh A, Sehgal A, Singh S, Chigurupati S et al (2021) Targeting probiotics in rheumatoid arthritis. Nutrients 13(10):3376

Cancer IAfRo. (2012) A review of human carcinogens. IARC Monogr Eval Carcinog Risks Hum 100:1–441

Cavallo S (2020) Immune-mediated genesis of multiple sclerosis. J Transl Autoimmun 3:100039

Chen J, Chia N, Kalari KR, Yao JZ, Novotna M, Paz Soldan MM et al (2016) Multiple sclerosis patients have a distinct gut microbiota compared to healthy controls. Sci Rep 6(1):1–10

Chervin CS, Gajewski T (2020) Microbiome-based interventions: therapeutic strategies in cancer immunotherapy. Immuno-Oncol Technol 8:12–20

Chetty A, Blekhman R (2024) Multi-omic approaches for host-microbiome data integration. Gut Microbes 16(1):2297860

Coradduzza D, Bo M, Congiargiu A, Azara E, De Miglio MR, Erre GL et al (2023) Decoding the microbiome's influence on rheumatoid arthritis. Microorganisms 11(9):2170

Cristofori F, Indrio F, Miniello VL, De Angelis M, Francavilla R (2018) Probiotics in celiac disease. Nutrients 10(12):1824

Croci S, D'Apolito LI, Gasperi V, Catani MV, Savini I (2021) Dietary strategies for management of metabolic syndrome: role of gut microbiota metabolites. Nutrients 13(5):1389

Cubillos-Ruiz A, Guo T, Sokolovska A, Miller PF, Collins JJ, Lu TK et al (2021) Engineering living therapeutics with synthetic biology. Nat Rev Drug Discov 20(12):941–960

Dehinsilu J, Sergaki C, Amos G, Fontana V, Pirmohamed M (2023) The interplay between the microbiome and colonic immune system in checkpoint inhibitor therapy. Front Microbiomes 2:1061193

Derosa L, Routy B, Thomas AM, Iebba V, Zalcman G, Friard S et al (2022) Intestinal Akkermansia muciniphila predicts clinical response to PD-1 blockade in patients with advanced non-small-cell lung cancer. Nat Med 28(2):315–324

Desai MS, Seekatz AM, Koropatkin NM, Kamada N, Hickey CA, Wolter M et al (2016) A dietary fiber-deprived gut microbiota degrades the colonic mucus barrier and enhances pathogen susceptibility. Cell 167(5):1339–1353. e21

El Haddad L, Mendoza JF, Jobin C (2022) Bacteriophage-mediated manipulations of microbiota in gastrointestinal diseases. Front Microbiol 13:1055427

Emencheta SC, Olovo CV, Eze OC, Kalu CF, Berebon DP, Onuigbo EB et al (2023) The role of bacteriophages in the gut microbiota: implications for human health. Pharmaceutics 15(10):2416

Fan S, Zhang Z, Zhao Y, Daglia M, Zhang J, Zhu Y et al (2023) Recent advances in targeted manipulation of the gut microbiome by prebiotics: from taxonomic composition to metabolic function. Curr Opin Food Sci 49:100959

Fernández-Veledo S, Vendrell J (2019) Gut microbiota-derived succinate: friend or foe in human metabolic diseases? Rev Endocr Metab Disord 20(4):439–447

Fugger L, Jensen LT, Rossjohn J (2020) Challenges, progress, and prospects of developing therapies to treat autoimmune diseases. Cell 181(1):63–80

Fusco W, Lorenzo MB, Cintoni M, Porcari S, Rinninella E, Kaitsas F et al (2023) Short-chain fatty-acid-producing bacteria: key components of the human gut microbiota. Nutrients 15(9):2211

Gasaly N, Hermoso MA, Gotteland M (2021) Butyrate and the fine-tuning of colonic homeostasis: implication for inflammatory bowel diseases. Int J Mol Sci 22(6):3061

Gavzy SJ, Kensiski A, Lee ZL, Mongodin EF, Ma B, Bromberg JS (2023) Bifidobacterium mechanisms of immune modulation and tolerance. Gut Microbes 15(2):2291164
Gebrayel P, Nicco C, Al Khodor S, Bilinski J, Caselli E, Comelli EM et al (2022) Microbiota medicine: towards clinical revolution. J Transl Med 20(1):111
Gill PA, Inniss S, Kumagai T, Rahman FZ, Smith AM (2022) The role of diet and gut microbiota in regulating gastrointestinal and inflammatory disease. Front Immunol 13:866059
Goldbaum AA (2022) Role of the gut microbiota, diet, and obesity in colorectal cancer risk. Purdue University Graduate School
Goldenberg SD, Merrick B (2021) The role of faecal microbiota transplantation: looking beyond Clostridioides difficile infection. Ther Adv Infect Dis 8:2049936120981526
Guarino MPL, Altomare A, Emerenziani S, Di Rosa C, Ribolsi M, Balestrieri P et al (2020) Mechanisms of action of prebiotics and their effects on gastro-intestinal disorders in adults. Nutrients 12(4):1037
Haghikia A, Jörg S, Duscha A, Berg J, Manzel A, Waschbisch A et al (2015) Dietary fatty acids directly impact central nervous system autoimmunity via the small intestine. Immunity 43(4):817–829
Hasan N, Yang H (2019) Factors affecting the composition of the gut microbiota, and its modulation. PeerJ 7:e7502
He J, Zhang P, Shen L, Niu L, Tan Y, Chen L et al (2020) Short-chain fatty acids and their association with signalling pathways in inflammation, glucose and lipid metabolism. Int J Mol Sci 21(17):6356
He R, Chen J, Zhao Z, Shi C, Du Y, Yi M et al (2023) T-cell activation Rho GTPase-activating protein maintains intestinal homeostasis by regulating intestinal T helper cells differentiation through the gut microbiota. Front Microbiol 13:1030947
Hitch TC, Hall LJ, Walsh SK, Leventhal GE, Slack E, de Wouters T et al (2022) Microbiome-based interventions to modulate gut ecology and the immune system. Mucosal Immunol 15(6):1095–1113
Horta-Baas G, Sandoval-Cabrera A, Romero-Figueroa MS (2021) Modification of gut microbiota in inflammatory arthritis: highlights and future challenges. Curr Rheumatol Rep 23:1–14
Hou K, Wu Z-X, Chen X-Y, Wang J-Q, Zhang D, Xiao C et al (2022) Microbiota in health and diseases. Signal Transduct Target Ther 7(1):1–28
Hrncir T (2022) Gut microbiota dysbiosis: triggers, consequences, diagnostic and therapeutic options. MDPI:578
Hu J, Mesnage R, Tuohy K, Heiss C, Rodriguez-Mateos A (2024) (Poly) phenol-related gut metabotypes and human health: an update. Food Funct 15(6):2814–2835
Jing L, Zhang H, Xiang Q, Shen L, Guo X, Zhai C et al (2022) Targeting trimethylamine N-oxide: a new therapeutic strategy for alleviating atherosclerosis. Front Cardiovasc Med 9:864600
Kang X, Lau HC-H, Yu J (2024) Modulating gut microbiome in cancer immunotherapy: harnessing microbes to enhance treatment efficacy. Cell Rep Med 5(4)
Kaźmierczak-Siedlecka K, Roviello G, Catalano M, Polom K (2021) Gut microbiota modulation in the context of immune-related aspects of Lactobacillus spp. and Bifidobacterium spp. in gastrointestinal cancers. Nutrients 13(8):2674
Khan I, Ullah N, Zha L, Bai Y, Khan A, Zhao T et al (2019) Alteration of gut microbiota in inflammatory bowel disease (IBD): cause or consequence? IBD treatment targeting the gut microbiome. Pathogens 8(3):126
Kim CH (2023) Complex regulatory effects of gut microbial short-chain fatty acids on immune tolerance and autoimmunity. Cell Mol Immunol 20(4):341–350
Kumar A, Ali A, Kapardar RK, Dar GM, Nimisha A et al (2023) Implication of gut microbes and its metabolites in colorectal cancer. J Cancer Res Clin Oncol 149(1):441–465
Kwoji ID, Aiyegoro OA, Okpeku M, Adeleke MA (2023) 'Multi-omics' data integration: applications in probiotics studies. NPJ Sci Food 7(1):25
Li W, Ma Z (2020) FBA ecological guild: trio of firmicutes-bacteroidetes alliance against actinobacteria in human oral microbiome. Sci Rep 10(1):287

Li H-Y, Zhou D-D, Gan R-Y, Huang S-Y, Zhao C-N, Shang A et al (2021) Effects and mechanisms of probiotics, prebiotics, synbiotics, and postbiotics on metabolic diseases targeting gut microbiota: a narrative review. Nutrients 13(9):3211

Li S, Zhu S, Yu J (2023) The role of gut microbiota and metabolites in cancer chemotherapy. J Adv Res

Liu L, Huh JR, Shah K (2022a) Microbiota and the gut-brain-axis: implications for new therapeutic design in the CNS. EBioMedicine:77

Liu Y, Zhang S, Zhou W, Hu D, Xu H, Ji G (2022b) Secondary bile acids and tumorigenesis in colorectal cancer. Front Oncol 12:813745

Llewellyn A, Foey A (2017) Probiotic modulation of innate cell pathogen sensing and signaling events. Nutrients 9(10):1156

López-Gómez L, Alcorta A, Abalo R (2023) Probiotics and probiotic-like agents against Chemotherapy-Induced Intestinal mucositis: a narrative review. J Person Med 13(10):1487

Ma Y, Chen H, Lan C, Ren J (2018) Help, hope and hype: ethical considerations of human microbiome research and applications. Prot Cell 9(5):404–415

Mahmoudi M, Kheder RK, Faraj TA, Abdulabbas HS, Esmaeili S-A (2023) Impacts of Porphyromonas gingivalis periodontitis on rheumatoid arthritis autoimmunity. Int Immunopharmacol 118:109936

Mangalam AK, Murray J (2019) Microbial monotherapy with Prevotella histicola for patients with multiple sclerosis. Expert Rev Neurother 19(1):45–53

Mazhar SF, Afzal M, Almatroudi A, Munir S, Ashfaq UA, Rasool M et al (2020) The prospects for the therapeutic implications of genetically engineered probiotics. J Food Qual 2020(1):9676452

Meijers BK, Claes K, Bammens B, de Loor H, Viaene L, Verbeke K et al (2010) p-Cresol and cardiovascular risk in mild-to-moderate kidney disease. Clin J Am Soc Nephrol 5(7):1182–1189

Mejía-Pitta A, Broset E, de la Fuente-Nunez C (2021) Probiotic engineering strategies for the heterologous production of antimicrobial peptides. Adv Drug Deliv Rev 176:113863

Menafra D, Proganò M, Tecce N, Pivonello R, Colao A (2024) Diet and gut microbiome: impact of each factor and mutual interactions on prevention and treatment of type 2, type 1, and gestational diabetes. Human. Nutr Metab:200286

Mo C, Lou X, Xue J, Shi Z, Zhao Y, Wang F et al (2024) The influence of Akkermansia muciniphila on intestinal barrier function. Gut Pathog 16(1):41

Mörbe UM, Jørgensen PB, Fenton TM, von Burg N, Riis LB, Spencer J et al (2021) Human gut-associated lymphoid tissues (GALT); diversity, structure, and function. Mucosal Immunol 14(4):793–802

Mousa WK, Chehadeh F, Husband S (2022) Microbial dysbiosis in the gut drives systemic autoimmune diseases. Front Immunol 13:906258

Mu Q, Kirby J, Reilly CM, Luo XM (2017) Leaky gut as a danger signal for autoimmune diseases. Front Immunol 8:598

Napolitano M, Fasulo E, Ungaro F, Massimino L, Sinagra E, Danese S et al (2023) Gut dysbiosis in irritable bowel syndrome: a narrative review on correlation with disease subtypes and novel therapeutic implications. Microorganisms 11(10):2369

Ott SJ, Waetzig GH, Rehman A, Moltzau-Anderson J, Bharti R, Grasis JA et al (2017) Efficacy of sterile fecal filtrate transfer for treating patients with Clostridium difficile infection. Gastroenterology 152(4):799–811. e7

Ou S, Wang H, Tao Y, Luo K, Ye J, Ran S et al (2022) Fusobacterium nucleatum and colorectal cancer: from phenomenon to mechanism. Front Cell Infect Microbiol 12:1020583

Padoan A, Musso G, Contran N, Basso D (2023) Inflammation, autoinflammation and autoimmunity in inflammatory bowel diseases. Curr Issues Mol Biol 45(7):5534–5557

Pan Q, Guo F, Huang Y, Li A, Chen S, Chen J et al (2021) Gut microbiota dysbiosis in systemic lupus erythematosus: novel insights into mechanisms and promising therapeutic strategies. Front Immunol 12:799788

Papoutsoglou G, Tarazona S, Lopes MB, Klammsteiner T, Ibrahimi E, Eckenberger J et al (2023) Machine learning approaches in microbiome research: challenges and best practices. Front Microbiol 14:1261889

Passmore IJ, Letertre MP, Preston MD, Bianconi I, Harrison MA, Nasher F et al (2018) Para-cresol production by Clostridium difficile affects microbial diversity and membrane integrity of Gram-negative bacteria. PLoS Pathog 14(9):e1007191

Pizzo F, Maroccia Z, Hammarberg Ferri I, Fiorentini C (2022) Role of the microbiota in lung cancer: insights on prevention and treatment. Int J Mol Sci 23(11):6138

Plaza-Díaz J, Solís-Urra P, Rodríguez-Rodríguez F, Olivares-Arancibia J, Navarro-Oliveros M, Abadía-Molina F et al (2020) The gut barrier, intestinal microbiota, and liver disease: molecular mechanisms and strategies to manage. Int J Mol Sci 21(21):8351

Portincasa P, Bonfrate L, Vacca M, De Angelis M, Farella I, Lanza E et al (2022) Gut microbiota and short chain fatty acids: implications in glucose homeostasis. Int J Mol Sci 23(3):1105

Pratt C, Campbell MD (2020) The effect of Bifidobacterium on reducing symptomatic abdominal pain in patients with irritable bowel syndrome: a systematic review. Probiot Antimicrob Prot 12(3):834–839

Principi N, Cozzali R, Farinelli E, Brusaferro A, Esposito S (2018) Gut dysbiosis and irritable bowel syndrome: the potential role of probiotics. J Infect 76(2):111–120

Pugliese A (2017) Autoreactive T cells in type 1 diabetes. J Clin Invest 127(8):2881–2891

Ramos GP, Papadakis KA (eds) (2019) Mechanisms of disease: inflammatory bowel diseases. Mayo Clinic Proceedings. Elsevier

Ranjbar R, Vahdati SN, Tavakoli S, Khodaie R, Behboudi H (2021a) Immunomodulatory roles of microbiota-derived short-chain fatty acids in bacterial infections. Biomed Pharmacother 141:111817

Ranjbar M, Salehi R, Haghjooy Javanmard S, Rafiee L, Faraji H, Jafarpor S et al (2021b) The dysbiosis signature of Fusobacterium nucleatum in colorectal cancer-cause or consequences? A systematic review. Cancer Cell Int 21:1–24

Rauf A, Khalil AA, Rahman U-U, Khalid A, Naz S, Shariati MA et al (2022) Recent advances in the therapeutic application of short-chain fatty acids (SCFAs): an updated review. Crit Rev Food Sci Nutr 62(22):6034–6054

Reyes-Castillo Z, Valdés-Miramontes E, Llamas-Covarrubias M, Muñoz-Valle JF (2021) Troublesome friends within us: the role of gut microbiota on rheumatoid arthritis etiopathogenesis and its clinical and therapeutic relevance. Clin Exp Med 21:1–13

Rhodes R (2016) Ethical issues in microbiome research and medicine. BMC Med 14(1):156

Roy S, Dhaneshwar S (2023) Role of prebiotics, probiotics, and synbiotics in management of inflammatory bowel disease: current perspectives. World J Gastroenterol 29(14):2078

Russo E, Nannini G, Dinu M, Pagliai G, Sofi F, Amedei A (2020) Exploring the food-gut axis in immunotherapy response of cancer patients. World J Gastroenterol 26(33):4919

Sadeghpour Heravi F, Hu H (2023) Bifidobacterium: host–microbiome interaction and mechanism of action in preventing common gut-microbiota-associated complications in preterm infants: a narrative review. Nutrients 15(3):709

Safari Z, Gérard P (2019) The links between the gut microbiome and non-alcoholic fatty liver disease (NAFLD). Cell Mol Life Sci 76:1541–1558

Schepici G, Silvestro S, Bramanti P, Mazzon E (2019) The gut microbiota in multiple sclerosis: an overview of clinical trials. Cell Transplant 28(12):1507–1527

Shah AB, Baiseitova A, Zahoor M, Ahmad I, Ikram M, Bakhsh A et al (2024) Probiotic significance of Lactobacillus strains: a comprehensive review on health impacts, research gaps, and future prospects. Gut Microbes 16(1):2431643

Shukla V, Singh S, Verma S, Verma S, Rizvi AA, Abbas M (2024) Targeting the microbiome to improve human health with the approach of personalized medicine: latest aspects and current updates. Clin Nutr ESPEN

Si J, Vázquez-Castellanos JF, Gregory AC, Decommer L, Rymenans L, Proost S et al (2022) Long-term life history predicts current gut microbiome in a population-based cohort study. Nat Aging 2(10):885–895

Singha B, Rawat BS, Venkataraman R, Nair T, Rosenn EH, Soni V (2023) Gut microbiome associated dysbiosis: limited regimens and expanding horizons of phage therapy. Aspects Mol Med:100029

Stern EK, Brenner DM (2018) Gut microbiota-based therapies for irritable bowel syndrome. Clin Transl Gastroenterol 9(2):e134

Tan J, McKenzie C, Potamitis M, Thorburn AN, Mackay CR, Macia L (2014) The role of short-chain fatty acids in health and disease. Adv Immunol 121:91–119

Tang H-Y, Jiang A-J, Wang X-Y, Wang H, Guan Y-Y, Li F et al (2021) Uncovering the pathophysiology of irritable bowel syndrome by exploring the gut-brain axis: a narrative review. Ann Transl Med 9(14)

Thomas MS, Fernandez ML (2021) Trimethylamine N-oxide (TMAO), diet and cardiovascular disease. Curr Atheroscler Rep 23:1–7

Valdes AM, Walter J, Segal E, Spector TD (2018) Role of the gut microbiota in nutrition and health. BMJ 361

Vandenplas Y, Carnielli V, Ksiazyk J, Luna MS, Migacheva N, Mosselmans J et al (2020) Factors affecting early-life intestinal microbiota development. Nutrition 78:110812

Varga A, Makszin L, Bufa A, Sipos D, Kása P, Pál S et al (2023) Efficacy of lyophilised bacteria-rich faecal sediment and supernatant with reduced bacterial count for treating patients with Clostridioides difficile Infection – A novel method for capsule faecal microbiota transfer. Front Cell Infect Microbiol 13:1041384

Venhorst J, van der Vossen JM, Agamennone V (2022) Battling enteropathogenic clostridia: phage therapy for Clostridioides difficile and Clostridium perfringens. Front Microbiol 13:891790

Vimal J, Himal I, Kannan S (2020) Role of microbial dysbiosis in carcinogenesis & cancer therapies. Indian J Med Res 152(6):553–561

Vinelli V, Biscotti P, Martini D, Del Bo' C, Marino M, Meroño T et al (2022) Effects of dietary fibers on short-chain fatty acids and gut microbiota composition in healthy adults: a systematic review. Nutrients 14(13):2559

Vongsavath T, Rahmani R, Tun KM, Manne V (2024) The use of fecal microbiota transplant in overcoming and modulating resistance to anti-PD-1 therapy in patients with skin Cancer. Cancers 16(3):499

Wahlström A, Sayin SI, Marschall H-U, Bäckhed F (2016) Intestinal crosstalk between bile acids and microbiota and its impact on host metabolism. Cell Metab 24(1):41–50

Wang Y, Zhang S, Borody TJ, Zhang F (2022) Encyclopedia of fecal microbiota transplantation: a review of effectiveness in the treatment of 85 diseases. Chin Med J 135(16):1927–1939

Watson RR, Preedy VR (2019) Dietary interventions in liver disease: foods, nutrients, and dietary supplements. Academic

Wei H, Yue Z, Han J, Chen P, Xie K, Sun Y et al (2024) Oral compound probiotic supplements can improve the quality of life for patients with lung cancer during chemotherapy: a randomized placebo-controlled study. Thoracic Cancer 15(2):182–191

Wensel CR, Pluznick JL, Salzberg SL, Sears CL (2022) Next-generation sequencing: insights to advance clinical investigations of the microbiome. J Clin Invest 132(7)

Wu J, Li Q, Fu X (2019) Fusobacterium nucleatum contributes to the carcinogenesis of colorectal cancer by inducing inflammation and suppressing host immunity. Transl Oncol 12(6):846–851

Xin Y, Liu C-G, Zang D, Chen J (2024) Gut microbiota and dietary intervention: affecting immunotherapy efficacy in non–small cell lung cancer. Front Immunol 15:1343450

Yang M-L, Sodré FM, Mamula MJ, Overbergh L (2021) Citrullination and PAD enzyme biology in type 1 diabetes–regulators of inflammation, autoimmunity, and pathology. Front Immunol 12:678953

Yang J, Yang H, Li Y (2023) The triple interactions between gut microbiota, mycobiota and host immunity. Crit Rev Food Sci Nutr 63(33):11604–11624

Yao S, Zhao Z, Wang W, Liu X (2021) Bifidobacterium longum: protection against inflammatory bowel disease. J Immunol Res 2021(1):8030297
Yu Z-K, Xie R-L, You R, Liu Y-P, Chen X-Y, Chen M-Y et al (2021) The role of the bacterial microbiome in the treatment of cancer. BMC Cancer 21:1–14
Yue B, Luo X, Yu Z, Mani S, Wang Z, Dou W (2019) Inflammatory bowel disease: a potential result from the collusion between gut microbiota and mucosal immune system. Microorganisms 7(10):440
Zhang Y, Gao X, Gao S, Liu Y, Wang W, Feng Y et al (2023a) Effect of gut flora mediated-bile acid metabolism on intestinal immune microenvironment. Immunology 170(3):301–318
Zhang B, Yang L, Ning H, Cao M, Chen Z, Chen Q et al (2023b) A matching strategy to guide donor selection for ulcerative colitis in fecal microbiota transplantation: meta-analysis and analytic hierarchy process. Microbiol Spectr 11(1):e02159–e02121
Zhao Q, Chen Y, Huang W, Zhou H, Zhang W (2023) Drug-microbiota interactions: an emerging priority for precision medicine. Signal Transduct Target Ther 8(1):386
Zhou S, Cui Y, Zhang Y, Zhao T, Cong J (2023a) Fecal microbiota transplantation for induction of remission in Crohn's disease: a systematic review and meta-analysis. Int J Color Dis 38(1):62
Zhou Y, Chen Y, He H, Peng M, Zeng M, Sun H (2023b) The role of the indoles in microbiota-gut-brain axis and potential therapeutic targets: a focus on human neurological and neuropsychiatric diseases. Neuropharmacology:109690

Exploring Clinical Applications: Precision Medicine in Gastrointestinal Disorders, Microbiota-Driven Cancer Therapies and Innovative Interventions for Neurological Disorders-Insights from Case Studies

Aanchal Misra and Sarthak Sharma

Abstract This chapter explores into real-world applications of precision medicine in gastrointestinal illnesses, microbiota-driven cancer therapy and advanced interventions for neurological conditions using extensive case studies. Precision medicine is revolutionising gastrointestinal health by tailoring treatments to individuals' genetic, environmental and microbiomic characteristics, allowing for therapies that target specific pathophysiological pathways rather than merely managing symptoms. The role of the microbiota in cancer care is becoming more well recognised, with personalised microbiome modulation improving immunotherapy efficacy and lowering adverse effects. Case studies show that microbiota-targeted techniques have the potential to alter therapeutic strategies in gastrointestinal oncology, resulting in safer and more efficient therapies for difficult cancer cases. Precision medicine approaches in neurology, such as gene editing, neurostimulation and artificial intelligence (AI)-guided diagnostics, are showing promise in treating difficult conditions including autism, amyotrophic lateral sclerosis (ALS) and epilepsy. By targeting the underlying causes of disease, tailored medicines that make use of pharmacogenetics, cell-based interventions and integrative technologies enhance patient outcomes. This study emphasises the importance of personalised genetic insights and high-density, responder-guided technologies in maximising treatment strategies.

The difficulties of clinical translation are discussed in the study's conclusion, with particular attention paid to accessibility, cost and ethical concerns. By focus-

A. Misra (✉)
Department of Pharmaceutical Analysis, Axis Institute of Pharmacy, Kanpur, Uttar Pradesh, India

S. Sharma
Department of Pharmacology, ISF College of Pharmacy, Moga, Punjab, India

S. K. Singh et al. (eds.), *Microbiota Profiling for Precision Medicine*, Interdisciplinary Biotechnological Advances,
https://doi.org/10.1007/978-981-95-2104-3_11

sing on these high-impact areas, this study highlights how precision medicine might advance personalised healthcare by providing patients with neurological and gastrointestinal illnesses with better, more individualised therapy options.

Keywords Precision medicine · Gastrointestinal disorders · Microbiota · Cancer therapy · Neurological disorders · Gene editing · Neurostimulation · Pharmacogenetics · Personalised healthcare · Case studies · AI-driven diagnostics · Individualised treatment

1 1. Introduction to the Concept of Precision Medicine

Precision medicine (PM) represents a transformative approach to healthcare, personalising treatment strategies based on an individual's genetic makeup, environmental influences and lifestyle factors. At the molecular level, the human genome demonstrates substantial structural diversity, leading to distinct profiles in gene expression, immune responses and pharmacogenomic landscapes. This genomic complexity continually adapts to maintain cellular homeostasis amidst varied internal and external pressures, with DNA replication requiring significant fidelity across approximately 7 billion base pairs in the diploid genome. However, environmental factors introduce selective pressures that can accumulate mutations, underscoring the intricate relationship between genetic determinants and environmental influences in disease development (Marques et al. 2024).

A global perspective on precision medicine is essential to understand how genetic variability influences disease presentation and impacts healthcare strategies worldwide (Khoury et al. 2020). Recognising the diversity of genomic and demographic variables, as well as epidemiological and ethnic considerations, is vital for effective patient care. The disparity in healthcare infrastructure and management approaches across regions, ranging from India to other global nations, demands a multidisciplinary approach that integrates various healthcare modalities, fostering collaboration between researchers and practitioners in PM implementation. This holistic model advocates for drug repurposing and dose optimisation to maximise therapeutic efficacy while simultaneously promoting proactive lifestyle modifications that reduce disease risk (Rebbeck et al. 2022).

Environmental factors substantially influence disease susceptibility, with genetic predisposition accounting for only about 10% of disease aetiology. For example, most cancers, which remain a significant global health issue, are attributed to modifiable lifestyle and environmental factors rather than inherited genetic mutations (Borja et al. 2024). Preventative measures, including tobacco cessation, healthy dietary habits, moderated alcohol consumption, sun protection and regular physical activity, play a critical role in minimising exposure-related genomic instability. Precision medicine leverages these insights to promote preventive strategies, emphasising the 'prevention is better than cure' philosophy.

In neurology, PM has spurred advancements in molecular biology, genetic engineering and neurostimulation, revolutionising care for conditions with complex pathophysiology, such as epilepsy, Alzheimer's disease, amyotrophic lateral sclerosis (ALS), spinal cord injuries (SCIs) and autism spectrum disorder (ASD). While traditional therapies offer some symptom management, they often fall short of addressing disease mechanisms and leave significant unmet needs. Recent innovations in biotechnology have led to treatments targeting specific molecular pathways linked to inflammation and neurodegeneration (Ju and Beyeler 2021). For instance, medications modulating pathways like Sirtuin 1 (SIRT-1), mammalian target of rapamycin (mTOR) and nuclear factor erythroid 2-related factor 2 (NRF2) show promise in slowing disease progression and offering neuroprotection for traditionally incurable conditions.

Gene editing technologies like Clustered Regularly Inter-spaced Short Palindromic Repeats (CRISPR) have created new avenues for treating genetic mutations directly, impacting both hereditary neurodegenerative diseases and disorders resulting from unique genetic abnormalities (Aslam et al. 2024). Cell-based therapies, such as stem cell transplantation, are expanding options for conditions requiring brain repair and regeneration. Neurostimulation techniques, including vagus nerve stimulation (VNS), deep brain stimulation (DBS) and transcranial magnetic stimulation (TMS) offer promising alternatives to pharmacology for managing symptoms in treatment-resistant neurological and psychiatric disorders, enhancing quality of life (QoL) for patients with conditions like chronic pain, epilepsy and major depressive disorder.

The personalised approach in neurology tailors treatment based on patients' unique genetic profiles, disease biomarkers and clinical presentations, enhancing therapeutic efficacy and minimising side effects. This strategy allows clinicians to better manage complex neurological conditions where treatment is complicated by symptom variability and disease progression. Complementary therapies, including cognitive behavioural therapy (CBT), mindfulness and the ketogenic diet, are increasingly used alongside traditional treatments, supporting patients' mental and physical health.

2 The Role of Microbiota in Gastrointestinal (GI) Disorders and Cancer Therapies

Gastrointestinal Tract (GIT)

The GI tract, also known as the alimentary or digestive canal, is an essential, uninterrupted component of the digestive system. It connects the oral cavity with the anus. The oesophagus, stomach and intestines are the major digestive organs that make up the GI tract. The oesophagus, stomach and intestines will digest, absorb nutrients and excrete metabolic waste products as faeces. The term 'gastrointestinal' refers to the functions and illnesses associated with the stomach and intestines (National Institute of Diabetes and Digestive and Kidney Diseases (NIDDK) 2024).

The gastrointestinal system is anatomically separated into two parts: the upper GI tract, which comprises the mouth, pharynx, oesophagus and stomach and the lower GI tract, which contains the small intestine, large intestine, rectum and anus. The tongue, salivary glands, pancreas, liver and gallbladder are accessory organs that supplement the digestive system by producing enzyme secretions and performing other digestive processes that aid in nutrient absorption and assimilation. This sophisticated system's coordinated mechanical and biochemical processes assure food digestion, allowing for the extraction of energy and other necessary nutrients required to maintain physiological equilibrium.

Microbiota

The 'microbiota' is a complex colony of microorganisms that inhabit ecological niches within and on multicellular creatures like plants. It contains a wide number of microbial taxa classified as bacteria, archaea, protists, fungi and viruses, each of which can form commensal, mutualistic or harmful associations. The host microbiota is intimately related to their physiological balance because it plays significant functions in immunological integrity, hormonal regulation and metabolic homeostasis (Vemuri et al. 2020).

The gastrointestinal system is inhabited by trillions of microorganisms, mostly bacteria, that make up the human gut microbiota, a vast and intricate community. These microbes participate in vital physiological processes through their symbiotic relationship with the host. In particular, the phrase 'gut microbiome' describes the unique ecological niche that these microbes occupy, which is typified by significant interindividual variety as well as significant intraindividual fluctuation across time.

Helicobacter pylori is the common name for a Gram-negative, microaerophilic bacterium that colonises the gastric mucosa and infects more than half of the world's population. The clinical results of *H. pylori* infection show significant variation, including host genetic variables, immunological responses and environmental influences. The majority of colonised people are asymptomatic, while approximately 20% develop stomach mucosal diseases. There are peptic ulcer disease (10–15% of cases), stomach adenocarcinoma (1–3%) and mucosa-associated lymphoid tissue (MALT) lymphoma, which are more common among the elderly (Ali and AlHussaini 2024).

In contrast, the absence of *H. pylori* colonisation has been linked to increased susceptibility to diseases such as gastro-oesophageal reflux disease (GERD), oesophageal carcinoma, diabetes mellitus and asthma, implying a complex interaction between microbial colonisation and disease susceptibility (Durazzo et al. 2021). Antibiotics along with proton pump inhibitors are the first line of treatment for eradicating *H. pylori* infection. However, new research, rising rates of antibiotic resistance among *H. pylori* strains and a lack of an effective vaccination limit therapeutic options for this infection.

3 Precision Medicine in Gastrointestinal Disorders

3.1 Overview of Precision Medicine

Definition and Importance Precision medicine, sometimes referred to as personalised medicine, is a cutting-edge approach to medical care and prevention that consider each person's unique genetic, environmental and lifestyle characteristics. Precision medicine seeks to customise treatments according to each patient's unique needs and traits, in contrast to the conventional 'one-size-fits-all' approach (Collins and Varmus 2015). Its potential to increase overall treatment efficiency, decrease side effects and improve results through the use of targeted medicines makes it significant in contemporary healthcare (National Research Council 2011).

Precision medicine has emerged as a result of advances in genomic technologies and a better understanding of the genetic underpinnings of diseases (Zerhouni 2005). Healthcare professionals can determine the most successful treatment options for certain patient populations by combining genomes, proteomics and metabolomics data. Precision medicine in the context of gastrointestinal illnesses allows for more precise diagnosis, personalised treatment strategies and better management of these complicated conditions (Betz et al. 2023).

3.2 Role in GI Disorders

Precision medicine has a significant impact on managing gastrointestinal illnesses like Crohn's disease, ulcerative colitis and IBS.

Crohn's Disease and Ulcerative Colitis
Chronic inflammation of the gastrointestinal system is a hallmark of both diseases, which are subtypes of inflammatory bowel disease (IBD). Disease progression and susceptibility are greatly influenced by genetic predisposition (Zhang et al. 2023). For example, genetic markers like interleukin-23R (IL-23R) and NOD2 (Nucleotide-binding Oligomerization Domain-containing protein 2) have been linked to Crohn's disease, enabling more individualised treatment plans that may involve biologic medicines that target particular inflammatory pathways (Baumgart 2017). Based on patient-specific genetic profiles, tailored therapies such as integrin inhibitors and anti-TNF (tumour necrosis factor) medicines have been developed in ulcerative colitis as a result of the identification of genetic variations (Arruebo et al. 2011).

Irritable Bowel Syndrome (IBS)
Even while IBS is frequently viewed as a functional condition rather than an inflammatory one, new findings about the gut microbiota and its genetic relationships have made it possible to use more precise diagnosis and treatment methods (Black et al.

2022). Knowing each person's unique gut microbiota composition might help advise dietary changes or probiotic use based on their individual needs.

Genetic, Epigenetic and Molecular Markers

Genetic Markers: Advanced genomic research continue to reveal different genetic variants linked to diverse GI problems. Polymorphisms in genes related to immune response, such as HLA genes, are useful in the diagnosis and treatment of IBD (Juan 2019).

Epigenetic Factors: Environmental influences can alter gene expression without changing the underlying DNA sequence, influencing illness patterns in gastrointestinal disorders (Su 2015). Epigenetic changes can be important in determining the onset and severity of illnesses such as IBD and IBS.

Molecular Markers: Biomarkers such as cytokines, proteomics and metabolomics profiles provide information on disease mechanisms and potential treatment targets (The Integrative Human Microbiome Project 2019).

For example, increased inflammatory cytokine levels in patients may signal the necessity for aggressive treatment options or therapy change based on responsiveness.

3.3 Advancements in Diagnostic Tools

Biomarker-Based Diagnostics

Biomarkers are crucial for the early detection and monitoring of gastrointestinal illnesses. These biological indicators, which can be obtained from blood, faeces or tissue samples, aid in the diagnosis of diseases and the prediction of treatment outcomes. In inflammatory bowel disease (IBD), specific inflammatory biomarkers such as C-reactive protein (CRP) and faecal calprotectin are utilised to determine disease activity and remission status (Kalla et al. 2016). Furthermore, the integration of advanced proteome and metabolomic profiles enables the discovery of novel biomarkers that improve diagnostic accuracy and risk stratification (Paananen and Fortino 2019).

Table 1 explains key biomarkers and their roles in the diagnosis and management of gastrointestinal disorders.

Figure 1 illustrates the steps taken for diagnosis based on biomarker testing in patients with GI disorders, including decision points for further testing or treatment.

Genetic Testing and Its Implications for Personalised Therapy in GI Disorders

In order to diagnose and customise treatments for gastrointestinal problems, genetic testing has completely changed the process. Clinicians are better able to anticipate therapy effectiveness and disease vulnerability by recognising particular genetic combinations. For example, determining the likelihood of responsiveness to specific medications in individuals with IBD can be done by testing for genetic markers like NOD2 (Kayali et al. 2024). In order to enable early management and heightened

Table 1 Common biomarkers in gastrointestinal disorders

Biomarker	Disorder	Role in diagnosis/Treatment
C-reactive protein (CRP)	Inflammatory bowel disease (IBD)	Indicates inflammation
Faecal calprotectin	Inflammatory bowel disease (IBD)	Monitors disease activity
Lactoferrin	Inflammatory bowel disease (IBD)	Differentiates between IBD and IBS

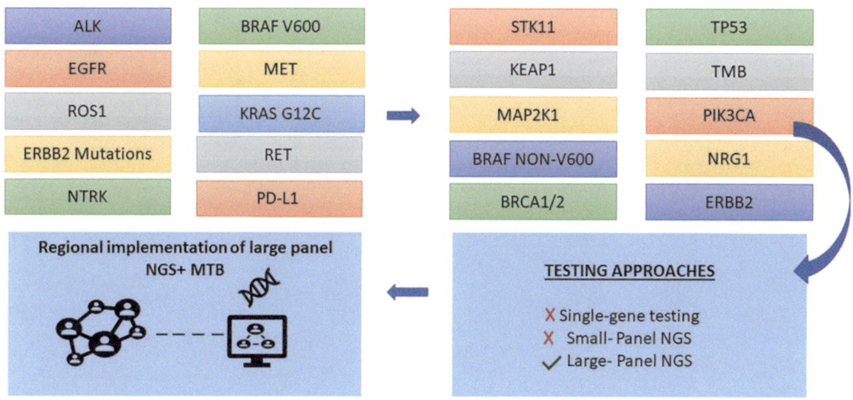

Fig. 1 Flowchart of the diagnostic process using biomarkers in GI disorders

surveillance, genetic testing can also help with screening for hereditary syndromes such as Lynch syndrome and familial adenomatous polyposis (FAP) (Collaborative Group of the Americas on Inherited Gastrointestinal Cancers (CGA-IGC) annual virtual meeting, November 13–15 2021). Beyond just diagnosing problems, these tests enable doctors to select the best individualised treatments, which improves patient care.

3.4 *Therapeutic Strategies*

Precision medicine has made it possible to develop personalised medications that consider an individual's genetic and molecular characteristics. By analysing a patient's genetic makeup, healthcare providers can choose medications that are more likely to be beneficial while reducing the chance of side effects. For example, individuals with specific mutations in the IL-23 receptor may benefit from tailored biologics that inhibit this pathway, resulting in more efficient healthcare resource allocation (Vuyyuru et al. 2023). Furthermore, the use of pharmacogenomics allows clinicians to change drug dosages based on the patient's genetic susceptibility to metabolise specific drugs differently, improving therapy efficacy and lowering toxicity (Tamsi and Falus 2012).

Table 2 Tailored therapies in gastrointestinal disorders

Genetic profile	Personalised therapy	Disorder
NOD2 mutations	Anti-TNF agents (e.g., infliximab)	Crohn's disease
IL-23 receptor variants	IL-23 inhibitors (e.g., guselkumab)	IBD

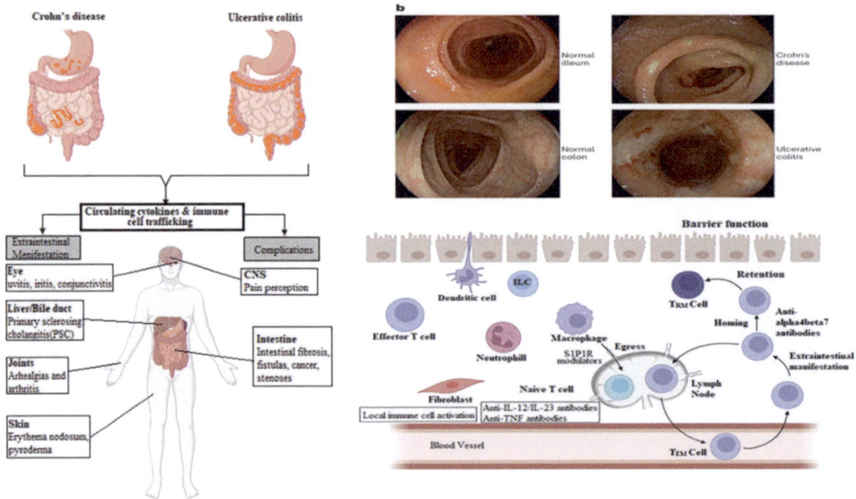

Fig. 2 Mechanism of action of biologics in IBD

Table 2 outlines the examples of personalised treatment strategies based on genetic profiles for managing gastrointestinal disorders.

Figure 2 shows how biologics work in inflammatory pathways, targeting specific immune responses in diseases such as Crohn's disease and ulcerative colitis (strategies for targeting cytokines in inflammatory bowel disease).

Emerging Treatments (e.g., Biologics, Small-Molecule Inhibitors) Precision medicine has paved the path for personalised therapeutics based on a person's genetic and molecular characteristics. By analysing a patient's genetic makeup, healthcare providers can choose medications that are more likely to be beneficial while minimising the risk of side effects. Patients with specific mutations in the IL-23 receptor, for example, may benefit from targeted biologics that inhibit this pathway, resulting in more efficient healthcare resource allocation (Tallarico et al. 2022). Furthermore, the application of pharmacogenomics allows clinicians to change drug dosages based on the patient's genetic susceptibility to metabolise specific drugs differently, increasing therapy efficacy and lowering toxicity (Sandborn et al. 2017).

Table 3 Emerging treatments for GI disorders

Treatment type	Examples	Mechanism of action
Biologics	Infliximab, adalimumab	Anti-TNF agents
Small-molecule inhibitor	Tofacitinib	JAK inhibitor
Microbiome modulation	Faecal microbiota transplantation	Restores healthy gut microbiota

Table 3 shows overview of innovative therapeutic approaches for gastrointestinal disorders, including biologics, small-molecule inhibitors and microbiome modulation, with examples and their mechanisms of action.

4 Microbiota in Gastrointestinal Health and Disease

4.1 Gut Microbiota Composition

The gastrointestinal (GI) system depends on a wide variety of microorganisms, mostly bacteria, that make up the human gut microbiome. Age, nutrition, genetics and environmental circumstances all affect the composition of the microbiome. Important bacterial phyla are usually found in a balanced gut microbiome, including:

Firmicutes These are known for digesting dietary fibres and generating short-chain fatty acids (SCFAs), which are helpful to gut health and metabolism.

Bacteroidetes These bacteria are important for breaking down complex carbs, which improves nutrition absorption.

Actinobacteria These bacteria help to synthesise vitamins and maintain gut homeostasis through immunomodulatory actions.

Table 4 outlines the key microbial phyla associated with gut health and their respective functions. Studies have shown that a diverse microbiome favours health, supporting various physiological functions including digestion, immune response and protection against pathogens (Sommer and Backhed 2013).

4.2 Role in Gastrointestinal Diseases

Figure 3 illustrates a comparative analysis of gut microbiota composition in healthy individuals versus those with dysbiosis.

Dysbiosis, or an imbalance in the gut microbiota, has been related to a variety of gastrointestinal conditions, most notably inflammatory bowel disease (IBD) and colorectal cancer (CRC).

Table 4 Key microbial phyla associated with gut health and their respective functions

Microbial Phylum	Key functions	Associated health benefits
Firmicutes	Fermentation of dietary fibres Production of short-chain fatty acids (SCFAs) Energy extraction from food	Improves gut barrier integrity Regulates immune response Reduces inflammation
Bacteroidetes	Degradation of complex carbohydrates Synthesis of essential vitamins	Enhances nutrient absorption Maintains gut microbiota diversity
Actinobacteria	Metabolisation of fibres Synthesis of vitamins (e.g., B12)	Supports gut health Contributes to immune modulation
Proteobacteria	Involved in short-chain fatty acid production	Weakened abundance may indicate dysbiosis; contributes to inflammation
Fusobacteria	Potential role in inflammation and immune response	Associated with colorectal cancer when overrepresented
Verrucomicrobia	Known for the presence of beneficial bacteria like Akkermansia muciniphila	Supports gut barrier function and metabolic health

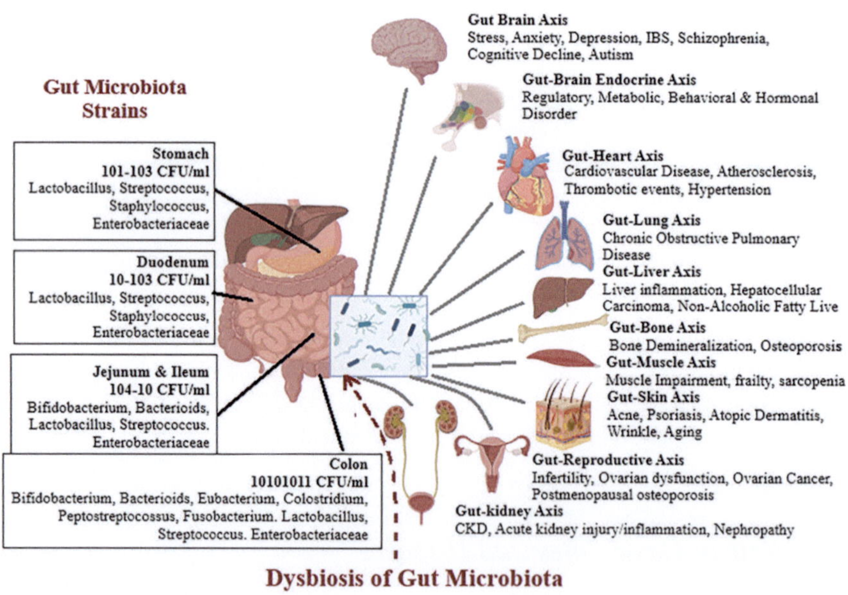

Fig. 3 Comparative analysis of gut microbiota composition in healthy individuals versus those with dysbiosis

Inflammatory Bowel Disease (IBD)

IBD, including Crohn's disease and ulcerative colitis, causes persistent inflammation of the gastrointestinal tract. Patients frequently show reduced microbial diversity, with an increase in pathogenic bacteria and a decrease in beneficial microorganisms (M'Koma 2018). This shift may impair intestinal barrier function, increase inflammation and contribute to disease development.

Colorectal Cancer (CRC)

Dysbiosis may raise the risk of colorectal cancer (CRC), according to research. Fusobacterium nucleatum has been linked to carcinogenesis by altering immune responses and generating cancer-causing metabolites (Cao et al. 2024). The method of action may entail altering epithelial cell metabolism and interacting with the host's immune system, resulting in an environment favourable to cancer formation.

Mechanisms of Interaction Between Microbiota and Host Immune Responses

The link between gut microbiota and host immune responses is complicated and reciprocal, regulating both gut health and overall immunity. The following are the primary processes by which bacteria interact with and affect host immune responses:

1. Immune System Education

Immune Cell Maturation: Commensal bacteria are required for immune system development and maturation. Gut microbiota promotes the development of naive T cells into numerous subsets, including regulatory T cells (Tregs), which are critical for maintaining immunological tolerance to commensal microorganisms and limiting excessive inflammatory responses (Round and Mazmanian 2009).

Cytokine Production: The gut microbiota helps immune cells produce cytokines, which influence both innate and adaptive immunity. For example, certain bacterial species can increase the synthesis of IL-10, a cytokine that helps regulate inflammation (Belkaid and Hand 2014).

2. Barrier Function Enhancement

Intestinal Epithelial Integrity: Commensal microorganisms enhance the production of tight junction proteins, which increases the permeability of the intestinal barrier. A healthy microbiome improves this barrier by preventing infections and chemicals from entering the bloodstream, lowering the risk of systemic inflammation (Virk et al. 2024).

Production of SCFAs: Short-chain fatty acids (SCFAs), such as butyrate generated by fermentation of dietary fibres by gut bacteria, serve as energy sources for colonocytes and play a protective function in gut barrier integrity (Varada et al. 2022).

3. Immune System Modulation

Regulatory Effects on Immunity: Some gut bacteria can regulate immune cell activity. Lactobacillus and Bifidobacterium can improve dendritic cell function, promote Treg formation and maintain immunological homeostasis (Catalkaya et al. 2020).

Dysbiosis and Immune Disorders: Dysbiosis, or an imbalance in the microbiota, can cause incorrect immune responses and is linked to a variety of disorders including inflammatory bowel disease, allergies and autoimmune diseases. Dysbiosis may change the synthesis of metabolites and immunological signalling molecules, which contributes to inflammation and disease progression (Kaur et al. 2022).

4. Metabolite Production and Immune Function

Signalling Molecules: Microbiota generate metabolites that communicate with host immune cells by acting as signalling molecules. For instance, by activating particular receptors on immune cells, bile acids generated by gut microorganisms might modify immunological responses (Loh et al. 2024).

Pathogen Inhibition: The microbiota can increase host defences against infections by outcompeting pathogenic organisms for resources and inducing the epithelium to produce antimicrobial peptides (Quinn 2017).

4.3 Precision Microbiota Therapeutics

The development of precision microbiota therapies has been accelerated by the growing understanding of the function of the gut microbiome in both health and disease. These strategies seek to use and alter the microbiome to enhance health outcomes, especially when it comes to chronic illnesses and dysbiosis-influenced ailments. Probiotics, prebiotics, faecal microbiota transplantation (FMT) and microbiome-based diagnostics are the main modalities in precision microbiota therapy listed below.

1. Probiotics

Probiotics are live bacteria that help the host's health when given in sufficient quantities. They can aid in the restoration of microbial equilibrium, particularly during illness or antibiotic-induced dysbiosis. Probiotics have shown promise in a number of areas:

Gut Health: Irritable bowel syndrome (IBS), inflammatory bowel disease (IBD) and antibiotic-associated diarrhoea can all be prevented and treated with probiotics by enhancing intestinal barrier function and modulating immune responses (Kamada et al. 2013).

Systemic Health: Probiotics are being researched for their effects on immunological regulation, metabolic health and mental health (the gut-brain axis), in addition to their advantages for the gastrointestinal tract (Stolfi et al. 2024).

2. Prebiotics

Prebiotics are indigestible food components that specifically promote the development and function of good gut flora. Complex carbohydrates like fructooligosaccharides and inulin make up the majority of them.

Supportive Role: Prebiotics can increase the synthesis of SCFAs, which nourish colonocytes and preserve the integrity of the gut barrier, by serving as substrates for good bacteria (Slavin 2013).

Impact on Metabolism: Prebiotics may help control obesity and diabetes since they have also been connected to better metabolic outcomes, including decreased body weight and increased insulin sensitivity (AlMarzooqi et al. 2024).

3. Faecal Microbiota Transplants (FMT)

The goal of FMT is to re-establish a balanced microbiome by transferring stool from a healthy donor to a recipient. Due to its efficacy, this treatment has become well-known, especially in:

Clostridium difficile Infection: The recurrent *C. difficile* infections have been successfully treated with FMT; studies have reported cure rates of more than 90% (Li et al. 2022).

Broader Applications: Although FMT is mostly used to treat *C. difficile*, studies are being conducted to determine whether it can also be used to treat irritable bowel syndrome, IBD and even mental health issues by way of the gut–brain axis (Merrick et al. 2023).

4. Microbiome-Based Diagnostics

Diagnostic methods to examine the makeup and function of the microbiome are being developed as the connection between the microbiome and different diseases is better recognised. This may result in individualised treatment strategies, such as:

Biomarkers for Disease: Certain microbial signatures have been found to be associated with the occurrence of diseases such as metabolic syndrome, inflammatory bowel disease and several types of cancer (Gopalakrishnan et al. 2018).

Tailored Precision Medicine: Employing microbiome data, medical professionals can design personalised therapies, such as advising specified probiotics or dietary adjustments, to restore microbial balance and address specific health conditions.

Table 5 summarises various microbiota modulation therapies, their mechanisms, clinical applications and notable examples or findings.

Table 5 Overview of microbiota modulation therapies

Therapeutic modality	Description	Clinical applications	Examples/Findings
Probiotics	Live microorganisms that confer health benefits	Antibiotic-associated diarrhoea IBD IBS	Lactobacillus for diarrhoea in children Bifidobacterium for IBS relief
Prebiotics	Non-digestible food ingredients that stimulate beneficial bacteria	Gut-barrier enhancement Metabolic health	Fructooligosaccharides improve gut health and SCFA production
Faecal microbiota transplants (FMT)	Transplantation of faecal matter from a healthy donor	Recurrent *C. difficile* infection IBD	High success rates (90%) in treating recurrent *C. difficile*
Microbiome-based diagnostics	Techniques to analyse microbiome composition to guide treatment	Personalised medicine Disease biomarker identification	Microbial signatures correlate with obesity and IBD

5 Microbiota-Driven Cancer Therapies

The intricate relationship between gut bacteria and cancer treatments has drawn more attention in recent years. Significant inter-individual heterogeneity in medication metabolism results from this relationship's bidirectional interaction, in which bacteria in the gut metabolically alter a variety of pharmacological drugs. Across a variety of illness states, these changes can significantly impact the therapeutic effectiveness and side effects of pharmaceutical therapies (Zhao et al. 2023; Pandey et al. 2023). On the other hand, systemic therapies have the potential to drastically change the gut microbiota's makeup and activity, which could affect patient outcomes (Shang et al. 2020).The link between cancer and the gut bacteria has created new opportunities for treatment approaches. This section emphasises microbiome-targeted therapy methods and talks about how microbiota affects the course of cancer.

5.1 *Microbiota and Cancer Development*

Dysbiosis and Inflammation
Chronic inflammation, a major factor in the development of cancer, is associated with dysbiosis, or an imbalance in the gut microbiota. For instance, research has demonstrated that some bacteria, like Fusobacterium nucleatum, are more prevalent in colorectal cancers and may stimulate the formation of tumours by activating inflammatory pathways (Shang et al. 2020).

Metabolite Production
The production of metabolites, such as short-chain fatty acids (SCFAs), by the gut microbiota may have an impact on the development of cancer. By encouraging apoptosis in cancer cells, butyrate, for instance, has been shown to have protective benefits against colon cancer (Shang et al. 2020).

Immune System Interaction
The growth of tumours and the effectiveness of treatments may be impacted by the microbiome's modulation of systemic immune responses. The efficacy of immunotherapies may be impacted by some gut bacteria that strengthen anti-tumour immunity (Zhao et al. 2023).

5.2 *Microbiota-Targeted Cancer Therapies*

Probiotics and Prebiotics
Probiotics may improve the efficacy of cancer treatments by assisting in the restoration of microbial equilibrium. They have been demonstrated to improve patient well-being by reducing gastrointestinal toxicity associated with chemotherapy (Yarahmadi and Afkhami 2024).

Faecal Microbiota Transplants (FMT)
FMT may be useful as a cancer treatment. FMT has been shown to change cancer patients' microbiomes, which may improve their reactions to immunotherapy (Zhou et al. 2022).

Targeted Antibiotic Therapy
By lowering the number of harmful bacteria while maintaining helpful microorganisms, the strategic use of antibiotics may improve immunotherapy results. To preserve microbial variety, this needs to be managed carefully (Pandey et al. 2023).

6 Challenges and Future Directions

Growing recognition of the ubiquitous function of the microbiota in influencing human health and disease, particularly in oncogenesis, dominates the current biomedical research landscape (Belkaid and Hand 2014; Zhao et al. 2023). We still lack a thorough knowledge of mechanisms, though. This change signals the urgent need to look into how the microbiota contributes to carcinogenesis and to change treatment paradigms to incorporate microbiome-directed strategies (Gopalakrishnan et al. 2018; Kandalai et al. 2023).

In fact, given the abundance of unanswered problems in the subject, we must further deepen our understanding of the intricate relationships. The most crucial challenge in the upcoming years will be to translate these preclinical discoveries into clinical settings (Rahman et al. 2022). The ability to forecast the effects of microbiota on human health will become increasingly important as precision medicine advances (Shang et al. 2020).

6.1 Integration of Precision Medicine with Microbiota Research Challenges in Integrating Omics Data (Genomics, Microbiomics, etc.) into Clinical Practice

In precision medicine, treatment regimens are tailored to each patient's particular genetic, environmental and behavioural characteristics. Several challenges arise when integrating omics data, especially in the fields of genomics and microbiomics:

6.1.1 Data Heterogeneity

Description Diverse data types (genomic sequences, microbiome profiles and clinical outcomes) from different research may be difficult to integrate due to variations in data collection, processing and presentation formats. Attempts to make insightful comparisons and inferences between research are hampered by this variability (Kandalai et al. 2023).

Impact Clinicians may encounter difficulties obtaining a comprehensive understanding of a patient's microbiome and its possible effects on treatment due to inconsistent data formats, which can cause issues with data integration tools.

6.1.2 Interpretation of Data

Description The biological implications of differences in microbiome composition are still not well known. Distinguishing between helpful and detrimental microbial communities and their functional implications necessitates advanced data analysis techniques (Pandey et al. 2023).

Impact Clinicians may struggle to make treatment decisions based on microbiome data in the absence of established guidelines on how specific microbial compositions affect drug metabolism and clinical outcomes.

6.1.3 Personalised Approaches

Description Individual microbiome-based treatments are still in their early stages. Technological and methodological limitations frequently limit the ability to identify the ideal microbiome for improving therapeutic effects (Shang et al. 2020).

Impact This problem causes gaps in treatment procedures, thereby limiting the efficacy of medicines and thorough patient care.

Challenges	Description
Data heterogeneity	Variability in data formats and methods complicates integration
Interpretation of data	Understanding clinical relevance of microbiome interactions is complex
Personalised approaches	Difficulty in developing tailored interventions based on microbiome profiles

6.2 Regulatory and Ethical Considerations Addressing Personalised Treatments and Patient Privacy Issues

The intersection of microbiome research with clinical applications raises regulatory and ethical concerns, including personalised treatments and patient privacy concerns.

6.2.1 Personalised Treatments

Description Implementing microbiome therapeutics in clinical settings necessitates rigorous clinical studies that can evaluate safety, efficacy and optimal dosage. To achieve inclusive outcomes, these trials must include a wide range of human microbiomes (Zhao et al. 2023).

Impact Developing guidelines for evaluating microbiome-based therapies is critical to avoiding premature clinical adoption that may not be universally applicable.

6.2.2 Patient Privacy

Description Genetic privacy is a problem when it comes to the collection and use of microbiome data. Patients need to be reassured that sensitive information about their microbiome will be safeguarded (Yarahmadi and Afkhami 2024).

Impact To preserve patient confidence and promote involvement in microbiome-related research, strict data protection procedures must be put in place.

6.2.3 Ethical Implications of Microbiome Alterations

Description Informed consent presents ethical challenges when a patient's microbiome is altered by treatments like faecal microbiota transplantation (FMT) or probiotics. Potential hazards, advantages and the long-term effects of changes to the microbiome must be explained to patients (Zhou et al. 2022).

Impact Preventing potential harm to patients by ensuring ethical rules are followed is crucial for responsible research and therapeutic development.

Areas of concern	Description
Personalised treatments	Regulatory hurdles for conducting rigorous clinical trials
Patient privacy	Need for regulations to safeguard sensitive microbiome data
Ethical implications	Questions about consent and long-term health effects of modifications

6.3 Future Trends in Precision Medicine and Microbiota Therapies Emerging Technologies (e.g., CRISPR, AI-Based Diagnostics) and Their Future Impact

It is anticipated that the field of microbiome medicine will change quickly because to new research methods, technology advancements and the developing. The following technologies are available:

6.3.1 CRISPR

Description CRISPR technology provides novel ways to modify particular genes in microbial populations. This can either eradicate harmful microorganisms or increase the presence of helpful ones (Pandey et al. 2023).

Impact CRISPR has the potential to significantly improve the treatment of health conditions associated with the microbiota and increase the effectiveness of cancer treatments by accurately modifying microbial functions.

6.3.2 AI-Based Diagnostics

Description The analysis of extensive microbiome data can be aided by artificial intelligence. Patterns in microbiome diversity that are associated with treatment outcomes can be found using AI algorithms (Pandey et al. 2023).

Impact Compared to conventional techniques, the use of AI tools can expedite diagnostics and more effectively customise intervention plans.

6.3.3 Prospective Clinical Trials and Translational Research

Description Clinical studies that pinpoint certain microbial fingerprints linked to treatment outcomes will be the subject of more and more future study. This will facilitate the development of more potent medicines that target the microbiome and the comprehension of their mechanisms (Zhao et al. 2023).

Impact Improved translational research can help close the gap between basic microbiome research and clinical practice, which will help microbiome therapies become more widely accepted.

Future trends	Description
CRISPR	Potential use in modifying beneficial microbial strains for better outcomes
AI-based diagnostics	Enhanced predictive models for treatment responses based on microbiome data
Clinical trials	Focus on identifying microbial signatures linked to patient outcomes

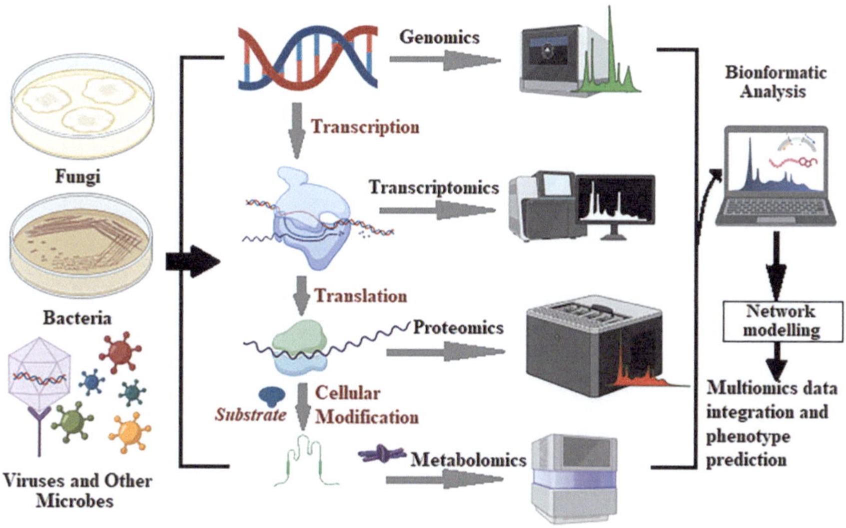

Fig. 4 Integration of omics data in microbiota research

Figure 4 is a flowchart illustrating how various forms of omics data (genomics, microbiomics, transcriptomics and metabolomics) can be integrated for personalised treatment strategies.

Figure 5 showcases the various relational frameworks for microbiome research with Indigenous communities.

7 Neuropharmacological Advancements

7.1 Pathway-Targeted Drug Development

Neuropharmacology advances rely significantly on the targeting of specific molecular and cellular pathways involved in neurodegeneration, neuroinflammation and cellular stress responses. Drugs that target pathways such as SIRT-1, mTOR and NRF2 are showing promise in clinical trials.

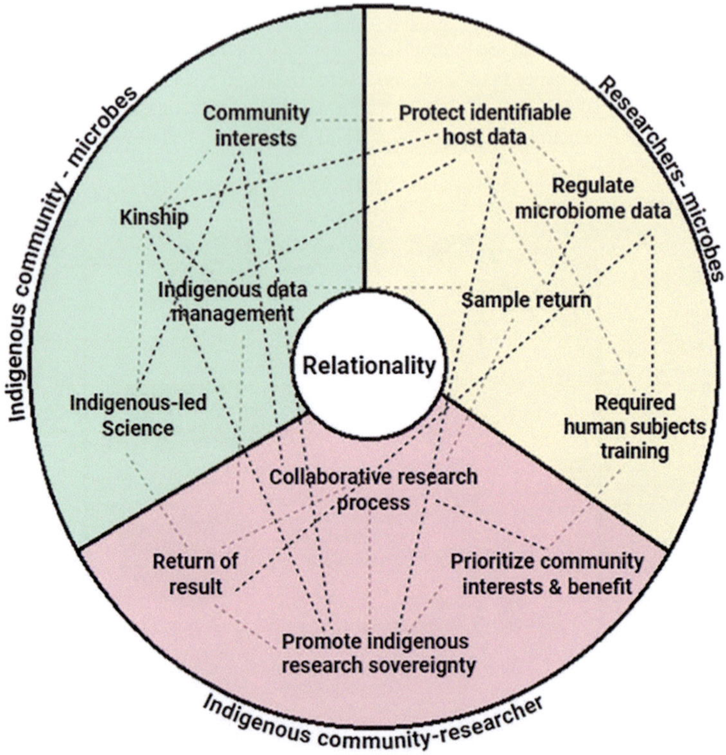

Fig. 5 A relational framework for microbiome research with Indigenous communities

7.2 SIRT-1 Activation in Neuroprotection for Alzheimer's Disease

Sirtuin 1 (SIRT-1), a NAD+-dependent deacetylase, is known for its ability to regulate cellular stress response, inflammation and neuronal survival. SIRT-1 is a promising therapeutic target for Alzheimer's disease (AD) due to its capacity to lower amyloid-beta (Aβ) accumulation, improve neuronal resilience and slow neurodegeneration (Guarente et al. 2018).

Mechanisms of Neuroprotection

SIRT-1 improves neuroprotection by diverting APP (Amyloid Precursor Protein) processing away from the amyloidogenic pathway, which causes the neurotoxic Aβ plaques associated with Alzheimer's disease (O'Brien et al. 2005) found that non-amyloidogenic processing promotes the production of neuroprotective soluble APPα over damaging Aβ fragments. SIRT-1 also aids in the fight against oxidative stress and inflammation, both of which play important roles in Alzheimer's disease. SIRT-1 promotes a neurodegeneration-resistant brain environment by activating antioxidant genes and suppressing proinflammatory responses.

Clinical Impact on Alzheimer's Disease

Reduction in Amyloid-Beta Accumulation: SIRT-1 activation lowers amyloid-beta accumulation via affecting APP processing and improving Aβ elimination. This activity may halt AD progression, providing neuroprotection through both reduced plaque formation and lower neuronal stress (Guarente et al. 2018).

Enhanced Neuronal Survival: SIRT-1 activation promotes synaptic plasticity and neuronal survival, both of which are critical for cognitive function. SIRT-1 has demonstrated promise in protecting neurones from AD-related degeneration by promoting mitochondrial health and cellular energy balance, which can enhance memory and learning processes (Gao et al. 2010).

Therapeutic Potential of SIRT-1 Activators

In AD models, substances such as resveratrol, a polyphenol that activates SIRT-1, have demonstrated neuroprotective effects. Research suggests that resveratrol may improve mitochondrial function, reduce inflammation and lower Aβ levels, underscoring its potential as a component of a multifaceted Alzheimer's disease treatment strategy (Porquet et al. 2013).

Table 6 lists the different substances that are known to target the SIRT1 pathway and details their interactions with SIRT1, indicating whether they are inhibitors or activators. Signals identify the illnesses for which these substances are being used or studied. Additionally, it summarises the main conclusions drawn from preclinical or clinical research on the efficacy of each medication.

Table 6 Summary of SIRT1 pathway-targeted drugs and clinical outcomes

Drug/Agent	Mechanism of action	Indications	Clinical outcomes
Resveratrol	SIRT1 activator, anti-inflammatory	Alzheimer's disease	Improved cognitive function and reduced amyloid-beta levels; increased SIRT1 expression
SRT1720	Selective SIRT1 activator	Metabolic disorders	Enhanced glucose metabolism and insulin sensitivity; decreased markers of inflammation
SIRT1 inhibitor Ex-527	SIRT1 inhibitor	Various cancer types	Induced apoptosis in cancer cells; reduced tumour growth in preclinical models
Nicotinamide	SIRT1 inhibitor (high doses)	Neurological disorders	Decreased neuroprotection; potential worsening of neurodegenerative conditions with long-term high dose.
Fisetin	SIRT1 activator, antioxidant	Neurodegenerative diseases	Cognitive improvement in mouse models; enhanced neuronal survival and function
Quercetin	SIRT1 activator, anti-inflammatory	Age-related diseases	Improved mobility and cognitive function in elderly population; upregulation of SIRT1 expression

7.3 Case Study: mTOR Pathway in Epilepsy

Neural activity, metabolism and cell development are all significantly impacted by the mammalian target of rapamycin (mTOR) pathway. Many neurological disorders, including epilepsy, are associated with dysregulation of this pathway, particularly in individuals who have mTOR signalling-related genetic abnormalities. mTOR pathway inhibitors have demonstrated encouraging outcomes in refractory epilepsy, a condition in which seizures are not responsive to conventional therapies.

Mechanism of Action By stabilising synaptic signalling and avoiding neuronal over excitation, which is a defining feature of epileptic activity, mTOR pathway inhibitors like rapamycin lower neuronal hyperactivity (Swiech et al. 2008). These inhibitors lessen the frequency and intensity of seizures by altering mTOR signalling, especially when mutations cause the system to become overactive.

Clinical Outcome

According to clinical research, mTOR inhibitors can greatly enhance seizure control in epileptic patients who have mTOR pathway mutations, including those found in tuberous sclerosis complex (TSC) (Krueger et al. 2013). In these situations, even in patients who had not responded to other antiepileptic medications, mTOR inhibitor treatment has resulted in decreased seizure frequency, enhanced quality of life and increased seizure stability.

Figure 6 shows how the mTOR pathway is activated and how it affects neuronal excitability and seizure frequency. This figure illustrates key elements, including synaptic plasticity, neuronal development, and the therapeutic advantages of mTOR inhibition.

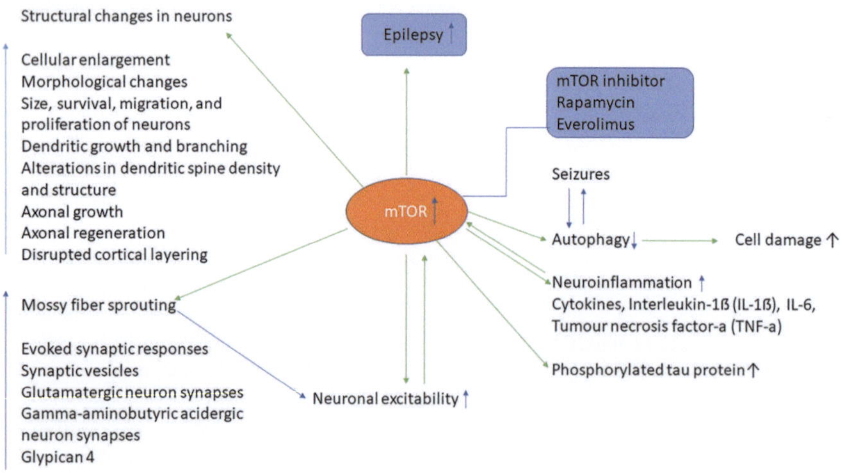

Fig. 6 Mechanism of mTOR inhibition in epilepsy

7.4 Case Study: NRF2 in Parkinson's Disease

A transcription factor called nuclear factor erythroid 2-related factor 2 (NRF2) is essential for cellular defence against oxidative stress because it controls the expression of antioxidant proteins. Oxidative stress and mitochondrial dysfunction play a major role in neuronal degeneration in Parkinson's disease (PD), which is characterised by the gradual loss of dopaminergic neurons in the substantia nigra. As a result, NRF2 pathway activation has become a viable treatment approach for PD management (Lastres-Becker et al. 2016).

Mechanism of NRF2 in Neuroprotection

Antioxidant enzymes like glutathione peroxidase and superoxide dismutase, which counteract free radicals and lessen oxidative damage in neurones, are expressed more when NRF2 is activated. NRF2 activation keeps mitochondrial function intact and stops the accumulation of harmful reactive oxygen species (ROS) that cause neurodegeneration in Parkinson's disease (PD) mice (Johnson et al. 2008). Furthermore, NRF2 reduces neuroinflammation, a major aspect of Parkinson's disease pathophysiology, by regulating inflammatory responses in glial cells.

Table 7 provides listings of substances that are known to support or activate the NRF2 pathway along with an explanation of how each one works with the pathway to produce its effects.

Table 7 Summary of NRF2-targeting agents in Parkinson's disease research

Agent	Mechanism of action	Indications	Clinical outcomes
BDNF (Brain-derived neurotrophic factor)	Activates NRF2, promotes neuronal survival	Parkinson's disease	Enhanced motor function and neuroprotection in preclinical models; improved quality of life in early-phase trials
L-theanine	NRF2 activator, antioxidant	Parkinson's disease	Increased NRF2 levels and reduced oxidative stress; improvement in motor deficits observed in animal models
Sulforaphane	NRF2 activator, detoxifier	Parkinson's disease	Significant neuroprotective effects; reduction in disease progression and motor symptoms in preliminary studies
Curcumin	NRF2 activator, anti-inflammatory	Neurodegenerative disorders	Improved cognition and reduced neuroinflammation in animal models; limited human clinical data showing symptom improvement
Tert-butylhydroquinone (tBHQ)	NRF2 activator	Neurodegenerative diseases	Protective effects against neurotoxins; improvement in behavioural tests in rodent models
Astaxanthin	Antioxidant, enhances NRF2 pathway activation	Neurodegenerative diseases	Showed significant protection against neuronal cell death; improvement in motor coordination in animal models

8 Gene and Cell-Based Therapies

The development of gene and cell-based therapies has transformed the therapeutic landscape for neurological illnesses by addressing the underlying genetic reasons. These treatments have the ability to change or repair faulty genes, providing hope for previously incurable illnesses. CRISPR (Clustered Regularly Inter-spaced Short Palindromic Repeats) has emerged as a game-changing technique because of its precision, efficiency and versatility in targeting specific genes.

8.1 CRISPR and Gene Editing

CRISPR-based gene editing technique enables targeted genome alterations, which is particularly useful in treating genetic neurological illnesses. CRISPR allows for the precise addition, deletion or repair of genetic information by directing the CRISPR-associated protein (Cas9) to specific places in the DNA via a guide RNA sequence (Doudna and Charpentier 2014). This method has showed enormous promise in targeting genes responsible for illnesses such as Huntington's disease, fragile X syndrome and amyotrophic lateral sclerosis. The capacity to directly target disease-causing mutations may result in long-term functional gains, taking symptomatic treatment to possibly curative levels.

8.2 Case Study: CRISPR in ALS

Amyotrophic lateral sclerosis (ALS), a neurological illness, is frequently associated with abnormalities in the super oxide dismutase 1 (SOD1) gene. Mutant SOD1 protein aggregates contribute to motor neurone degeneration, which is characteristic of ALS disease. For years, therapy options were restricted to controlling symptoms rather than addressing the underlying genetic cause.

Intervention

Researchers have successfully employed CRISPR technology to target SOD1 mutations in ALS animal models. Researchers were able to directly edit and knock down mutant SOD1 in these mice using CRISPR-Cas9, lowering protein aggregation and so alleviating neuronal injury. This technique has enhanced motor function, increased survival and reduced neurodegeneration in ALS mice, providing potential insights into future clinical applications (Smythe et al. 2024).

Figure 7 effectively summarises the CRISPR mechanism in gene editing, detailing both its components and processes involved in modifying genetic material accurately and efficiently.

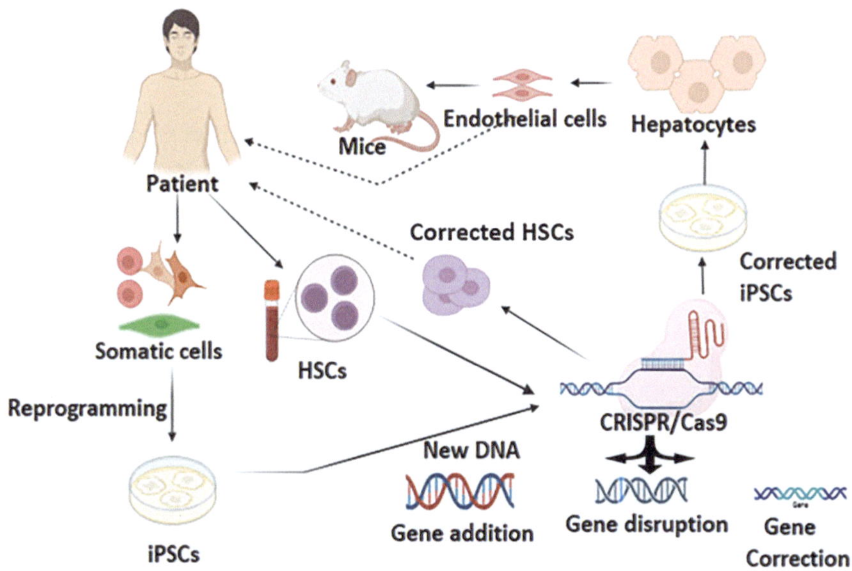

Fig. 7 Diagram of CRISPR mechanism in gene editing

8.3 *Stem Cell-Based Therapies*

When it comes to treating neurological conditions, especially those that involve the loss or damage of neurones, such spinal cord injuries and neurodegenerative diseases, stem cell-based therapies offer a revolutionary new approach. Because of their exceptional capacity to develop into a variety of cell types, such as neurones and glial cells, stem cells can be used to replace damaged cells and aid in the central nervous system's (CNS) functional recovery.

Mechanism of Stem Cell Therapy

Stem cell treatment is the transplantation of stem cells that can develop into specific neural cell types, assisting in the regeneration, repair and functional recovery of injured brain tissues. Several types of stem cells, including embryonic stem cells (ESCs), induced pluripotent stem cells (iPSCs) and mesenchymal stem cells (MSCs), have been studied for their restorative potential in neurological diseases. Following implantation, these cells can migrate to damage sites, develop into neurones or supporting glial cells and produce neurotrophic factors that aid in the survival and repair of existing neurones (Cummings et al. 2005).

Clinical Impact on Spinal Cord Injuries and Neurodegenerative Diseases

Spinal Cord Injury (SCI)

Stem cell therapy for spinal cord injuries can restore motor function by stimulating neuronal regeneration and bridging disrupted neural networks. Clinical trials of stem cell transplantation in SCI patients have indicated partial restoration of motor and sensory functioning, albeit results vary depending on injury severity and treatment timing (Tetzlaff et al. 2011).

Neurodegenerative Disorders

In disorders like Parkinson's and Alzheimer's disease, where gradual neuronal loss causes symptoms, stem cell therapy has showed promise in preclinical tests by replacing damaged neurones and controlling neuroinflammation. For example, iPSC-derived dopaminergic neurones have been utilised to replace degenerating neurones in Parkinson's disease, resulting in improved motor function in animal models (Takahashi et al. 2007).

Table 8 outlines the type of stem cell-based intervention, clarifies the source of the stem cells employed in the intervention and summarises significant findings and improvements reported in clinical or preclinical investigations.

Table 8 Stem cell-based interventions for neurological disorders

Intervention	Type of stem cells	Indications	Mechanism of action	Clinical outcomes
Mesenchymal stem cells	Adult stem cells from bone marrow	Amyotrophic lateral sclerosis (ALS)	Promote neuronal survival, reduce inflammation, secrete neurotrophic factors	Improved motor function and prolonged survival in early-phase trials.
Neural stem cells	Neural stem cells from foetal tissue	Parkinson's disease	Differentiate into dopaminergic neurons, secrete growth factors	Significant improvements in motor function and quality of life in small trials
Induced pluripotent stem cells (iPSCs)	Reprogrammed somatic cells	Spinal cord injury	Differentiate into neurons and glial cells to promote regeneration	Partial recovery of function and improved sensory/motor abilities in rodent models.
Umbilical cord blood stem cells	Haematopoietic stem cells	Multiple sclerosis	Improve immune response, promote myelination in the CNS	Reduced disease activity in MS patients; improvements in magnetic resonance imaging (MRI) findings
Neuroprogenitor cells	Progenitor cells from the brain	Stroke	Support neurogenesis, reduce apoptosis in injured neurons	Improved functional recovery and reduced neurological deficits in small studies
Adipose-derived stem cells	Adult stem cells from adipose tissue	Traumatic brain injury	Anti-inflammatory effects, support for neurogenesis	Enhanced recovery of cognitive functions and decreased brain oedema in limited trials

9 Neurostimulation Techniques

In order to treat a variety of neurological and psychiatric disorders, particularly those that don't respond to traditional treatments, neurostimulation techniques have become crucial. These strategies use either invasive or non-invasive techniques to apply electrical or magnetic stimulation to alter neuronal activity in certain brain regions. Transcranial magnetic stimulation (TMS), deep brain stimulation (DBS) and vagus nerve stimulation (VNS) are common neurostimulation procedures that have demonstrated promise in the treatment of conditions like epilepsy, depression and chronic pain.

9.1 Overview of Neurostimulation Techniques

1. Transcranial Magnetic Stimulation (TMS) TMS is a non-invasive method of stimulating brain nerve cells using magnetic fields. TMS is administered by a coil applied to the scalp, which creates tiny electric currents in specific brain regions, including the prefrontal cortex, to alter cortical excitability. It is mostly used to treat depression that is resistant to treatment.

2. Deep Brain Stimulation (DBS) DBS is an invasive procedure in which electrodes are surgically implanted in particular brain areas. DBS is a recommended treatment for disorders including Parkinson's disease and refractory epilepsy because these electrodes alter aberrant neural activity by delivering controlled electrical impulses.

3. Vagus Nerve Stimulation (VNS) The vagus nerve, which runs from the brain to the belly, is stimulated by implanting a device during a minimally invasive surgery called vagus nerve stimulation (VNS). VNS has demonstrated effectiveness in lowering the symptoms of depression and refractory epilepsy by altering the neural networks linked to mood and seizure control.

9.2 Case Study: TMS for Depression

Mechanism

The dorsolateral prefrontal cortex is one of the brain regions linked to mood regulation where TMS acts by modifying cortical excitability. It has been demonstrated that TMS improves depression symptoms, especially in cases that are resistant to therapy, by enhancing neuronal connections and changing neurotransmitter levels in this area (Perera et al. 2016).

Table 9 Comparative outcomes in TMS for depression and chronic pain

Outcome	Depression	Chronic pain
Treatment efficacy	60–70% response rate in major depressive disorder (MDD)	30–50% reduction in pain severity reported
Duration of effect	Sustained remission observed up to 6 months post-treatment	Pain relief effects may last 3–6 weeks after treatment
Session frequency	Typically 5 days per week for 4–6 weeks	Two to three sessions per week for 4–8 weeks
Patient tolerance	Generally well-tolerated with few side effects	Some patients report discomfort at the stimulation site
Improvement in quality of life	Significant improvement in QoL measures (e.g., QIDS)	Improvement in QoL and functionality following pain reduction
Long-term follow-up	Decreased relapse rates among responders	Less consistent long-term pain relief among responders

Outcomes

According to clinical research, TMS can dramatically lessen depressed symptoms while having few adverse effects, such a slight headache or soreness in the scalp. TMS is well-tolerated and offers a non-pharmacological alternative for depressed people who don't react to conventional antidepressants. According to meta-analyses, TMS helps between 50% and 60% of patients with treatment-resistant depression with their depressive symptoms (Lefaucheur et al. 2014).

Table 9 compares the outcomes of transcranial magnetic stimulation (TMS) in the treatment of depression and chronic pain. It outlines the efficacy, duration of effects, session frequency, patient tolerance, quality of life improvements and long-term follow-up for each condition.

9.3 *Case Study: DBS in Refractory Epilepsy*

Mechanism

Deep brain stimulation sends precise electrical impulses to particular brain regions linked to seizure activity, like the thalamic anterior nucleus, in patients with intractable epilepsy. By stabilising abnormal neuronal firing patterns and lowering hyperexcitability, these impulses can lower the frequency of seizures (Fisher et al. 2010).

Outcomes

In patients with medication-resistant epilepsy, DBS has demonstrated great potential in lowering seizure frequency. According to studies, DBS can reduce seizures by over 50% in most patients over time, providing an alternative for people whose epilepsy cannot be controlled with medicine alone (Salanova et al. 2015). Although there may be dangers associated with DBS, including as surgical problems and device-related infections, it has shown long-term efficacy and safety.

Table 10 VNS efficacy in epilepsy

Patient population	Duration of VNS therapy	Reduction in seizure frequency	Improvement in quality of life	Adverse effects
89 patients with refractory epilepsy	12 months	45% median reduction; 20% achieved >70% reduction	57% reported improved quality of life	Hoarseness, infection risk, vagus nerve irritation
121 patients with refractory seizures	24 months	Average reduction of 31% seizure frequency	Significant improvements in daily living activities	Nausea, pain at stimulation site
100 patients with drug-resistant epilepsy	5 years	54% reported >50% reduction; 15% achieved >90% reduction	Enhanced emotional and mental well-being	Drowsiness, difficulty swallowing
30 patients with focal epilepsy	6 months	Seizure frequency decreased by 36%	Positive effects on mood reported	Voice changes, sleep disturbances
50 patients with refractory epilepsy	12 months	40% of patients achieved >50% reduction	70% of patients reported general well-being	Minor surgical complications, weight gain

9.4 VNS for Epilepsy

Refractory epilepsy, or situations in which seizures continue despite medication, can be effectively managed by vagus nerve stimulation (VNS). The vagus nerve receives controlled electrical impulses from the VNS, which helps control the aberrant brain activity linked to seizures. Its effectiveness in lowering seizure frequency and improving the quality of life for epileptic patients has been shown in numerous trials.

Table 10 presents the efficacy of vagus nerve stimulation (VNS) in treating patients with refractory epilepsy. It summarises the reduction in seizure frequency, improvements in quality of life and common adverse effects across different patient populations and therapy durations.

10 Personalised Medicine in Neurology

In neurology, personalised medicine refers to a change towards more customised treatment plans based on the distinct genetic and biological characteristics of each patient. By tailoring medicines to the unique genetic, biochemical and clinical traits of each patient, this strategy seeks to maximise therapeutic results. For genetically based neurological disorders including epilepsy and autism spectrum disorder (ASD), where focused therapy is becoming made possible by genetic profiling and biomarker analysis, personalised medicine has shown promise.

10.1 The Role of Genetic Profiling

In personalised neurology, genetic profiling is essential because it might reveal gene changes that may make a person more susceptible to certain neurological disorders or influence how they react to particular therapies. This method looks for mutations or polymorphisms linked to a patient's neurological condition by analysing their genetic data. For example, certain types of epilepsy have been linked to genetic variations in the SCN1A gene, which enables medical professionals to choose more potent therapies that target the underlying genetic mechanism directly (Ottman et al. 2010). Furthermore, genetic profiling plays a crucial role in detecting possible drug–gene interactions, improving the efficacy and safety of therapeutic interventions.

10.2 Case Study: Genetic-Based Treatment in Epilepsy

Context
Certain genetic alterations in epilepsy affect both the response to treatment and the likelihood of seizures. Numerous epilepsy syndromes have been linked to gene mutations in SCN1A, KCNQ2 and DEPDC5, allowing medical professionals to modify treatment plans appropriately.

Intervention
Neurologists can recommend tailored therapies for patients with genetic epilepsy syndromes by using genetic screening. Because of their distinct genetic profile, patients with SCN1A-linked Dravet syndrome, for instance, may benefit from different drugs such stiripentol or clobazam, which are more effective than sodium-channel blockers (Harkin et al. 2007). When opposed to a one-size-fits-all treatment plan, this individualised approach improves seizure control and minimises side effects.

10.3 Applications in Autism Spectrum Disorder (ASD)

Personalised medicine is also advancing therapy for autism spectrum disorder (ASD), with genetic and biomarker analysis used to personalise intervention techniques to each child's unique biological and neurological profile. Autism is a complicated neurodevelopmental disease with a strong genetic base, and genetic variations in genes including SHANK3, CHD8 and NRXN1 are linked to ASD symptoms (Satterstrom et al. 2020). Genetic testing and biomarker analysis enable physicians to create tailored interventions to improve behavioural and cognitive outcomes in people with ASD.

Personalised therapy regimens that target distinct genetic pathways can address the individual neurobiological demands of ASD patients. For example, certain metabolic and mitochondrial dysfunctions seen in ASD can be addressed with dietary changes or supplements matched to the patient's metabolic profile, resulting in improved symptoms and overall quality of life.

10.4 Rare Neurogenetic Disorders

Rare neurogenetic illnesses, which are frequently caused by mutations in specific genes, pose considerable problems due to their distinct and complex pathologies. However, recent advances in genetic therapeutics have created new opportunities for controlling these illnesses, allowing for personalised medications that target the specific genetic abnormalities involved. Unlike traditional therapy, these methods seek to alter the disease's underlying aetiology, with promising developmental and functional effects.

Tailored Genetic Therapies for Rare Disorders
Tailored genetic therapies, such as gene substitution, gene editing and antisense oligonucleotide (ASO) therapy, have been developed to target the unique genetic alterations seen in uncommon neurogenic illnesses. For example, gene replacement therapy has been very effective in disorders such as spinal muscular atrophy (SMA), where an injectable medication, onasemnogene abeparvovec, inserts a functioning copy of the SMN1 gene, significantly improving motor outcomes and life expectancy in infants (Mendell et al. 2017).

Improved Developmental Outcomes
Clinical trials have demonstrated that these tailored methods can result in significant improvements in developmental outcomes for kids with uncommon diseases. ASO treatments, for example, are being explored in illnesses such as Batten disease and Rett syndrome, where preliminary studies show promise for symptom reduction and improved cognitive and motor functioning (Finkel et al. 2017). Furthermore, gene-editing methods such as CRISPR-Cas9 are being investigated in preclinical trials for illnesses such as Fragile X syndrome, with preliminary results suggesting the ability to reverse specific neurodevelopmental symptoms in animal models (BenDavid et al. 2024).

11 Complementary and Integrative Approaches

Complementary and integrative methods to neurology have received recognition as helpful supplements to established pharmacological treatments, particularly for chronic and treatment-resistant neurological illnesses. These techniques, which include non-pharmacological therapies including cognitive behavioural therapy

(CBT), mindfulness practices and dietary changes, seek to improve patient outcomes by addressing the larger physiological and psychological aspects that influence neurological health.

11.1 Non-Pharmacological Interventions

Non-pharmacological approaches are becoming increasingly popular in neurology to supplement normal treatments, particularly for chronic neurological diseases when traditional medicines may be insufficient. For example:

Cognitive Behavioural Therapy (CBT) Cognitive behavioural therapy (CBT) is used to treat neurological illnesses with psychological symptoms, such as persistent pain and anxiety from epilepsy or multiple sclerosis. CBT enables patients to recognise and change thinking and behaviour patterns that contribute to their symptoms, so improving overall coping mechanisms and quality of life.

Mindfulness-Based Practices Meditation helps alleviate stress, anxiety and sadness in neurological patients. Research suggests that these techniques can improve emotional well-being and resilience, potentially benefiting illnesses such as epilepsy, Parkinson's disease and chronic pain (Zeidan et al. 2014).

Dietary Modifications Certain diets, such as the ketogenic diet, are used to treat neurological diseases including epilepsy. The ketogenic diet, which is high in fat and low in carbohydrates, has been found to lower seizure frequency in people with treatment-resistant epilepsy by modifying brain metabolism (Neal et al. 2008).

11.2 Case Study: Ketogenic Diet for Epilepsy

Mechanism
A condition of ketosis is brought on by the ketogenic diet, which is high in fat and extremely low in carbs. In this state, the body uses fat as fuel instead of glucose. By lowering neuronal excitability and changing neurotransmitter levels, this metabolic change has an impact on how much energy the brain uses. According to Neal et al. (2008), these modifications are linked to a decrease in the frequency and intensity of seizures, especially in individuals with drug-resistant epilepsy.

Clinical Impact
According to clinical research, patients with treatment-resistant epilepsy may have a considerable decrease in seizures when following a ketogenic diet. Children and adults following a ketogenic diet may see a 50% or more decrease in the frequency of seizures, which can significantly enhance quality of life. For kids with epileptic encephalopathies who don't respond to medication, the ketogenic diet is very helpful (Kossoff et al. 2018).

Table 11 compares various dietary and pharmacological interventions used in the management of epilepsy, outlining their mechanisms of action, efficacy in seizure reduction and potential adverse effects.

Table 11 Comparison of dietary and pharmacological interventions in epilepsy

Intervention type	Intervention	Mechanism of action	Efficacy	Adverse effects
Dietary interventions	Ketogenic diet	Alters metabolism to produce ketones, providing an alternative fuel for the brain and reducing excitability	Significant reduction in seizure frequency in refractory epilepsy cases; ~50% reduction in seizures reported in many studies	Gastrointestinal issues, hypoglycaemia, dehydration
	Modified Atkins diet	Similar to ketogenic but less restrictive; encourages low-carb intake for seizure control	Effective for some patients with refractory epilepsy; good for long-term adherence	Weight loss, nutrient deficiency
	Low glycaemic index diet	Reduces glucose variability; hypothesised to stabilise neuronal activity	Some evidence of efficacy in seizure reduction	Not well-studied; potential nutritional insufficiencies
Pharmacological interventions	Carbamazepine	Stabilises the inactive state of sodium channels, reducing neuronal excitability	Effective for partial seizures; can reduce seizure frequency by ~50%	Dizziness, drowsiness, gastrointestinal effects
	Lamotrigine	Inhibits voltage-gated sodium channels, enhancing the refractory state of neurons	Broad-spectrum efficacy; effective for partial and general seizures	Rash, dizziness, sedation
	Levetiracetam	Modulates SV2A protein and inhibits synaptic vesicle release, reducing neurotransmitter release	Effective for a wide range of seizure types; typically well tolerated	Behavioural changes, fatigue
	Valproate	Increases GABA levels and inhibits excitatory neurotransmission	Highly effective for generalised seizures; effective for many types	Weight gain, liver toxicity, hair loss

12 Conclusion and Summary

The revolutionary potential of precision medicine and microbiome research in the treatment of neurological, gastric and cancer conditions is highlighted by this investigation. A paradigm shift in the diagnosis, management and treatment of diseases is signalled by the complex interrelationships among human health, the microbiome and precision medicine.

12.1 Advances in Microbiome Research and Precision Medicine

12.1.1 Developments in Precision Medicine and Microbiome Research

The expanding knowledge of microbial dynamics highlights how crucial a role they play in regulating immune responses and affecting the effectiveness of treatments. By combining genetic, environmental and microbiological data, precision medicine customises care, greatly enhancing patient outcomes. Probiotics, prebiotics and faecal microbiota transplantation are examples of microbiota modulation therapies that show significant clinical promise (Zhao et al. 2023), whereas omics technologies have transformed biomarker identification and disease classification in gastrointestinal illnesses. Likewise, the influence of the microbiota on immune checkpoint inhibitors and other cancer treatments has opened the door for customised cancer treatments based on the microbiome.

12.2 Challenges and Ethical Considerations

Even with these developments, problems still exist. The significance of responsible scientific advancement is highlighted by ethical considerations concerning gene editing, particularly germline changes (The Royal Society 2021) Precision therapy adoption is hampered by financial and practical obstacles to accessibility and price. Furthermore, ongoing study is necessary because of the individual heterogeneity in microbiome profiles and the difficulties in integrating and interpreting data.

12.3 The Role of Artificial Intelligence

AI presents previously unheard-of possibilities to improve individualised medical care. By facilitating early diagnosis and bolstering precision treatment approaches, artificial intelligence has promise for revolutionising the field of neurology. Large amounts of genetic, imaging and clinical data can be analysed by AI approaches to find trends and predict the course of an illness, providing crucial insights into complex neurological conditions. For instance, by examining minute alterations in MRI scans and biomarkers, machine learning algorithms can assist in the earlier diagnosis of neurodegenerative diseases like Alzheimer's or Parkinson's disease than conventional methods (Feng 2023).

AI can find patterns in diseases, forecast results and improve treatment strategies by examining enormous collections of genetic, imaging and clinical data. Although problems like data privacy and algorithm bias need to be addressed, these tools have the potential to improve neurological treatment precision and enable earlier diagnosis of neurodegenerative disorders.

12.4 Vision for the Future

The combination of precision medicine, microbiome research and AI provides a substantial advancement in patient-centred care. By embracing these breakthroughs, we may create personalised, effective and easily available therapies that improve outcomes and quality of life. To make this vision a reality, we must take a multidisciplinary strategy that includes promoting research, developing collaborations, tackling ethical and economic constraints and putting patients first.

In conclusion, there is a great deal of promise for medical advancements in the future as we understand the complex interactions among our microbiome, genetics and health. A new era of healthcare, characterised by individualised, comprehensive and revolutionary care for patients worldwide, is promised by the convergence of precision medicine, microbiome research and artificial intelligence.

References

Ali A, AlHussaini KI (2024) Helicobacter pylori: a contemporary perspective on pathogenesis, diagnosis and treatment strategies. Microorganisms 12(1):222. https://doi.org/10.3390/microorganisms12010222

AlMarzooqi E et al (2024) Deciphering the complex interplay of obesity, epithelial barrier dysfunction, and tight junction remodeling: unraveling potential therapeutic avenues. Obes Rev 25(8). https://doi.org/10.1111/obr.13766

Arruebo M et al (2011) Assessment of the evolution of cancer treatment therapies. Cancers 3(3):3279–3330. https://doi.org/10.3390/cancers3033279

Aslam A, Kumar D, Yadav K, Yadav A, Ahmad R (2024) CRISPR advancements in correcting protein misfolding: implications for neurodegenerative disorders. Gene Expr 23(4):309–335. https://doi.org/10.14218/GE.2024.00002

Baumgart DC (2017) Crohn's disease and ulcerative colitis: from epidemiology and immunobiology to a rational diagnostic and therapeutic approach. Springer. https://doi.org/10.1007/978-3-319-33703-6

Belkaid Y, Hand TW (2014) Role of the microbiota in immunity and inflammation. Cell 157(1):121–141. https://doi.org/10.1016/j.cell.2014.03.011
BenDavid E et al (2024) Emerging perspectives on prime editor delivery to the brain. Pharmaceuticals 17(6):763. https://doi.org/10.3390/ph17060763
Betz F et al (2023) Game changers in science and technology – now and beyond. Technol Forecast Soc Chang 193:122588. https://doi.org/10.1016/j.techfore.2023.122588
Black C et al (2022) British Society of Gastroenterology guidelines on the management of functional dyspepsia. Gut 71(9):1697–1723. https://doi.org/10.1136/gutjnl-2022-327737
Borja NA, Silva-Smith R, Calfa C, Sussman DA, Tekin M (2024) Triple primary cancers: an analysis of genetic and environmental factors. Cancer Prev Res 17(5):209–215. https://doi.org/10.1158/1940-6207.CAPR-23-0395
Cao Y et al (2024) Intratumoural microbiota: a new frontier in cancer development and therapy. Signal Transduct Target Ther 9(1). https://doi.org/10.1038/s41392-023-01693-0
Catalkaya G et al (2020) Interaction of dietary polyphenols and gut microbiota: microbial metabolism of polyphenols, influence on the gut microbiota, and implications on host health. Food Front 1(2):109–133. https://doi.org/10.1002/fft2.25
Collaborative Group of the Americas on Inherited Gastrointestinal Cancers (CGA-IGC) annual virtual meeting, November 13–15 (2021) 2022. Fam Cancer 21(4):489–555. https://doi.org/10.1007/s10689-021-00284-8
Collins FS, Varmus H (2015) A new initiative on precision medicine. N Engl J Med 372(9):793–795. https://doi.org/10.1056/NEJMp1500523
Cummings BJ et al (2005) Human neural stem cells differentiate and promote locomotor recovery in spinal cord-injured mice. Proc Natl Acad Sci 102(39):14069–14074. https://doi.org/10.1073/pnas.0507063102
Doudna JA, Charpentier E (2014) The new frontier of genome engineering with CRISPR-Cas9. Science 346(6213):1258096. https://doi.org/10.1126/science.1258096
Durazzo M, Adriani A, Fagoonee S, Saracco GM, Pellicano R (2021) Helicobacter pylori and respiratory diseases: 2021 update. Microorganisms 9(10):2033. https://doi.org/10.3390/microorganisms9102033
Feng T (2023) Applications of artificial intelligence to diagnosis of neurodegenerative diseases. Stud Health Technol Inform 308:648–655. https://doi.org/10.3233/SHTI230896
Finkel RS et al (2017) Diagnosis and management of spinal muscular atrophy: part 2: pulmonary and acute care; medications, supplements, and immunizations; other organ systems; and ethics. Neuromuscul Disord 27(11):1025–1036. https://doi.org/10.1016/j.nmd.2017.08.006
Fisher R et al (2010) Electrical stimulation of the anterior nucleus of thalamus for treatment of refractory epilepsy. Epilepsia 51(5):899–908. https://doi.org/10.1111/j.1528-1167.2010.02536.x
Gao J et al (2010) A novel pathway regulates memory and plasticity via SIRT1 and miR-134. Nature 466(7310):1105–1109. https://doi.org/10.1038/nature09271
Gopalakrishnan V et al (2018) Gut microbiome modulates response to anti-PD-1 immunotherapy in melanoma patients. Science 359(6371):97–105. https://doi.org/10.1126/science.aan4236
Guarente L, Mostoslavsky R, Kazantsev A (2018) Introductory review on sirtuins in biology, aging, and disease. Academic Press
Harkin LA et al (2007) The spectrum of SCN1A-related infantile epileptic encephalopathies. Brain 130(3):843–852. https://doi.org/10.1093/brain/awm002
Johnson JA et al (2008) The Nrf2–ARE pathway: an indicator and modulator of oxidative stress in neurodegeneration. Ann N Y Acad Sci 1147(1):61–69. https://doi.org/10.1196/annals.1427.036
Ju A, Beyeler A (2021) A new player in neural circuits of emotions. Nat Neurosci 24:1506–1507. https://doi.org/10.1038/s41593-021-00945-y
Juan AS (2019) Exome sequencing in gastrointestinal food allergy induced by multiple food protein. https://doi.org/10.4995/thesis/10251/134361
Kalla R et al (2016) Serum calprotectin: a novel diagnostic and prognostic marker in inflammatory bowel diseases. Am J Gastroenterol 111(12):1796–1805. https://doi.org/10.1038/ajg.2016.342

Kamada N, Seo SU, Chen GY, Núñez G (2013) Role of the gut microbiota in immunity and inflammatory disease. Nat Rev Immunol 13(5):321–335. https://doi.org/10.1038/nri3430
Kandalai S et al (2023) The human microbiome and cancer: a diagnostic and therapeutic perspective. Cancer Biol Ther 24(1):2240084. https://doi.org/10.1080/15384047.2023.2240084
Kaur H, Kaur G, Ali SA (2022) Dairy-based probiotic-fermented functional foods: an update on their health-promoting properties. Fermentation 8(9):425. https://doi.org/10.3390/fermentation8090425
Kayali S et al (2024) NOD2 and Crohn's disease clinical practice: from epidemiology to diagnosis and therapy, rewired. Inflamm Bowel Dis:izae075. https://doi.org/10.1093/ibd/izae075
Khoury MJ, Armstrong GL, Bunnell RL et al (2020) The intersection of genomics and big data with public health: opportunities for precision public health. PLoS Med 17(10):e1003373. https://doi.org/10.1371/journal.pmed.1003373
Kossoff EH et al (2018) Optimal clinical management of children receiving dietary therapies for epilepsy: updated recommendations of the International Ketogenic Diet Study Group. Epilepsia Open 3(2):175–192. https://doi.org/10.1002/epi4.12225
Krueger DA et al (2013) Everolimus treatment of refractory epilepsy in tuberous sclerosis complex. Ann Neurol 74(5):679–687. https://doi.org/10.1002/ana.23960
Lastres-Becker I et al (2016) Repurposing the NRF2 activator dimethyl fumarate as therapy against synucleinopathy in Parkinson's disease. Antioxid Redox Signal 25(2):61–77. https://doi.org/10.1089/ars.2015.6554
Lefaucheur JP et al (2014) Evidence-based guidelines on the therapeutic use of repetitive transcranial magnetic stimulation (rTMS). Clin Neurophysiol 125(11):2150–2206. https://doi.org/10.1016/j.clinph.2014.05.021
Li ZK et al (2022) Global trends in gut microbiota and Clostridioides difficile infection research: a visualized study. J Infect Public Health 15(7):806–815. https://doi.org/10.1016/j.jiph.2022.06.011
Loh JS et al (2024) Microbiota–gut–brain axis and its therapeutic applications in neurodegenerative diseases. Signal Transduct Target Ther 9(1). https://doi.org/10.1038/s41392-024-01743-1
M'Koma AE (2018) The multifactorial etiopathogeneses interplay of inflammatory bowel disease: an overview. Gastrointest Disord 1(1):75–105. https://doi.org/10.3390/gidisord1010007
Marques L, Costa B, Pereira M, Silva A, Santos J, Saldanha L, Silva I, Magalhães P, Schmidt S, Vale N (2024) Advancing precision medicine: a review of innovative in silico approaches for drug development, clinical pharmacology and personalized healthcare. Pharmaceutics 16(3):332. https://doi.org/10.3390/pharmaceutics16030332
Mendell JR et al (2017) Single-dose gene-replacement therapy for spinal muscular atrophy. N Engl J Med 377(18):1713–1722. https://doi.org/10.1056/NEJMoa1706198
Merrick B, Sergaki C, Edwards. (2023) Modulation of the gut microbiota to control antimicrobial resistance (AMR)—a narrative review with a focus on faecal microbiota transplantation (FMT). Infect Dis Rep 15(3):238–254. https://doi.org/10.3390/idr15030025
National Institute of Diabetes and Digestive and Kidney Diseases (NIDDK) (2024) Your digestive system and how it works. Retrieved from https://www.niddk.nih.gov
National Research Council (2011) Toward precision medicine: building a knowledge network for biomedical research and a new taxonomy of disease. National Academies Press
Neal EG et al (2008) The ketogenic diet for the treatment of childhood epilepsy: a randomised controlled trial. Lancet Neurol 7(6):500–506. https://doi.org/10.1016/S1474-4422(08)70092-9
O'Brien J, Ames D, McKeith I, Chiu E (2005) Dementia with Lewy bodies: and Parkinson's disease dementia. CRC Press
Ottman R et al (2010) Genetic testing in the epilepsies—report of the ILAE genetics commission. Epilepsia 51(4):655–670. https://doi.org/10.1111/j.1528-1167.2009.02429.x
Paananen J, Fortino V (2019) An omics perspective on drug target discovery platforms. Brief Bioinform 21(6):1937–1953. https://doi.org/10.1093/bib/bbz122
Pandey H et al (2023) Gut microbiota in colorectal cancer: biological role and therapeutic opportunities. Cancers 15(3):866. https://doi.org/10.3390/cancers15030866

Perera T et al (2016) The clinical TMS Society consensus review and treatment recommendations for TMS therapy for major depressive disorder. Brain Stimul 9(3):336–346. https://doi.org/10.1016/j.brs.2016.03.010
Porquet D et al (2013) Dietary resveratrol prevents Alzheimer's markers and increases life span in SAMP8. Age 35(5):1851–1865. https://doi.org/10.1007/s11357-012-9474-5
Quinn O (2017) Microbiota of an invasive wasp Vespula vulgaris and hymenopteran relatives: interpreting the microbiome. https://doi.org/10.26686/wgtn.17067914.v1
Rahman M, Islam M et al (2022) Microbiome in cancer: role in carcinogenesis and impact in therapeutic strategies. Biomed Pharmacother 149:112898. https://doi.org/10.1016/j.biopha.2022.112898
Rebbeck TR, Bridges JFP, Mack JW et al (2022) A framework for promoting diversity, equity, and inclusion in genetics and genomics research. JAMA Health Forum 3(4):e220603. https://doi.org/10.1001/jamahealthforum.2022.0603
Round JL, Mazmanian SK (2009) The gut microbiota shapes intestinal immune responses during health and disease. Nat Rev Immunol 9(5):313–323. https://doi.org/10.1038/nri2515
Salanova V et al (2015) Long-term efficacy and safety of thalamic stimulation for drug-resistant partial epilepsy. Neurology 84(10):1017–1025. https://doi.org/10.1212/WNL.0000000000001334
Sandborn WJ et al (2017) Tofacitinib as induction and maintenance therapy for ulcerative colitis. N Engl J Med 376(18):1723–1736. https://doi.org/10.1056/NEJMoa1606910
Satterstrom FK et al (2020) Large-scale exome sequencing study implicates both developmental and functional changes in the neurobiology of autism. Cell 180(3):568–584.e23. https://doi.org/10.1016/j.cell.2019.12.036
Shang F, Jiang X et al (2020) The inhibitory effects of probiotics on colon cancer cells: in vitro and in vivo studies. J Gastrointest Oncol 11(6):1224–1232. https://doi.org/10.21037/jgo-20-573
Slavin J (2013) Fiber and prebiotics: mechanisms and health benefits. Nutrients 5(4):1417–1435. https://doi.org/10.3390/nu5041417
Smythe L, Guliaeva I, Skuse D, Muntoni F (2024) Abstracts of the 17th UK Neuromuscular Translational Research Conference, 17th and 18th April 2024: Differences in executive function skills used during planning and navigation observed in boys with DMD. Journal of Neuromuscular Diseases 11(s1):DC05
Sommer F, Backhed F (2013) The gut microbiota—masters of host development and physiology. Nat Rev Microbiol 11(4):227–238. https://doi.org/10.1038/nrmicro2974
Stolfi F et al (2024) Omics approaches open new horizons in major depressive disorder: from biomarkers to precision medicine. Front Psychiatry 15. https://doi.org/10.3389/fpsyt.2024.1422939
Su LJ (2015) Environmental epigenetics and obesity: evidences from animal to epidemiologic studies. In: Su L, Chiang TC (eds) Environmental epigenetics, Molecular and integrative toxicology. Springer, pp 251–266. https://doi.org/10.1007/978-1-4471-6678-8_6
Swiech L et al (2008) Role of mTOR in physiology and pathology of the nervous system. Biochim Biophys Acta 1784(1):116–132. https://doi.org/10.1016/j.bbapap.2007.08.015
Takahashi K et al (2007) Induction of pluripotent stem cells from adult human fibroblasts by defined factors. Cell 131(5):861–872. https://doi.org/10.1016/j.cell.2007.11.019
Tallarico M et al (2022) Biologics for inflammatory bowel disease in clinical practice: a Calabria (southern Italy) prospective pharmacovigilance study. Pharmaceutics 14(11):2449. https://doi.org/10.3390/pharmaceutics14112449
Tamsi V, Falus A (2012) Genetic and epigenetic factors affecting cytochrome P450 phenotype and their clinical relevance. InTech eBooks. https://doi.org/10.5772/36451
Tetzlaff W et al (2011) A systematic review of cellular transplantation therapies for spinal cord injury. J Neurotrauma 28(8):1611–1682. https://doi.org/10.1089/neu.2009.1177
The Integrative Human Microbiome Project (2019) The Integrative Human Microbiome Project. Nature 569(7758):641–648. https://doi.org/10.1038/s41586-019-1238-8
The Royal Society et al (2021) Heritable human genome editing. National Academies Press

Varada VV et al (2022) Host-specific probiotics feeding influence growth, gut microbiota, and fecal biomarkers in buffalo calves. AMB Express 12(1). https://doi.org/10.1186/s13568-022-01460-4
Vemuri R, Shankar EM, Chieppa M, Eri R, Kavanagh K (2020) Beyond just bacteria: functional biomes in the gut ecosystem including virome, mycobiome, archaeome and helminths. Microorganisms 8(4):483. https://doi.org/10.3390/microorganisms8040483
Virk MS et al (2024) The anti-inflammatory and curative exponent of probiotics: a comprehensive and authentic ingredient for the sustained functioning of major human organs. Nutrients 16(4):546. https://doi.org/10.3390/nu16040546
Vuyyuru SK et al (2023) Targeting IL-23 for IBD: rationale and progress to date. Drugs 83(10):873–891. https://doi.org/10.1007/s40265-023-01882-9
Yarahmadi A, Afkhami H (2024) The role of microbiomes in gastrointestinal cancers: new insights. Front Oncol 13. https://doi.org/10.3389/fonc.2023.1344328
Zeidan F et al (2014) Neural correlates of mindfulness meditation-related anxiety relief. Soc Cogn Affect Neurosci 9(6):751–759. https://doi.org/10.1093/scan/nst041
Zerhouni EA (2005) Translational and clinical science—time for a new vision. N Engl J Med 353(15):1621–1623. https://doi.org/10.1056/NEJMp051100
Zhang X, Tang B, Guo J (2023) Parkinson's disease and gut microbiota: from clinical to mechanistic and therapeutic studies. Transl Neurodegener 12(1). https://doi.org/10.1186/s40035-023-00392-8
Zhao L-Y et al (2023) Role of the gut microbiota in anticancer therapy: from molecular mechanisms to clinical applications. Signal Transduct Target Ther 8(1):201. https://doi.org/10.1038/s41392-023-01406-7
Zhou P et al (2022) Microbiome in cancer: an exploration of carcinogenesis, immune responses, and immunotherapy. Front Immunol 13:877939. https://doi.org/10.3389/fimmu.2022.877939

The Next Frontier in Microbiome-Based Therapeutics: Advanced Clinical Applications

Thumpati Prasanth, K. P. Reddy, and Ravichandiran Velayutham

Abstract The human microbiome, a diverse community of microorganisms vital for digestion, immunity, and metabolism, is emerging as a powerful target for therapeutic intervention. This chapter examines the burgeoning field of microbiome-driven therapies, particularly focusing on the microbiome's impact on health and its role in personalized medicine. We explore how microbial imbalances contribute to diseases like inflammatory bowel disease (IBD), metabolic disorders, neurological conditions, and cancer, and discuss the latest advancements in therapeutic strategies. These include probiotics, prebiotics, fecal microbiota transplantation (FMT), and microbial-based drugs, while also acknowledging the challenges of clinical translation. Furthermore, we highlight the integration of microbiome profiling and metagenomic analysis in advancing precision medicine, enabling tailored treatments based on individual microbial signatures. This chapter aims to provide a comprehensive overview of current knowledge, address research gaps, and outline future directions in microbiome-based therapies, offering a valuable resource for clinicians, researchers, and policymakers seeking to leverage the microbiome for improved healthcare outcomes.

Keywords Microbiome · Personalized medicine · Precision medicine · Health and disease · Therapeutic strategies · Probiotics · Prebiotics · Fecal microbiota transplantation (FMT) · Synbiotics · Live biotherapeutic products (LBPs) · Postbiotics · Inflammatory bowel disease (IBD) · Microbiome profiling

---

T. Prasanth (✉)
Department of Natural Products, National Institute of Pharmaceutical Education and Research, Chunilal Bhawan, Kolkata, West Bengal, India

Department of Chemistry, Virginia Commonwealth University, Richmond, VA, USA

K. P. Reddy · R. Velayutham (✉)
Department of Natural Products, National Institute of Pharmaceutical Education and Research, Chunilal Bhawan, Kolkata, West Bengal, India

S. K. Singh et al. (eds.), *Microbiota Profiling for Precision Medicine*, Interdisciplinary Biotechnological Advances,
https://doi.org/10.1007/978-981-95-2104-3_12

1 Introduction

The human microbiome, the collection of microorganisms living in and on our bodies, has long been linked to health and disease (Young 2017). Dysbiosis, the imbalance of the microbiome, is increasingly understood to play a role in a wide range of health disorders. These include inflammatory bowel disease (IBD), neurological conditions like autism and Parkinson's disease (PD), metabolic disorders such as obesity and diabetes, and even certain cancers, highlighting the far-reaching impact of a disrupted microbial ecosystem on human health (Zhu et al. 2010). Attempts to harness "good" microbes trace back centuries to fermented foods and early probiotics, reflecting an intuition that modulating the microbiota could improve health (Fig. 1). Ancient records even describe stool-based remedies for diarrhea, a precursor to modern fecal microbiota transplantation (FMT). However, it is only in recent decades, with advances in DNA sequencing and microbiology, that microbiome research has exploded. Scientists revealed that the gut microbiome plays an integral role in digestion, immune development, and metabolism, and that disruptions (dysbiosis) can contribute to disease. This understanding spurred the development of microbiome-based therapeutics interventions aiming to restore or modify the microbial community to treat illness (Khanna and Pardi 2012; Kelly et al. 2021). Early successes with probiotics (live beneficial bacteria) and FMT in refractory infections validated the concept (Al-Fakhrany et al. 2024; Petrosino 2018; Ryu et al. 2021). Today, microbiome therapeutics encompass a broad arsenal: diet modifications, prebiotics (fibers that feed beneficial microbes), probiotics, synbiotics (combined pre/probiotics), phage therapy (viruses targeting bacteria), live biotherapeutic products (LBPs, defined live microbes used as drugs), and even microbiome-derived metabolites or "postbiotics." Recent technological leaps enable more targeted approaches from rationally designed microbial consortia to genetically engineered bacteria, marking the next frontier of advanced clinical applications (Rasul et al. 2022;

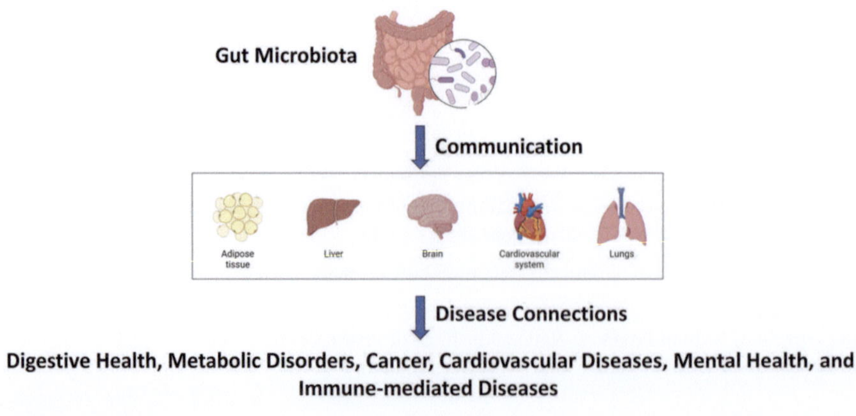

Fig. 1 Gut microbiota and disease connections

Sioson et al. 2021). Recent therapeutic developments are increasingly focused on the microbiome, exploring its potential to treat a range of diseases. In the past five years alone, significant progress has been made in understanding the mechanisms behind these interventions, leading to advancements in regulatory approval and commercialization. (Asma et al. 2024; Rhodes 2016; Ringel et al. 2012).

1.1 Unveiling Precision Approaches for Gastrointestinal Disorders

Irritable Bowel Syndrome (IBS): IBS, a functional gastrointestinal (GI) disorder characterized by chronic abdominal pain and altered bowel habits, has also been linked to microbiota composition. Some IBS patients exhibit lower microbial diversity or imbalances that correlate with symptoms (e.g., bloating). Clinical trials with probiotics have shown modest symptom relief. Certain *Bifidobacterium* and *Lactobacillus* strains can reduce IBS pain and stool irregularity in subsets of patients, although effects vary (Carroll et al., 2012 and Carroll et al., 2010 and Shaikh et. al., 2023).). FMT trials in IBS have yielded conflicting results: while a few studies reported global symptom improvement after FMT, others found no benefit over placebo (and one trial even noted placebo recipients did better than FMT by enema) (Wanget al. 2020). These discrepancies may relate to differences in donor material and the large placebo effect in IBS. Nonetheless, the *idea* of microbiome modulation for IBS persists (Degruttola et. al., 2016 and Baldelli et. al., 2021). Approaches like targeted antibiotics (rifaximin) for IBS-diarrhea work by partial microbiome alteration, and diets such as low- Fermentable Oligosaccharides, Disaccharides, Monosaccharides, and Polyols (FODMAP) fermentable fiber restriction) alleviate symptoms by reducing gut microbial fermentation (Di Vincenzo et al., 2024 and Jang et al., 2024 and Coates et al. 2023). Looking ahead, more personalized microbiome therapeutics (selecting interventions based on a patient's baseline microbiota profile) may yield better IBS outcomes. For now, microbiome therapy in IBS is experimental, but ongoing research seeks to validate if a "gut flora reset" can reliably calm the irritable gut (Pasta et. al., 2024 and Gros et. al., 2021).

Inflammatory Bowel Disease (IBD): IBD is a chronic inflammatory condition affecting the gastrointestinal tract (GIT), primarily in Crohn's disease (CD) and ulcerative colitis (UC), leading to damage to the intestinal lining (Fakhoury et. a., 2014). Recent studies have highlighted significant alterations in the gut microbiota of IBD patients, shedding light on its potential role in disease development and progression (Pittayanon et.al., 2020). One of the most consistent findings is the depletion and imbalance of the beneficial bacterial populations. A study by Sokol et al. (2008) demonstrated that one of the beneficial bacteria, *Faecalibacterium prausnitzii*, is significantly depleted in patients with IBD, particularly those with CD (Corthier et al. 2008). *F. prausnitzii* is known for producing short-chain fatty acids (SCFAs), particularly butyrate, which is crucial for maintaining intestinal barrier function and regulating immune responses (Nogal et.al., 2021 and Miquel et. al., 2013). Its depletion has been linked to increased intestinal permeability, which allows harmful substances and bacteria from the gut lumen to cross the epithelial barrier, leading to immune activation and chronic inflammation (Corthier et al. 2008). Another protective role of *F. prausnitzii* is that it can inhibit the secretion of pro-inflammatory cytokines, such as tumor necrosis factor-alpha (TNF-α) and interleukin-6 (IL-6), which are major contributors to the inflammatory cascade in IBD patients (Bowene et al. 2015). This ability to dampen inflammation underscores the importance of maintaining a balanced gut microbiota to prevent disease exacerbation (Chloé Mirsepasi-Lauridsen et al., 2019) and Kamali Dolatabadi et. al., 2021).

Clostridioides difficile Infection (CDI): The most clear-cut success of microbiome therapy has been in recurrent *C. difficile* infection. Antibiotic treatment for CDI often fails to prevent recurrence because it does not restore a healthy gut flora. FMT, which involves transferring stool from a healthy donor to the patient's gut, addresses the underlying dysbiosis. Across multiple studies, FMT has shown striking efficacy in recurrent CDI, with cure rates around 85–90% in many cases. For example, one randomized trial reported a 90% resolution of recurrent CDI with FMT versus 62.5% with standard therapy. Such outcomes, and supportive meta-analyses, firmly established FMT as a treatment for CDI that *outperforms* antibiotics alone. The benefits stem from restoring colonization resistance. FMT reintroduces a diverse microbiota that produces inhibitory compounds (like secondary bile acids and short-chain fatty acids) and outcompetes *C. difficile*, preventing its spore germination and toxin production. This success catalyzed the commercialization of standardized microbiome products for CDI. In *2022 and 2023, the Food and Drug Administration (FDA) granted the first-ever approvals for microbiome therapeutics*: a stool-derived enema product (REBYOTA™) and an oral capsule of purified spores (SER-109, brand name VOWST™) for the prevention of recurrent CDI. These *live biotherapeutic products* demonstrated significant efficacy (e.g., ~12% recurrence on treatment vs. ~40% with placebo in SER-109's Phase 3 trial). Their approval marks a milestone for the field, providing an off-the-shelf alternative to individualized FMT. Notably, both products underwent rigorous quality controls (donor screening, pathogen inactivation, standardized dosing) to address safety concerns that had limited FMT's broader use. CDI remains the flagship indication for microbiome therapy, transforming a once intractable recurring infection into a highly manageable condition.

1.2 Metabolic and Liver Disorders

Obesity and Metabolic Syndrome: Groundbreaking mouse studies in the 2000s first revealed that transferring gut microbiota from an obese donor to germ-free mice could transmit increased fat gain, suggesting a causal role of the microbiome in obesity (Colecchia et. al., 2014). Humans with obesity often harbor a different microbiome composition (e.g., lower diversity, more *Firmicutes* relative to *Bacteroidetes*) than lean individuals (Magne et al., 2020). These findings spurred trials to alter the microbiome as a therapy for metabolic disorders (Fusco et al., 2023 and Ghosh et.al., 2022). FMT from lean, healthy donors into individuals with metabolic syndrome has shown hints of metabolic benefit: a landmark study found that insulin sensitivity improved in obese recipients after FMT from lean donors, alongside increased gut microbial diversity (Ramírez-Pérez et. al., 2017). Similarly, in a small trial of obese patients with hypertension, FMT plus diet counseling led to more favorable microbiota changes and modest blood pressure improvements compared to diet alone. While these effects can be temporary (some studies report that metabolic improvements wane weeks after FMT unless sustained by diet/lifestyle), they provide proof-of-concept that the microbiome influences host metabolic pathways. Probiotic supplementation has also been examined: certain strains (e.g., *Bifidobacterium breve* or *Lactobacillus reuteri*) have been reported to assist weight loss or insulin sensitivity in pilot studies (Corbin et al. 2023, and Alcock et al., 2014). One trial in adolescents with obesity showed that a multi-strain probiotic improved metabolic parameters and gut hormone profiles relative to placebo. Additionally, diets rich in fermentable fiber (prebiotics) can boost short-chain fatty acid-producing gut bacteria, which in turn may reduce systemic inflammation and improve glucose metabolism. Overall, microbiome modulation for obesity/diabetes is still in early stages—weight loss outcomes are modest—but the *mechanistic insights* are strong. Microbes can influence energy harvest from diet, gut barrier integrity (translocation of endotoxin can promote insulin resistance), and production of metabolites like butyrate that affect adiposity. Future therapies might combine traditional approaches (diet, exercise, medications) with microbiome-targeted adjuvants to amplify metabolic benefits. For instance, one could envision a personalized probiotic regimen to enrich an individual's capacity to produce beneficial metabolites that promote leanness or insulin sensitivity.

Non-alcoholic Fatty Liver Disease (NAFLD): NAFLD—the accumulation of fat in the liver, often linked to obesity and diabetes—has a well-recognized gut–liver axis component. An altered gut microbiome may contribute to NAFLD progression by increasing intestinal permeability (allowing inflammatory bacterial products into portal circulation) and generating metabolites that affect liver fat metabolism. Patients with NAFLD commonly exhibit dysbiosis and gut barrier dysfunction. Initial clinical attempts to modulate the microbiome in NAFLD have yielded mixed results. In a 2020 randomized trial, obese NAFLD patients received either donor FMT or their own stool (placebo); FMT did *not* significantly improve liver fat or insulin resistance at 6 weeks Craven et al. (2020). Interestingly, however, the subset of FMT patients who had abnormal intestinal permeability at baseline showed a restoration of gut barrier integrity post-FMT. This suggests microbiome therapy might help by reducing gut leakiness and inflammation in NAFLD, even if hepatic fat changes take longer. Other small studies have reported that certain probiotics (e.g., *Lactobacillus* or *Bifidobacterium* strains) can lower liver enzymes and inflammatory markers in NAFLD patients, though substantial effects on liver fat are unproven (Kobyliak et al., 2018). Given NAFLD's rising prevalence and lack of approved pharmaceuticals, the microbiome is a tantalizing target: interventions like FMT, synbiotics, or engineered bacteria that produce short-chain fatty acids (which can reduce hepatic lipogenesis) are being explored. While no microbiome-based NAFLD therapy is yet established, ongoing trials aim to determine if altering the gut microbiota can slow progression of fatty liver to steatohepatitis. It may be that microbiome therapies will play a *supporting* role (improving gut-liver signaling and inflammation) in conjunction with diet and weight loss for NAFLD management.

1.3 Navigating Neurological Frontiers: Microbiota-Targeted Clinical Innovations

1.3.1 Neurological and Psychiatric Disorders (Gut–Brain Axis)

An exciting frontier is the *microbiota–gut–brain axis*, the bidirectional communication network between the gastrointestinal tract and the central nervous system. The gut microbiome can produce neurotransmitters (like GABA, serotonin precursors), modulate immune signals, and influence vagus nerve activity—all of which can affect brain function and behavior. Conversely, neurological conditions often coincide with GI abnormalities or dysbiosis, raising the question of causality. While research here is nascent, several applications are under investigation:

Autism Spectrum Disorder (ASD): Children with ASD frequently have gastrointestinal issues and distinct gut microbiota profiles compared to neurotypical peers. This has led to trials of microbiome modulation to mitigate ASD symptoms. FMT has been piloted in ASD under a protocol called "Microbiota Transfer Therapy." In early open-label studies, chronic FMT (over 7–8 weeks) resulted in not only improved GI symptoms (less diarrhea, constipation, abdominal pain) but also reported improvements in behavior and core ASD symptoms that persisted even 2 years after treatment in some cases. For example, Kang et al. (2017) found significant improvements in Autism Behavior Checklist scores after FMT, and a follow-up study noted that participants maintained improved GI function and had an average ~45% reduction in ASD symptoms 2 years post-FMT (Krajmalnik-Brown 2017). Other case series and observational studies similarly report enhanced social responsiveness and language in some ASD children following FMT.

However, these studies lacked placebo controls; a recent systematic review concluded that while FMT *appears* to improve ASD-associated metrics, rigorous randomized trials are needed. Probiotics and dietary interventions (like gluten/casein free diets or prebiotic fibers) have also been tried, with anecdotal benefits. The mechanism is thought to involve reducing gut dysbiosis and intestinal inflammation, thereby decreasing neuroinflammatory signaling. The gut microbiome may also affect levels of neuroactive compounds; for instance, certain bacterial strains can influence tryptophan metabolism, potentially affecting serotonin levels. As of now, microbiome therapy for ASD is experimental but promising—*so much so that multiple clinical trials are underway* (e.g., testing oral encapsulated FMT in ASD in double-blind fashion). This line of research represents a paradigm shift, treating a neurodevelopmental disorder by targeting the gut ecology.

Depression and Anxiety: Mental health conditions such as major depressive disorder (MDD) have been associated with altered gut microbiomes, including lower diversity and a pro-inflammatory microbial milieu (Grau-Del Valle et. al., 2023). Preclinical studies show that transplanting gut microbes from a depressed individual into rodents can induce depression-like behaviors in the animals. Conversely, introducing beneficial bacteria (so-called "psychobiotics") can elevate brain neurotransmitter levels and reduce stress hormones in animal models (Chen et al. 2021). These findings provide a rationale for microbiome-based adjunct therapies in mood disorders. A handful of small clinical studies have tested FMT or probiotic supplementation in people with depression. One open-label trial in patients with irritable bowel syndrome and comorbid depression observed that FMT led to improvements in both GI and depressive symptoms, along with increased quality of life (Braga et al. 2024). Probiotics such as *Bifidobacterium longum* and *Lactobacillus helveticus* have shown anxiolytic and antidepressant effects in pilot trials—for example, improving mood and sleep quality compared to placebo in mild-to-moderate depression. The proposed mechanisms include gut bacteria producing metabolites (e.g., butyrate, which has neuroactive and anti-inflammatory properties) and modulating systemic inflammation (elevated inflammatory cytokines are often linked to depression). While these early results are encouraging, evidence is not yet definitive. The *current landscape*, as summarized in a 2024 review, is that FMT appears safe in depression and can improve GI comorbidities and possibly mood, but larger controlled studies are needed to validate efficacy and optimal protocols. If confirmed, microbiome therapeutics could become novel adjuncts to antidepressants—a way to "reset" the gut–brain axis and enhance treatment response in refractory depression.

Parkinson's Disease (PD): Parkinson's disease is a neurodegenerative (Caradonna et al., 2024 and Boehme, et.al., 2023) disorder wherein motor symptoms (tremor, rigidity) often co-occur with constipation and gut microbiota alterations. Researchers have found that PD patients harbor distinct microbial communities, for instance, reduced levels of anti-inflammatory producers like Prevotella and an overabundance of opportunistic pathogens (Difruscolo et al. 2024). This dysbiosis is hypothesized to contribute to PD pathology via increased intestinal permeability and inflammation or even by generating neurotoxic molecules that travel via the vagus nerve. Given this link, small trials have examined FMT in PD. A recent placebo-controlled *randomized trial* (2024) assessed a single colonic FMT in PD patients but found *no significant clinical improvement* in motor symptoms compared to placebo, despite the procedure being safe. By contrast, another Phase 2 trial (GUT-PARFECT 2024) reported that nasojejunal FMT led to *mild but sustained improvements* in motor scores at 12 months in early-stage PD, relative to a placebo (autologous transplant) group. The discrepancy might be due to differences in FMT dosing or patient selection. Both studies underscore that any motor benefit is modest. However, some PD patients did report improved constipation and gut function post-FMT, which can indirectly better quality of life. Beyond FMT, researchers are exploring specific microbial metabolites that might protect dopaminergic neurons. For example, certain *Enterococcus* strains produce amyloid-like proteins that could influence α-synuclein aggregation in the gut—targeting such strains might slow PD progression. While microbiome therapy is *not yet an established PD treatment*, it remains an intriguing area. If nothing else, managing the GI aspects of PD via probiotics or FMT (to alleviate constipation and dysbiosis) could form part of a holistic care plan. Ongoing and future trials will clarify whether more intensive or targeted microbiome interventions can impact the neurological course of Parkinson's disease (Cox et al. 2023).

1.4 Revolutionizing Cancer Treatment: Microbiota-Driven Strategies

1.4.1 Oncological Applications

The relationship between the microbiome and cancer is twofold: microbes can influence cancer development (e.g., *Helicobacter pylori* in gastric cancer, *Fusobacterium* in colorectal cancer), and intriguingly, they can also affect responses to cancer treatment (Kim and Lee et al., 2022 and White and Sears et. al., 2024). Recent studies have uncovered that the gut microbiome can modulate the efficacy and toxicity of various anticancer therapies (Ritapa et al. 2023, 2024; Panda et al. 2019; Fatma et al. 2024). This has opened up new avenues to use microbiome-based therapeutics to improve cancer outcomes:

Cancer Immunotherapy (Checkpoint Inhibitors): Immune checkpoint inhibitors (ICIs) like anti–PD-1 or anti–CTLA-4 antibodies have revolutionized cancer treatment, but only a subset of patients respond to these drugs. Beginning around 2017, seminal studies showed that responders to ICIs often have distinct gut microbiota compositions compared to non-responders (Binayak et al. 2024; Biswa et al. 2023; Priya et al. 2018; Prasanth et al. 2022; Thumpati et al. 2025). For example, melanoma patients who responded to PD-1 blockade had a higher abundance of *Ruminococcaceae* and *Bifidobacterium* and greater microbiome diversity than those with progressive disease. In contrast, antibiotic use (which perturbs the gut flora) before immunotherapy was associated with poorer survival in multiple cancers. These observations suggest the microbiome primes the immune system for anti-tumor activity. To directly test this, two ground-breaking clinical trials in 2021 took *fecal microbiota from cancer patients who responded to immunotherapy and transplanted it into patients who had not responded.* In metastatic melanoma, FMT combined with anti–PD-1 therapy led to clinical responses (tumor regression or disease stabilization) in a subset of previously resistant patients. In one study, ~6 out of 15 refractory melanoma patients achieved disease control after receiving FMT from a responder donor and resuming anti–PD-1 therapy. These proof-of-concept trials demonstrated that altering the gut microbiome could *convert an immunotherapy non-responder into a responder in some cases.* Mechanistically, FMT was found to increase infiltration of anti-cancer immune cells into the tumor microenvironment and reduce immunosuppressive factors. Essentially, a favorable microbiome "reprogrammed" the patient's immune system to better attack cancer. Building on these insights, companies are developing defined consortia of bacteria as adjuvants for immunotherapy. For instance, a consortia of *Akkermansia* and *Enterococcus* (bacteria linked to good ICI responses) is being tested to see if oral administration can improve outcomes in checkpoint inhibitor-treated patients. While still experimental, the ability to manipulate the microbiome to enhance immunotherapy represents a novel strategy to increase response rates. In the near future, we may screen patients' microbiomes as biomarkers for immunotherapy success and offer microbiome modulation (via diet, probiotics, or FMT) to those with "unfavorable" microbiota in order to boost their cancer therapy (White and Sears et al., 2024 and Doocey et al., 2022).

Chemotherapy and Other Cancer Therapies: The gut microbiota also influences conventional cancer treatments, affecting both how well they work and how severe the side effects are. A classic example is the chemotherapy drug cyclophosphamide: it was discovered that cyclophosphamide causes certain gut bacteria to translocate into lymphoid tissue, where they stimulate a specific immune response that contributes to the drug's anti-tumor effect. Mice lacking gut bacteria (germ-free mice or those given antibiotics) had impaired anti-cancer responses to cyclophosphamide, highlighting that an intact microbiome is needed for optimal efficacy. Similarly, the efficacy of platinum chemotherapy (like oxaliplatin) against colon cancer can be enhanced by microbial metabolites—butyrate produced by commensal bacteria was shown to improve oxaliplatin's tumor-killing ability in experimental models. On the other hand, some microbes can cause *resistance* to chemotherapy (Parida and Sharma et. al., 2021). For instance, certain gut and tumor-associated bacteria can metabolize the chemotherapeutic gemcitabine into an inactive form, rendering the treatment less effective. Eliminating those bacteria (or inhibiting their enzyme) restores drug potency. The microbiome also plays a role in chemotherapy side effects: GI mucositis and diarrhea induced by drugs like irinotecan are exacerbated by bacterial beta-glucuronidase activity in the gut co-administration of an inhibitor or using probiotics to modulate these enzymes has been shown to reduce such toxicity (Sextonet al., 2022). Recognizing these influences, researchers are looking at microbiome interventions to support cancer patients (Talapko et al. 2024). This could mean giving antibiotics or specific probiotics to shape the microbiome during cancer therapy. For example, in leukemia patients receiving stem cell transplants, maintaining a diverse gut microbiome via probiotic supplementation has correlated with lower rates of graft-vs-host disease and better survival, whereas broad antibiotics that decimate the microbiome have the opposite effect (Santacroce et. al., 2023). Another emerging modality is *microbial ecosystem therapeutics*. To mitigate therapy side effects, one company is evaluating a live biotherapeutic to prevent immunotherapy-associated colitis by introducing strains that promote regulatory immunity (McCune et al. 2024). In summary, oncology is embracing the idea that the microbiome is a modifiable factor in treatment success. By *targeting the gut microbiota*, we might enhance patients' tolerance to aggressive therapy and improve tumor responses, truly an advanced clinical application of microbiome science (Table 1).

1.5 Major Recent Advancements (Last 5 Years)

The past 5 years have been transformative for microbiome therapeutics, transitioning many concepts from bench to bedside:

First-in-Class Approvals: As noted, 2022–2023 saw the *first regulatory approvals of microbiome-based drugs*. The FDA approval of REBYOTA® (a fecal microbiota suspension) and VOWST® (SER-109 oral spores) for recurrent CDI is a watershed moment. These approvals required demonstrating quality manufacturing of live microbes, consistent potency, and clinical efficacy in Phase 3 trials. They validate that live biotherapeutics can meet pharmaceutical standards and pave the way for approvals in other indications. Companies like *Seres*, *Rebiotix/Ferring*, and *Vedanta* pioneered this path, showing that industry can harness consortia of commensal bacteria as a new therapeutic class.

Breakthrough Clinical Trials: Beyond CDI, numerous key trials were completed. In immuno-oncology, the 2021 FMT trials in melanoma (described above) offered *proof-of-concept* that modifying the microbiome can overcome drug resistance. In metabolic disease, a 2019 trial of *Akkermansia muciniphila* (a gut bacterium depleted in obesity) showed that a pasteurized form of this bacterium improved insulin sensitivity and reduced cholesterol in overweight individuals, marking one of the first trials of a novel commensal as a therapy. In 2020, a randomized trial in type 2 diabetes reported that a high-fiber prebiotic diet reshaped the gut microbiome and led to better glycemic control compared to standard diet, linking targeted microbiota changes to clinical endpoints. Additionally, *engineered bacteria* started entering the clinic (discussed more below). For example, Synlogic Inc. conducted Phase 1 trials of an engineered *E. coli* Nissle capable of consuming phenylalanine in patients with phenylketonuria, demonstrating safety and preliminary efficacy in lowering blood phenylalanine. This was among the first human trials of a synthetic biology-based microbe for a metabolic disorder.

Table 1 Key bacterial strains and clinical implications of microbiome-related disorders

Disorder	Bacterial strains	Alterations	Clinical implications	References
Irritable bowel syndrome (IBS)	*Bifidobacterium* spp., *Lactobacillus genera* (e.g., *Lacticaseibacillus*, *Limosilactobacillus*), *Methanobrevibacter smithii, Escherichia coli.*	Decreased *Bifidobacterium* and members of the reclassified *Lactobacillus* genera: increased *Methanobrevibacter* and *E. coli*	Increased gas production, altered motility, and inflammation	Evangelia, et.al., (2021)
Inflammatory bowel disease (IBD)	*Faecalibacterium prausnitzii, Roseburia* spp., *Eubacterium rectale, Akkermansia muciniphila, Escherichia coli*	Depletion of anti-inflammatory species (*F. prausnitzii, Roseburia*); overgrowth of pathogenic *E. coli*	Loss of gut barrier integrity, chronic inflammation	Ha et. al., (2023)
Clostridioides difficile infection	*Clostridioides difficile*, reduced *Bacteroidetes* and *Firmicutes* diversity	Overgrowth due to disrupted microbiota (e.g., post-antibiotics)	Severe diarrhea and colitis	Latifi-Navid et al. (2021)
Helicobacter pylori infection	*Helicobacter pylori*	Colonizes the stomach lining, reduces protective microbial diversity	Gastritis, ulcers, increased gastric cancer risk	Shariati, et. al., (2021)
Colorectal cancer (CRC)	*Fusobacterium nucleatum, Bacteroides fragilis, Escherichia coli*	Enrichment of *F. nucleatum* and *B. fragilis*	Promotes tumorigenesis via inflammation and DNA damage	Li et. al., (2022)
Diverticulitis	*Bacteroides fragilis, Escherichia coli, Enterococcus* spp.	Altered microbial diversity increases inflammation	Pain, fever, abscess formation	Xu et. al., (2020)
Metabolic disorders	*Akkermansia muciniphila, Bacteroidetes* spp., *Firmicutes* spp.	Reduced *Akkermansia*; altered *Firmicutes/Bacteroidetes* ratio	Obesity, insulin resistance, increased inflammation	Akobeng, et. al., (2020)
Celiac disease	*Bifidobacterium* spp., *members of the reclassified Lactobacillus genera (e.g., Lacticaseibacillus, Limosilactobacillus), increased Enterobacteriaceae.*	Reduced beneficial bacteria; increased pathogenic strains	Triggers inflammatory responses in the gut	Marasco et. al., (2016)

(continued)

Table 1 (continued)

Disorder	Bacterial strains	Alterations	Clinical implications	References
Autism spectrum Disorders (ASD)	*Bacteroides* spp., *Clostridium spp.*, *Prevotella* spp.	Decreased *Prevotella*; increased *Clostridium*	Altered gut–brain axis signaling, behavioral symptoms	Ho et. al., (2020)
Cardiovascular diseases	Members of the reclassified *Lactobacillus genera* (e.g., *Lacticaseibacillus, Limosilactobacillus*), *Bifidobacterium* spp., *Firmicutes*	Increased trimethylamine-N-oxide (TMAO)-producing species	Links to atherosclerosis and hypertension	Ding et. al., (2017)

Emerging Technologies: The microbiome field has embraced cutting-edge technologies to develop next-generation therapeutics. *Synthetic Biology* has been a game changer—researchers can now genetically program bacteria to perform specific tasks. We've seen engineered probiotics that secrete anti-inflammatory molecules for IBD, bacteria that light up (diagnostically) in the presence of tumor metabolites, and gut microbes modified to produce neurotransmitters for neuropsychiatric indications. One striking example is an engineered *E. coli* that produces a STING agonist to stimulate anti-tumor immunity: Synlogic's SYNB1891 was designed to release cyclic dinucleotides in the tumor microenvironment to boost immune cell activity, and it entered a Phase 1 trial in solid tumor patients in 2020. Parallel advances in *bacteriophage (phage) therapy* are providing tools to selectively eliminate pathogenic or resistant strains without broad antibiotics. For instance, phages targeting *C. difficile* or multi-drug-resistant *E. coli* have been engineered to enhance their killing efficacy and safety, and some are in early clinical testing for recurrent infections. Another emerging technology is *consortia design* using computational models—rather than relying on donor stool, companies are using in silico modeling to predict which combination of cultured bacteria might recreate the functions of a healthy microbiome. These defined consortia (e.g., VE303, a mix of *Firmicutes* spores for CDI) have shown promising Phase 2 results.

Microbiome Diagnostics and Biomarkers: Alongside therapies, recent years brought advances in identifying microbiome *biomarkers* to guide treatment. For example, a patient's gut microbiome profile can potentially predict their response to immunotherapy or likelihood of developing complications like graft-vs-host disease. Companies are investing in microbiome sequencing diagnostics a "stool test" to personalize medicine. A notable 2022 study used machine learning on microbiome data to stratify IBS patients into subgroups that responded differently to a probiotic versus placebo, illustrating precision medicine via microbiome profiling. These diagnostic tools will help identify who might benefit most from a given microbiome intervention, thereby improving clinical trial success rates and patient outcomes.

Overall, from 2018 to 2023, the field moved from intriguing correlations to tangible interventions. The success of FMT in new domains (immunotherapy, etc.), the entrance of synthetic biotics to clinical trials, and the first approvals all signal that microbiome therapeutics are maturing. There is also a notable influx of investment—many biotech startups in this arena went public or struck big pharma partnerships (e.g., Nestlé with Seres, AstraZeneca with Caelus Health) in the last 5 years. This momentum is propelling the field into a new phase of larger trials and expanded indications.

1.6 Regulatory and Commercialization Aspects

Bringing microbiome-based therapies to market poses unique regulatory and commercialization challenges (Ma et. al., 2018 and Daloiso et. al., 2015). Both the U.S. FDA and European Medicines Agency (EMA) have been grappling with how to classify and evaluate these living medicines:

Regulatory Classification: In the United States, the FDA has defined *live biotherapeutic products (LBPs)* as biological products that contain live organisms applicable to disease treatment or prevention, excluding vaccines. Early on, FMT was used under enforcement discretion (essentially tolerated without full approval) for CDI. Now with specific products, the FDA requires rigorous Investigational New Drug (IND) applications and clinical trials akin to traditional drugs. The FDA's approval of two LBPs for CDI in 2022–2023 confirms that a regulatory pathway exists. These approvals were treated similarly to biologic drugs (licensed via a Biologics License Application). In the European Union (EU), there is not yet a formally approved microbiome drug, but EMA has issued guidance that such products would be regulated as medicinal products or possibly as Advanced Therapy Medicinal Products (ATMPs) if they involve substantially manipulated cells/tissues. Defining the "active ingredient" of a microbiome therapy can be tricky—is it each strain? the whole consortium? or their collective metabolic output? Regulators have required detailed characterization of the microbial components, consistency of manufacturing (genomic stability of strains, etc.), and absence of unwanted contaminants or genes (like antibiotic resistance genes). The *regulatory bar is high*, but the recent approvals provide a roadmap. Ongoing dialogue between industry and regulators is refining standards for potency assays (how do you measure the dose of a microbial therapy?), quality control (sterility, purity of cultures), and efficacy endpoints.

Safety Considerations: Safety is paramount since we are introducing live organisms into patients. Regulatory agencies require extensive pathogen screening of donor-derived products to prevent transmission of infections. This became evident in 2019 when the FDA issued alerts after two patients contracted drug-resistant *E. coli* infections from an FMT that had not been adequately tested—one patient died. As a result, stringent donor screening for fecal transplants (including for multidrug-resistant organisms, viruses like severe acute respiratory syndrome coronavirus 2 [SARS-CoV-2], etc.) has been mandated. Companies have responded by developing processes to further inactivate potential pathogens (e.g., ethanol treatment in SER-109 to enrich spores and kill vegetative bacteria). Another safety concern is the theoretical risk of triggering unwanted immune reactions or transferring antibiotic resistance genes via the microbiome intervention. So far, clinical trials of defined LBPs have reported side effects mostly limited to mild GI symptoms, and no serious systemic infections attributable to the products, which are reassuring. However, long-term safety monitoring is in place (since microbes can theoretically colonize and persist). Regulators also consider *ecosystem risks*—for example, could a live microbe drug disrupt the native microbiome in negative ways or spread to others? These concerns drive careful design of containment features in engineered strains (like auxotrophic killswitches that prevent them from surviving outside the target environment). Overall, while microbiome therapies are generally well-tolerated, gaining regulatory approval requires robust safety data and risk mitigation strategies, given the novelty of using live consortia.

Manufacturing and Quality Control: Unlike conventional small molecules, growing and formulating a complex mix of bacteria at industrial scale are challenging. Early FMT was essentially a crude transplant of donor stool. For a commercial product, companies must produce standardized batches—this involves culturing multiple anaerobic species in specialized bioreactors, then combining and preserving them (often via freeze-drying). Ensuring each dose has the right composition and viability is non-trivial. Some bacteria are oxygen-sensitive and hard to lyophilize (freeze-dry) without losing efficacy. Innovative solutions, such as encapsulating spores (which are naturally hardy, as in SER-109) or formulation with protective cryoprotectants, have been key to making shelf-stable products. The FDA and EMA expect Good Manufacturing Practice (GMP) compliance—meaning quality controls at every step: verifying strain identity (using genomic sequencing), testing for contaminants, and assays to confirm functionality (e.g., the ability of the product to produce a known metabolite or engraft in a fecal sample). Companies like four-dimensional (4D) Pharma and Vedanta Biosciences have built manufacturing facilities dedicated to live biotherapeutics, reflecting the specialized needs. From a regulatory view, consistency is crucial—each lot should perform similarly. This has driven some developers away from whole-stool products to defined strain mixes that can be more tightly controlled. Indeed, one of the FDA-approved products (VOWST) contains a defined consortium of spore-forming *Firmicutes* grown from stool isolates, not stool itself. This approach eases standardization. As the industry matures, we anticipate clearer guidelines on release criteria for microbiome therapeutics—for example, how to measure total live cells and how to ensure lot-to-lot genomic consistency. The manufacturing know-how developed for probiotics and fermented foods provides a foundation, but therapeutic-grade products face higher scrutiny.

Market and Commercialization Challenges: Bringing a microbiome drug to market involves more than regulatory approval; there are economic and practical considerations. One challenge has been *intellectual property (IP)*—since many commensal microbes are naturally occurring, companies have focused on patenting specific formulations, delivery methods, or uses. There's also competition from the "DIY" space: for years, patients could receive FMT via stool banks or even do at-home transplants. Convincing payers and providers to adopt a commercial product that might be more expensive than donor stool is a real concern. However, standardized products offer reliability and safety that ad-hoc FMT lacks, which is a strong selling point. *Cost* of goods for live biotherapeutics could be high, given the complex manufacturing, but with scale and improved techniques (like continuous culture or synthetic biology approaches that simplify consortia), costs should come down. We are witnessing the first steps—Ferring's REBYOTA and Seres' VOWST are now being marketed to hospitals for CDI. Their success (or hurdles) will shape commercial models. Another issue is *physician and patient acceptance*: For example, some patients may be uncomfortable with the idea of "processed stool" treatments. Education is key to normalize microbiome therapies as just another form of biologic medicine. From a regulatory standpoint, post-marketing surveillance is being required to track long-term outcomes. Both FDA and EMA will likely update guidelines as more products seek approval, possibly creating a dedicated framework for microbiome therapeutics. In summary, while challenges exist, the path to market for microbiome-based therapies is clearer than ever. Regulatory agencies have shown openness, provided safety and efficacy are demonstrated. With CDI products as trailblazers, companies are now advancing pipeline candidates in IBD, allergy, oncology, and more, armed with lessons learned about meeting regulatory standards and proving real-world value.

1.7 Mechanistic Insights and Emerging Therapeutic Strategies

A deeper understanding of how the microbiome interacts with the host is driving innovation in therapy design. Here, we outline key mechanisms by which microbes influence health, and how these are being harnessed or modulated in cutting-edge interventions:

Microbiome–Host Interactions: The microbiome impacts the host through multiple pathways—metabolic, immune, and neural. For instance, gut bacteria ferment dietary fibers to produce short-chain fatty acids (SCFAs) like acetate, propionate, and butyrate. SCFAs serve as energy sources for colon cells, strengthen the gut barrier, and have systemic anti-inflammatory effects. Butyrate, in particular, fosters regulatory immune cells and can even influence the blood–brain barrier integrity and neuroinflammation. An imbalance in SCFA production has been implicated in conditions from IBD to depression. Another example is bile acids:

gut microbes convert primary bile acids into secondary bile acids, which can signal through host receptors (like Farnesoid X Receptor (FXR) and Takeda G protein-coupled Receptor 5 (TGR5)) to regulate metabolism and immune homeostasis. In CDI, a low level of secondary bile acids (due to microbiome disruption) fails to suppress *C. difficile* spore germination restoring those microbes restores colonization resistance. The microbiome also synthesizes vitamins (vitamin K and B vitamins) and aids mineral absorption. Critically, it "trains" the immune system—certain commensals induce anti-inflammatory regulatory T-cells (e.g., *Bacteroides fragilis* produces a polysaccharide that calibrates T-cell balance) and others stimulate immunoglobulin A (IgA) antibody production in the gut mucosa. This immune education is why germ-free animals have underdeveloped immune structures. In the context of cancer, specific gut bacteria can prime tumor-infiltrating myeloid cells or T-cells to be more effective when immunotherapy is given. In the gut–brain axis, microbes can produce or modulate neurotransmitters (90% of serotonin is made in the gut, influenced by microbial metabolites), and can affect stress reactivity via cortisol modulation. All these insights emphasize that the microbiome acts as an "organ" of its own, producing signals that shape physiology. Therapeutically, this means interventions can target these pathways—for example, giving fiber (prebiotic) to boost SCFA producers for colonic health, or using a consortium of bile-acid-converting bacteria to prevent pathogen colonization. It also means that unintended microbiome changes (like from antibiotics) can have wide ripple effects on immunity and metabolism.

Engineered Probiotics and Synthetic Biology: One of the most exciting advances in the creation of *engineered microbial therapeutics* designer probiotics endowed with new functions (. Using synthetic biology tools, scientists can modify the genome of a bacterium to, for example, secrete a specific drug at the disease site or consume a harmful metabolite. We have already seen this with Synlogic's strains for metabolic disorders. The SYNB1618 *E. coli* Nissle was engineered to express two non-native enzymes (PAL and LAAD) that break down phenylalanine, aiming to treat phenylketonuria. In clinical testing, it successfully lowered blood phenylalanine levels, offering hope for a novel therapy beyond dietary restriction. Similarly, an engineered strain SYNB8802 carries oxalate-degrading genes from *Oxalobacter formigenes*, designed to treat kidney stone disease by reducing urinary oxalate. In oncology, bacteria like *Salmonella* and *E. coli* have been engineered to selectively grow within tumors (taking advantage of the low-oxygen tumor core) and deliver cancer-killing payloads. For example, the SYNB1891 strain mentioned earlier delivers a STING agonist locally in the tumor to stimulate an immune attack. Engineered microbes can also act as "smart" delivery vehicles: one can program bacteria to release a drug only under certain conditions (e.g., when they sense inflammation markers). To ensure safety, synthetic biologists build in kill switches or nutritional dependencies so that if the bacteria stray beyond their target environment, they cannot survive. This field of *live therapeutic GMOs* (genetically modified organisms) is rapidly evolving, with multiple clinical-stage programs. Engineered probiotics could address diseases that were hard to tackle with conventional drugs—for instance, a microbe that continuously secretes a short-lived molecule right in the gut mucosa could treat localized inflammation without systemic side effects. There are efforts to engineer bacteria to mop up excess ammonia in cirrhosis patients (to prevent hepatic encephalopathy) and to produce dopamine precursors in the gut for Parkinson's. As our ability to program microbes grows, we could effectively turn bacteria into microscopic "pharmacies" operating inside the patient.

Phage Therapy and Microbiome Editing: Another innovative approach involves bacteriophages—viruses that infect bacteria. Phage therapy itself is not new (it has been used for decades in Eastern Europe to treat bacterial infections), but applying it to modulate the microbiome is novel. Scientists are developing phage cocktails that can selectively knock down deleterious bacteria in the gut while sparing the rest of the community. For example, targeting *Enterobacteriaceae* that produce pro-inflammatory metabolites could be a strategy in ulcerative colitis. There is also interest in *using phages as vectors for gene editing in the microbiome*. Recent research has shown it's possible to deliver CRISPR/Cas systems via phages to selectively kill bacteria carrying specific genes (like antibiotic resistance genes or virulence factors). This way, one could *edit out* the undesirable traits in the microbiome without disturbing others. Though not yet in the clinic, this "precision editing" of microbiota is a tantalizing concept for the future—a way to remove just the pathogen or just the toxin-encoding gene from a complex community. Such tools could address the specificity challenge that antibiotics could never solve.

Microbiome-Derived Metabolites (Postbiotics): Not all microbiome therapies involve live organisms. There is growing interest in *postbiotics*—administering beneficial molecules that the microbiome produces, thereby bypassing the need to change the microbes themselves. Many clinical effects of microbes are mediated by metabolites like SCFAs, tryptophan metabolites, and conjugated linoleic acids. Companies are isolating or synthesizing these and testing them as drugs or supplements. For example, 2,3-butanediol, a metabolite from certain commensals, has been tested for its immune-modulatory effects. Urolithin A, a microbial metabolite from pomegranate, is being developed as a therapy to improve muscle function in aging. In the realm of GI infections, *secondary bile acids* or analogues might be given to patients after antibiotic treatment to prevent *C. difficile* relapse by chemically inhibiting spore germination (mimicking what a healthy microbiome would do). While these approaches don't "restore" the microbiome, they can compensate for its dysfunction. They also typically have fewer regulatory hurdles (since they're defined chemical entities) and potentially a better safety profile (no live components). We may see a hybrid model where a patient receives a live therapeutic plus a metabolite booster that helps the engraftment or function of that microbiome therapy.

Personalized Nutrition and Microbiome: A simpler but powerful intervention is using diet to shape one's microbiome for therapeutic benefit. Precision nutrition studies (like the 2015 Zeevi et al. study) have shown that individuals respond differently to the same foods depending on their microbiome. This has led to algorithms that suggest tailored diets to improve blood sugar or other parameters by leveraging one's microbiome composition. Additionally, certain specialized diets are being used as therapy: For example, a high-fiber diet to promote anti-inflammatory microbes in IBD, or ketogenic diets that alter the gut flora in ways that may help epilepsy. While diet is a blunt tool compared to drugs, it remains one of the most accessible ways to alter the microbiome and can be a first-line adjunct in many conditions. The future might integrate microbiome data into nutritional counseling—for example, recommending prebiotic-rich foods to a patient whose microbiome analysis shows low butyrate producers, thereby boosting their levels naturally.

In summary, mechanistic research has underscored that microbe–host interactions are complex but targetable. Whether by adding beneficial functions (engineered LBPs), removing harmful members (phages/CRISPR), or supplying key metabolites (postbiotics), a toolbox of novel interventions is emerging. Synthetic biology stands out as a particularly revolutionary area, effectively allowing us to *redesign* the microbiome to better serve human health. Each strategy comes with its considerations—live engineered bugs need safety containment, phages need to avoid immune neutralization, and metabolites need efficient delivery to target site—but rapid progress is being made on all these fronts. The mechanistic insights ensure that new therapies are grounded in a deep understanding of cause and effect in the microbiome realm, improving the odds of clinical success.

1.8 Future Directions and Conclusion

Microbiome-based therapeutics are poised to become a mainstream component of medicine in the next decade. Looking forward, several trends and future directions are evident:

Expanded Clinical Indications: Thus far, most success has been in GI-related illnesses, but the *next frontier* will see microbiome therapies for systemic and even CNS diseases. Ongoing trials are exploring LBPs in conditions like asthma (to induce tolerance), malnutrition (to correct microbiota immaturity in children), autoimmune diseases like type 1 diabetes, and even as adjuncts in Alzheimer's disease. If early signals pan out, the pipeline of indications will broaden considerably. Each new disease area will require tailored approaches—the bacteria that help in one context may differ in another—underscoring the need for extensive microbiome profiling and mechanistic studies disease by disease.

Personalized Microbiome Medicine: One size may not fit all when it comes to microbiome interventions. We anticipate a shift toward personalized or stratified therapy. Prior to treatment, patients might be categorized by their microbiome "enterotype" or specific microbial signatures. This could guide which therapy to use—for example, only patients lacking certain butyrate producers might be given a supplement targeting that deficiency. Machine learning and artificial intelligence (AI) are being harnessed to find patterns in vast microbiome data that correlate with treatment response (Past, present, and future of microbiome-based therapies—PMC). Advanced AI algorithms can integrate microbiome sequences with clinical data to predict outcomes and suggest optimal interventions. In fact, researchers are already talking about *"3rd generation" microbiome therapies designed with AI*, where complex microbial consortia are computationally optimized for a given individual's needs (Past, present, and future of microbiome-based therapies—PMC) (Past, present, and future of microbiome-based therapies—PMC). This vision aligns with the broader trend of precision medicine, making therapies more effective and avoiding unnecessary treatments for those unlikely to benefit.

Integration with Mainstream Therapies: Rather than treating the microbiome in isolation, future practice will likely integrate microbiome modulation with conventional treatments. For example, cancer patients might receive a microbiome adjuvant along with chemotherapy to protect their gut and immune function during treatment. Psychiatric patients might get a psychobiotic alongside antidepressants to enhance mood regulation. In surgery or critical care, prophylactic microbiome management (like giving beneficial strains to intensive care unit [ICU] patients) could reduce hospital-acquired infections and complications. As evidence accumulates, clinical guidelines will begin to incorporate microbiome status as a factor in decision-making ("check the microbiome before prescribing X drug" could become a norm if strong drug–microbe interactions are known (The Use of Fecal Microbiota Transplant in Overcoming and Modulating Resistance to Anti-PD-1 Therapy in Patients with Skin Cancer—PMC)). We may also see regulatory requirements to evaluate how new pharmaceuticals impact the microbiome, to catch potential issues early.

Improved Delivery Systems: A practical aspect for the future is developing better delivery methods for microbiome therapeutics. Oral capsules work for GI delivery, but for targeting other body sites (like the lung or skin), novel formulations are needed. Researchers are exploring encapsulating bacteria in protective coatings to ensure they reach the colon intact, or embedding them in gels/films for skin applications (think of a probiotic wound dressing to prevent infection). Another concept is *bioreactors in a pill*—devices that could release microbes gradually or only when triggered by a gut condition. Ensuring that introduced microbes can effectively engraft (take up residence) in the host microbiome is an area of active investigation, as persistence might be needed for long-term benefit in chronic diseases. Conversely, for some uses transient presence is enough, and then designing bugs that self-limit and clear after releasing their therapeutic payload may be preferable.

Safety and Ethical Oversight: With increasing power to manipulate the microbiome comes responsibility. Long-term surveillance of patients who receive live microbiome products will inform safety guidelines—for instance, could altering someone's microbiome have intergenerational effects (if a mother's microbiome is permanently changed, what about her child's initial colonization)? As these therapies become more widespread, ethical considerations will also emerge, such as equitable access (microbiome therapies could be expensive at first) and avoidance of "hyper personalization" where one's microbial data might be used discriminatorily. Another future concern might be *microbiome hacking* or unregulated use—already we see Internet forums where patients discuss DIY FMT for conditions like autism or Crohn's disease, sometimes with risky practices. The medical community will need to provide safe, regulated options to meet this demand and discourage unsafe experimentation.

In conclusion, microbiome-based therapeutics have moved from a fringe idea to a validated treatment modality with broad potential. We now recognize the microbiome as a crucial mediator of health, one that can be leveraged to prevent and treat disease. The past 5 years delivered concrete advances—landmark trials, new drug approvals, and mechanistic breakthroughs—that set the stage for rapid expansion of this field. The concept of treating disease by "rethinking our bacteria" is revolutionary: instead of solely targeting human cells or pathogens, we are now targeting the supportive community within us. This paradigm shift opens numerous possibilities for conditions previously difficult to manage. Still, challenges remain in optimizing these therapies, ensuring safety, and integrating them into clinical practice. Multidisciplinary collaboration between microbiologists, clinicians, computational scientists, and regulatory experts will be key to fully realize microbiome therapeutics' promise. The *next frontier* will likely bring smarter, more personalized interventions—from designer microbes to AI-guided microbiome edits—that can restore homeostasis in disease-specific ways. Such therapies could transform management of inflammatory diseases, metabolic syndromes, neurodegeneration, and even enhance outcomes in cancer and beyond. In essence, harnessing our microbial allies represents a new frontier of medicine, one that is steadily advancing from theory to practice. The coming years will undoubtedly see microbiome interventions becoming a routine part of the therapeutic arsenal, heralding a new era of symbiotic medicine where healing involves not just the patient, but their microbial partners as well.

References

Akobeng AK, Singh P, Kumar M, Al Khodor S (2020) Role of the gut microbiota in the pathogenesis of coeliac disease and potential therapeutic implications. Eur J Nutr 59:3369–3390

Alcock J, Maley CC, Aktipis CA (2014) Is eating behavior manipulated by the gastrointestinal microbiota? Evolutionary pressures and potential mechanisms. BioEssays 36:940–949

Al-Fakhrany OM, Elekhnawy E (2024) Next-generation probiotics: the upcoming biotherapeutics. Mol Biol Rep 51:505

Bakhti SZ, Latifi-Navid S (2021) Interplay and cooperation of helicobacter pylori and gut microbiota in gastric carcinogenesis. BMC Microbiol 21:258

Baldelli V, Scaldaferri F, Putignani L, Del Chierico F (2021) The role of enterobacteriaceae in gut microbiota dysbiosis in inflammatory bowel diseases. Microorganisms 9:697

Binayak L, Chaudhuri R, Prasanth T, Burkhart I, Schwalbe H, Dash J (2024) Guanosine-based hydrogel as a supramolecular scaffold for template-assisted macrocyclization. Chem Commun 60:34

Biswa RS, Ranjan JS, Beriha SK, Mahanta CS, Jena BB, Prasanth T, Samanta L, Satapathy R, Dash BP (2023) Synthesis and anticancer properties of dendritic glycoconjugates containing multiple o-carborane clusters. New J Chem 47:10
Boehme M, Guzzetta KE, Wasen C, Cox LM (2023) The gut microbiota is an emerging target for improving brain health during ageing. Gut Microbiome 4:e2
Braga JD, Thongngam M, Kumrungsee T (2024) Gamma-aminobutyric acid as a potential postbiotic mediator in the gut–brain axis. npj Sci Food 8:16
Caradonna E, Nemni R, Bifone A, Gandolfo P, Costantino L, Giordano L, Mormone E, Macula A, Cuomo M, Difruscolo R et al (2024) The Brain-gut Axis, an important player in Alzheimer and Parkinson disease: a narrative review. J Clin Med 13:4130
Carroll IM, Chang YH, Park J, Sartor RB, Ringel Y (2010) Luminal and mucosal-associated intestinal microbiota in patients with diarrhea-predominant irritable bowel syndrome. Gut Pathog 2:19
Carroll IM, Ringel-Kulka T, Siddle JPf, Ringel Y (2012) Alterations in composition and diversity of the intestinal microbiota in patients with diarrhea-predominant irritable bowel syndrome. Neurogastroenterol Motil 24(521–530):e248
Chen Y, Xu J, Chen Y (2021) Regulation of neurotransmitters by the gut microbiota and effects on cognition in neurological disorders. Nutrients 13:2099
Chloé Mirsepasi-Lauridsen H, Vallance BA, Krogfelt KA, Petersen AM (2019) Escherichia Coli Pathobionts associated with inflammatory bowel disease. Clin Microbiol Rev 32:e00060–e00018
Coates MD, Clarke K, Williams E, Jeganathan N, Yadav S, Giampetro D, Gordin V, Smith S, Vrana K, Bobb A et al (2023) Abdominal pain in inflammatory bowel disease: an evidence-based. Multidisciplinary Review Crohn's Colitis 360(5):otad055
Corbin KD, Carnero EA, Dirks B, Igudesman D, Yi F, Marcus A, Davis TL, Pratley RE, Rittmann BE, Krajmalnik-Brown R et al (2023) Host-diet-gut microbiome interactions influence human energy balance: a randomized clinical trial. Nat Commun 14:3161
Craven LJ et al (2020) Allogenic fecal microbiota transplantation in patients with nonalcoholic fatty liver disease improves abnormal small intestinal permeability: a randomized control trial. Am J Gastroenterol 115:1055–1065
Daloiso V, Minacori R, Refolo P, Sacchini D, Craxi L, Gasbarrini A, Spagnolo AG (2015) Ethical aspects of fecal microbiota transplantation (FMT). Eur Rev Med Pharmacol Sci 19:3173–3180
Degruttola AK, Low D, Mizoguchi A, Mizoguchi E (2016) Current understanding of dysbiosis in disease in human and animal models. Inflamm Bowel Dis 22:1137–1150
Di Vincenzo F, Del Gaudio A, Petito V, Lopetuso LR, Scaldaferri F (2024) Gut microbiota, intestinal permeability, and systemic inflammation: a narrative review. Intern Emerg Med 19:275–293
Ding HT, Taur Y, Walkup JT (2017) Gut microbiota and autism: key concepts and findings. J Autism Dev Disord 47:480–489
Doocey CM, Finn K, Murphy C, Guinane CM (2022) The impact of the human microbiome in tumorigenesis, cancer progression, and biotherapeutic development. BMC Microbiol 22:53
Evangelia L, Karanasou EA, Gazouli M (2021) Gut microbiome and gastrointestinal disorders. In: Maria G, Theodoropoulos G (eds) Gut microbiome-related diseases and therapies. Springer International Publishing, Cham, pp 41–91. ISBN 978-3-030-59642-2
Fakhoury M, Negrulj R, Mooranian A, Al-Salami H (2014) Inflammatory bowel disease: clinical aspects and treatments. J Inflamm Res 7:113–120
Fatma K, Prasanth T, Panda D, Ravichandiran V, Dash J (2024) Selective recognition of c-KIT 1 G-Quadruplex by structural tuning of heteroaromatic scaffolds and side chains. ACS Med Chem Lett 15:3
Festi D, Schiumerini R, Eusebi LH, Marasco G, Taddia M, Colecchia A (2014) Gut microbiota and metabolic syndrome. World J Gastroenterol 20:16079–16094
Fusco W, Lorenzo MB, Cintoni M, Porcari S, Rinninella E, Kaitsas F, Lener E, Mele MC, Gasbarrini A, Collado MC et al (2023) Short-chain fatty-acid-producing bacteria: key components of the human gut microbiota. Nutrients 15:2211

Ghosh TS, Shanahan F, O'Toole PW (2022) The gut microbiome as a modulator of healthy ageing. Nat Rev Gastroenterol Hepatol 19:565–584
Grau-Del Valle C, Fernandez J, Sola E, Montoya-Castilla I, Morillas C, Banuls C (2023) Association between gut microbiota and psychiatric disorders: a systematic review. Front Psychol 14:1215674
Gros M, Gros B, Mesonero JE, Latorre E (2021) Neurotransmitter dysfunction in irritable bowel syndrome: emerging approaches for management. J Clin Med 10:3429
Han Z, Min Y, Pang K, Wu D (2023) Therapeutic approach targeting gut microbiome in gastrointestinal infectious diseases. Int J Mol Sci 24:15654
Ho LKH, Tong VJW, Syn N, Nagarajan N, Tham EH, Tay SK, Shorey S, Tambyah PA, Law ECN (2020) Gut microbiota changes in children with autism spectrum disorder: a systematic review. Gut Pathog 12:6
Jang JH, Jang SY, Ahn S, Oh JY, Yeom M, Ko SJ, Park JW, Kwon SK, Kim K, Lee IS et al (2024) Chronic gut inflammation and dysbiosis in IBS: unraveling their contribution to atopic dermatitis progression. Int J Mol Sci 25:2753
Kamali Dolatabadi R, Feizi A, Halaji M, Fazeli H, Adibi P (2021) The prevalence of adherent-invasive Escherichia Coli and its association with inflammatory bowel diseases: a systematic review and meta-analysis. Front Med 8:730243
Kang D, Adams J, Gregory A, Gregory A, Borody T, Chittick L, Chittick L, Fasano A, Khoruts A, Geis E, Maldonado J, McDonough-Means S, Pollard E, Roux S, Roux S, Sadowsky M, Lipson K, Sullivan M, Sullivan M, Caporaso J, Krajmalnik-Brown R (2017) Microbiota transfer therapy alters gut ecosystem and improves gastrointestinal and autism symptoms: an open-label study. Microbiome 5:10
Kelly CR, Yen EF, Grinspan AM, Kahn SA, Atreja A, Lewis JD, Moore TA, Rubin DT, Kim AM, Serra S et al (2021) Fecal microbiota transplantation is highly effective in real-world practice: initial results from the FMT National Registry. Gastroenterology 160:183–192.e3
Khanna S, Pardi DS (2012) Clostridium difficile infection: new insights into management. Mayo Clin Proc 87:1106–1117
Kim J, Lee HK (2022) Potential role of the gut microbiome in colorectal cancer progression. Front Immunol 12:807648
Kobyliak N, Abenavoli L, Mykhalchyshyn G, Kononenko L, Boccuto L, Kyriienko D, Dynnyk O (2018) A multi-strain probiotic reduces the fatty liver index, cytokines and aminotransferase levels in NAFLD patients: evidence from a randomized clinical trial. J Gastrointestin Liver Dis 27(1):41–49
Li S, Liu J, Zheng X, Ren L, Yang Y, Li W, Fu W, Wang J, Du G (2022) Tumorigenic bacteria in colorectal cancer: mechanisms and treatments. Cancer Biol Med 19:147–162
Lu Y, Yuan X, Wang M, He Z, Li H, Wang J, Li Q (2022) Gut microbiota influence immunotherapy responses: mechanisms and therapeutic strategies. J Hematol Oncol 5:47
Ma Y, Chen H, Lan C, Ren J (2018) Help, hope and hype: ethical considerations of human microbiome research and applications. Protein Cell 9:404–415
Magne F, Gotteland M, Gauthier L, Zazueta A, Pesoa S, Navarrete P, Balamurugan R (2020) The Firmicutes/Bacteroidetes ratio: a relevant marker of gut dysbiosis in obese patients? Nutrients 12:1474
Marasco G, Di Biase AR, Schiumerini R, Eusebi LH, Iughetti L, Ravaioli F, Scaioli E, Colecchia A, Festi D (2016) Gut microbiota and celiac disease. Dig Dis Sci 61:1461–1472
Markowiak P, Śliżewska K (2017) Effects of probiotics, prebiotics, and synbiotics on human health. Nutrients 9:1021
McCune E, Sharma A, Johnson B, O'Meara T, Theiner S, Campos M, Heditsian D, Brain S, Gilbert JA, Esserman L et al (2024) Gut and oral microbial compositional differences in women with breast cancer, women with ductal carcinoma in situ, and healthy women. mSystems 9:e0123724
Miquel S, Martín R, Rossi O, Bermúdez-Humarán LG, Chatel JM, Sokol H, Thomas M, Wells JM, Langella P (2013) Faecalibacterium Prausnitzii and human intestinal health. Curr Opin Microbiol 16:255–261

Miquel S, Leclerc M, Martin R, Chain F, Lenoir M, Raguideau S, Hudault S, Bridonneau C, Northene T, Bowene B et al (2015) Identification of metabolic signatures linked to anti-inflammatory effects of Faecalibacterium Prausnitzii. MBio 6:e00300–e00315

Nogal A, Valdes AM, Menni C (2021) The role of short-chain fatty acids in the interplay between gut microbiota and diet in cardio-metabolic health. Gut Microbes 13:1897212

Panda D, Saha P, Chaudhuri R, Prasanth T, Ravichandiran V, Dash J (2019) A competitive pull-down assay using G-quadruplex DNA-linked magnetic nanoparticles to determine specificity of G-quadruplex ligands. Anal Chem 91:12

Parida S, Sharma D (2021) The microbiome and cancer: creating friendly neighborhoods and removing the foes within A C. Cancer Res 81:790–800

Pasta A, Formisano E, Calabrese F, Plaz Torres MC, Bodini G, Marabotto E, Pisciotta L, Giannini EG, Furnari M (2024) Food intolerances, food allergies and IBS: lights and shadows. Nutrients 16:265

Petrosino JF (2018) The microbiome in precision medicine: the way forward. Genome Med 10:12

Pittayanon R, Lau JT, Leontiadis GI, Tse F, Yuan Y, Surette M, Moayyedi P (2020) Differences in gut microbiota in patients with vs. without inflammatory bowel diseases: a systematic review. Gastroenterology 158:930–946.e1

Prasanth T, Chakraborti G, Mandal T, Ravichandiran V, Dash J (2022) Cycloaddition of N-sulfonyl and N-sulfamoyl azides with alkynes in aqueous media for the selective synthesis of 1,2,3-triazoles. Green Chem 24:911

Priya D, Gumpu MB, Thumpati P, Rayappan JBB, Ravichandiran V, Pazhani GP, Veerapandian M (2018) Chemically synthesized butein and butin: optical, structure and electrochemical redox functionality at electrode interface. J Photochem Photobiol B Biol 182:122–129

Ramírez-Pérez O, Cruz-Ramón V, Chinchilla-López P, Méndez-Sánchez N (2017) The role of the gut microbiota in bile acid metabolism. Ann Hepatol 16:S21–S26

Rasul MF, Hussen BM, Salihi A, Ismael BS, Jalal PJ, Zanichelli A, Jamali E, Baniahmad A, Ghafouri-Fard S, Basiri A et al (2022) strategies to overcome the main challenges of the use of CRISPR/Cas9 as a replacement for cancer therapy. Mol Cancer 21:64

Rhodes R (2016) Ethical issues in microbiome research and medicine. BMC Med 14:156

Ritapa R, Prasanth T, Dash J (2023) Expanding the toolbox of target directed bio-orthogonal synthesis: in situ direct macrocyclization by DNA templates. Angew Chem Int Ed 135:7

Ritapa C, Prasanth T, Biswas D, Mandal S, Dash J (2024) Combating multidrug-resistance in S. pneumoniae: a G-quadruplex binding inhibitor of efflux pump and its bio-orthogonal assembly. Nar Mol Med 1:2

Ryu G, Kim H, Koh A (2021) Approaching precision medicine by tailoring the microbiota. Mamm Genome 32:206–222

Santacroce L, Passarelli PC, Azzolino D, Bottalico L, Charitos IA, Cazzolla AP, Colella M, Topi S, Godoy FG, D'Addona A (2023) Oral microbiota in human health and disease: a perspective. Exp Biol Med 248:1288–1301

Sexton RE, Uddin MH, Bannoura S, Khan HY, Mzannar Y, Li Y, Aboukameel A, Al-Hallak MN, Al-Share B, Mohamed A et al (2022) Connecting the human microbiome and pancreatic cancer. Cancer Metastasis Rev 41:317–331

Shaikh SD, Sun N, Canakis A, Park WY, Weber HC (2023) Irritable bowel syndrome and the gut microbiome: a comprehensive review. J Clin Med 12:2558

Shariati A, Razavi S, Ghaznavi-Rad E, Jahanbin B, Akbari A, Norzaee S, Darban-Sarokhalil D (2021) Association between colorectal cancer and Fusobacterium Nucleatum and Bacteroides Fragilis bacteria in Iranian patients: a preliminary study. Infect Agent Cancer 16:41

Sioson VA, Kim M, Joo J (2021) Challenges in delivery systems for CRISPR-based genome editing and opportunities of nanomedicine. Biomed Eng Lett 11:217–233

Sokol H, Pigneur B, Watterlot L, Lakhdari O, Bermúdez-Humarán LG, Gratadoux JJ, Blugeon S, Bridonneau C, Furet JP, Corthier G, Grangette C et al (2008) Faecalibacterium Prausnitzii is an anti-inflammatory commensal bacterium identified by gut microbiota analysis of Crohn disease patients. Proc Natl Acad Sci USA 105:16731–16736

Talapko J, Erić S, Meštrović T, Stipetić MM, Juzbašić M, Katalinić D, Bekić S, Muršić D, Flam J, Belić D, Lešić D et al (2024) The impact of oral microbiome dysbiosis on the aetiology, pathogenesis, and development of oral cancer. Cancer 16:2997

Thumpati P, Rai SN, Prajapati C, Ramakrishna K, Singh SK (2025) Targeting c-MYC G-Quadruplexes for cancer treatment by small molecule. SciPharm 93(1):6

Tiwari A, Ika Krisnawati D, Susilowati E, Mutalik C, Kuo TR (2024) Next-generation probiotics and chronic diseases: a review of current research and future directions. J Agric Food Chem 18:27679–27700

Waheed R, Farooq AZ, Asma L (2024) Regulatory considerations for microbiome-based therapeutics. In: Khurshid M, Akash MSH (eds) Human microbiome: techniques, strategies, and therapeutic potential. Singapore, Springer Nature, pp 657–689. ISBN 978-981-97-3790-1

Wang L, Alammar N, Singh R, Nanavati J, Song Y, Chaudhary R, Mullin GE (2020) Gut microbial dysbiosis in the irritable bowel syndrome: a systematic review and meta-analysis of case-control studies. J Acad Nutr Diet 120:565–586

White MT, Sears CL (2024) The microbial landscape of colorectal cancer. Nat Rev Microbiol 22:240–254

Xu Y, Wang N, Tan HY, Li S, Zhang C, Feng Y (2020) Function of Akkermansia muciniphila in obesity: interactions with lipid metabolism, immune response and gut systems. Front Microbiol 11:219

Young VB (2017) The role of the microbiome in human health and disease: an introduction for clinicians. BMJ 356:j831

Zhu B, Wang X, Li L (2010) Human gut microbiome: the second genome of human body. Protein Cell 1:718–725

Navigating the Landscape: Regulatory Frameworks and Ethical Considerations in Microbiota-Based Precision Medicine

Samaneh Hatami and Seyed Ali Issazadeh

Abstract This chapter investigates the convergence of ethical considerations and regulatory frameworks within the rapidly developing area of microbiota-based precision medicine. As microbiome therapies draw more attention, it is essential to comprehend the consequences of their development and commercialization. Ethical considerations including health disparities and informed consent are of utmost importance, especially considering the disproportionate representation of specific ancestries in current research, which could worsen inequities in access to healthcare. However, the regulatory perspective also has many problems to consider concerning efficacy and safety-not just of fecal microbiota transplantation, but also that of other microbiome interventions. This chapter also discusses the role of the microbiome in affecting pharmacokinetics and variability in drug response, and identifies the need for multi-dimensional strategies that incorporate both microbiota and genetic signals. Eventually, responsible advancement of precision medicine based on microbiota will depend upon the resolution of these multi-dimensional challenges.

Keyword Ethical considerations · Microbiota · Regulatory frameworks · FMT

1 Introduction

This has been characterized by the dramatic rise in human microbiome-associated scientific knowledge and research in the recent past, facilitated by an era of unprecedented development of high-efficiency molecular technologies (Ma et al. 2018).

S. Hatami (✉)
Department of Food Science and Technology, Faculty of Agriculture, Ferdowsi University of Mashhad, Mashhad, Iran

S. Ali Issazadeh
Department of Food Science and Technology, Quchan Branch, Islamic Azad University, Quchan, Iran

S. K. Singh et al. (eds.), *Microbiota Profiling for Precision Medicine*,
Interdisciplinary Biotechnological Advances,
https://doi.org/10.1007/978-981-95-2104-3_13

The microbiome comprises the collective genomes of the human host and the diverse viruses and bacteria residing on the skin, membranes, mucous, gastrointestinal tract, and other organs of the human body (Sanyal et al. 2024).

The Human Microbiome Project was an initiative of the National Institute of Health, which in 2007 began funding research proposals with the aim of using technological advances for a thorough description of the microbial communities inhabiting the human body and exploring the relationships between microbiota and the human body, its influence on human health and disease, physiology, development, nutrition, and immunity (Olteanu et al. 2024). Insight into the microbiome is likely to revolutionize medical practice. However, it has the potential to also impact our juridical and social order and alter some of our fundamental ethical considerations relative to biomedical and medical research. Because of this, consideration of some of the legitimate, social, and ethical issues emerging from human microbiome research is required in order to advise researchers engaging in such research activities and participants who might undergo its results (Rhodes 2016). Examination of problems from the perspective of microbiome research offers a new lens through which we will be able to perceive our identity, our role throughout the society, and our adherence to each other. For instance, human microbiome research necessitates a change in perspective from regarding germs as adversaries to be destructed to conceiving Personal identity (Honey 2008). Research in the human microbiome, often referred to as our second genome, stands at a threshold that will make great strides toward a better understanding of identity, normality, health, and disease. It should catalyze a revolution in the management of public health and clinical practices and in the manufacture and distribution of commercial products claiming health effects and disease prevention. Such products would range from personalized diets, prebiotics, and probiotics to other microbial-related interventions (Shukla et al. 2024).

For instance, the diagnostic criteria, which are commonly utilized for vaginal microbiota, partly rely on evaluating the abundance of *Lactobacillus morphotypes* as an indicator of "healthiness"; however, a research study reported that this strain was not detected in the vagina of 25% of healthy women. This research emphasizes the need for a deeper understanding of a "healthy" and "normal" vaginal ecosystem, focusing on its functionality instead of its composition (Sherrill-Mix et al. 2020). Another important example is the growing use of fecal microbiota transplantation (FMT) as a novel therapeutic approach, which risks being recognized as a panacea for various diseases, leading to misuses as more websites have emerged promoting domestic DIY FMT kits as a form of "self-treatment" (Shi et al. 2017). A high level of caution is crucial for interventions based on microbiome to ensure that risk-benefit assessment and application indications are specific. We may have to rethink the human body as an ecosystem and consider each person to be a superorganism, instead of regarding it as just a single person (Ribas et al. 2023).

The importance of this new perception of ourselves has great implications for the approach and conduct regarding property, privacy, and identity questions associated with the human microbiome-for instance, how integral is the microbiome to our concept of self? (Rose and Dening 2023). To what degree do people have ownership over their microbiome considering that some of its sources, like feces, are typically regarded as waste? How does our information on microbiome influence our perception of health and disease? Who can benefit when the microbial profile of an individual is distinct and may have commercial value? How can we regulate microbiome-based products, and what evidence is needed to support health claims for probiotic products such as yogurt? Are we allowed to permit the agencies of public security to access the biobanks that collect and store the samples of human microbiota that could include distinct and recognizable information about donors? (Ma et al. 2018).

There could be one or more optimum microbial profiles. Enhancing an individual's microbiome in the pursuit of better-than-usual health should not be a purely medical or technical issue. It is already largely accepted across scientific scholarship that the microbiome lies at the heart of our health, both psychi and physical, indeed in health and illness generally. The question is, to what extent is the microbiome doing these things? (Carabotti 2015).

The gut–brain axis (GBA) is a two-way, very complex system of communication between the gut and the brain through which homeostasis is maintained. Recent findings indicate that gut microbiota, made up of trillions of microorganisms within the gastrointestinal system, occupies a position where it can play a significant role in the modulation of the gut–brain axis (GBA) (Ahmed and Hens 2022). The intestinal microbiota differs from person to person; basically, each forms during early life and depends on multiple variables: genetic, use of antibiotics, diet, lifestyle, and environment. Gut microbiota exerts significant immunological effects and has a symbiotic interaction with the immune system of the host to respond to antigens of the microbiota and various metabolic products. Indeed, recent research has also documented that gut microbiota influences the brain via neural, immune, and endocrine pathways, leading to the overarching concept of a microbiota gut–brain axis (MGB) (Afzaal et al. 2022).

As a result, the microbiome effect on mental health is also addressed along with its implications. Microbiome research in psychiatry offers valuable insights, particularly concerning the ethical challenges and the potential of intervention based on the microbiome (Mörkl et al. 2020).

Although our knowledge about human microbiome and its influences on the human health and disease has considerably been enhanced in recent years, robust evidence has still remained restricted. Moreover, the intricate results of research works are usually published by social networks as "exciting," "attractive," "groundbreaking" and even "miraculous," and people, even though misinformed, try to apply the most recent "science on human microbiome" in their health choices while neglecting the unpredicted long-term risks (Ostendorf et al. 2020).

There is growing excitement and effort to manipulate or modify individual microbiome in the quest for a "healthy" microbiome; nevertheless, no one knows the definition of a healthy microbiome. Furthermore, there is a possibility that such definition does not exist, which also raises ethical concerns, because these variations could influence the surrounding community or society. Consequently, it is essential to take into account these social and ethical factors at the outset of the research on human microbiome in order to be active and perfectly ready for confronting challenges (Ma et al. 2018). In this chapter, we are going to present a comprehensive overview of the ethical issues associated with the wider legal, social, and ethical dimensions of human microbiome research and its clinical uses. We concentrate on (1) human microbiome and personal identity; (2) safety, privacy, and risks; (3) informed consent in microbiome-related interventions such as FMT and vaginal seeding; (4) biobanks; (5) public health concerns; (6) application process of investigational new drug (IND); (7) current regulatory environment for FMT; (8) challenges in development and commercialization; and (9) ethical guidelines for in-vivo microbiota research. In the entire discussion, we also present some suggestion for the sustainable and healthy advancement of human microbiome research.

These ethical concerns, particularly in novel microbiome research, underscore the necessity of a cautious and accountable approach to psychiatric research and interventions. Tackling these problems is essential for progressing the field while safeguarding patients' well-being and rights.

In Fig. 1, these relationships are clearly showed and each aspect is discussed separately to ensure the practices responsible for the microbiome research.

2 Personal Identity

"I" used to be a straightforward expression that anyone was able to understand its meaning. Currently, according to our scientific understanding, "I" signifies not only my identity as a moral parameter and the focus of my consciousness but also encompasses my genotype, phenotype, and microbiome which include organisms that are not a part of me and many of them are transient. Depending on the context, various notions of "me" can be pertinent (Ascienzo et al. 2022).

From an ethical point of view, only certain humans can qualify as moral people, as we merely can hold some people accountable for their decisions (for example not young children). As a victim of disease, my body represents my identity, but as a carrier of disease, my identity is formed by my microbiome which affects my body health; at the same time, its impacts are partially specified by the collective properties of the microorganisms that constitute it (Gordon et al. 2005). A certain microorganism may positively influence one person while producing adverse effects on another. These variations will make it more challenging to establish a definition for the word "normal." While identity and personhood have always been complex terms, as we begin to look at ourselves as a blend of us and others, we need to reconceptualize normalcy, personal identity, and their implications (Freytag and Lister 2020). Our comprehension of the human microbiome and how it interacts with the human body also influences the way we conceive personal identity and personhood; the latter is typically defined by vital and distinguishing traits. Therefore, if diet, genes, and microorganisms affect our sensitivity and resistance to illness and our responses to treatments, all of them can contribute to our identity (Conlon and Bird 2014; Chen et al. 2015). Moreover, the microbiome of every

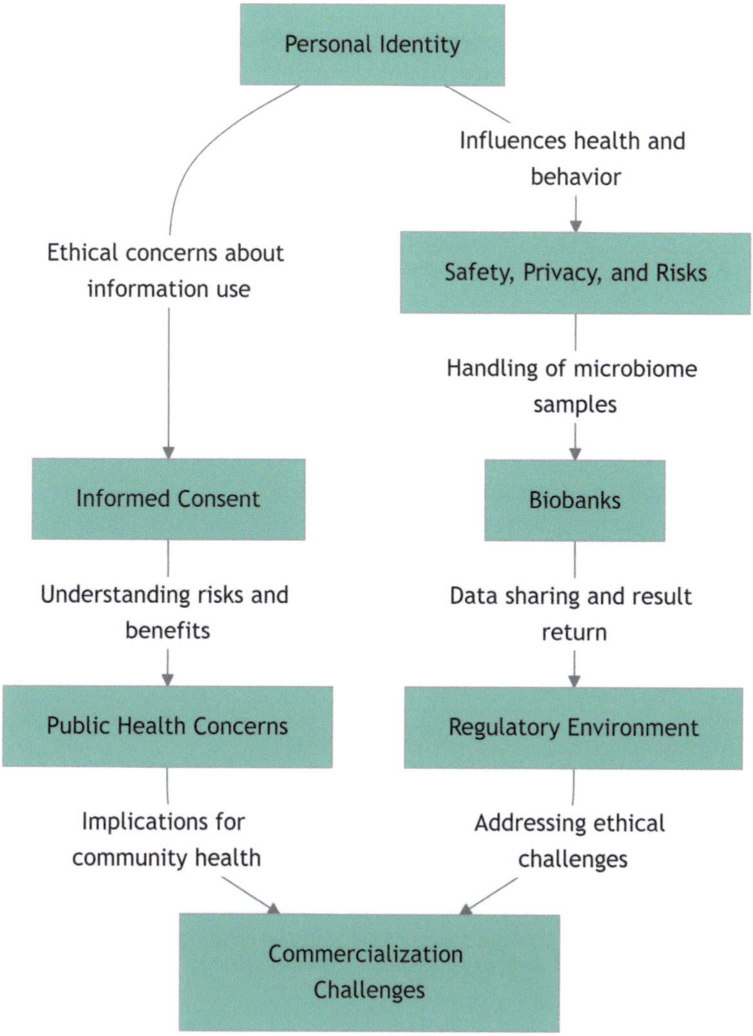

Fig. 1 Ethical dimensions of human microbiome research

person is distinct (Karvonen et al. 2022). In this regard, the microbiome could be integrated into our understanding of personal identity. Many experts believe that we must consider human to be a superorganism made up of the human body in addition to a variety of microorganisms residing in it. At the same time, they have approached the implication of identity problems in human microbiome from various prospects. As noted by Ahmed and Hens (2022), the individual and common perceptions of self and personal identity may be influenced by the rise of research projects on human microbiome, "the features of our microbiome are features of ourselves" (Ahmed and Hens 2022). On the other hand, from a philosophical standpoint, they also noted that discoveries associated with the human microbiome will not significantly impact the numerical criteria for personal identity as time goes by. In other words, it is possible that a person's social identity or psychology will remain consistent over time even if their microbiome undergoes significant alterations, for example, through FMT. Conversely, from an evolutionary and ecological point of view, Schneider et al. (2024) argued that the human microbiome has significant implications for our conception of humanity: "the shared evolutionary destiny of humans and their symbiotic bacteria has favored mutualistic interactions crucial for human health, and genetic or ecological alterations that disrupt this shared destiny could lead to illness" (Schneider et al. 2024). Aligned with this perspective, Al-Hwsali stated that "the coexistence of human and the microbiome reveals that the human evolution encompasses more than just the human history" (Al-Hwsali et al. 2023). Ultimately, from the viewpoint of selfhood and individuality, it has been suggested that the single Homo Sapiens is not actually the true biological human, as the true biological human is a super-human defined as the combination of the organism and its microbiome (Seekatz 2020).

Beyond the realm of philosophy, "personal identity" typically pertains to specific characteristics to which an individual feels a particular sense of connection or ownership. In this context, an individual's personal identity comprises the characteristics that are believed to "define him/her as a person" or "shape who he/she is." The literature demonstrated the ability of microbiome-based results to reveal group affiliation and more personal data associated with nationality, race, ethnicity, and even socio-economic circumstances. Additionally, microorganisms could offer insights into the human ancestry (Dauda et al. 2023).

The microorganisms making up human microbiota coevolve with humans and engage in intricate interactions with their hosts; furthermore, they are able to be transmitted vertically. For instance, it is proposed that *Helicobacter pylori* (HP) can serve as an indicator for migrations and ancestry (Dominguez-Bello and Blaser 2011), implying that the individuals sharing a specific strain of HP might have the same ancestor. At the same time, there is significant skepticism about such highly simplified classification, as it likely ignores the intricacy of the human microbiome and could lead to new types of stigma and stereotype.

3 Risks, Safety, and Privacy

Most studies include social, physical, or psychological risks to some extent. Research ethics mandates the potential risks of harm related to taking part in research should be weighed against the predicted social advantages. Azria et al. asserted research projects should have a desirable risk–benefit ratio, signifying that research teams should investigate the entire types of harms engaged, compare them with potential benefits, and make sure the benefits have a higher weight than the risks (Azria et al. 2024). Studies on the human microbiome are full of unresolved clinical, microbiological, and social issues, making it challenging, if not impossible, to balance potential risks and benefits. Special care must be provided in the area of the clinical uses of microbial research.

Generally, the boundaries of privacy align with physical borders. They correspond to personal enclosed areas like my body, my home, inside my bedroom, thoughts, and diary. Except for few cases, nobody is allowed to come into my privacy without my consent, and government intrusions necessitate strong justification. Confidentiality and privacy differ from one another (Rhodes 2016). Confidentiality is a crucial professional duty for healthcare providers and certain other professionals such as accountants, lawyers, and clerics. In such professions, an artificial area is established where information is protected. In this area, information shared by the individuals seeking professional services can be disclosed to others, if necessary, to advance the client's interests. Beyond this area, the shared information can only be revealed if the client agrees. Therefore, in medicine, patients anticipate that their medical history, prognosis, and diagnosis will be shared solely among the clinicians who require it for care provision (Mantel and Fowler 2020).

Additionally, patients have a reasonable expectation that their information will not be revealed. It is essential for medical practitioners to ensure confidentiality, as it enables patients to disclose their information comfortably. While the terms "privacy" and "confidentiality" are often used interchangeably and their differences overlooked, distinguishing between them may be crucial in human microbiome research (Shirazi and Shekhani 2021).

When we investigate confidentiality and privacy from the standpoint of microbiome research, we must be careful what information requires protection and why. Many debates on research ethics strongly advocate for handling research specimens with privacy care; however, this question is raised if stringent restrictions on sharing the samples are logical. Gaining more knowledge on the human microbiome will cause further privacy concerns with significant consequences for personal freedom. Microbiome constitutes about 95% of my feces, and my microbiome narrates my experiences and the people I have communicated with. Are the law enforcement officials permitted to have the same access to my microbiome as they do to my fingerprint records? (Rhodes 2016; Cernava et al. 2022).

3.1 Case Study: From Fecal Microbiota Transplantation to Vaginal Seeding Fecal Microbiota Transplantation (FMT)

FMT is known as the most efficient therapy for recurrent *Clostridium difficile* infection (rCDI) and is being investigated for a variety of other disorders outside the gastrointestinal system. The literature suggests that FMT is able to lower the reliance on antibiotics, thus reducing economic costs related to conventional antibiotic treatment. On the other hand, there are concerns about the potential transmission of

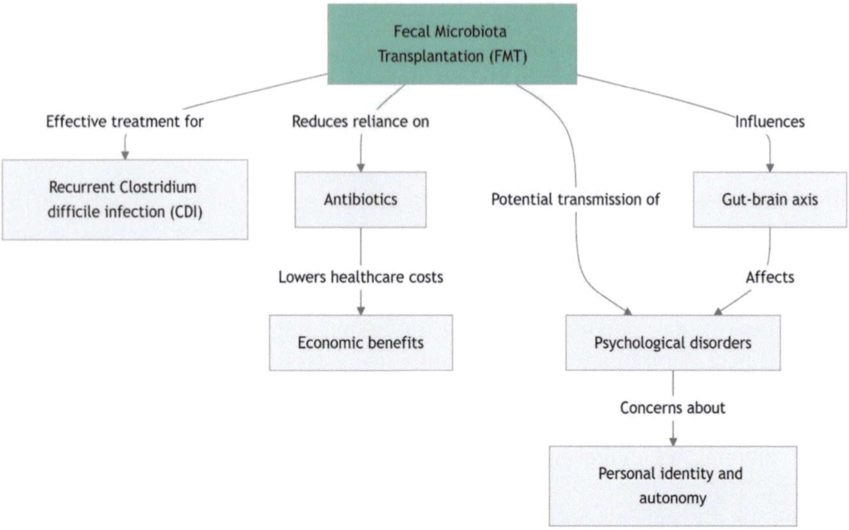

Fig. 2 Interconnections of fecal microbiota transplantation (FMT) and its implications

psychological disorders including anxiety and depression via FMT, which brings up ethical issues associated with personal identity and autonomy (Shi et al. 2017; Sun et al. 2023; Fig. 2).

3.2 *Vaginal Seeding*

Vaginal seeding refers to the transfer of a mother's vaginal fluids to a newborn's eyes, mouth, and skin, especially for infants delivered by cesarean section. The purpose of this practice is to replicate the microbiome exposure that occurs in vaginally-born babies, which potentially decreases the risks of asthma and other immune-related disorders. This method became popular after a research project conducted in 2010, which declared that the seeded infants' microbiomes were more similar to the vaginally-born ones than those delivered through C-section. Vaginal seeding is very dangerous and carries a risk of transmitting pathogens like *C. trachomatis* and herpes simplex virus, associated with neonatal infection (Anthoulaki et al. 2023).

3.3 *Ethical Considerations*

Both vaginal seeding and FMT have some safely and ethical COLONY which should be proactively addressed. The American College of Obstetricians and Gynecologists views vaginal implants as "unnecessary" and even "extremely hazardous," and their use is recommended only within confirmed research protocols until more safety data are available (Grigoryan et al. 2020).

4 Property

All the following undermentioned various insights consider the term "property" only as a philosophical discussion on the legal, ethical, and social problems revolving around the aspect of ownership.

These debates also encompass deposits in sample banks and biobanks (Gründer 2021; O'Keefe et al. 2020). Does human being possess property rights over their microbiome as their bodies host it? In specific conditions, the government might require interventions in (e.g., vaccination) or applications of (e.g., mandatory military service) our body. Does the state have the right to require us to dedicate our microbiome or genome samples to research projects or public health initiatives? If the findings of a microbiome research project are commercially valuable, who is entitled to the profits? If an individual's microbiome is distinctive and has commercial value, who has ownership of it? While there have not been any regulations related to the human microbiome thus far, we have seen a number of legal decisions regarding the genetic research and biological samples associated with the human microbiome (Rhodes 2016).

5 Public Health

Throughout history, diseases caused by microorganisms have posed major risks to public health. Protection and promotion of public health necessitates disease monitoring, tracking, and data gathering on disease epidemics and mortalities. Research on the human microbiome has changed our conventional understanding and societal practices from conceiving that all bacteria are dangerous and must be destructed to finding out that most of the bacteria residing in our body are beneficial for us rather than harmful. Moreover, the excessive use of antimicrobials can in fact be detrimental to our health, with potentially lasting effects. Research has shown that the adaptation of the human gut microbiome to a modern diet, high in dairy products and simple sugars, could have a dramatic effect on the expansion and persistence of global obesity epidemic (Mitra et al. 2019; Stapleton et al. 2024). This change in perspective underscores the necessity for a balanced insight into diet and hygiene, encouraging the incorporation of varied whole foods that can promote a healthy microbiome and potentially alleviate health problems associated with obesity.

Human microbiomes do not function independently, but rather are continuously evolving and affected by environmental factors including diet, consumption of antibiotics, and the microbiomes belonging to the family members and other people (Salerno et al. 2023). Based on discussion regarding public health and microbiome research ethical implications, we focus on three major points: first, as previously mentioned by Salerno, the technologies of microbiome aiming to impact individuals, may produce notable and unpredicted effects on families, communities, and the entire public health (Salerno et al. 2023). We must recognize that individuals can serve as disease potential victims and carriers (DeWane and Bercovitch 2021). There is growing evidence that microorganisms are capable of being transmitted and exchanged among people who are near each other, including members of a family and even participants in the same sport activities (Hale 2021; Yaji et al. 2024). The potential of microbial transmission to continuously influence microbial profiles (particularly in children) results in significant autonomy implications. This implies that people's microbiomes could change due to the health choices of others, potentially in contrast to their will (or even without informing them) in ways that can be harmful to them (Salerno et al. 2023).

Second, it is essential to pay attention to the danger of microbiome cross-contamination among diverse species. In their book entitled "Beasts of the Earth: Animals, Humans and Disease," Torrey and Yolken highlighted the unplanned outcome of the decision of a poultry company to enhance efficacy and price by making a change in chickens' microbiomes to remove a risk to their health. Unintentionally, this created a risk for human consumers (Canavan 2023; Al-Hwsali et al. 2023). This example serves as an apparent illustration of the complicated interactions between the microbiomes of various species, and how attempts to generate profit can lead to unforeseen outcomes (Al-Hwsali et al. 2023).

Third, from a public health viewpoint, microbiome-based interventions could be developed to improve and support community and public health. For instance, if a specific probiotic strain is identified to eliminate *H. pylori* (a bacterium that can cause stomach cancer), introducing this probiotic strain into the food supply or water system could potentially lower the incidence and suffering related to stomach cancer (Kononova et al. 2022; Bai et al. 2023). At the same time, there is also evidence suggesting that the reduction in *H. pylori* might be linked to the increase in esophagus cancer, and the individuals without this bacterium are more possible to suffer from asthma, childhood skin allergies, and hay fever (Chen et al. 2022). This example emphasizes the intricacy of the role and function of bacteria; they can be both detrimental and beneficial. The interventions of public health typically restrict the freedom and decision making of people. Understanding how public health issues are able to justify the restrictions of individual freedom, how information obtained from microbiome research could be employed to develop effective policies of public health, and how to assess the potential benefits to the public against possible expenses are all crucial issues to address prior to implementing public policies. In addition, public interventions that influence newborns' microbiome, children, and other susceptible communities, must take a precautionary approach by thoroughly examining any unintended long-term impacts. Moreover, research on the human microbiome presents opportunities to tackle the global issues pertaining to infectious diseases and antibiotic resistance. For instance, the literature has indicated that consumption of antibiotics might be circumvented if FMT were applied as an initial treatment for recurrent CDI instead of being regarded as the last resort (Shi et al. 2017), and this decreases economic expenses in comparison to standard antibiotic treatment (Parnell et al. 2021; Gweon et al. 2015). Edge claimed that gaining "a deeper knowledge about the mechanisms behind colonization resistance and other interactions between microbiomes and pathogens could disclose new approaches for the treatment or prevention of infections" (Edge et al. 2020). Tosh and McDonald proposed that the programs of infection inhibition should be restructured to take advantage of the increasing knowledge about the microbiome protective functions and to retain and restore a balanced indigenous microbiome (Tosh and McDonald 2012). An adequate policy of public health requires a comprehensive evaluation of the harms, risks, and benefits that impact the whole society.

Comparable challenges emerge in public health and the research on it (Perlman 2008). Since microbiome has an effect on several infectious diseases, safeguarding the public health necessitates disease monitoring. Gathering data on the outbreaks, births, and deaths of different diseases is vital for the development and execution of efficient disease inhibition, outbreak management, and disaster response strategies. The efficient actions of public health rely on information; nevertheless, collecting individual's information during an emergency might not permit a pre-assessment by an Institutional Review Board or the opportunity for informed consent (Wan and Wang 2022).

Most clinical research studies subject merely their own participants to harms and risks. However, since research projects on the human microbiome result in the investigation of phage (viral), probiotic, and prebiotic (bacterial) treatments, the risks to both research and public participants are important factors to consider. When we consider modifying the human microbiome by altering naturally-occurring groups of microorganisms, we should be aware that individuals can be both carriers and victims of disease (Yuliana et al. 2021). Overall, clinical research into the human microbiome poses a dual risk: first, not only to individuals participating but also to the greater public health. Interactive voice response (IVR) As our understanding and utilization of probiotic, prebiotic, and phage therapies continue to rise, we must consider what implications an altered microbiome might have not only for the individual research subjects but also for humanity as a whole. Recognition of people both as vectors of disease and as victims of infection underlines the complexity of such interventions and invokes for rigorous risk assessments. Future studies should focus more on ethical implications and those regarding public health associated with microbiome manipulation to ensure safety and well-being of research participants and the entire population as a whole (Segura Benedicto 2021).

6 Biobanks

6.1 *Overview of Microbiome Biobanks*

As the research on human microbiome progresses, it is unavoidable that numerous biobanks comprise a series of microbiota samples necessary for genomic research. These biobanks containing human microbiota and the corresponding genetic data have the potential to be a precious health resource. Just like the biobanks that retain the tissue samples of humans, such as the UK Biobank, the gathered samples and the relevant clinical data stored in the human microbiome biobank differ according to the objectives and scope of the biobank, in addition to the characteristic of the research being carried out (Shi et al. 2017). For instance, the "American Gut" project is an open-access, community-based initiative aimed at mapping the microbial diversity of the American society and investigating how lifestyle and diet could influence health via every individual's gut microbiome. Disease-specific biobanks, such as the cystic fibrosis one in Canada, are aimed at making the sharing of data and samples easier for research focused on understanding the relationship between microbial dynamics and disease progression in the lungs of both adult and pediatric patients (Editorial Board 2016). The Multi-Omic Microbiome Study-Pregnancy Initiative (MOMS-PI) is a collaborative project conducted at Settle Children's and funded by the National Institutes of Health (NIH). The purpose of this project is to elucidate the effects of the vaginal microbiome on pregnancy, pregnancy-based disorders, as well the establishment of the infant microbiome. By applying the "multi-omics" approaches, the MOMS-PI project aims to gather samples from a cohort of 2000 pregnant women during their pregnancy period to investigate how the microbiome influences preterm birth risk and to elucidate the temporal alterations in the pregnancy microbiome over time.

6.2 *Stool Banks*

Among the different types of biobanks, "stool banks" have emerged as both a controversial and promising option in recent years. They have demonstrated remarkable potential in curing patients suffering from intestinal disorders, in particular those with CDI (Van Nood et al. 2013; Boyle et al. 2015). Stool banks have also played a role in the progression of translational research on the human gut microbiome (Jamshidi et al. 2023). The U.S. Food and Drug Administration (FDA) defines a stool bank as "an establishment that gathers, processes, and maintains FMT products for distribution to other medical facilities, healthcare professionals, or other organizations for therapeutic purposes or clinical research." The advantages of stool banks are clear, as shown by our latest survey on gastroenterologists in China, which indicated most of the participants in the survey strongly supported the development of a stool bank. They stated that it "not only alleviates the burden of handling stool samples, but also mitigates privacy concerns" (Shi et al. 2017). To date, an increasing number of stool banks have been founded throughout the world, with OpenBiome being the first to establish in Massachusetts, the United States, in 2012, which has supplied stool for 13,000 FMT across the United States and six other countries. Furthermore, other stool banks exist which have been launched by various research organizations and clinics, including the Taymount Clinic in the United Kingdom, Advancing Bio in the United States, Melbourne FMT in Australia, the Netherlands Donor Faeces Bank (NDFB), the Chinese FMT Bank in China, and the Asian Microbiota Bank in Hong Kong. Basically, such banks are necessary for medical research. Furthermore, the insights accomplished from such banks, along with the information from patients and donors, contribute to the optimization of the therapeutic applications of stool samples. However, utilization of highly efficient methods will also allow for the recognition and analysis of all the microorganisms linked to the human gut along with the host genetics. In addition, it aids in finding the way in which these factors help to alterations in the gut microbial ecosystem (Shi et al. 2017; Rackerby et al. 2020).

It is too simple to understand that everyone likes to prevent diseases and their associated burdens, and when prevention is not possible, each individual seeks effective and efficient ways to address them. The availability of such treatments will need considerable scientific progress which requires translational research and fundamental science encompassing biobanking, sample storage, and research on human subject. As an example, improvement in cancer therapy through individualized medicine needs research on samples obtained from diverse groups of patients suffering from various types of cancer (Asgiani et al. 2021; McCarty et al. 2005). The growing demand for additional research and broader collaboration in this field brings up concerns regarding the regulations and management of sample banks and biobanks. Who can have access to the samples; what data can be disclosed; in what projects can the samples be used; and who is liable for making these decisions?

Although both FMT and vaginal seeding present promising opportunities for health interventions and microbiome research, they also raise considerable safety, ethical, and regulatory concerns which need to be accurately addressed. Some of them are outlined below.

7 Investigational New Drug (IND) Application Process

In order to carry out human clinical trials involving a biological or experimental drug in the United States, it is essential to submit an IND application. For the application of FMT, a researcher or clinician has the option to submit one of the INDs below (Wu et al. 2020):

(a) Patients who do not meet the criteria of an approved research protocol or for whom there is no authorized research protocol could qualify to file an emergency IND if the clinical conditions do not allow for the submission of an IND.

(b) Experimental medications which are potent for treating severe diseases in clinical research could be submitted for expanded access through a therapeutic IND. This option must be utilized by physicians and scientists intending to treat several participants suffering from rCDI.
(c) FMT submitted as a research IND may be provided or distributed under the direct surveillance of researchers (with or without their assistants). This approach should be applied if the data collected from patients involved in the protocol are going to be utilized for further research (Thanush et al. 2023).

8 Current FMT Regulatory Environment

The regulatory landscape for fecal microbiota transplantation (FMT) varies considerably across countries and regions, reflecting diverse approaches to the classification and management of this emerging therapeutic intervention. This section provides a comparative overview of these regulatory frameworks and shows how each jurisdiction defines, restricts, or permits the use of FMT. This comparative analysis highlights the challenges and complexities of standardizing FMT practices globally, given the unique ethical, safety, and clinical considerations of each regulatory body.

(a) United States The Food and Drug Administration (FDA) classifies FMT as an unapproved biological medication. Although this is typically the most restrictive designation, the FDA has considered an exception for the application of FMT in the treatment of rCDI provided that the doctor obtains informed consent. This distinctive enforcement discretion policy highlights the challenges faced by regulatory agencies in balancing classification with patient access. The first randomized controlled trial by Van Nood et al. (2013), published in the *New England Journal of Medicine*, demonstrated that fecal microbiota transplantation (FMT) is significantly more effective than vancomycin for treating recurrent Clostridium difficile infection, highlighting the need for robust ethical considerations in donor selection and safety protocols. In May 2013, the FDA organized a public workshop to discuss the scientific and regulatory issues of FMT. In this workshop, the FDA categorized FMT as an unapproved biological medication. It is essential for researchers to submit an IND application prior to administering FMT to patients for any purpose, and the treatment may solely be provided within the context of clinical trials or under emergency conditions (Gupta et al. 2021). The IND requirement raised concerns among doctors, organizations of patient advocacy, medical communities, and researchers who believed that an alternative regulatory approach was necessary for patients suffering from rCDI. In response, the FDA announced a discretionary enforcement approach in July 2013, underscoring that this is a temporary approach while they continue to further examine the issue. As years passed, the FDA has revised its FMT rules through some draft guidelines (Verbeke et al. 2017) which were made available for public comment even though none of them has been implemented. The FDA suggested modifying the enforcement discretion policy in March 2014, stipulating that both doctors and patients should be acquainted with the stool donor and the whole screening procedures should be monitored by the doctor directing the FMT. By performing this plan, hospitals are permitted to establish their own stool banks while family and friends could directly contribute to independent public stool banks such as OpenBiome in Cambridge, MA (Scheeler 2019). A coalition of medical professionals and patient activists expressed concerns regarding the potential effects of the suggested policy on patient access. In March 2016, the FDA published a new version of the draft guidelines for public comment. Based on its draft guideline, the FDA advised health professionals to utilize public stool banks under an IND for the treatment of rCDI. Since the period for public comment on the proposal ended in May 2016, no draft or final guidelines have been published. Clinicians and healthcare advocates believe that the FDA had better create a balance between accessibility and patient safety by re-classifying FMT as a Human Cell, Tissue, or Cellular- or Tissue-Based Product (HCT/P). Although the HCT/P framework regulates various therapeutic applications in different ways, the classification of biological therapeutic products necessitates a complex exemption for access to patients with rCDI (Thanush et al. 2023).

(b) Canada Health Canada mandates the submission of clinical trial applications (CTAs) for investigating FMT, as it is classified as a state-of-the-art experimental biological medicine. An interim policy has been established by Health Canada that permits FMT for patients with rCDI under certain circumstances. The application criteria of Health Canada are much more strict and limiting compared to those in the United States. Health Canada requires doctors to receive informed consent, sets criteria for recordkeeping, and prepares a list of microbiome-related and infectious diseases for which donors have to test negative. Moreover, either the clinician or the patient involved in the patient's care should be familiar with the donor. If these criteria are not met, patients with rCDI can undergo FMT merely under a CTA (Scheeler 2019; Thanush et al. 2023).

(c) European Union The EMA has transferred the responsibility of decision making to its member states. The Competent Authorities on Substances of Human Origin Expert draw the conclusion that feces do not meet the criteria for being included in the European Human Tissue Directive 2004/23/EC and cannot be classified as HCT/P at the European-level group of 2012 (Baunwall et al. 2021). Similar to that in the United States, it was founded on a stringent interpretation of the directive provisions. While FMT consists of both bacterial and human cells, it falls outside the directive's scope, because the active ingredient is conceived to be bacterial. As a result, the member countries can regulate FMT in any manner that they find it appropriate.

(d) Belgium Stool products used in FMT are regarded as human body components, comparable to human tissues or cells. The FAMHP adopted the SHC's suggestions in October 2018 following the updates in law made by the Belgian legislators in

December 2008, that permitted stool to be categorized as a HCT/P (Scheeler 2019; Thanush et al. 2023).

(e) France FMT has been classified as an experimental drug by the Agency National de Securite du Medicament et des Produits (ANSM). A pharmacy or hospital which produces FMT could utilize it in an authorized clinical experiment or when it is produced in accordance with Article L.5121-1 of the French Code of Public Health. FMT must be regarded as the last option, applied to severe or rare situations where usual treatments are not successful, and no alternative treatments are available (Scheeler 2019; Thanush et al. 2023).

(f) United Kingdom Medications and healthcare products are monitored by the Medicines and Healthcare Products Regulation Agency (MHRA). Several options available for patients to access FMT have been outlined in the position paper of MHRA since June 2015. Pharmacies can prepare FMT for patients in three ways: based on a prescription for a special patient (referred to as the "magistral option"), for the patients of the pharmacy (known as the "officinal formula"), or as a part of clinical trials. In line with the "Specials" framework, a doctor is allowed to prescribe FMT for a certain patient to be applied under direct supervision, according to their needs (Scheeler 2019; Thanush et al. 2023).

(g) Australia Patients are able to access FMT in four ways, which is presently classified as an unauthorized medical product by the Therapeutic Goods Administration (TGA). If their prescribing doctor considers it essential in specific rare instances, patients are allowed to receive FMT via the Special Access Scheme. The TGA's Authorized Prescriber Scheme enables doctors curing a particular category of patients to request blanket approval to administer FMT to the whole patients for a certain treatment indication. Patients wishing to import therapeutic products for personal uses or take part in a clinical research project involving FMT must obtain permission from the TGA which is considering alterations in the regulations surrounding FMT. In October 2018, stakeholders convened to explore the potential for the safe and therapeutic modification of the FMT rules while ensuring its long-term availability (Thanush et al. 2023). All of the above are summarized for better understanding in Table 1.

9 Challenges in Development and Commercialization

While great potential exists with microbiome-based therapies, their development and commercialization bear unique challenges. Intellectual property, clinical trial design, and regulatory frameworks are intricately connected with defining the feasibility and rapidity of innovation within this new field.

Table 1 Comparative regulatory approaches to fecal microbiota transplantation (FMT) across regions

n	Region	Regulatory body	Classification	Access conditions	Sources
1	United States	FDA	Unapproved biological medication	Requires IND application; discretionary enforcement for rCDI; calls for reclassification as HCT/P	Verbeke et al. (2017), Scheeler (2019), Thanush et al. (2023)
2	Canada	Health Canada	Experimental biological medicine	Requires CTAs; informed consent; strict donor screening criteria	Scheeler (2019), Thanush et al. (2023)
3	European Union	EMA	N/A	Member states regulate independently; does not meet HCT/P criteria	Baunwall et al. (2021)
4	Belgium	FAMHP	HCT/P	Stool products classified as human body components	Scheeler (2019), Thanush et al. (2023)
5	France	ANSM	Experimental drug	Used as a last resort in severe cases; must comply with public health regulations	Scheeler (2019), Thanush et al. (2023)
6	United Kingdom	MHRA	N/A	Access through prescriptions, officinal formulas, or clinical trials	Scheeler (2019), Thanush et al. (2023)
7	Australia	TGA	Unauthorized medical product	Access via Special Access Scheme, Authorized Prescriber Scheme, or clinical trials	Thanush et al. (2023)

N/A not applicable

(a) The Intellectual Property Paradox Absence of intellectual property or limited patent rights might be a disincentive to investment in the research and study of microbiomes. To illustrate, patent laws exclude from patentability living body parts or materials occurring naturally (Sabatelli et al. 2017).

(b) Considerations for Trial Design and Clinical Development Product development related to the microbiome is a very complex task, insofar as an individual's microbiome is highly variable in its disrupted state, and the exact composition of bacteria clinically administered, as in the case of FMT, cannot be clinically predicted. Many of the factors connected with these studies are hard to control. Moreover, adequate patient selection and accurate patient selection in clinical trials can prove highly problematic.

(c) Regulatory Frameworks Regulatory systems have not kept pace with microbiome treatments, and it has remained unclear how these treatments will be evaluated, although the FDA has issued draft guidance and signaled its openness to consultation with industry (Belizário et al. 2018). It is to be seen if this will result in the establishment of a new regulatory direction for the microbiome-based therapies inside the regulatory bodies. The FDA has granted orphan drug status for rCDI to both agents in Phase III. The manufacturers will most likely sell the agents at a premium price in order to recoup their investment in development. Microbiome-based therapies, and in particular those for CDI, do hold promise for areas of high unmet need and of great public health interest. It is further obvious that FDA is quite anxious to work with the industry in commercializing FMTs. Whatever the result may be, microbiome-based therapies for CDI will lead the way to additional indications, bringing a new class of drugs and approaches to therapy to the clinic (Stallhofer et al. 2024).

10 Guidelines on Ethics in In Vivo Microbiota Research

Ethics considerations and guidelines by context: In-vivo microbiota research involves looking into microbial communities in living organisms through animal models very frequently. Below are some ethical considerations and guidelines according to the given contexts and their summary is given in Table 2.

10.1 Models Under Germ-Free Conditions

It normally employs germ-free animal models as a way to study microbiota interactions without the introduction of outside microorganisms. This consideration provides the possibility of more clearly grasping the basic function of microbiota in health and disease. The ethical considerations entail treating such models humanely and ensuring that their application should be justified in terms of benefits that may accrue from the studies (Foster and Clarke 2024; Aghighi and Salami 2023).

10.2 Minimizing Animal Suffering

Ethical guidelines underscore the necessity of reducing distress and pain in animal studies. This involves application of the least invasive techniques available and making sure that all procedures conducted on the animals are essential for accomplishing the research goals. The contexts demonstrate that in-vivo research should be structured to reach the highest scientific output while lowering harm (Karas and Silverman 2007).

10.3 Transplantation Procedures

In research projects involving FMT, such as transplanting human feces into germ-free mice, the donors' permission and the origin of the fecal material should be included in the ethical considerations. The possible benefits and risks to the animal subjects should be accurately assessed (Nooij and Terveer 2022).

10.4 Conformity to Regulations

Animal participant studies should employ necessary standards set both locally and internationally, dealing with animal welfare. This includes the obtaining of requisite permissions from ethics committees or boards of institutional review before conducting such research (Olsson et al. 2022).

Table 2 Ethical guidelines for in vivo microbiota research

Ethical guideline	Description	Sources
Germ-free models	In the presented work, use germ-free animal models to study the interactions between microbiota without any external microbial influence. The animals must be treated humanely, in line with the justification of research benefits.	Aghighi and Salami (2023)
Minimize animal suffering	Apply the 3Rs: Replacement, Reduction, and Refinement in order to reduce pain and distress, whenever similar nông are conducted on animals. Application of least invasive techniques should be employed.	Karas and Silverman (2007)
Transplantation procedures	For FMT, consent from the donor is needed and risks to animal subjects must be weighed against benefits.	Nooij and Terveer (2022)
Regulatory compliance	Observe local and international regulations concerning animal welfare and permissions from ethics committees.	Olsson et al. (2022)
Future innovations	Address the emergence of ethical challenges concerning the manipulation of microbiota, especially for human applications.	Ma et al. (2018)
In vitro alternatives	In addition, in vitro fermentation models allow for the avoidance of ethical considerations involved with live animal testing.	Gao et al. (2023)
Precision medicine	Informed consent regarding gut microbiota profiling should also be provided, bearing in mind potential discrimination based on microbiota data.	Ahmed and Hens (2022)
Protocol standardization	Develop standardized protocols for microbiota research to ensure safety and efficacy.	Perrot and Horn (2022)
Transparency in research	Ensure transparency in research methodologies and findings to inspire trust through ethical integrity.	Ahmed and Hens (2022)
Community engagement	Engage the community in giving their views, particularly in research design studies which affect public health.	Perrot and Horn (2022)

10.5 Future Innovations and Ethical Considerations

Treatments using microbiota are developing, and ethical discussions shall be developed in tandem to address the emerging challenges, especially with regard to human applications and outcomes of the manipulation of microbiota. In summary, Research Oversight Norms (RONs) that pertain to ethical issues in in-vivo microbiota research emphasize humane approaches to animal subjects, adherence to regulations, and transparency in research studies, thus enabling scientific knowledge development with particular respect for animal welfare (Ma et al. 2018).

10.6 Simulation of Models of Fermentation

As Gao et al. stated, 2023, in-vitro simulated fermentation of intestinal microbiota can be conducted without the ethical restrictions involved in the mechanism of animal models. In this way, it provides a chance for researchers to test the compatibility of the microbiota easily without the involvement of ethical concerns using live animals, and it is more acceptable for bioethical points of view for preliminary tests. Gao et al. (2023) discussed the development of a variety of problems in microbiota precision medicine, including ethics consideration.

10.6.1 Precision Medicine and Individual Differences

Thus, the introduction provides an overview of personalized therapies with a focus on precision medicine. It seeks to ensure minimal adverse responses while ensuring high effectiveness resulting from individual differences in response to ingested food and medications based on gut microbiota variations. But these raise a host of ethical concerns, like the consent from the individuals regarding their gut microbiota profile and possible discrimination upon the basis of one's gut microbiota profile, which needed more detailed discussion (Ahmed and Hens 2022).

10.6.2 Absence of Established Protocols

Complexity of microbiota communities, together with the lack of harmonized procedures for their precise control, is indeed another major problem. This uncertainty raises ethical questions concerning the efficiency and safety of microbiota-based treatments, besides liability issues from scientists who must make sure their findings do not contribute to deleterious practices or misunderstanding within a clinical setting (Perrot and Horn 2022).

11 Future Direction for Research

The conclusion this study reached was that, although the field is particularly promising, the methods developed so far are not sufficient to allow the development of medication and food targeting microbiota. The above places a premium on the need for ethical frameworks guiding further research to ensure that progress in microbiota precision medicine occurs with responsibility vis-à-vis the potential effects it may exert on society. In the final analysis, though this chapter has shed light on the ethical landscape of microbiota precision medicine, deep perception requires further research—particularly with a focus on guidelines and frameworks in this new area.

12 Conclusion

While there is considerable potential for the rapidly growing area of microbiome research to improve personalized medicine, this area also raises many regulatory, ethical, and scientific challenges that will need to be pondered carefully. The complex interaction of a person's distinctive microbiome with so many different health outcomes signals a deep level of detail in understanding microbial interaction and how they influence immune function, drug metabolism, and disease susceptibility. In moving forward within this arena, consideration of regulatory frameworks and ethical issues by the biomedical community will be essential to the full expression of the potential of treatments targeting the microbiome. The way forward will demand interdisciplinary collaboration on the design and implementation of treatments involving the microbiome, based on valid scientific scholarship and moral integrity. Addressing such challenges will ultimately enhance the microbiome interventions, not only in terms of safety and efficiency but will also open ways to new methodology of intervention, which may revolutionize healthcare and improve patient outcomes by bringing a complete change in our perspective towards human health.

CRediT Authorship Contribution Statement

Author Contribution:

- *Samaneh Hatami* was involved in writing—review, editing, supervision, resources, project administration, designed pictures, investigation, and conceptualization.

Seyed Ali Issazadeh designed pictures and contributed in writing—review and editing, writing original draft, and resources.

Declaration of Competing Interest The authors declare that they have no known competing financial interests or personal relationships that could have appeared to influence the work reported in this paper.

References

Afzaal M, Saeed F, Shah YA, Hussain M, Rabail R, Socol CT, Hassoun A, Pateiro M, Lorenzo JM, Rusu AV, Aadil RM (2022) Human gut microbiota in health and disease: unveiling the relationship. Front Microbiol 13:999001. https://doi.org/10.3389/fmicb.2022.999001

Aghighi F, Salami M (2023) Manipulating gut microbiota in germ-free and antibiotic-treated animal models in research with therapeutic goals: advantages and disadvantages, pp 83–107. https://doi.org/10.22541/au.169956180.01909034/v1

Ahmed E, Hens K (2022) Microbiome in precision psychiatry: an overview of the ethical challenges regarding microbiome big data and microbiome-based interventions. AJOB Neurosci 13(4):270–286

Al-Hwsali A, Alsaadi B, Abdi N, Khatab S, Alzubaidi M, Solaiman B, Househ M (2023) Scoping review: legal and ethical principles of artificial intelligence in public health. In: Studies in health technology and informatics. IOS Press

Anthoulaki X, Oikonomou E, Bothou A, Papanikolopoulou S, Nikolettos K, Damaskos C et al (2023) Comparison of gut microbiome in neonates born by caesarean section and vaginal seeding with gut microbiomes of neonates born by caesarean section without vaginal seeding and neonates born by vaginal delivery. Mater Sociomed 35(3):234

Ascienzo S, Sprang G, Royse D (2022) "My bad experiences are not the only things shaping me anymore": thematic analysis of youth trauma narratives. J Child Adolesc Trauma 15(3):741–753. https://doi.org/10.1007/s40653-021-00431-4

Asgiani P, Suryawati C, Agushybana F (2021) A literature review: security aspects in the implementation of electronic medical records in hospitals. Media Ilmu Kesehatan 10(2):161–166. https://doi.org/10.30989/mik.v10i2.561

Azria E, Haaser T, Schmitz T, Froeliger A, Bouchghoul H, Madar H, Pineles BL, Sentilhes L (2024) The ethics of induction of labor at 39 weeks in low-risk nulliparas in research and clinical practice. Am J Obstet Gynecol 230(3S):S775–S782. https://doi.org/10.1016/j.ajog.2023.07.037

Bai X-F, Tian D, Wang T-Y, Shu J-C, He Y-J, Zhu M-J (2023) The impact of probiotics on gut microbiota in the eradication of Helicobacter pylori infection: a systematic review. Eur Rev Med Pharmacol Sci 27(14):6736–6743. https://doi.org/10.26355/eurrev_202307_33144

Baunwall SMD, Terveer EM, Dahlerup JF, Erikstrup C, Arkkila P, Vehreschild MJ, Ianiro G, Gasbarrini A, Sokol H, Kump PK, Satokari R, De Looze D, Vermeire S, Nakov R, Brezina J, Helms M, Kjeldsen J, Rode AA, Kousgaard SJ et al (2021) The use of faecal microbiota transplantation (FMT) in Europe: a Europe-wide survey. Lancet Reg Health Eur 9:100181. https://doi.org/10.1016/j.lanepe.2021.100181

Belizário JE, Faintuch J, Garay-Malpartida M (2018) Gut microbiome dysbiosis and immunometabolism: new frontiers for treatment of metabolic diseases. Mediat Inflamm 2018(1):2037838

Boyle ML, Ruth-Sahd LA, Zhou Z (2015) Fecal microbiota transplant to treat recurrent Clostridium difficile infections. Crit Care Nurse 35(2):51–64. https://doi.org/10.4037/ccn2015356

Canavan B (2023) Historical literature related to zoonoses and pandemics. Isis 114(S1):S104–S142

Carabotti M, Scirocco A, Maselli MA, Severi C (2015) The gut-brain axis: interactions between enteric microbiota, central and enteric nervous systems. Ann Gastroentol 28(2):203

Cernava T, Rybakova D, Buscot F, Clavel T, McHardy AC, Meyer F et al (2022) Metadata harmonization–standards are the key for a better usage of omics data for integrative microbiome analysis. Environ Microbiome 17(1):33

Chen J, Li Y, Tian Y, Huang C, Li D, Zhong Q, Ma X (2015) Interaction between microbes and host intestinal health: modulation by dietary nutrients and gut-brain-endocrine-immune axis. Curr Protein Pept Sci 16(7):592–603

Chen Y, Zhan X, Wang D (2022) Association between Helicobacter pylori and risk of childhood asthma: a meta-analysis of 18 observational studies. J Asthma 59(5):890–900

Conlon MA, Bird AR (2014) The impact of diet and lifestyle on gut microbiota and human health. Nutrients 7(1):17–44

Dauda B, Molina SJ, Allen DS, Fuentes A, Ghosh N, Mauro M, Neale BM, Panofsky A, Sohail M, Zhang SR, Lewis ACF (2023) Ancestry: how researchers use it and what they mean by it. Front Genet 14:1044555. https://doi.org/10.3389/fgene.2023.1044555

DeWane ME, Bercovitch L (2021) The ethical responsibilities of physicians during epidemics. In: Dermatoethics. Springer International Publishing, pp 267–275

Dominguez-Bello MG, Blaser MJ (2011) The human microbiota as a marker for migrations of individuals and populations. Annu Rev Anthropol 40(1):451–474

Edge TA, Baird DJ, Bilodeau G, Gagné N, Greer C, Konkin D et al (2020) The Ecobiomics project: advancing metagenomics assessment of soil health and freshwater quality in Canada. Sci Total Environ 710:135906

Editorial Board (2016) Antimicrob Agents Chemother 60(1). https://doi.org/10.1128/aac.masthead.60-1

Foster J, Clarke G (2024) Chapter 5 – Microbiota influence behavior—work in animal models. In: Foster J, Clarke G (eds) Microbiota brain axis. Academic Press

Freytag S, Lister R (2020) schex avoids overplotting for large single-cell RNA-sequencing datasets. Bioinf (Oxf) 36(7):2291–2292. https://doi.org/10.1093/bioinformatics/btz907

Gao X, Zhao J, Chen W, Zhai Q (2023) Food and drug design for gut microbiota-directed regulation: current experimental landscape and future innovation. Pharmacol Res 194:106867

Gordon JI, Ley RE, Wilson R, Mardis E, Xu J, Fraser CM, Relman DA (2005) Extending our view of self: the human gut microbiome initiative (HGMI). National Human Genome Research Institute

Grigoryan Z, Shen MJ, Twardus SW, Beuttler MM, Chen LA, Bateman-House A (2020) Fecal microbiota transplantation: uses, questions, and ethics. Med Microecol 6:100027

Gründer G (2021) Psychedelics: a new treatment paradigm in psychiatry? Pharmacopsychiatry 54(4):149–150. https://doi.org/10.1055/a-1520-5020

Gupta S, Mullish BH, Allegretti JR (2021) Fecal microbiota transplantation: the evolving risk landscape. Am J Gastroenterol 116(4):647–656. https://doi.org/10.14309/ajg.0000000000001075

Gweon T-G, Lee KJ, Kang D, Park SS, Kim KH, Seong H, Ban T-H, Moon SJ, Kim JS, Kim SW (2015) A case of toxic megacolon caused by Clostridium difficile infection and treated with fecal microbiota transplantation. Gut Liver 9(2):247–250. https://doi.org/10.5009/gnl14152

Hale VL (2021) mSphere of influence: drivers of host-associated microbial community structure and change. mSphere 6(2):e00248-21. https://doi.org/10.1128/msphere.00248-21

Honey K (2008) Good bugs, bad bugs: learning what we can from the microorganisms that colonize our bodies. J Clin Invest 118(12):3817–3817

Jamshidi P, Farsi Y, Nariman Z, Hatamnejad MR, Mohammadzadeh B, Akbarialiabad H et al (2023) Fecal microbiota transplantation in irritable bowel syndrome: a systematic review and meta-analysis of randomized controlled trials. Int J Mol Sci 24(19):14562

Karas A, Silverman J (2007) Pain and distress. IACUC Handb 2:241–286

Karvonen KL, Menjívar-López JS, Brissett D, McBride D, Olveda R, Cribb Fabersunne C, Low O, Cronin M, Argueza BR (2022) A resident-led initiative to advance diversity, equity, inclusion, and antiracism in a pediatrics residency program. Acad Pediatr 22(3):360–364. https://doi.org/10.1016/j.acap.2021.09.020

Kononova AG, Kolbasnikov SV, Beldiev SN (2022) The role of probiotics in Helicobacter pylori eradication therapy regimens: current evidence of effectiveness. Eksperimental'naia i Klinicheskaia Gastroenterologiia [Exp Clin Gastroenterol] 4:160–168. https://doi.org/10.31146/1682-8658-ecg-200-4-160-168

Ma Y, Chen H, Lan C, Ren J (2018) Help, hope and hype: ethical considerations of human microbiome research and applications. Protein Cell 9(5):404–415

Mantel J, Fowler L (2020) Thinking outside the silos: information sharing in medical–legal partnerships. J Leg Med 40(3–4):369–389. https://doi.org/10.1080/01947648.2020.1854135

McCarty CA, Wilke RA, Giampietro PF, Wesbrook SD, Caldwell MD (2005) Marshfield Clinic Personalized Medicine Research Project (PMRP): design, methods and recruitment for a large population-based biobank. Pers Med 2(1):49–79

Mitra R, Das L, Das SK (2019) Gut microbiomes and their impact on human health. In: Microbial diversity in ecosystem sustainability and biotechnological applications: volume 1. Microbial diversity in normal & extreme environments. Springer, Singapore, pp 355–385

Mörkl S, Wagner-Skacel J, Lahousen T, Lackner S, Holasek SJ, Bengesser SA et al (2020) The role of nutrition and the gut-brain axis in psychiatry: a review of the literature. Neuropsychobiology 79(1):80–88

Nooij S, Terveer EM (2022) Evaluating donor microbiome before fecal microbiota transplantation: reply. Gastroenterology 162(3):994–995

O'Keefe RJ, Tuan RS, Lane NE, Awad HA, Barry F, Bunnell BA, Colnot C, Drake MT, Drissi H, Dyment NA, Fortier LA, Guldberg RE, Kandel R, Little DG, Marshall MF, Mao JJ, Nakamura N, Proffen BL, Rodeo SA et al (2020) American Society for bone and Mineral Research-Orthopaedic Research Society joint Task Force report on cell-based therapies – secondary publication. J Orthop Res 38(3):485–502. https://doi.org/10.1002/jor.24485

Olsson IAS, Nielsen BL, Camerlink I, Pongrácz P, Golledge HD, Chou JY et al (2022) An international perspective on ethics approval in animal behaviour and welfare research. Appl Anim Behav Sci 253:105658

Olteanu G, Ciucă-Pană M-A, Busnatu, Ştefan S, Lupuliasa D, Neacşu SM, Mititelu M, Musuc AM, Ioniță-Mîndrican C-B, Boroghină SC (2024) Unraveling the microbiome–human body axis: a comprehensive examination of therapeutic strategies, interactions and implications. Int J Mol Sci 25(10):5561. https://doi.org/10.3390/ijms25105561

Ostendorf S, Müller SM, Brand M (2020) Neglecting long-term risks: self-disclosure on social media and its relation to individual decision-making tendencies and problematic social-networks-use. Front Psychol 11:543388. https://doi.org/10.3389/fpsyg.2020.543388

Parnell JM, Nicholson MR, Kellermayer R, Kahn SA (2021) Pediatric fecal microbiota transplantation in recurrent Clostridioides difficile. Pediatr Ann 50(12):e515–e521. https://doi.org/10.3928/19382359-20211111-01

Perlman D (2008) Public health practice vs research: implications for preparedness and disaster research review by State Health Department IRBs. Disaster Med Public Health Prep 2(3):185–191

Perrot A, Horn R (2022) Health professionals and scientists' views on genome-wide NIPT in the French public health system: critical analysis of the ethical issues raised by prenatal genomics. PLoS One 17(11):e0277010. https://doi.org/10.1371/journal.pone.0277010

Rackerby B, Kim HJ, Dallas DC, Park SH (2020) Understanding the effects of dietary components on the gut microbiome and human health. Food Sci Biotechnol 29:1463–1474

Rhodes R (2016) Ethical issues in microbiome research and medicine. BMC Med 14(1):156

Ribas MP, García-Ulloa M, Espunyes J, Cabezón O (2023) Improving the assessment of ecosystem and wildlife health: microbiome as an early indicator. Curr Opin Biotechnol 81:102923. https://doi.org/10.1016/j.copbio.2023.102923

Rose T, Dening KH (2023) Who am I? Identity, person-centred care and dementia. Nurs Older People 35(5):22–27. https://doi.org/10.7748/nop.2023.e1446

Sabatelli AD, Vincent NG, Puleo DE (2017) Patenting the microbiome: trends, challenges and insights. Pharm Pat Anal 6(6):273–282

Salerno J, Coughlin SS, Goodman KW, Hlaing WM (2023) Current ethical and social issues in epidemiology. Ann Epidemiol 80:37–42. https://doi.org/10.1016/j.annepidem.2023.02.001

Sanyal S, Nigam K, Singh S, Lohani P, Dwivedi M (2024) Pathophysiological and clinical potential of human microbiome: microbe-based therapeutic insights. Curr Pharm Biotechnol 26:2082–2096. https://doi.org/10.2174/0113892010314433240823113111

Scheeler A (2019) Where stool is a drug: international approaches to regulating the use of fecal microbiota for transplantation. J Law Med Ethics 47(4):524–540

Schneider E, Leigh S-J, Lynch CMK, Hilbert A, Clarke G, Higgs S, Cryan JF (2024) Microbiota-gut-brain axis in binge-eating disorder: towards microbiome-based therapies. Neurosci Appl 3:104088. https://doi.org/10.1016/j.nsa.2024.104088

Seekatz AM (2020) mSphere of influence: translating gut microbiome studies to benefit human health. mSphere 5(4):1–3. https://doi.org/10.1128/msphere.00592-20

Segura Benedicto A (2021) Ética, Salud Pública y COVID-19: algunas consideraciones (elementales) impopulares. Rev Bioética Derecho (51):209–224

Sherrill-Mix S, Connors K, Aldrovandi GM, Brenchley JM, Boucher C, Bushman FD, Collman RG, Dandekar S, Klatt NR, Lagenaur LA, Paredes R, Tachedjian G, Turpin JA, Landay AL, Ghosh M (2020) A summary of the fifth annual virology education HIV microbiome workshop. AIDS Res Hum Retrovir 36(11):886–895. https://doi.org/10.1089/aid.2020.0121

Shi J, Wang Y, He J, Li P, Jin R, Wang K, Xu X, Hao J, Zhang Y, Liu H, Chen X, Wu H, Ge Q (2017) Intestinal microbiota contributes to colonic epithelial changes in simulated microgravity mouse model. FASEB J 31(8):3695–3709. https://doi.org/10.1096/fj.201700034r

Shirazi B, Shekhani S (2021) Patient's expectations of privacy and confidentiality in Pakistan: a mixed-methods study. J Pak Med Assoc 71(2(A)):537–539. https://doi.org/10.47391/jpma.888

Shukla V, Singh S, Verma S, Verma S, Rizvi AA, Abbas M (2024) Targeting the microbiome to improve human health with the approach of personalized medicine: latest aspects and current updates. Clin Nutr ESPEN 63:813–820. https://doi.org/10.1016/j.clnesp.2024.08.005

Stallhofer J, Steube A, Katzer K, Stallmach A (2024) Microbiota-based therapeutics as new standard-of-care treatment for recurrent Clostridioides difficile infection. Visc Med 40(2):82–91

Stapleton S, Welch G, DiBerardo L, Freeman LR (2024) Sex differences in a mouse model of diet-induced obesity: the role of the gut microbiome. Biol Sex Differ 15(1):5. https://doi.org/10.1186/s13293-023-00580-1. PMID: 37961721; PMCID: PMC10635401

Sun C, Zhou X, Guo T, Meng J (2023) The immune role of the intestinal microbiome in knee osteoarthritis: a review of the possible mechanisms and therapies. Front Immunol 14:1168818

Thanush D, Basavaraj HC, Gowrav MP (2023) Current regulation and initial considerations for successful development and commercialization of microbiome therapies. Adv Gut Microbiome Res 2023(1):6657515

Tosh PK, McDonald LC (2012) Infection control in the multidrug-resistant era: tending the human microbiome. Clin Infect Dis 54(5):707–713

Van Nood E, Vrieze A, Nieuwdorp M, Fuentes S, Zoetendal EG, de Vos WM et al (2013) Duodenal infusion of donor feces for recurrent Clostridium difficile. N Engl J Med 368(5):407–415

Verbeke F, Janssens Y, Wynendaele E, De Spiegeleer B (2017) Faecal microbiota transplantation: a regulatory hurdle? BMC Gastroenterol 17:1–11

Wan S, Wang J (2022) A sequence obfuscation method for protecting personal genomic privacy. Front Genet 13:876686

Wu C, Lee SL, Taylor C, Li J, Chan YM, Agarwal R et al (2020) Scientific and regulatory approach to botanical drug development: a US FDA perspective. J Nat Prod 83(2):552–562

Yaji ME, Sar TT, Aernan PT (2024) Relationship, association and distribution of Helicobacter pylori infection in endoscopy patients at a Tertiary Hospital in Makurdi, Benue State, Nigeria to the symptoms of nausea and vomiting. World J Adv Res Rev 21(2):790–795. https://doi.org/10.30574/wjarr.2024.21.2.0391

Yuliana R, Alfiniyah C, Windarto (2021) Stability analysis of SIVS epidemic model with vaccine ineffectiveness. AIP Conf Proc 2329:040008

Challenges and Future Perspectives in Microbiota-Based Precision Medicine: Standardization of Profiling Methods, Ethical Considerations and the Potential for Pre-emptive Healthcare

Chandrabhan Prajapati, Sachchida Nand Rai, Ghanshyam R. Parmar, and Santosh Kumar Singh

Abstract Trillions of microbes majorly bacteria dwell in various parts of the human body, and play crucial roles in health and illness as well. About equal proportion of microbe cells compared with human cells estimated" with "With an estimated population of almost the same size as human cells, trillions of microorganisms, primarily bacteria, live in different regions of the human body and are essential to both health and illness. The microbiota of newborns is initially acquired via their mothers, and it continues to be altered by nursing. Adulthood brings the microbiome into equilibrium, and as people get older, its microbiota diversity can decrease and inflammation may rise, making them more vulnerable to disease. There are various factors including genetic, environmental, lifestyle and others that can alter the human microbiota, and this imbalance in human microbiota called dysbiosis is responsible for various metabolic and chronic diseases. Recently, the whole microbial genome has been sequenced using whole-genome shotgun-based metagenomics, which provides both functional along with taxonomic information. Research on the human microbiome is currently in its third phase. Personalized medicine is the main emphasis of this phase, which uses each person's unique microbiome profile to customize medicinal and nutritional treatments. Technological developments in areas such as artificial intelligence (AI), machine learning, as well as high-throughput culturomics are essential for managing big datasets and detecting important signals in the microbiome. Enhancing clinical trial strategies and controlling the hype towards microbiome research by highlighting strong, data-driven methodolo-

C. Prajapati · S. N. Rai · S. K. Singh (✉)
Centre of Experimental Medicine and Surgery, Institute of Medical Sciences, Banaras Hindu University, Varanasi, Uttar Pradesh, India
e-mail: santosh.singh01@bhu.ac.in

G. R. Parmar (✉)
Department of Pharmacy, Sumandeep Vidyapeeth Deemed to be University Piparia Waghodia, Vadodara, Gujarat, India

S. K. Singh et al. (eds.), *Microbiota Profiling for Precision Medicine*, Interdisciplinary Biotechnological Advances,
https://doi.org/10.1007/978-981-95-2104-3_14

gies are a few of the problems faced by the area. In this chapter, we focus on the challenges related to standardization in microbiota profiling methods, ethical issues and quality control data along with the significance of human microbiota research in integrated personalized medicines.

Keywords Gut microbiota · Microbiota profiling · Metabolic disorders · Neurodegenerative disorders · Precision medicine · Probiotic

1 Introduction

Microorganisms like bacteria, viruses and fungi live in several parts of the body, including the mouth, skin, pulmonary tract, stomach and other body parts as well. They are collectively referred to as the microbiota of the human body. The habitat in which they reside is called the microbiome. In the human body, most microorganisms are beneficial; however, if conditions go against balance, these may become dangerous. The microbiome is essential for proper digestion, immunological system control, pathogen defence and general metabolism (Dekaboruah et al. 2020). About 30 trillion cells belonging to humans make up the human body; however, there are also more than 100 trillion gut-associated bacterial cells throughout it (Dupont et al. 2020). Trillions of microorganisms, comprising bacteria, viruses, as well as fungi, live in the gastrointestinal system and make up the human gut microbiota. These microorganisms support immune system performance, digestion, food absorption and pathogen defence (Gomaa 2020). A balanced and diversified microbial population indicates an advantageous microbiome; dysbiosis, on the other hand, is an imbalance in the microbiota of the human body that has been related to several illnesses. Appropriate interactions between human cells and the microbiota are essential for maintaining general health and wellbeing (Schippa and Conte 2014). Dysbiosis of microbiota affects glucose metabolism as well as insulin sensitivity regarding metabolic diseases, which in turn leads to obesity and type 2 diabetes via altered energy harvesting and storage (Moreno-Indias et al. 2014). In cancer, *Helicobacter pylori*-induced alterations in the stomach microbiota raise the risk of gastric cancer, whereas gut dysbiosis can cause colorectal cancer by inflaming the colon and producing carcinogenic factors (Gao et al. 2017; Kumar et al. 2021). Dysbiosis has a neurological impact on gut–brain signalling and neuroinflammation, which aggravates Parkinson's as well as Alzheimer's diseases (Denman et al. 2023; Singh et al. 2020). Microbiota abnormalities are linked to autoimmune disorders such as rheumatoid arthritis along with inflammatory bowel disorder, or IBD, which cause aberrant immune responses and persistent inflammation (Forbes et al. 2016). Furthermore, dysbiosis impacts lipid metabolism along with liver inflammation, which exacerbates asthma, and pulmonary dysfunction as well, while gut–microbiota imbalance can also promote non-alcoholic fatty liver dysfunction. Preventing and managing disease requires an understanding of and commitment to the microbiota balance (Tokuhara 2021). A new area of study called 'microbiota-based precise medicine' uses the intricate relationships that exist between the host metabolism as well as the human microbiome to provide individualistic medical interventions and therapies. By offering personalized treatment plans that take into account each person's particular microbiological makeup, this strategy holds the potential to completely transform the healthcare industry (S. Wang et al. 2024a). There are multiple techniques used presently regarding human microbiota profiling, and while participating each method has its advantages and disadvantages (Kuczynski et al. 2012). Based on the remarkably stable 16S ribosomal RNA (rRNA) sequence of the gene, the sequencing of 16S rRNA genes is frequently used to identify and categorize bacteria within a population. Although it lacks functional insights, this technique yields taxonomic information only (Buetas et al. 2024). The whole microbial genome is sequenced using whole-genome shotgun-based metagenomics, which provides both functional along with taxonomic information (Purushothaman et al. 2022). Additional data on microorganism's gene expression, protein synthesis and metabolic activities are provided by other methods such as meta-transcriptomics techniques, meta-proteomics, as well as metabolomics respectively (Beura et al. 2022; Prajapati et al. 2025a). Furthermore, the discovery of microbial biomarkers linked to health and illness has been made easier by developments like bioinformatics along with computational tools, which have made it possible to integrate and evaluate challenging microbiome data (Marcos-Zambrano et al. 2021). The legal and moral utilization of microbiota knowledge is critical, given the sensitivity associated with private health information along with the possible consequences for people and communities. Microbiota research ethics frameworks need to cover things like ownership, privacy of information, as well as informed permission. The breadth of the study, any possible risks or rewards and the uses and sharing of the participants' data are all explained in detail to the participants. Ethical standards also take into account how advantages from microbiome research should be distributed fairly, so that advances in precision medicine assist a variety of communities and avoid aggravating already-existing health inequities (Safarlou et al. 2023). The relationship between humans and their microbiomes is symbiotic, where both parties benefit from their coexistence (Malard et al. 2021). Advances in understanding this relationship hold promise for innovative treatments and preventive strategies. For example, probiotics and prebiotics are being explored to restore healthy microbiota, while faecal microbiota transplantation (FMT) has successfully treated certain infections and gastrointestinal disorders (Cammarota et al. 2014). The intricate interplay between human cells and the microbiome is crucial for maintaining health and preventing disease. As research continues to uncover the depths of this relationship, it opens new openings for enhancing human health through microbiome-targeted therapies and personalized medicine (Ratiner et al. 2024). However, several challenges of microbiome-based precision medicine include guaranteeing data privacy, gaining informed permission, regulating the multifaceted nature of microbiota interactions, resolving individual fluctuations and fairly dispersing benefits. The precise analysis, interpretation and integration of microbiome data into healthcare settings are hampered by technological issues as well. This chapter discusses the significance of precision medicine as well as the difficulties and potential applications of this field of study.

2 Historical Overview Concerning Human Microbiota Exploration and Therapeutic Implications

Microbiota study benefited greatly from the groundbreaking work of Élie Metchnikoff, a Russian immunologist as well as a zoologist. His theory, which laid the groundwork for current probiotics, was that pathogenic gut bacteria create toxins that accelerate aging and that consuming good bacteria, like those in yogurt, may extend life and improve health (Georgieva et al. 2022). Metchnikoff, who made the discovery of phagocytosis, highlighted the microbe's role in the gastrointestinal tract and other interior environments of the body play in both health and sickness. His opinions on the influence of gut microbiota in health gained prominence and he was awarded with the Nobel Prize in 1908 in Physiology or Medicine for his pioneering work on immunity (Metchnikoff 2004; Podolsky 2012). Another pioneer of microbiota study was the German-Austrian paediatrician as well as bacteriologist 'Theodor Escherich'. His most recognizable discovery, in the year 1885, was the bacterium *Escherichia coli*, or *E. coli*. It turned out to be an essential model organism for research on the physiology as well as genetics of bacteria. Escherich's research played a crucial part in comprehending the functions and variety of the gut microbiota. He comprehended that healthy bacteria survive in the gastrointestinal tract and play a crucial role in digestion as well as immunity (Friedmann 2006; Geurtsen et al. 2022). In the beginning of the twentieth century, German physician as well as bacteriologist Alfred Nissle pioneered important advances in probiotic medications. He obtained and identified a harmless strain of *Escherichia coli* referred as *E. coli* Nissle in the year 1917, through a stool sample of soldier who suffered from a serious incident of diarrhoea at that time. One of the earliest strains of probiotics to be utilized in medicine was this one. Early proof that altering the gut microbiota may improve health was provided by Nissle's study, which demonstrated that certain strains of gut bacteria may also be utilized to treat gastrointestinal disorders (Nissle 1925; Sonnenborn 2016). In the early nineteenth century, the germ hypothesis of disease revolutionized medicine by claiming that microbes are the fundamental cause of numerous ailments. It replaced pre-existing notions like spontaneous creation as well as miasma theory to establish scientific knowledge of infection, and it was pioneered through scientists like Louis Pasteur as well as Robert Koch. Significant advancements in cleanliness, immunizations and medicines were made possible by this change, which significantly decreased the prevalence of infectious diseases. The germ hypothesis improved our ability to fight a variety of diseases by highlighting the role that bacteria play in both health and sickness. It also established the foundation for innovative microbiology as well as investigations exploring the human microbiome (Chen et al. 2019). Human microbiome research has progressed via crucial strategies, increasing our comprehension of human–microorganism interactions. In the early twentieth century, breakthroughs in microscopy along with staining methods enabled researchers to see microbial diversity, showing the development of intricate microbial communities (Bartholomew and Mittwer 1952; Burdon 1946; Porter 1976; Zernike 1955). The field of microbiota study underwent a revolutionary shift with the introduction of 16S rRNA sequencing of genes, which allowed for high-throughput screening and revealed an unexpected number of microorganisms, around the last decades of the twentieth century (Woo et al. 2008). Metagenomics along with subsequent multi-omics methods offered information about microbial functions as well as interactions (Zhang et al. 2019) (Fig. 1). Artificial intelligence (AI) and other computational tool advancements have also enhanced data analysis along with early illness prediction (Ahmed et al. 2020; Chen and Shen 2021). The promise of the microbiome in personalized medicine as well as disease prevention was highlighted by clinical applications such as faecal microbiota transplantation, probiotics, as well as prebiotics, which have demonstrated therapeutic advantages (Kashyap et al. 2017). Human microbiota research has recently been centred on elucidating its complex relationships with the immune system as well as the gut–brain axis, highlighting the microbiota's critical role in both health and illness (Bhalla et al. 2024). As a result, studies on the human microbiota have progressed from primitive bacterial isolation to modern multi-omics, underscoring its therapeutic promise for the immune system as well as gut–brain interaction.

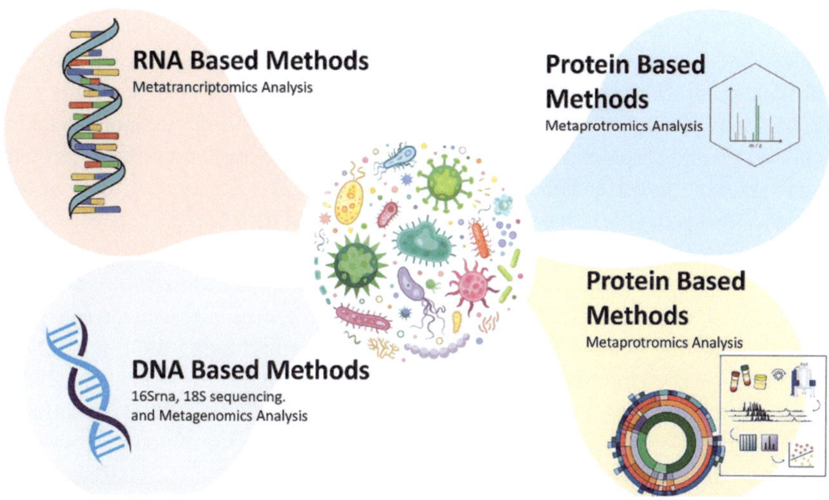

Fig. 1 Overview on multi-omics approaches for microbiome analysis. This figure highlights RNA-based methods (metatranscriptomics analysis), DNA-based methods (16S rRNA, 18S sequencing and metagenomics analysis) and protein-based methods (metaproteomics analysis). These methodologies provide comprehensive insights into microbial structure, function and activity in diverse environments

3 Health Problems Caused by Microbiota Imbalances in the Human Body

Dysbiosis, or deviations in the microbiota, may result in a number of serious negative effects on human health. Dysbiosis in the digestive system is associated with diseases including irritable bowel syndrome (IBS) as well as inflammatory bowel disorder (IBD). Unbalanced microbiota in IBS may lead to gas as well as irritants, which can cause pain in the abdomen and change bowel patterns. IBD is characterized by persistent inflammation brought on by pathogenic bacteria, which also cause ulcerative colitis as well as Crohn's disease (Lopetuso et al. 2018). Dysbiosis is also linked to metabolic diseases as type 2 diabetes as well as obesity. Some bacteria can increase the amount of calories that are extracted from meals, which can cause weight gain. Other bacteria can interfere with insulin sensitivity as well as glucose metabolism, which can cause diabetes. These problems may be made worse by the imbalance, which can result in persistent low-level inflammation (Scheithauer et al. 2020; Turnbaugh et al. 2006). Dysbiosis in the mouth can lead to tooth decay as well as periodontal conditions. Because acid-producing microbes can create cavities by demineralizing enamel on teeth, a high prevalence of harmful bacteria in the oral cavity can induce gum inflammation followed by decay of the teeth (Costalonga and Herzberg 2014). The gut–lung axis defines a mutual relationship among the lungs as well as the gut, impacted by immune system responses, microbial metabolites, along with microbial motility. Lung functioning can be impacted by dysbiosis in the stomach, which can change immunological responses and cause inflammation. Diseases including lung fibrosis, asthma, food allergies, chronic pulmonary disease, as well as respiratory infections are associated with this imbalance. The synthesis of cytokines can alter valuable short-chain fatty acids that are depleted, and pathogenic bacterial components are translocated, besides other understood mechanisms (Zhang et al. 2020). Dysbiosis can cause pro-inflammatory cytokine levels to rise and chronic inflammation, which can aggravate heart failure, coronary artery disease, hypertension, as well as myocardial infarction. Consequently, decreased levels of short-chain fatty acids also referred to as SCFAs can lessen anti-inflammatory benefits, while harmful metabolites such as trimethylamine-N-oxide (TMAO) encourage plaque development in arteries as well. Furthermore, dysbiosis raises the risk of cardiovascular disease by influencing insulin resistance as well as lipid metabolism. Comprehending these associations highlights the possibility of microbiota-focused treatments to enhance cardiovascular health (Wang et al. 2023). The gut–brain axis influences mental health illnesses such as depressive and anxious disorders. Dysbiosis in this context impacts the generation of neurotransmitters along with the stress response (Appleton 2018; Vamanu and Rai 2021). Recent findings show that microbial imbalances might be related to digestive issues in autism spectrum conditions (Li et al. 2024; Lin et al. 2024; Phan et al. 2024). Neuroinflammation and the generation of toxic metabolites that impair brain function are a few ways in which dysbiosis is linked to neurodegenerative disorders like Alzheimer's, Parkinson's disease, as well as amyotrophic lateral sclerosis (Birla et al. 2019; Rai et al. 2019; Villavicencio-Tejo et al. 2024). Microbiota imperfections have also been related to autoimmune illnesses including multiple sclerosis as well as rheumatoid arthritis. Increased gut permeability brought on by dysbiosis can cause components of bacteria to pass through into the bloodstream triggering immunological reactions, which can result in long-term inflammation along with autoimmune conditions (Sadeghpour Heravi 2024). Skin conditions including psoriasis as well as eczema are associated with human microbe dysbiosis. The imbalance of microbes impairs the skin barrier as well as induces inflammation in eczema by increasing pathogens and decreasing beneficial microorganisms. Unbalanced in the human microbiota can cause or accelerate psoriasis by promoting inflammatory markers (Ryguła et al. 2024). Inflammation of the follicles can be caused by the overgrowth of *Cutibacterium acnes*, which results from dysbiosis (Cavallo et al. 2022). Dysbiosis of the stomach and skin affects rosacea and leads to both localized and systemic swelling (Sánchez-Pellicer et al. 2024). Recent investigation indicates that vaginal microbiota may be associated with a variety of reproductive health issues, such as heightened vulnerability to infection, difficulties during pregnancy and delayed vaginal tissue healing. However, the specific mechanisms behind these unfavourable consequences remain ongoing (Ottinger et al. 2024). Therefore, dysbiosis affects health by impairing immune system regulation, which results in autoimmunity as well as chronic inflammation. Toxins as well as metabolites produced by harmful bacteria cause tissue damage along with promoting illness. Unbalanced microbiota can lead to deficits or excesses in the absorption of nutrients and can also weaken the integrity of the intestinal barrier, which can let dangerous chemicals enter the circulation. These problems can be lessened by maintaining a healthy microbiota through food, probiotics, as well as lifestyle decisions. On the other hand, based on particular imbalances, therapy and preventative measures have to be adapted (Pires et al. 2024).

4 Advancement with Timelines Regarding Human Microbiota Profiling Methods

Frederick Sanger's invention of the chain-termination technique for DNA sequencing, called Sanger sequencing, in the year 1977 marked a major achievement in the 1970s. This technique is revolutionizing the area of genetic analysis by selectively including chain-terminating dideoxynucleotides throughout DNA replication (Sanger et al. 1977). Additional breakthroughs were made in the 1980s when Applied Biosystems introduced the ABI 370, the initial computerized DNA sequencer, in the year 1986. Large-scale genetic initiatives were possible due to the substantial acceleration of analysis of DNA sequences brought about by this machine, which advanced the Sanger sequencing procedure (Barba et al. 2014; Trainor 1990). The Human Genome Project, also known as HGP, was started in the 1990s. This global scientific project, which made extensive use of Sanger sequencing technology, attempted to map and comprehend every gene throughout the human genome (Watson 1990). The entire genome for *Haemophilus influenzae* Rd. was sequenced during 1995 by J. Craig Venter along with collaborators using whole-genome shotgun DNA sequencing, making it the initial genome of a live creature to be fully sequenced (Fleischmann et al. 1995). The effort known as the Human Genome Project as well as Celera Genomics published a preliminary manuscript regarding the human genome during 2001, demonstrating the potential of massive DNA sequencing projects in the 2000s (Brown 2000). Increasingly parallel sequencing was made possible in 2005 by the advancement of next-generation sequencing or NGS innovations, such as Roche 454, Illumina (Solexa) as well as ABI SOLiD, which significantly increased throughput and lowered costs (Van Dijk et al. 2014). An important step towards personalized genomics was taken in 2008 when James Watson, a co-discoverer regarding the DNA doubled helix, experienced personal genome sequenced employing NGS methods (Wheeler et al. 2008). The decade of the 2010s saw the introduction of nanopore sequencing by Oxford Nanopore Technology (ONT) in the year 2011. With the use of this technique, DNA strands may be directly sequenced by tracking variations with electrical conductivity as they go through the single nanopore (Lu et al. 2016). With the ability to precisely modify DNA sequences, the finding and subsequent development regarding Clustered Regularly Interspaced Short Palindromic Repeats (CRISPR)-CRISPR-associated protein 9 (Cas9) as an effective genome-editing tool during 2012 has important consequences for genomics as well as genetic research (Peng et al. 2016). Massive genomic investigations became more viable in 2015 when Illumina announced the HiSeq X Ten, which can sequence around 18,000 human genomes annually for about $1000 per genome (Xiao et al. 2016). Sequencing technology was important in combating the coronavirus disease 2019 (COVID-19) pandemic in the early 2020s. The COVID-19 virus, severe acute respiratory syndrome coronavirus 2 (SARS-CoV-2), was identified and tracked in 2020 because of fast sequencing, which also made vaccine development possible (Krammer 2020). The first full sequencing of the human genome was revealed by the Telomere-to-Telomere (T2T) group in 2021, covering gaps regarding the Human Genome Project along with providing an improved reference understanding human genetics (Nurk et al. 2022). The National Institutes of Health, also known as NIH, initiated the Human Microbiome Project (HMP) in the year 2008. Its main objective was to describe the many microbial communities that exist within and around the human body as well as to comprehend the functions that these communities play in both health and illness (Peterson et al. 2009). An initial description of these kinds of microbes was the main objective of the 2013 completion of the HMP's first phase. This stage produced a large amount of scientific information regarding the bacteria, viruses, as well as fungi that inhabit different parts of the human body (Aagaard et al. 2013). After the completion of the initial phase, the Integrated Human Microbiome Project (iHMP called HMP2) began in 2014. By examining how the microbiome affects health and illness across time and in various populations, this subsequent stage attempted to expand on the previous phase's results (Huang et al. 2014). After HMP2 was finished in 2019, researchers gained more understanding of how the human microbiome remains dynamic and how it affects a range of illnesses, including diabetes, inflammatory bowel disorders, pregnancy, as well as premature delivery. With the help of the Human Microbiome Project, our knowledge of the intricate relationships that exist between humans as well as their microbial ecosystems has greatly increased, opening the door to novel medication diagnostics along with treatment strategies (Voth and Khanna 2020). Culturomics has considerably revolutionized human microbiota exploration by allowing the growth of previously uncultivable microorganisms under a variety of culture conditions and using high-throughput approaches. This method may produce a wealth of new microbiological species, contributing to a deeper comprehension of the complexity and variety of the human microbiota. Culturomics has considerably revolutionized human microbiota exploration by allowing the growth of previously uncultivable microorganisms under a variety of culture conditions and using high-throughput approaches. This strategy may produce an immense number of new microbiological species, contributing to a deeper comprehension of the diversity as well as the complexity of the microbiota found in humans (Song et al. 2024). Researchers may carry out in-depth genetic along with functional investigations of these microorganisms, providing understanding of how they participate in health and illness, by combining culturomics with sequencing technology. Furthermore, culturomics tracks how the microbiota changes in response to different stimuli and aids in advancing the research of microbiota-based therapeutics like faecal microbiota transplantation as well as probiotics, opening the door to personalized medicine and novel therapeutic approaches (Abou Chacra et al. 2024; Dubourg et al. 2018; Pfleiderer et al. 2013; F. Wang et al. 2024b). The ability to see microbe communities in extreme detail along with high resolution has contributed to advances in technologies for imaging such as confocal laser scanning microscopy, also known as CLSM as well as optical coherence tomography, or OCT, precious for studying the human microbiome. OCT enables real-time, non-invasive visualization of in vivo microbe habitats and biofilms, exposing their dynamic behaviours. With the use of three-dimensional, multicoloured imaging provided by CLSM, population of microbes quantitative assessments and in-depth structural analysis are made possible. These technological advancements further the research of the microbiota's involvement in health and illness by enhancing our comprehension of the ecology of microbes, host–microbiota relationships and the invention of targeted therapeutics (Darvin 2023; Grohmann and Vaishampayan 2017). Research on the human microbiome is currently in its third phase. Personalized medicine is the main emphasis of this phase, which uses each person's unique microbiome profile to customise medicinal and nutritional treatments. Technological developments in areas such as AI, machine learning, as well as high-throughput culturomics are essential for managing big datasets and detecting important signals in the microbiome. Enhancing clinical trial strategies and controlling the hype towards microbiome research by highlighting strong, data-driven methodologies are a few of the problems facing the area (McGuinness et al. 2024). Sequencing of amplicons, largely focusing on the 16S ribosomal RNA (rRNA) gene, continues to be a central tool in microbial community analysis; this strategy takes advantage of the variable portions of the 16S rRNA gene to identify bacterial species and generate detailed community descriptions. This method has been optimized to amplify hypervariable regions, for instance, V3–V4, in order to improve the taxonomic classification across the phylogenetic tree (Grogan et al. 2019; Szóstak et al. 2022). On the other hand, the metagenomic shotgun sequencing is more comprehensive and provides extra comprehensive description of the microbial separators of a sample by whole food content. This method offers strain resolution and functional characterization and can accommodate a broad spectrum of microbial forms such as bacteria, archaea, fungi, viruses and more, while providing a unified synopsis of both phyletic and functional diversity, offering a true picture of microbial communities with functional capabilities encoded in the genomes.

Variability in microbiota profiling has been targeted by the International Human Microbiome Standards project (IHMS) project with the establishment of the standards; however, adoptions remain sporadic, and many researchers continue to use different protocols that provide fundamentally different results from the same sample (Szóstak et al. 2022). In view of these disparities and with the purpose of increasing data credibility, a number of standardization procedures have been recommended. High-throughput samples and standard sample prep methods are essential since their variation affects DNA quality and necessitates mechanical lysis. The specific sequencing strategies like the use of tagmentation-based library prep kits can be regarded to have improved the reproducibility (Kool et al. 2022). The comparison of results to be obtained from bioinformatics pipelines requires that the pipelines have been made more consistent by having similar tools and taxonomic databases used wherever they are. Retrospective comparison of samples prepared differently and analysed under identical conditions in different laboratories calls for ideology and establish standardized protocols. In addition, the training in best practice materials and standardized methods is crucial to encouraging other researchers to adopt high-quality techniques and increase the repeatability of microbiome research (Bharti and Grimm 2021; Roume et al. 2023). As shown in Fig. 1, RNA-based (metatranscriptomics), DNA-based (16S rRNA, 18S sequencing, metagenomics), and protein-based (metaproteomics) techniques offer thorough insights into the composition, function, and activity of microorganisms in a variety of settings.

5 Challenges in Human Microbiota Profiling Methods

Human microbiota profiling encounters considerable computational hurdles, such as dealing with large datasets that demand a lot of processing power along with storage. These datasets require sophisticated algorithms and reliable infrastructure for processing and analysis, which are frequently costly and technically challenging (Bharti and Grimm 2021; Cammarota et al. 2020; Navas-Molina et al. 2017). The heterogeneity and noise underlying microbiome data make it difficult to build reliable models for prediction with machine learning as well as artificial intelligence (AI), while connecting microbiome datasets together with additional clinical and genetic data adds difficulty because of the different formats and the integrity of the data (Abavisani et al. 2024; Marcos-Zambrano et al. 2021). The human microbiome varies greatly between individuals due to factors such as genetics, lifestyle, surroundings and nutrition. This makes it challenging to identify universal indicators and broad-based treatments. The constantly evolving character of the microbiome necessitates resource-intensive as well as logistically difficult longitudinal investigations because it alters as time passes in response to different factors. Furthermore, it is challenging to identify causal relationships to generate specific therapies due to the complex interactions that exist between various microbial species as well as between microorganisms and the host (Kodikara et al. 2022). Sample collection along with laboratory analysis provides a number of obstacles for human microbiota profiling. Microbe community profiles may vary significantly depending on the procedures used for collection, involving swabs, biopsies and faecal samples. Strict processes are necessary in order to avoid contamination, along with proper storage and transportation, which are essential to prevent changes in the microbial numbers and varieties (Short et al. 2021). Even as the technologies for microbiota profiling have advanced, it still remains a major issue to arrive at a standard technique. Heterogeneity of sample preparation techniques, sequence generation technologies and data analysis workflows contributes valuable noise that compromises the kinetic of the studies. For instance, variations in homogenization periods during DNA extraction have been shown to affect diversity measures from microbial samples by a large margin. Furthermore, the decision of library preparation kits and sequencing technologies adds extra sources of variation thus making it extremely challenging to obtain standardized data and quality values which are consistent when making comparable conclusions (Kool et al. 2022; Szóstak et al. 2022). Varying sequencing sites and methodologies might produce different findings in the lab; in addition DNA extraction techniques can create biases. Variability is also impacted by the various steps involved in data handling and evaluation, which are determined by the algorithms as well as reference databases used. It is important to establish uniform techniques across these phases to guarantee precise and consistent microbiome profiling (Nearing et al. 2021) (Fig. 2). Using trustworthy sample and sequencing techniques is essential for appropriate microbiome profiling. Such procedures can become biased and have issues with repeatability when they are variable. Although culturomics has made microbial culture more advanced, it is still laborious and might not fully capture the richness of the human microbiome. Although it challenges with precise genome assembly as well as annotation, metagenomics offers extensive genetic information. Metatranscriptomics, metaproteomics, along with metabolomics multifunctional profiling provide valuable information but need comprehensive reference databases along with advanced analytical methods (Fondi and Liò 2015; Prajapati et al. 2025b). The utilization of microbiome data raises ethical and legal questions, especially when it comes to sharing and storing private health information. It's critical to establish precise rules and legal frameworks. Furthermore, the inconsistency in microbiome research methodology, such as protocols for data processing, sequencing, as well as sampling, impedes the capacity to compare studies along with applying research findings in clinical settings (Ugwu et al. 2024) (Fig. 2).

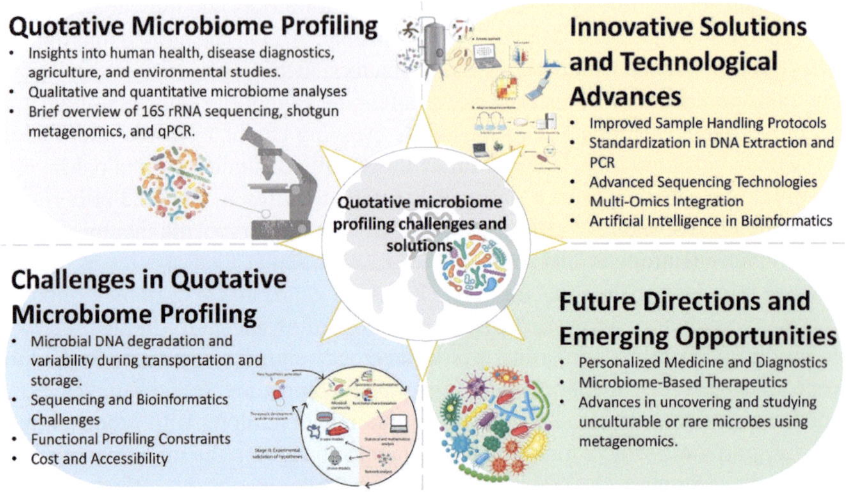

Fig. 2 Overview of quantitative microbiome profiling, highlighting its applications, challenges, innovative solutions and future opportunities. This figure displays the key aspects include insights into health and environmental research, technological advances such as multi-omics and artificial intelligence, challenges in sample handling and bioinformatics and emerging opportunities in personalized medicine and microbiome-based therapeutics

6 Protecting Standardization and Effectiveness in Human Microbiota Studies: Ethical and Regulatory Aspects

Research on the human microbiome has not only enormous potential to advance medicine, but it also raises ethical issues that must be carefully considered in terms of consent procedures, privacy rights and minimizing harm (McGuire et al. 2008). Receiving informed permission is crucial and necessitates thorough explanations of the goals, methods, risks and advantages of the study. Participant consent and withdrawal policies must be made clear to participants by researchers. For vulnerable groups, including youngsters and people who are mentally impaired, more attention needs to be provided for making sure that their involvement remains authentically free and ethically acceptable (Ma et al. 2018). Anonymizing participant information as well as safeguarding it from unwanted access is necessary to protect their privacy. To combat breaches as well as misuse, strong security of information procedures and unambiguous data sharing regulations are crucial. It is also essential to make sure that receivers follow precise privacy guidelines and that provided information can be de-identified (O'Doherty et al. 2016). Microbiota exploration should be preceded by a comprehensive risk assessment that recognizes potential hazards. Employing less disruptive techniques and making sure research design adheres to scientific standards are two strategies to reduce such risks. Every study should be reviewed by an independent ethics system, or the institutional review board (IRB) to make sure that participant damage is kept to a minimum and ethical criteria are fulfilled (Aagaard et al. 2013). Regulations pertaining to human microbiome research have been put in place to guarantee adherence to moral guidelines, uphold quality assurance procedures and fulfil reporting obligations. These rules should be crucial for maintaining data integrity, protecting participant well-being and encouraging transparency within scientific studies (Marques et al. 2019). In microbiota investigations, compliance with local, state, federal and global regulations seems crucial. This involves following the rules established by regulatory agencies including the EMA (European Medicines Agency), the United States FDA (Food and Drug Administration), along with other appropriate authorities (Edelstein et al. 2015; Komala et al. 2023). The Good Clinical Practice (also called GCP) principles as well as the Declaration of Helsinki, serving as a framework to conduct ethically acceptable and scientifically legitimate research using human beings, must be followed by researchers (Bruggeman et al. 2024). Putting strong quality control procedures into action has become essential to guaranteeing the reliability and authenticity of microbiota research. For the collecting, processing, storing and analysis of samples, standard operating procedures, or SOPs, have to be created and maintained properly. Programmes for quality assurance, such as expertise testing and routine audits, aid in upholding high standards across the whole research process. To guarantee accuracy as well as consistency regarding the results they report, researchers should additionally use equipment and procedures that have been verified (Annaratone et al. 2021; Broadhurst et al. 2018). Strict reporting criteria have been adopted to maintain accountability as well as transparency throughout microbiota research; researchers must register their trials with public databases such as ClinicalTrials.gov along with provide comprehensive protocols. Accurate and timely reporting of results needs to be done, including positive as well as negative effects (Zarin et al. 2011). Regular progress information and notification regarding any unfavourable occurrences or protocol violations are required by ethical regulatory organizations, like IRBs (institutional review boards) or to committees on ethics. Protecting data integrity as well as safeguarding participant privacy requires the use of appropriate data management procedures, such as secure storage, restricted access, along with information anonymization (Dupras and Bunnik 2021; Prajapati et al. 2025c). To improve reproducibility and promote scientific advancement, researchers need to abide by standard guidelines regarding data sharing along with transparency, especially the FAIR (Findable, Accessible, as well as Interoperable, and Reusable) rules. In addition to submitting comprehensive protocols as well as informed consent applications to regulatory bodies including ethical review boards, investigators must get all relevant regulatory authorizations before beginning the research study (Cernava et al. 2022). Certifications could also be required to access particular biological materials as well as research locations. Researchers are accountable for keeping correct records and getting ready for evaluations, but post-approval, ongoing monitoring as well as verification of compliance by regulatory authorities in addition to ethical committees involving visits to the facility, audits, especially inspections guarantee conformity to authorized protocols as well as standards (Müller et al. 2020). Considering these regulatory factors guarantees a high standard legal study on human microbiota that conforms with all applicable regulations. This increases field comprehension and promotes assurances regarding scientific research.

7 Key Factors Regulating Human Microbiota

According to recent estimations, the proportion of microbial cells versus human cells inside the human body measures approximately 1:1, containing 39 trillion microbe cells versus 30 trillion cells belonging to humans. This disproves the previous assumption that germs prevail over human cells with a ratio of 10:1. The microbiome is a very varied group comprising various species, mostly bacteria, that differ greatly depending on the part of the body they are in, including the mouth, skin cells as well as the gastrointestinal tract (Nazaroff 2019; Sender et al. 2016). The human microbiota's makeup and functioning are influenced by a number of variables, such as age, atmosphere, genetics, antibiotic use, nutrition, along with lifestyle (Flandroy et al. 2018). Foods rich in nutrients, particularly fibre, encourage the growth of good bacteria which generate the short-chain fatty acids found to be vital for a healthy gut (Koh et al. 2016). The human microbiota could be positively modulated by probiotics that contain active beneficial bacteria, as well as prebiotics, which are indigestible dietary components. Probiotics are found in foods like yogurt, kefir, as well as fermented fruits and vegetables, whereas prebiotics are found in foodstuffs including garlic, onions and bananas (Khan et al. 2023). Antibiotics can destroy both good and harmful strains of bacteria, disrupting the microbial equilibrium for a short period of time or even longer. Bacteria with resistance to antibiotics can arise as a result of overuse or misconduct, which can change the resilience along with balance of an individual's microbiome (Langdon et al. 2016). Excessive impeccable hygiene could decrease microbial diversity, increasing the possibility of allergies as well as autoimmune disorders. On the other hand, diseases and dysbiosis can also result from improper hygiene (Bach 2021; Gupta et al. 2020). The microbiome can be affected by genetic factors, such as inherited factors as well as epigenetic alterations. Genetic predispositions to particular microbial communities have been found in individuals; in addition environmental influences can change gene expression, influencing the microbiome throughout generations (Benson et al. 2010; El-Sayed et al. 2021; Ussar et al. 2015). The microbiota of newborns can initially acquire via their mothers, and it continues to be altered by nursing. Adulthood brings the microbiome into equilibrium, and as people get older, its microbiota diversity can decrease and inflammation may rise, making them more vulnerable to disease (Rinninella et al. 2019). The microbiome can be affected by pollution, prolonged exposure with regard to environmental contaminants, along with geographic location. In general, the microbiome in rural locations appears more diverse than that in towns and cities. Pollutants as well as chemicals may damage the microbiome by eliminating beneficial bacteria while encouraging dangerous ones (Jin et al. 2017). Prolonged stress can negatively impact the gut microbiota via the gut–brain axis, causing alterations including gut motility, the release of chemicals, along with permeability. IBS, also known as irritable bowel syndrome, along with additional gastrointestinal problems may be exacerbated by the aforementioned. Unbalances may result from physical stress on the microbiota, such as those caused by disease or hard exercise (Karl et al. 2018). All of these factors interconnect in complex manners to establish the human microbiome, impacting general health as well as disease risk. All of these factors interconnect in complex manners to establish the human microbiome, impacting general health as well as disease risk. Comprehending these factors can aid in formulating tactics to preserve a robust microbiome along with improving outcomes for medicine.

8 Human Microbiota Research Drives Innovations in Pre-emptive Medicine for Healthcare Integration

Numerous internal human niches, including the mouth cavity, skin, gut, urogenital systems, peripheral blood and even tumours, include microbiomes that can be used as biomarkers to predict various illnesses. As a result, these microbial communities provide insightful data for the earlier identification, precise diagnosis and management of a range of human illnesses (Ratiner et al. 2024) (Fig. 2). There are three basic approaches to using the microbiome in personalized medicinal therapies. Supplementing with microorganisms, such as faecal microbiota transplantation (FMT), prebiotics, as well as next-generation probiotics, improves dysbiosis and generates advantageous metabolites (Al-Fakhrany and Elekhnawy 2024). Utilizing RNA-guided nucleases, predator bacteria, as well as bacteriophages, microbe suppression targeted and destroyed pathogenic microorganisms along with their genomes (Citorik et al. 2014; Matias et al. 2024). Recent approaches to maintaining the human microbiome include metabolite control, which involves halting the manufacture of harmful metabolites, altering pathways and supplementing beneficial metabolites as well. These methods seek to successfully cure illnesses connected to the microbiome as well as restore the balance of microorganisms. Chemotherapy efficacy is greatly impacted by the microbiota, which may both enhance and inhibit drug activity. For instance, the drug gemcitabine can become inactive by *Mycoplasma hyorhinis*, which results in medication failure and tumour growth. In contrast, certain bacteria can increase the effectiveness of oxaliplatin by producing tumour necrosis factor (TNF) along with reactive oxygen species. The response of the microbiome towards combined chemotherapies such as FOLFIRINOX can also be affected by the metabolites of bacteria, such as indole acetic acid (IAA), which hinder autophagy within tumour cells and stimulate apoptotic death. Furthermore, via boosting memory T cell traffickers, producing interleukin-12 (IL-12), as well as promoting tumour suppression, the bacteria *Akkermansia muciniphila* increases the effectiveness of anti-Programmed Death-1 (PD-1) antibodies. The results of cancer treatment may thus be improved by microbiome-targeted therapies (Ratiner et al. 2024). According to a recently published study, there are several ways in which the gut microbiota affects brain function and plays a role in neurodegenerative illnesses. LPS and other microbiome-derived chemicals get across the intestinal barrier, stimulating immune cells to produce proinflammatory cytokines like TNF-α, IL-1β, as well as IL-6 and decreasing the ability of monocytes to phagocytose amyloid proteins, which result in amyloid accumulation and neuroinflammation. Dysbiosis stimulates Th1 and Th17-mediated proinflammatory T cells, which exacerbates inflammation. By interacting with microglia and passing across an impaired blood–brain barrier, these microorganisms can exacerbate neuroinflammation as well (Belizário et al. 2018). Furthermore, intestinal bacteria activate the vagus nerve, which reduces brain-derived neurotrophic factor (BNDF) and transports amyloids into the brain. Their by-products also pass the blood–brain barrier together with harm neurons as well as microglia, while promoting neuroinflammation along with amyloid aggregation (Megur et al. 2020; Ranuh et al. 2019). The human microbiome plays a critical role in personalized dietary counselling as well as stress management for pre-emptive medicine, helping to preserve health and warn against illness. Through the use of microbiome profiling, specific dietary recommendations which boost beneficial bacteria along with improving gut health can potentially be made, providing information regarding metabolic predispositions as well as digestive health. By controlling lipid profiles, blood glucose levels, as well as weight, personalized dietary strategies lower the possibility of metabolic disorders (de Toro-Martín et al. 2017).

This has been made possible by improvement in techniques used in identifying microbiome such as next-generation sequencing that help researchers pinpoint specific microbes associated with specific disease states making diagnosis possible and specific treatments initiated (Kashyap et al. 2017). Compositional alterations in the microbiome have been linked to diseases including IBD for which specific microbiome signatures can be used to characterize disease state and response to therapy. This shows the importance of the microbiome as a biomarker for the concept of personalized therapy depending on a person's microbial signature (Huang et al. 2023; Zhao et al. 2023). Furthermore, the pharmacomicrobiomics, the relation between the gut microbiota and pharmacotherapy, show that variations in microbial composition can affect drug metabolism and efficacy and enhance the therapeutic index potential via modulation of the gut flora (Hou et al. 2022). New approaches as FMT, probiotics and diet targeting the disrupted microbiota will therefore open the new ways to the prevention and treatment of related disorders. In addition, knowing the processes through which gut microbes influence systemic diseases, including cardiovascular diseases, diabetes and cancer, opens new avenues for prevention based on microbial health. Including microbiota profiling into clinical practice may change the approach from traditional treatment to preventing diseases at their source (Wu et al. 2022). Unfolding comparative approaches that remain relevant to annotating the developed constellations of the microbiome connecting to host physiology, there is a general agreement that further pursuit of such knowledge will be critical for building a better and more accurate medical future concerning the multiple human populations. In fact, microbiota profiling appears tremendously promising for pre-emptive medicine that advances not only individual ways of prevention and treatment but also redefines the way that health and illness factors are conceived in an increasingly complex biological environment (Kashyap et al. 2017; Nagpal et al. 2014).

The gut microbiota's impact on responses to stress along with mental health can be seen by the microbiota–gut–brain pathway. Psychobiotics, along with stress-reduction practices like yoga as well as meditation, can help balance the bacteria in the gut and improve mental health as well. This all-encompassing method combines nutritional and mental health techniques to promote general health and avoid illness (Pedroza Matute and Iyavoo 2023). Sustainable approaches to regulate the human microbiota, including herbal medicines, fermented goods, along with plant-based diets, can improve gut functioning and general well-being. Certain herbs, notably chicory root, function as prebiotics, encouraging the growth of desirable intestinal bacteria, whereas others, such as garlic as well as turmeric, possess antibacterial and anti-inflammatory properties (Carrera-Quintanar et al. 2018; Perović et al. 2021; Prajapati et al. 2024a, b; Tian et al. 2024). Probiotics from fermented foods, such as yogurt as well as kimchi, boost nutrient absorption along with diversity of gut bacteria, consequently improving immune function. In addition to lowering inflammation and preserving the balance of microbes, a plant-based diet that is high in fibre and phytochemicals can promote gut health as well. Combining these strategies boosts immunity, maintains intestinal health and promotes overall health over a long period of time (Shah et al. 2023). Table 1 highlights various innovative approaches

Table 1 Innovations in pre-emptive medicine through human microbiota research

Field of innovation	Description	Examples	References
Microbiota-Based Diagnostics	Using microbial signatures as biomarkers to diagnose diseases	Detection of colorectal cancer via gut microbiome profiling	Herlo et al. (2024); Zhou et al. (2020)
Probiotics and Psychobiotics	Live microorganisms that confer mental and physical health benefits	Probiotic supplementation to alleviate anxiety and depression	Bermúdez-Humarán et al. (2019); Freijy et al. (2023)
Microbiota-Driven Immunotherapy	Leveraging gut microbiota to enhance immune responses	Faecal microbiota transplantation (FMT) to boost cancer immunotherapy	Brusnic et al. (2024)
Nutritional Modulation of Microbiota	Dietary strategies to alter microbiota composition and function	Prebiotics like inulin to increase beneficial gut bacteria	Druart et al. (2014); Tawfick et al. (2022)
Microbiome-Based Therapeutics	Developing drugs targeting microbiota for specific diseases	Small-molecule drugs modulating microbial metabolites	Bishai and Palm (2021); Hussain et al. (2023)
Early-Life Microbiota Interventions	Shaping the microbiome during early developmental stages	Probiotics in infant formulas to promote immune health	Vandenplas et al. (2020); Yao et al. (2021)

and strategies, such as microbiota-based interventions, personalized treatments and sustainable practices, which drive advancements in pre-emptive medicine and healthcare integration through microbiota research. Future studies on the role of human microbiota along with genetic predispositions in regulating health might provide revolutionary discoveries for personalized treatment. Examining the ways in which genetic variants affect the structure and functioning of the microbiota may identify underlying risk factors for illnesses such as type 2 diabetes, obesity, as well as inflammatory disorders. Comprehending these connections might result in personalized measures, including treatments that modify the microbiota, enhanced according to personal genetic characteristics. Preventive measures might be improved by investigating interactions along with environmental impacts regarding microbiota at the beginning of life. Through the use of genetic and microbiome data, precision medicine can improve health results and disease preventive measures by predicting earlier health risks along with personalizing treatments (Fan and Pedersen 2021).

9 Conclusion and Future Prospects

The human microbiota, comprised of bacteria, viruses, as well as fungi residing throughout various body areas, plays some fundamental role in good health as well as illnesses. Furthermore, the 1:1 ratio of microbiota to human cells demonstrates the symbiotic relationship among microbiota as well as human health. Beneficial bacteria help to maintain a healthy environment by assisting in nutrition intake, pathogen defence, along with immune system modulation. On the other hand, dysbiosis—an imbalance in the composition of the microbiota—can result in a number of illnesses, including digestive problems, neurological ailments like Alzheimer's or Parkinson's disease, in addition to metabolic illnesses like diabetes and excess body weight. The microbiota of newborns is initially acquired via their mothers, and it continues to be altered by nursing. Adulthood brings the microbiome into equilibrium, and as people get older, its microbiota diversity can decrease and inflammation may rise, making them more vulnerable to disease. The profiling of human microbiome has been transformed by the shift from conventional sequencing to sophisticated nanostructure-based sequencing technologies. These developments have greatly improved our comprehension of the dynamics along with the intricate nature of microbial communities. Particularly with regard to nanotechnology, there is potential for improved accessibility, speed and accuracy in microbiome studies, which might lead to personalized treatment strategies with specific medications. There are a number of obstacles that human microbiota profiling techniques must overcome, such as complex analysis of data, biases with sequencing, as well as sample processing. To safeguard participants, ethical factors including informed consent along with information privacy are essential. To guarantee accuracy and reliable outcomes, standardization and quality control—which include strict validation, standardized processes and adherence to legal requirements—are crucial.

Microbiome investigations regarding healthcare and personalized medicine should become more reliable and useful if these issues are resolved and ethical as well as quality standards are maintained. Future research on the human microbiome should concentrate on holistic lifestyle including sustainability approaches, as well as the optimization of probiotics, prebiotics, plant-based chemicals, faecal microbiota transplantation, synbiotics, along with microbiota-targeted medications. Future research on the correlation between genetic predispositions and the human microbiome may transform personalized medicine by identifying risk factors and developing tailored therapies that improve health outcomes along with disease prevention.

Acknowledgments SKS lab would like to acknowledge funding from the Institution of Eminence Incentive Grant Scheme, Trans-Disciplinary Research Scheme, Department of Science and Technology, SEED grant, New Delhi and funding from Institute of Medical Sciences intramural grant, Banaras Hindu University, Varanasi, India. CP & SNR would like to acknowledge CSIR (09/0013(12277)/2021-EMR-I) and ICMR (3/1/3/196/Neuro/2021-NCD-I) for providing fellowship assistance respectively.

Disclosure Statement The author declares no conflict of interest.

References

Aagaard K, Petrosino J, Keitel W, Watson M, Katancik J, Garcia N et al (2013) The Human Microbiome Project strategy for comprehensive sampling of the human microbiome and why it matters. FASEB J 27(3):1012

Abavisani M, Foroushan SK, Ebadpour N, Sahebkar A (2024) Deciphering the gut microbiome: the revolution of artificial intelligence in microbiota analysis and intervention. Curr Res Biotechnol 7:100211

Abou Chacra L, Benatmane A, Iwaza R, Ly C, Alibar S, Armstrong N et al (2024) Culturomics reveals a hidden world of vaginal microbiota with the isolation of 206 bacteria from a single vaginal sample. Arch Microbiol 206(1):20

Ahmed Z, Mohamed K, Zeeshan S, Dong X (2020) Artificial intelligence with multi-functional machine learning platform development for better healthcare and precision medicine. Database 2020:baaa010

Al-Fakhrany OM, Elekhnawy E (2024) Next-generation probiotics: the upcoming biotherapeutics. Mol Biol Rep 51(1):505

Annaratone L, De Palma G, Bonizzi G, Sapino A, Botti G, Berrino E et al (2021) Basic principles of biobanking: from biological samples to precision medicine for patients. Virchows Arch 479:233–246

Appleton J (2018) The gut-brain axis: influence of microbiota on mood and mental health. Integr Med 17(4):28

Bach J-F (2021) Revisiting the hygiene hypothesis in the context of autoimmunity. Front Immunol 11:615192

Barba M, Czosnek H, Hadidi A (2014) Historical perspective, development and applications of next-generation sequencing in plant virology. Viruses 6(1):106–136

Bartholomew JW, Mittwer T (1952) The gram stain. Bacteriol Rev 16(1):1–29

Belizário JE, Faintuch J, Garay-Malpartida M (2018) Gut microbiome dysbiosis and immunometabolism: new frontiers for treatment of metabolic diseases. Mediat Inflamm 2018(1):2037838

Benson AK, Kelly SA, Legge R, Ma F, Low SJ, Kim J et al (2010) Individuality in gut microbiota composition is a complex polygenic trait shaped by multiple environmental and host genetic factors. Proc Natl Acad Sci 107(44):18933–18938

Bermúdez-Humarán LG, Salinas E, Ortiz GG, Ramirez-Jirano LJ, Morales JA, Bitzer-Quintero OK (2019) From probiotics to psychobiotics: live beneficial bacteria which act on the brain-gut axis. Nutrients 11(4):890

Beura S, Kundu P, Das AK, Ghosh A (2022) Metagenome-scale community metabolic modelling for understanding the role of gut microbiota in human health. Comput Biol Med 149:105997

Bhalla D, Dinesh S, Sharma S, Sathisha GJ (2024) Gut-brain axis modulation of metabolic disorders: exploring the intertwined neurohumoral pathways and therapeutic prospects. Neurochem Res 49(4):847–871

Bharti R, Grimm DG (2021) Current challenges and best-practice protocols for microbiome analysis. Brief Bioinform 22(1):178–193

Birla H, Rai SN, Singh SS, Zahra W, Rawat A, Tiwari N et al (2019) Tinospora cordifolia suppresses neuroinflammation in parkinsonian mouse model. NeuroMolecular Med 21:42–53

Bishai JD, Palm NW (2021) Small molecule metabolites at the host-microbiota interface. J Immunol 207(7):1725–1733

Broadhurst D, Goodacre R, Reinke SN, Kuligowski J, Wilson ID, Lewis MR, Dunn WB (2018) Guidelines and considerations for the use of system suitability and quality control samples in mass spectrometry assays applied in untargeted clinical metabolomic studies. Metabolomics 14:1–17

Brown K (2000) The human genome business today. Sci Am 283(1):50–55

Bruggeman A, Vandendriessche C, Hamerlinck H, De Looze D, Tate DJ, Vuylsteke M et al (2024) Safety and efficacy of faecal microbiota transplantation in patients with mild to moderate Parkinson's disease (GUT-PARFECT): a double-blind, placebo-controlled, randomised, phase 2 trial. EClinicalMedicine 71:102563

Brusnic O, Onisor D, Boicean A, Hasegan A, Ichim C, Guzun A et al (2024) Fecal microbiota transplantation: insights into colon carcinogenesis and immune regulation. J Clin Med 13(21):6578

Buetas E, Jordán-López M, López-Roldán A, D'Auria G, Martínez-Priego L, De Marco G et al (2024) Full-length 16S rRNA gene sequencing by PacBio improves taxonomic resolution in human microbiome samples. BMC Genomics 25(1):310

Burdon KL (1946) Fatty material in bacteria and fungi revealed by staining dried, fixed slide preparations. J Bacteriol 52(6):665–678

Cammarota G, Ianiro G, Bibbò S, Gasbarrini A (2014) Gut microbiota modulation: probiotics, antibiotics or fecal microbiota transplantation? Intern Emerg Med 9:365–373

Cammarota G, Ianiro G, Ahern A, Carbone C, Temko A, Claesson MJ et al (2020) Gut microbiome, big data and machine learning to promote precision medicine for cancer. Nat Rev Gastroenterol Hepatol 17(10):635–648

Carrera-Quintanar L, Lopez Roa RI, Quintero-Fabián S, Sánchez-Sánchez MA, Vizmanos B, Ortuño-Sahagún D (2018) Phytochemicals that influence gut microbiota as prophylactics and for the treatment of obesity and inflammatory diseases. Mediat Inflamm 2018(1):9734845

Cavallo I, Sivori F, Truglio M, De Maio F, Lucantoni F, Cardinali G et al (2022) Skin dysbiosis and Cutibacterium acnes biofilm in inflammatory acne lesions of adolescents. Sci Rep 12(1):21104

Cernava T, Rybakova D, Buscot F, Clavel T, McHardy AC, Meyer F et al (2022) Metadata harmonization–standards are the key for a better usage of omics data for integrative microbiome analysis. Environ Microbiome 17(1):33

Chen G, Shen J (2021) Artificial intelligence enhances studies on inflammatory bowel disease. Front Bioeng Biotechnol 9:635764

Chen CC, Chen YN, Liou JM, Wu MS, Taiwan Gastrointestinal D, Helicobacter C (2019) From germ theory to germ therapy. Kaohsiung J Med Sci 35(2):73–82

Citorik RJ, Mimee M, Lu TK (2014) Sequence-specific antimicrobials using efficiently delivered RNA-guided nucleases. Nat Biotechnol 32(11):1141–1145

Costalonga M, Herzberg MC (2014) The oral microbiome and the immunobiology of periodontal disease and caries. Immunol Lett 162(2):22–38

Darvin ME (2023) Optical methods for non-invasive determination of skin penetration: current trends, advances, possibilities, prospects, and translation into in vivo human studies. Pharmaceutics 15(9):2272

de Toro-Martín J, Arsenault BJ, Després J-P, Vohl M-C (2017) Precision nutrition: a review of personalized nutritional approaches for the prevention and management of metabolic syndrome. Nutrients 9(8):913

Dekaboruah E, Suryavanshi MV, Chettri D, Verma AK (2020) Human microbiome: an academic update on human body site specific surveillance and its possible role. Arch Microbiol 202:2147–2167

Denman CR, Park SM, Jo J (2023) Gut-brain axis: gut dysbiosis and psychiatric disorders in Alzheimer's and Parkinson's disease. Front Neurosci 17:1268419

Druart C, Alligier M, Salazar N, Neyrinck AM, Delzenne NM (2014) Modulation of the gut microbiota by nutrients with prebiotic and probiotic properties. Adv Nutr 5(5):624S–633S

Dubourg G, Baron S, Cadoret F, Couderc C, Fournier P-E, Lagier J-C, Raoult D (2018) From culturomics to clinical microbiology and forward. Emerg Infect Dis 24(9):1683

Dupont HL, Jiang Z-D, Dupont AW, Utay NS (2020) The intestinal microbiome in human health and disease. Trans Am Clin Climatol Assoc 131:178

Dupras C, Bunnik EM (2021) Toward a framework for assessing privacy risks in multi-omic research and databases. Am J Bioeth 21(12):46–64

Edelstein CA, Kassam Z, Daw J, Smith MB, Kelly CR (2015) The regulation of fecal microbiota for transplantation: an international perspective for policy and public health. Clin Res Regul Aff 32(3):99–107

El-Sayed A, Aleya L, Kamel M (2021) The link among microbiota, epigenetics, and disease development. Environ Sci Pollut Res 28:28926–28964

Fan Y, Pedersen O (2021) Gut microbiota in human metabolic health and disease. Nat Rev Microbiol 19(1):55–71

Flandroy L, Poutahidis T, Berg G, Clarke G, Dao M-C, Decaestecker E et al (2018) The impact of human activities and lifestyles on the interlinked microbiota and health of humans and of ecosystems. Sci Total Environ 627:1018–1038

Fleischmann RD, Adams MD, White O, Clayton RA, Kirkness EF, Kerlavage AR et al (1995) Whole-genome random sequencing and assembly of Haemophilus influenzae Rd. Science 269(5223):496–512

Fondi M, Liò P (2015) Multi-omics and metabolic modelling pipelines: challenges and tools for systems microbiology. Microbiol Res 171:52–64

Forbes JD, Van Domselaar G, Bernstein CN (2016) The gut microbiota in immune-mediated inflammatory diseases. Front Microbiol 7:1081

Freijy TM, Cribb L, Oliver G, Metri N-J, Opie RS, Jacka FN et al (2023) Effects of a high-prebiotic diet versus probiotic supplements versus synbiotics on adult mental health: the "Gut Feelings" randomised controlled trial. Front Neurosci 16:1097278

Friedmann HC (2006) Escherich and Escherichia. Adv Appl Microbiol 60:133–196

Gao R, Gao Z, Huang L, Qin H (2017) Gut microbiota and colorectal cancer. Eur J Clin Microbiol Infect Dis 36:757–769

Georgieva M, Georgiev K, Hvarchanova N (2022) Probiotics: past, present, and future challenges. In: Probiotics in the prevention and management of human diseases. Elsevier, pp 431–448

Geurtsen J, de Been M, Weerdenburg E, Zomer A, McNally A, Poolman J (2022) Genomics and pathotypes of the many faces of Escherichia coli. FEMS Microbiol Rev 46(6):fuac031

Gomaa EZ (2020) Human gut microbiota/microbiome in health and diseases: a review. Antonie Van Leeuwenhoek 113(12):2019–2040

Grogan MD, Bartow-McKenney C, Flowers L, Knight SAB, Uberoi A, Grice EA (2019) Research techniques made simple: profiling the skin microbiota. J Invest Dermatol 139(4):747–752

Grohmann E, Vaishampayan A (2017) Techniques in studying biofilms and their characterization: microscopy to advanced imaging system in vitro and in situ. In: Biofilms in plant and soil health. Wiley, Hoboken, pp 215–230

Gupta V, Kumar R, Sood U, Singhvi N (2020) Reconciling hygiene and cleanliness: a new perspective from human microbiome. Indian J Microbiol 60(1):37–44

Herlo L-F, Salcudean A, Sirli R, Iurciuc S, Herlo A, Nelson-Twakor A et al (2024) Gut microbiota signatures in colorectal cancer as a potential diagnostic biomarker in the future: a systematic review. Int J Mol Sci 25(14):7937

Hou K, Wu Z-X, Chen X-Y, Wang J-Q, Zhang D, Xiao C et al (2022) Microbiota in health and diseases. Signal Transduct Target Ther 7(1):135

Huang B, Fettweis JM, Brooks JP, Jefferson KK, Buck GA (2014) The changing landscape of the vaginal microbiome. Clin Lab Med 34(4):747–761

Huang G, Khan R, Zheng Y, Lee P-C, Li Q, Khan I (2023) Exploring the role of gut microbiota in advancing personalized medicine. Front Microbiol 14:1274925

Hussain A, Patwekar U, Mongad DS, Shouche YS (2023) Strategizing the human microbiome for small molecules: approaches and perspectives. Drug Discov Today 28(2):103459

Jin Y, Wu S, Zeng Z, Fu Z (2017) Effects of environmental pollutants on gut microbiota. Environ Pollut 222:1–9

Karl JP, Hatch AM, Arcidiacono SM, Pearce SC, Pantoja-Feliciano IG, Soares JW (2018) Effects of psychological, environmental and physical stressors on the gut microbiota. Front Microbiol 9:372026

Kashyap PC, Chia N, Nelson H, Segal E, Elinav E (2017) Microbiome at the frontier of personalized medicine. Mayo Clin Proc 92(12):1855–1864

Khan FF, Sohail A, Ghazanfar S, Ahmad A, Riaz A, Abbasi KS et al (2023) Recent innovations in non-dairy prebiotics and probiotics: physiological potential, applications, and characterization. Probiotics Antimicrob Proteins 15(2):239–263

Kodikara S, Ellul S, Lê Cao K-A (2022) Statistical challenges in longitudinal microbiome data analysis. Brief Bioinform 23(4):bbac273

Koh A, De Vadder F, Kovatcheva-Datchary P, Bäckhed F (2016) From dietary fiber to host physiology: short-chain fatty acids as key bacterial metabolites. Cell 165(6):1332–1345

Komala MG, Ong SG, Qadri MU, Elshafie LM, Pollock CA, Saad S (2023) Investigating the regulatory process, safety, efficacy and product transparency for nutraceuticals in the USA, Europe and Australia. Foods 12(2):427

Kool J, Tymchenko L, Shetty S, Fuentes S (2022) Standardization of laboratory practices for the study of the human gut microbiome. bioRxiv, 2022-2011. https://doi.org/10.1101/2022.11.10.515556

Krammer F (2020) SARS-CoV-2 vaccines in development. Nature 586(7830):516–527

Kuczynski J, Lauber CL, Walters WA, Parfrey LW, Clemente JC, Gevers D, Knight R (2012) Experimental and analytical tools for studying the human microbiome. Nat Rev Genet 13(1):47–58

Kumar S, Patel GK, Ghoshal UC (2021) Helicobacter pylori-induced inflammation: possible factors modulating the risk of gastric cancer. Pathogens 10(9):1099

Langdon A, Crook N, Dantas G (2016) The effects of antibiotics on the microbiome throughout development and alternative approaches for therapeutic modulation. Genome Med 8:1–16

Li H, Huang S, Jing J, Yu H, Gu T, Ou X et al (2024) Dietary intake and gastrointestinal symptoms are altered in children with Autism Spectrum Disorder: the relative contribution of autism-linked traits. Nutr J 23(1):27

Lin P, Zhang Q, Sun J, Li Q, Li D, Zhu M et al (2024) A comparison between children and adolescents with autism spectrum disorders and healthy controls in biomedical factors, trace elements, and microbiota biomarkers: a meta-analysis. Front Psych 14:1318637

Lopetuso LR, Petito V, Graziani C, Schiavoni E, Paroni Sterbini F, Poscia A et al (2018) Gut microbiota in health, diverticular disease, irritable bowel syndrome, and inflammatory bowel diseases: time for microbial marker of gastrointestinal disorders. Dig Dis 36(1):56–65

Lu H, Giordano F, Ning Z (2016) Oxford nanopore MinION sequencing and genome assembly. Genomics Proteomics Bioinform 14(5):265–279
Ma Y, Chen H, Lan C, Ren J (2018) Help, hope and hype: ethical considerations of human microbiome research and applications. Protein Cell 9(5):404–415
Malard F, Dore J, Gaugler B, Mohty M (2021) Introduction to host microbiome symbiosis in health and disease. Mucosal Immunol 14(3):547–554
Marcos-Zambrano LJ, Karaduzovic-Hadziabdic K, Loncar Turukalo T, Przymus P, Trajkovik V, Aasmets O et al (2021) Applications of machine learning in human microbiome studies: a review on feature selection, biomarker identification, disease prediction and treatment. Front Microbiol 12:634511
Marques FZ, Jama HA, Tsyganov K, Gill PA, Rhys-Jones D, Muralitharan RR et al (2019) Guidelines for transparency on gut microbiome studies in essential and experimental hypertension. Hypertension 74(6):1279–1293
Matias LLR, da Silva Chaves Damasceno KSF, Pereira AS, Passos TS, Morais AHDA (2024) Innovative biomedical and technological strategies for the control of bacterial growth and infections. Biomedicine 12(1):176
McGuinness AJ, Stinson LF, Snelson M, Loughman A, Stringer A, Hannan AJ et al (2024) From hype to hope: considerations in conducting robust microbiome science. Brain Behav Immun 115:120–130
McGuire AL, Colgrove J, Whitney SN, Diaz CM, Bustillos D, Versalovic J (2008) Ethical, legal, and social considerations in conducting the Human Microbiome Project. Genome Res 18(12):1861–1864
Megur A, Baltriukienė D, Bukelskienė V, Burokas A (2020) The microbiota–gut–brain axis and Alzheimer's disease: neuroinflammation is to blame? Nutrients 13(1):37
Metchnikoff II (2004) The prolongation of life: optimistic studies. Springer Publishing Company
Moreno-Indias I, Cardona F, Tinahones FJ, Queipo-Ortuño MI (2014) Impact of the gut microbiota on the development of obesity and type 2 diabetes mellitus. Front Microbiol 5:190
Müller H, Dagher G, Loibner M, Stumptner C, Kungl P, Zatloukal K (2020) Biobanks for life sciences and personalized medicine: importance of standardization, biosafety, biosecurity, and data management. Curr Opin Biotechnol 65:45–51
Nagpal R, Yadav H, Marotta F (2014) Gut microbiota: the next-gen frontier in preventive and therapeutic medicine? Front Med 1:15
Navas-Molina JA, Hyde ER, Sanders JG, Knight R (2017) The microbiome and big data. Curr Opin Syst Biol 4:92–96
Nazaroff WW (2019) Embracing microbes in exposure science. J Expo Sci Environ Epidemiol 29(1):1–10
Nearing JT, Comeau AM, Langille MGI (2021) Identifying biases and their potential solutions in human microbiome studies. Microbiome 9(1):113
Nissle A (1925) Weiteres über grundlagen und praxis der mutaflorbehandlung. Dtsch Med Wochenschr 51(44):1809–1813
Nurk S, Koren S, Rhie A, Rautiainen M, Bzikadze AV, Mikheenko A et al (2022) The complete sequence of a human genome. Science 376(6588):44–53
O'Doherty KC, Virani A, Wilcox ES (2016) The human microbiome and public health: social and ethical considerations. Am J Public Health 106(3):414–420
Ottinger S, Robertson CM, Branthoover H, Patras KA (2024) The human vaginal microbiota: from clinical medicine to models to mechanisms. Curr Opin Microbiol 77:102422
Pedroza Matute S, Iyavoo S (2023) Exploring the gut microbiota: lifestyle choices, disease associations, and personal genomics. Front Nutr 10:1225120
Peng R, Lin G, Li J (2016) Potential pitfalls of CRISPR/Cas9-mediated genome editing. FEBS J 283(7):1218–1231
Perović J, Šaponjac VT, Kojić J, Krulj J, Moreno DA, García-Viguera C et al (2021) Chicory (Cichorium intybus L.) as a food ingredient–nutritional composition, bioactivity, safety, and health claims: a review. Food Chem 336:127676

Peterson J, Garges S, Giovanni M, McInnes P, Wang L, Schloss JA et al (2009) The NIH human microbiome project. Genome Res 19(12):2317–2323

Pfleiderer A, Lagier JC, Armougom F, Robert C, Vialettes B, Raoult D (2013) Culturomics identified 11 new bacterial species from a single anorexia nervosa stool sample. Eur J Clin Microbiol Infect Dis 32:1471–1481

Phan J, Calvo DC, Nair D, Jain S, Montagne T, Dietsche S et al (2024) Precision synbiotics increase gut microbiome diversity and improve gastrointestinal symptoms in a pilot open-label study for autism spectrum disorder. mSystems 9(5):e00503-00524

Pires L, González-Paramás AM, Heleno SA, Calhelha RC (2024) The role of gut microbiota in the etiopathogenesis of multiple chronic diseases. Antibiotics 13(5):392

Podolsky SH (2012) Metchnikoff and the microbiome. Lancet 380(9856):1810–1811

Porter JR (1976) Antony van Leeuwenhoek: tercentenary of his discovery of bacteria. Bacteriol Rev 40(2):260–269

Prajapati C, Kumar D, Ambastha V, Singh SK (2024a) Fungal endophytes as potential anticancer candidate over synthetic drugs: latest development and future prospects. In: Endophytic fungi: the hidden sustainable jewels for the pharmaceutical and agricultural industries. Springer, pp 27–56

Prajapati C, Rai SN, Mishra SS, Singh SK, Singh AK (2024b) Natural sources of drugs for neurodegenerative diseases. In: Neurodegenerative diseases. CRC Press, pp 120–134

Prajapati C, Rai SN, Singh AK, Chopade BA, Singh Y, Singh SK et al (2025a) An update of fungal endophyte diversity and strategies for augmenting therapeutic potential of their potent metabolites: recent advancement. Appl Biochem Biotechnol 197:1–68

Prajapati C, Rai SN, Singh AK, Rajak N, Garg N, Singh SK (2025b) Untargeted metabolomics and bioactivities assessment of Xylaria ellisii, an endophytic fungus isolated from the leaf of the plant Acorus calamus Linn. Curr Pharm Des 31:1781

Prajapati C, Tripathi PN, Sood S, Rai SN, Mishra SS, Singh SK, Tiwari AK (2025c) Intellectual property rights in neuroprotective biomaterials. In: Biomaterials and neurodegenerative disorders. Springer, pp 251–269

Purushothaman S, Meola M, Egli A (2022) Combination of whole genome sequencing and metagenomics for microbiological diagnostics. Int J Mol Sci 23(17):9834

Rai SN, Zahra W, Singh SS, Birla H, Keswani C, Dilnashin H et al (2019) Anti-inflammatory activity of ursolic acid in MPTP-induced parkinsonian mouse model. Neurotox Res 36:452–462

Ranuh R, Athiyyah AF, Darma A, Risky VP, Riawan W, Surono IS, Sudarmo SM (2019) Effect of the probiotic Lactobacillus plantarum IS-10506 on BDNF and 5HT stimulation: role of intestinal microbiota on the gut-brain axis. Iran J Microbiol 11(2):145

Ratiner K, Ciocan D, Abdeen SK, Elinav E (2024) Utilization of the microbiome in personalized medicine. Nat Rev Microbiol 22(5):291–308

Rinninella E, Raoul P, Cintoni M, Franceschi F, Miggiano GAD, Gasbarrini A, Mele MC (2019) What is the healthy gut microbiota composition? A changing ecosystem across age, environment, diet, and diseases. Microorganisms 7(1):14

Roume H, Mondot S, Saliou A, Le Fresne-Languille S, Doré J (2023) Multicenter evaluation of gut microbiome profiling by next-generation sequencing reveals major biases in partial-length metabarcoding approach. Sci Rep 13(1):22593

Ryguła I, Pikiewicz W, Grabarek BO, Wójcik M, Kaminiów K (2024) The role of the gut microbiome and microbial dysbiosis in common skin diseases. Int J Mol Sci 25(4):1984

Sadeghpour Heravi F (2024) Gut microbiota and autoimmune diseases: mechanisms, treatment, challenges, and future recommendations. Curr Clin Microbiol Rep 11(1):18–33

Safarlou CW, Jongsma KR, Vermeulen R, Bredenoord AL (2023) The ethical aspects of exposome research: a systematic review. Exposome 3(1):osad004

Sánchez-Pellicer P, Eguren-Michelena C, García-Gavín J, Llamas-Velasco M, Navarro-Moratalla L, Núñez-Delegido E et al (2024) Rosacea, microbiome and probiotics: the gut-skin axis. Front Microbiol 14:1323644

Sanger F, Nicklen S, Coulson AR (1977) DNA sequencing with chain-terminating inhibitors. Proc Natl Acad Sci 74(12):5463–5467

Scheithauer TPM, Rampanelli E, Nieuwdorp M, Vallance BA, Verchere CB, Van Raalte DH, Herrema H (2020) Gut microbiota as a trigger for metabolic inflammation in obesity and type 2 diabetes. Front Immunol 11:571731

Schippa S, Conte MP (2014) Dysbiotic events in gut microbiota: impact on human health. Nutrients 6(12):5786–5805

Sender R, Fuchs S, Milo R (2016) Revised estimates for the number of human and bacteria cells in the body. PLoS Biol 14(8):e1002533

Shah AM, Tarfeen N, Mohamed H, Song Y (2023) Fermented foods: their health-promoting components and potential effects on gut microbiota. Fermentation 9(2):118

Short MI, Hudson R, Besasie BD, Reveles KR, Shah DP, Nicholson S et al (2021) Comparison of rectal swab, glove tip, and participant-collected stool techniques for gut microbiome sampling. BMC Microbiol 21:1–9

Singh SS, Rai SN, Birla H, Zahra W, Rathore AS, Singh SP (2020) NF-κB-mediated neuroinflammation in Parkinson's disease and potential therapeutic effect of polyphenols. Neurotox Res 37:491–507

Song HS, Kim YB, Kim JY, Roh SW, Whon TW (2024) Advances in culturomics research on the human gut microbiome: optimizing medium composition and culture techniques for enhanced microbial discovery. J Microbiol Biotechnol 34(4):757

Sonnenborn U (2016) Escherichia coli strain Nissle 1917—from bench to bedside and back: history of a special Escherichia coli strain with probiotic properties. FEMS Microbiol Lett 363(19):fnw212

Szóstak N, Szymanek A, Havránek J, Tomela K, Rakoczy M, Samelak-Czajka A et al (2022) The standardisation of the approach to metagenomic human gut analysis: from sample collection to microbiome profiling. Sci Rep 12(1):8470

Tawfick MM, Xie H, Zhao C, Shao P, Farag MA (2022) Inulin fructans in diet: role in gut homeostasis, immunity, health outcomes and potential therapeutics. Int J Biol Macromol 208:948–961

Tian Y, Jian T, Li J, Huang L, Li S, Lu H et al (2024) Phenolic acids from Chicory roots ameliorate dextran sulfate sodium–induced colitis in mice by targeting TRP signaling pathways and the gut microbiota. Phytomedicine 128:155378

Tokuhara D (2021) Role of the gut microbiota in regulating non-alcoholic fatty liver disease in children and adolescents. Front Nutr 8:700058

Trainor GL (1990) DNA sequencing, automation, and the human genome. Anal Chem 62(5):418–426

Turnbaugh PJ, Ley RE, Mahowald MA, Magrini V, Mardis ER, Gordon JI (2006) An obesity-associated gut microbiome with increased capacity for energy harvest. Nature 444(7122):1027–1031

Ugwu OP-C, Alum EU, Okon MB, Obeagu EI (2024) Mechanisms of microbiota modulation: implications for health, disease, and therapeutic interventions. Medicine 103(19):e38088

Ussar S, Griffin NW, Bezy O, Fujisaka S, Vienberg S, Softic S et al (2015) Interactions between gut microbiota, host genetics and diet modulate the predisposition to obesity and metabolic syndrome. Cell Metab 22(3):516–530

Vamanu E, Rai SN (2021) The link between obesity, microbiota dysbiosis, and neurodegenerative pathogenesis. Diseases 9(3):45

Van Dijk EL, Auger H, Jaszczyszyn Y, Thermes C (2014) Ten years of next-generation sequencing technology. Trends Genet 30(9):418–426

Vandenplas Y, Carnielli VP, Ksiazyk J, Luna MS, Migacheva N, Mosselmans JM et al (2020) Factors affecting early-life intestinal microbiota development. Nutrition 78:110812

Villavicencio-Tejo F, Olesen MA, Navarro L, Calisto N, Iribarren C, García K et al (2024) Gut-brain axis deregulation and its possible contribution to neurodegenerative disorders. Neurotox Res 42(1):4

Voth E, Khanna S (2020) The Integrative Human microbiome project: a mile stone in the understanding of the gut microbiome. Expert Rev Gastroenterol Hepatol 14(8):639–642
Wang W, Zhu L-J, Leng Y-Q, Wang Y-W, Shi T, Wang W-Z, Sun J-C (2023) Inflammatory response: a crucial way for gut microbes to regulate cardiovascular diseases. Nutrients 15(3):607
Wang S, Ju D, Zeng X (2024a) Mechanisms and clinical implications of human gut microbiota-drug interactions in the precision medicine era. Biomedicine 12(1):194
Wang F, Yu L, Ren Y, Zhang Q, He S, Zhao M et al (2024b) An optimized culturomics strategy for isolation of human milk microbiota. Front Microbiol 15:1272062
Watson JD (1990) The human genome project: past, present, and future. Science 248(4951):44–49
Wheeler DA, Srinivasan M, Egholm M, Shen Y, Chen L, McGuire A et al (2008) The complete genome of an individual by massively parallel DNA sequencing. Nature 452(7189):872–876
Woo PCY, Lau SKP, Teng JLL, Tse H, Yuen KY (2008) Then and now: use of 16S rDNA gene sequencing for bacterial identification and discovery of novel bacteria in clinical microbiology laboratories. Clin Microbiol Infect 14(10):908–934
Wu H, Chen X, Zhang S, Li J (2022) Gut microbiota, the potential biological medicine for prevention, intervention and drug sensitization to fight diseases. Nutrients 14(20):4220
Xiao W, Wu L, Yavas G, Simonyan V, Ning B, Hong H (2016) Challenges, solutions, and quality metrics of personal genome assembly in advancing precision medicine. Pharmaceutics 8(2):15
Yao Y, Cai X, Ye Y, Wang F, Chen F, Zheng C (2021) The role of microbiota in infant health: from early life to adulthood. Front Immunol 12:708472
Zarin DA, Tse T, Williams RJ, Califf RM, Ide NC (2011) The ClinicalTrials.gov results database—update and key issues. N Engl J Med 364(9):852–860
Zernike F (1955) How I discovered phase contrast. Science 121(3141):345–349
Zhang X, Li L, Butcher J, Stintzi A, Figeys D (2019) Advancing functional and translational microbiome research using meta-omics approaches. Microbiome 7:1–12
Zhang D, Li S, Wang N, Tan H-Y, Zhang Z, Feng Y (2020) The cross-talk between gut microbiota and lungs in common lung diseases. Front Microbiol 11:301
Zhao Q, Chen Y, Huang W, Zhou H, Zhang W (2023) Drug-microbiota interactions: an emerging priority for precision medicine. Signal Transduct Target Ther 8(1):1–27
Zhou Z, Ge S, Li Y, Ma W, Liu Y, Hu S et al (2020) Human gut microbiome-based knowledgebase as a biomarker screening tool to improve the predicted probability for colorectal cancer. Front Microbiol 11:596027

MIX
Papier aus verantwortungsvollen Quellen
Paper from responsible sources
FSC® C105338

If you have any concerns about our products,
you can contact us on
ProductSafety@springernature.com

In case Publisher is established outside the EU,
the EU authorized representative is:
Springer Nature Customer Service Center GmbH
Europaplatz 3, 69115 Heidelberg, Germany

Printed by Libri Plureos GmbH
in Hamburg, Germany